ANNUAL REVIEW
OF ENERGY AND
THE ENVIRONMENT

ANNUAL REVIEW OF ENERGY AND THE ENVIRONMENT

VOLUME 16, 1991

JACK M. HOLLANDER, *Editor*

University of California at Berkeley

ROBERT H. SOCOLOW, *Associate Editor*

Princeton University

DAVID STERNLIGHT, *Associate Editor*

Pasadena, California

ANNUAL REVIEWS, INC.   4139 EL CAMINO WAY   P.O. BOX 10139   PALO ALTO, CALIFORNIA 94303-0897

ANNUAL REVIEWS INC.
Palo Alto, California, USA

International Standard Serial Number: 1056–3466
International Standard Book Number: 0–8243–2316-5

Annual Review and publication titles are registered trademarks of Annual Reviews Inc.

⊗ The paper used in this publication meets the minimum requirements of American National Standard for Information Sciences—Permanence of Paper for Printed Library Materials, ANZI Z39.48-1984.

Annual Reviews Inc. and the Editors of its publications assume no responsibility for the statements expressed by the contributors to this *Review*.

Typesetting by Kachina Typesetting Inc., Tempe, Arizona; John Olson, President; Janis Hoffman, Typesetting Coordinator; and by the Annual Reviews Inc. Editorial Staff

PREFACE

The explosive events of 1991 in the Middle East were a reminder of the crucial role of energy, and oil in particular, in the geopolitics of the region and the world. While the Gulf War was not exclusively a war over control of oil supplies, the intensity of the West's reaction to the Iraqi occupation of Kuwait demonstrated the great importance that many oil-consuming nations attach to security of oil supply from the Gulf states.

The full magnitudes of the political, economic, and environmental consequences of the Gulf War will only become known as these impacts slowly unfold in the future. Already apparent, however, is that the war's environmental damages are huge. Iraq's deliberate dumping of millions of barrels of oil directly into the Gulf dwarfs any previous accidental oil spill including the *Exxon Valdez* spill off the coast of Alaska. Damage to the fragile ecosystems of the Gulf is only beginning to be assessed. Even more grotesque was the Iraqi torching of the Kuwaiti oil wells, an act of environmental sabotage that surpassed even the most pessimistic predictions. The smoke and fallout from these fires are having serious health effects in the region, and the economic costs of recovery and rehabilitation of the oil fields will be immense. Fortunately, the fires will probably not affect the Earth's climate, because the smoke does not rise high enough in the atmosphere to produce appreciable "nuclear winter" phenomena on a global scale. The reader of this series may expect to find reviews and analyses, in subsequent volumes, of the most significant energy- and environment-related aspects of the Gulf War.

Meanwhile, peaceful human activities, although less dramatic, are definitely of a scale that causes worldwide impacts. Growing attention is being focused on ways to mitigate the impacts of the so-called "greenhouse gases" on the world's climate system—even though the science of this enormously complex problem is still only partially understood. This volume contains two approaches to the question of mitigation. In his paper, D. Spencer argues that the developed countries must invest as much as $350 billion per year in efficiency improvements of energy end-use and generation systems, and must also subsidize the costs of energy conservation in the developing countries, if carbon emissions are to be kept to an acceptable level by mid-century. In their wide-ranging paper, A. and H. Lovins conclude that shrewd application of market forces, together with currently available or near-term high-technologies, can not only reduce carbon emissions enough to abate climate change but can save money as well.

(*continued*)   v

One strategy for carbon stabilization is the use of renewable biomass fuels, which do not contribute net carbon dioxide to the atmosphere. The review by Cook, Beyea, & Keeler discusses the potential for biomass fuels in the context of serious potential ecological consequences of large-scale production of biomass fuels, especially the risks to biological diversity.

The comprehensive review by J. Holdren of the environmental and safety characteristics of nuclear fusion energy is a welcome contribution to a subject that has not received wide public attention. Although the current state of fusion technology is still primitive in comparison to fission technology, significant environmental and safety advantages of fusion are already perceived, especially the lower consequences of severe reactor accidents and the smaller links of fusion technology to nuclear weapons production.

Because of its great abundance and accessibility, coal has the potential to remain a major fuel for as long as fossil fuels are used, provided that economically viable technologies can be implemented to burn coal "cleanly," i.e. with acceptable environmental impacts. This volume carries three reviews of clean coal technology (CCT). The paper by Rose, Labys, & Torries views the subject from the perspective of the large US government CCT program and includes an estimate of future US energy resource use, including coal. S. Alpert reviews in more technical detail the various CCTs, categorizing them in terms of removal of sulfur and nitrogen before, during, and after combustion, and also includes an industry projection of US electricity demand. S. Tavoulareas focuses his review on fluidized-bed combustion (FBC), one of the major CCTs, covering the evolution of FBC technology, the developmental status of various FBC designs, potential applications and market penetration, and issues associated with the commercialization of FBC technology.

Another paper dealing with an important energy supply issue is that of Kaufmann & Cleveland, who evaluate the costs and benefits of policies that seek to increase domestic supply. These authors conclude that the benefits of such policies do not justify either the economic or environmental costs that would be entailed.

Among the many advances in technology that have the potential to increase the efficiency of energy-using devices, none is more exciting than superconductivity. Although discovered in 1911, the phenomenon of superconductivity entered the realm of practicality only in 1987, when new superconducting materials were discovered that could be operated at easily obtainable cold temperatures. T. Schneider reviews the kinds and magnitudes of energy savings that will be available when this scientific breakthrough translates into engineering practice in the marketplace.

In an effort to increase the performance efficiency and environmental acceptability of their energy systems, many utilities are broadening their corporate purpose to include the provision of energy services rather than only energy supply. The paper by E. Hirst & C. Goldman reviews the new concepts of integrated utility resource planning, by which utilities (in this case, electric utilities) consider all resources, e.g. energy efficiency and load-management programs as alternatives to some power plants, environmental costs as well as direct economic costs, and uncertainties, risks, and public attitudes toward various resource portfolios and strategies. The paper by K. Lee reviews similar utility management issues from a different perspective—that of utilities in hydro-rich Washington state that suffered an unprecedented financial failure in 1983 because of overcommitment to nuclear power plants at a time of falling electricity demand, and their subsequent attempts to recover through an emphasis on energy efficiency and environmental objectives.

In keeping with the strong international character of this series, the current volume carries reviews of energy policies and outlooks in a number of countries and regions. These are: Asia and the Pacific, papers by Razavi & Fesharaki, and by Sathaye & Tyler; Brazil, by Geller & Zylbersztajn; Southern Africa, by Raskin & Lazarus; China, by Perlack, Russell, & Shen; Poland, by Z. Nowak; and the Soviet Union, by Y. Sinyak.

Finally, the editors are pleased to announce a new title for this series. Starting with this volume, *Annual Review of Energy* becomes *Annual Review of Energy and the Environment*. The new name reflects the increasing recognition of the strong linkages between energy and environmental issues. Reviews and analyses of these linkages have always been an integral part of the *Annual Review of Energy* and will continue to be so. The new name reflects a formal recognition of the breadth of the volume's constituency.

Annual Review of Energy and the Environment
Volume 16, 1991

CONTENTS

For the convenience of readers, a detachable order form/envelope is bound into the back of this volume.

Annu. Rev. Energy Environ. 1991. 16:1–23

CLEAN COAL TECHNOLOGY AND ADVANCED COAL-BASED POWER PLANTS

S. B. Alpert

Electric Power Research Institute, 3412 Hillview Avenue, Palo Alto, California 94303

KEY WORDS: coal technology, generation of electricity, power generation, advanced power plants.

Introduction

Clean Coal Technology is an arbitrary terminology that has gained increased use since the 1980s when the debate over acid rain issues intensified over emissions of sulfur dioxide and nitrogen oxides. In response to political discussions between Prime Minister Brian Mulroney of Canada and President Ronald Reagan in 1985, the US government initiated a demonstration program by the Department of Energy (DOE) on Clean Coal Technologies, which can be categorized as:

1. precombustion technologies wherein sulfur and nitrogen are removed before combustion,
2. combustion technologies that prevent or lower emissions as coal is burned, and
3. postcombustion technologies wherein flue gas from a boiler is treated to remove pollutants, usually transforming them into solids that are disposed of.

Typical technologies in these categories are shown in Table 1.

 The DOE Clean Coal Technology (CCT) program is being carried out with $2.5 billion of federal funds and additional private sector funds. By the end of

1

1056-3466/91/1022-0001$02.00

Table 1 Typical Clean Coal Technologies

Category	Technologies
Precombustion	Microbiological treatment, coal gasification, liquefaction of coal, pyrolysis and mild gasification, coal desulfurization, physical and chemical coal cleaning
Combustion	Pressurized and atmospheric fluidized bed, furnace sorbent injection, duct sorbent injection, combustion modification, slagging combustion
Postcombustion	Flue gas desulfurization, regenerative flue gas desulfurization, selective catalytic reduction, selective noncatalytic reduction

1989, 38 projects were under way or in negotiation (Figure 1 shows their locations). These projects were solicited in three rounds, known as Clean Coal I, II, and III, and two additional solicitations are planned by DOE.

Worldwide, about 100 clean coal demonstration projects are being carried out. Table 2 lists important requirements of demonstration plants based on experience with such plants. These requirements need to be met to allow a technology to proceed to commercial application with ordinary risk, and represent the principal reasons that a demonstration project is necessary when introducing new technology.

Status of Electricity Industry In the United States

In 1989, about 2.8 trillion kilowatt-hours of electricity were generated in the United States from the 600,000 MWe of capacity of plants that use coal, oil, gas, hydropower, nuclear, and renewable energy to satisfy a growing market for electric service. More than 50% was generated by using coal, which consumed most of US production of about one billion tons.

The public has come to depend on the electricity industry as it has the telephone industry. The energy supply system works very well to sustain the life-style the United States enjoys. A characteristic of a developing country such as Thailand today or Korea of the 1950s is an unreliable supply of power that is not sufficient or robust enough to meet the demand of customers. In Thailand electricity supply is allocated by a priority that first supplies industrial customers. Residential customers are often without electric power at night is some parts of the country. In Korea during the 1950s, electricity for residential users was shut off at night and represented an unreliable supply of energy. Hospitals and military bases were forced to have their own generators for primary supply because of frequent (almost daily) interruptions. As countries develop, low-cost, reliable electric supply becomes an expected service and is a benchmark of a modern industrialized nation.

On a per capita annual basis, about 10,000 kWh are used to support the US life-style. In contrast, in less developed countries the per capita electricity

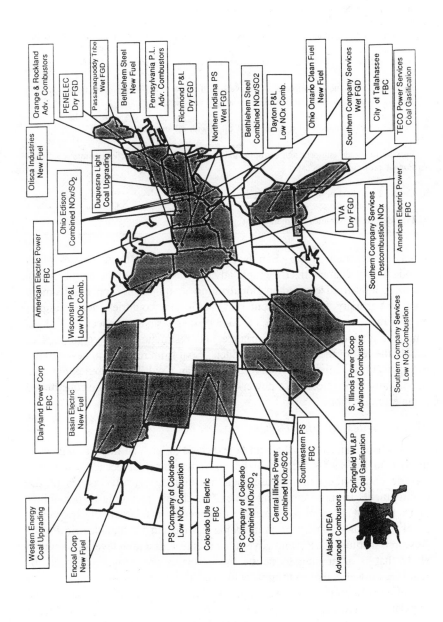

Table 2   Important requirements for demonstration plants

1. Involve technology/plant owner interested in commercial application in subsequent plants

2. Scale up from development program by about five times in size

3. Prove reliability and performance of innovative, "first of a kind" equipment and materials in sustained operations to establish plant reliability and performance

4. Include a single module of a commercial plant with all subsystems present

5. Provide full-scale engineering/environmental design basis for commercial plants, confirmation of projected plant costs

6. Certify equipment/designs of suppliers

7. Provide product samples for market test

supply is as small as 5% (Bangladesh) of that in the United States. This decade will be characterized by rapid advances as the rest of the world rapidly moves to industrialize based on globalization of equipment supply and a more open opportunity for investment. In Pacific rim countries, for example, growth rates of electric power of 20–30% per year are projected as these emerging nations rapidly industrialize and restructure their economies to supply the rest of the world with products. Coal from regional mines is likely to represent an important resource, and Clean Coal Technologies have an opportunity to be selected for power-generation plants that will be constructed in emerging nations.

Major Issues for the 1990s

Table 3 lists major issues that can be expected to have an impact on electric utilities in the 1990s. This will be an "environmental decade" in which electric companies will increase their attention to sensitive issues of the environment. In these emerging issues, the debate regarding responsibility and action that are required to ensure safe generation and use of electric power has begun. The industry program being carried out by the Electric Power Research Institute (EPRI) and government represents a high level of scientific study that is attempting to define the problems and provide an array of realistic, cost-effective and timely solutions. The problems are complex, with causes and effects that are not straightforward. The solutions often seem to be unpalatable, and costly for the public. For example, significantly reducing carbon dioxide emissions could increase electricity costs by a factor of two or more. Sensible strategies for sequestering carbon dioxide are not readily apparent despite the attention of some very talented scientists. The technological options still need definition based on credible scientific study.

Table 3   Major issues of the 1990s for the electricity industry

1. Increased environmental pressures by federal and state regulation agencies
   Clean Air Act provisions
   Electric and magnetic field effects
   Global warming
   Toxic and solid wastes

2. Capacity shortages and need for new plants

3. Fuel supply and price uncertainties
   Natural gas and petroleum

4. Increased competition from new players in the electricity business

During the 1990s, the pace of work on the science base of electricity will accelerate and the national and international debate on the environment can be expected to intensify. Unlike in the past, the debate is likely to take the form of global discussions of strategies, as the issues cross national boundaries and action on a multinational basis is required to arrive at solutions. Precipitous action by single nations is likely to result in competitive disadvantage, so all nations will need to act in concert if the issues are to be resolved.

Supply of Electricity

In the United States, electric capacity additions have fallen behind demand growth as the overcapacity in regions of the United States has been eroded. Electric utilities are now reluctant to commit to new capacity additions because of several factors including adverse rate decisions on nuclear and fossil plants brought on line in the 1980s, uncertainties with regard to fuel supply and cost, uncertainty about future environmental regulations, and siting issues.

The National Electric Reliability Council's 10-year summer peak forecast is shown in Figure 2. The consensus of the electric utility industry is that peak demand will grow at 2% per year over the next decade. This projected rate of growth is lower than the 2.7% per year average growth in peak electricity demand over the past decade in the United States.

Table 4 shows the actual and forecast growth in total electricity demand for the past three years. Actual expansion in electricity demand has exceeded projected demand of 2%. In 1990, overall demand grew by about 3%.

As shown in Figure 3, if demand continues to grow at about 2.5% per year, new capacity will need to be brought on line by the mid-1990s, and will therefore need to be sited and constructed quickly.

Table 5 shows the regions of the United States that are projected to have less than 17% capacity margin toward the end of this decade. Historically,

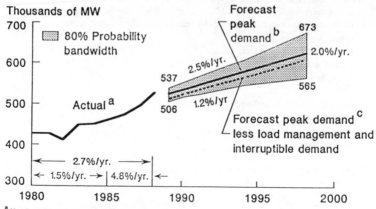

a Unadjusted for operator-controlled demand reductions, emergency operating procedures, or weather.
b Item 3a, line 03, April 1, 1989 IE-411 reports.
c Item 3a, line 06, April 1, 1989 IE-411 reports.

0548.09

Figure 2 Summer peak demands, 1980–1998 (1989–1998 forecast). Source: US National Electric Reliability Council (NERC)

reliability considerations indicate that 17% of a system's capacity represents a comfortable margin for ensuring reliability of service. As mentioned, the regions of the United States with the lowest reserve margins are the northeast and southeast.

In response to the need for new capacity, utilities plan major new plant commitments for gas turbine equipment from US and foreign suppliers that use low-cost natural gas and petroleum. Equipment supply has become increasingly globalized, with suppliers from Europe and Japan as well as the United States selling to the US market.

We can expect that if current projections and trends continue, US utilities

Table 4 Electricity demand in the United States

	1987		1988		1989	
	Actual	Forecast	Actual	Forecast	Actual	Forecast
Summer peak (1000 MW)	496	484	529	500	523	522
Growth, %	4.0		6.6		−1.1	
Winter peak (1000 MW)	448	444	467	457	496	476
Growth, %	4.4		4.1		6.5	
Total in millions of MWh	2644	2589	2769	2690	2850	2802
Growth, %	4.4		4.7		2.8	

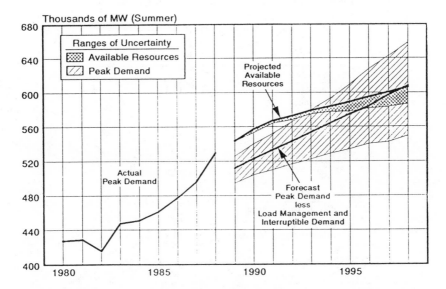

Figure 3 Actual peak demand, 1980–1988, and projected available resources and ranges of uncertainty, 1989–1998 forecast. Source: US National Electric Reliability Council (NERC)

will order enough gas turbine equipment to double their capacity of such equipment in this decade (to 100 gigawatts). The economically attractive large-size new combined-cycle equipment represents capacity increments of 200–225 MWe, but with a 50% lower cost than that of a pulverized coal plant. The overall efficiency of these systems is about 50% as compared with 35% for a pulverized coal plant. Equipment that uses natural gas has the added benefit of a fuel cost of about one-third the cost of petroleum fuels. Table 6 shows characteristics of gas turbines that are offered by manufacturers.

Major concerns regarding electric supply are listed in Table 7, and discussed below.

As mentioned, capacity additions have severely lagged market growth in the United States. As the capacity margin shrinks, reliability of supply may decrease, and "brownouts" that have been experienced during recent summer peak months are likely to become a permanent fact of life.

In order to cope with this situation, US utilities plan to install about 500 new gas turbines in this decade. This represents a challenge, since newly designed equipment is often unreliable. Experience in achieving high reliability with combustion turbines in the 1960s was poor. It is important that this poor experience not be repeated, so that electricity will continue to represent a reliable energy service.

Combustion turbines increase the electric industry's dependency on natural

Table 5 Regions of the United States with low capacity margins[a]

Summer peak	Summer capacity margin, %	
SPP (Southwest)		
Southeast	12.1	
West Central	16.6	
ERCOT (Texas)	16.8	
WSCC (West)		
Arizona–New Mexico	13.0	
Main (Mid America)		
Commonwealth Edison	13.0	
East Missouri	13.1	
Wisconsin–Upper Michigan	13.9	
NPCC (Northeast)		
New England Pool	8.3	(8.5)[b]
SERC		
Florida	13.5	(5.9)
Virginia Carolina	12.2	(14.2)
Southern	16.0	

[a] Anything under 17%
[b] ( ) = Winter

gas and oil. Recent experience regarding the erratic price of petroleum fuels indicates a need to be cautious of increased dependency on high-quality fuels that can suddenly escalate in price. The use of natural gas by the electricity industry is projected to rise from 3 trillion to about 6 trillion cubic feet per year over the next decade; there is concern about the price stability and availability of natural gas used for electricity generation.

As indicated, suppliers of combustion turbines represent domestic as well as foreign suppliers who are just entering the US market. Foreign electrical equipment suppliers are providing competition for historic domestic suppliers during the 1990s. The continued presence of suppliers is needed to service equipment throughout its long lifetime.

Because of the low level of new construction activity during the 1980s, engineering staffs at electric utilities and contracting organization were reduced. In addition, many suppliers of conventional equipment, e.g. valves, water treatment, etc went out of business. If the construction program of electric utilities is accelerated to install baseload coal and nuclear plants, the limited capabilities of US engineering staffs may cause delays, reduce quali-

Table 6 Characteristics of gas turbines

Manufacturer	Capacity (MW)	
	Simple cycle	Combined cycle
A. Current offerings		
General Electric	83	123
Westinghouse	104	145
Asea/Brown Boveri	73	121
Siemans KWU	103	155
B. New models (1990s)		
General Electric	150	225
Westinghouse/Mitsubishi	150	210
Asea/Brown Boveri	155	220
Siemans KWU	140	198
C. Advanced models (2000)		
Major manufacturers	200	270

ty, and require suppliers from outside the United States to enter the market.

Fossil Power Generation Options

Rapid deployment and commercialization of clean coal technologies over the next decade represents an opportunity for the United States to protect existing export coal markets and to expand coal utilization in both existing and new energy applications.

After 30 years of relative technological stagnation, diverse new technology options for fossil fuel–based power generation are coming into serious commercial consideration. These include:

1. combustion turbines and combined cycles for natural gas/oil firing,
2. state-of-the-art coal power plant (SOAPP),
3. advanced sulfur dioxide (SO_2) and nitrogen oxide (NO_x) control,

Table 7 Concerns regarding electric supply

1. Capacity additions lag market growth

2. Need to add ~500 new turbines in 10 years

3. Increased use of gas and oil by the electricity industry

4. Entry of foreign suppliers into US market

5. Engineering capability shortages

4. fluidized-bed combustion systems—both atmospheric and pressurized—primarily for the retrofit and repowering markets (AFBC and PFBC),
5. slagging combustors and gasification-combined-cycle power plants, for their emission control capabilities and long-term economic potential.

A brief description of the status of each of these technologies follows.

COMBUSTION TURBINE POWER PLANTS   Currently, US electric utilities own 54,000 MW of combustion turbine capacity representing more than 2000 separate units. Forecasts project that combustion turbine capacity will rise dramatically by the year 2000 as the result of significant planned additions by both utilities and nonregulated generators. The additional capacity is anticipated to range between 40,000 and 75,000 MW with initial emphasis on simple-cycle, peaking duty applications. Unit sizes range from a few hundred kilowatts for special power-generation applications to the more commonly considered 20–150-MW units operated by electric utilities. Unit sizes continue to grow as higher temperatures and great airflow are incorporated into new power generators and the opportunities for combined-cycle operation expand.

The upturn in the market for combustion turbines confirms the strategic value they represent in long-term supply planning. At the current delivered price for natural gas (about $2/MBtu), combustion turbine capacity's low capital cost ($300/kW for simple-cycle, $600/kW for combined-cycle), rapid modular installation, and high efficiency when operated in the combined-cycle mode gives it the lowest projected cost of electricity of any of the alternatives now available or expected in the next several years. EPRI's analysis indicates that gas prices could increase by 50–100% from 1990 levels before the life-cycle power cost of combined-cycle generation would exceed that of coal-fired baseload capacity built today. Thus, clean coal technology has strong competition and will be introduced slowly during the 1990s.

EPRI is undertaking a project to perform a comprehensive durability test program on the advanced gas turbines currently being offered to utilities. This project will involve extensive instrumentation and surveillance of these machines to allow needed improvements in design and operating characteristics before many of a given design are placed into service. The first such surveillance project will involve a General Electric MS7001F engine providing peaking service at a plant of the Potomac Electric Power Co.

EPRI is also considering concepts emerging for the next generation of gas turbines, which roughly follow a 10–12-year development cycle. Researchers and manufacturers envision combustion turbines offered early in the next century to be about 200 MW in rating, with firing temperatures around 2500°F (1370°C), a simple-cycle efficiency of 37%, and a combined-cycle unit efficiency of more than 50% (higher heating value—HHV). Improvements in thermodynamic cycles, using such techniques as intercooling and steam injection, are also being explored.

Combustion turbines, which can be installed and operating with minimal field construction within a two-year period, represent a modular, incremental approach to capacity expansion. As demands grow, additional combined-cycle units can be added. And if gas prices exceed the break-even point with coal-generated electricity, a utility could build or buy the output of a coal gasification plant, fueling the combustion turbine with clean syngas.

STATE-OF-THE-ART POWER PLANT (SOAPP)   The past decade has seen great improvements in pulverized coal (PC) plant technology. Recent technological advances in virtually all components of the PC generating plant have been widespread, from improved coal-cleaning processes to more integrated environmental control systems, and with sophisticated electronics improving plant controls at all stages.

EPRI has initiated a project to assemble detailed specifications of a state-of-the-art pulverized coal power plant (SOAPP), which can be used for baseload or cycling. SOAPP uses proven, state-of-the-art designs, materials, and components, which conform to environmental restrictions, and incorporates them into a clean, efficient, modular plant. Many SOAPP technologies are equally applicable to oil- and gas-fired steam plants and other power plant types such as IGCC and FBC (see later sections on these types). The SOAPP concept standardizes major components while providing options among subsystems and materials. Standard systems include the boiler, turbine, control system, and air heater (Figure 4).

The incorporation of SOAPP technology will make possible new plants with lower life-cycle costs than those of current plants. Capital costs for SOAPP are expected to be lower than for recently constructed plants (e.g. $1400/kW for a plant in the north central United States). Shorter construction times (below 40 months) for SOAPP will decrease the interest charged during construction. In addition, a 25% reduction in nonfuel operating and maintenance costs from present practice is expected. And improved components and designs in SOAPP will result in improved heat rates.

SOAPP specifications will be assembled from EPRI research and development (R&D) results and other worldwide technological advances. Flow and energy balances will be provided at the level of detail necessary for preliminary engineering. Lists of materials, cost details, and a construction schedule will be provided in sufficient detail for utility planning studies. After the specifications are published, demonstration projects are planned with utilities. The results of these demonstrations will provide a broad base of experience for actual application of SOAPP technology.

SOAPP will benefit utilities that need to build new capacity and are faced

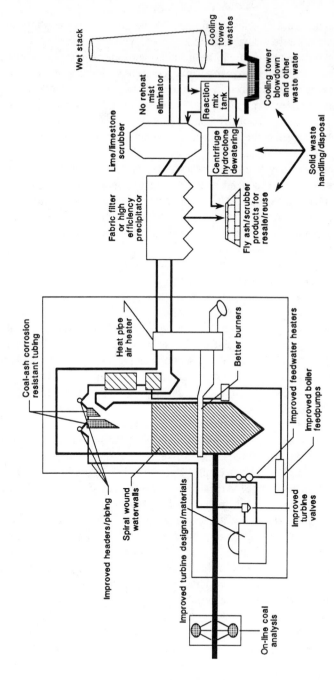

Figure 4 State-of-the-art power plant (SOAPP). SOAPP is an automated, cycling power plant that uses proven, state-of-the-art designs, materials, and components, including environmental control systems that meet or exceed anticipated environmental standards. An improved coal cleaning facility, although not a part of SOAPP, may be appropriate at some sites.

with choosing among a bewildering range of designs offered by different suppliers. It will provide design and cost benchmarks and guidelines for evaluating these options. SOAPP will serve the needs of US utilities in avoiding risk and preserving future generating options because of its flexibility, shorter construction times, and reduced operating and maintenance costs, and availability of operating experience gained worldwide.

ADVANCED SO$_2$ CONTROL   Advanced SO$_2$ control research goals include: (a) Improve flue gas desulfurization (FGD) reliability to $\geq$99% through simplification of systems. (b) Reduce capital and operating costs to less than 50% of historic wet FGD costs. (c) Minimize or eliminate solid wastes, and maximize the reuse of material from SO$_2$ control technologies. (d) Demonstrate 50–70% SO$_2$ dry injection technology at costs of $500–1000/ton of SO$_2$ removed.

Several advanced FGD processes are being or have been tested at large pilot and commercial scale. The CT-121 process (in which the flue gas is bubbled through a liquid absorber) has been pilot tested on high-sulfur coal at a 40 MW scale by EPRI. It is operating in Japan on 3% sulfur coal at a 250 MW scale, and EPRI will work with Southern Company Services to demonstrate an improved version at 100 MW under DOE's Clean Coal Technology Demonstration Program.

Another technology, high sulfur coal spray drying, adapts the successful concept currently used on 7000 MW of low-sulfur coal capacity. Two currently operating pilot plants are testing the performance on high-sulfur coal—a 10 MW spray dryer/ESP combination at Tennessee Valley Authority's Shawnee power plant, and a 4 MW spray dryer/fabric filter/combination at EPRI's High Sulfur Test Center near Buffalo, New York. Initial results indicate more than 90% removal of SO$_2$ can be obtained. If these results are confirmed at a larger scale, spray dry FGD would be a potential competitor with wet scrubbing for plants fueled with medium- to high-sulfur coal applications.

SORBENT INJECTION   Furnace Sorbent Injection is being demonstrated at two US power plants burning high-sulfur coal: Ohio Edison's Edgewater Plant, and Richmond Power and Light's Whitewater Valley station. Based on results from these two demonstrations, a third demonstration by the US Environmental Protection Agency (EPA) is now being planned at the Yorktown Station of Virginia Power.

Sorbent injection—both in furnace postfurnace (duct)—is characterized by moderate sulfur capture (40–70% SO$_2$ removal), simplicity of design and operation, low capital cost, and attendant savings in levelized removal costs. An issue in all sorbent injection processes, furnace or duct, is the increase in the quantities and the altered characteristics of solid wastes for disposal.

Disposal costs could increase from the presence of unreacted sorbent, which will need to be handled.

EPRI is updating its FGD cost estimates, which will evaluate a wide variety of FGD processes. Preliminary results show that for new plants, the experience gained and the advances made in part through EPRI R&D appear to have reversed the cost escalation that has characterized wet FGD in the past.

ADVANCED NO_x CONTROL   R&D to reduce NO_x emissions from stationary combustion sources in the United States is being applied for retrofit modifications of conventional pulverized-coal-fired boilers. Twelve demonstrations will be conducted involving low-NO_x burners (LNBs), overfire air (OFA), and reburning on each of the four types of conventional pulverized-coal-fired boilers in sizes ranging from 40 to 800 MW. Six NO_x-related Clean Coal II demonstrations represent a financial commitment of approximately $85 million by government and private industry.

Experience to date with these technologies shows NO_x reductions up to 50%, with little, if any, impact on boiler operation and maintenance. Capital costs are in the range of $10–30/kW. The first retrofit low-NO_x burner demonstration with overfire air, on a 320 MW tangential-fired boiler of Kansas Power and Light, showed that while 50% NO_x reduction was achievable at high load, the reduction was much less at low load factors. In addition, the overfire air system was responsible for the largest fraction (about 75%) of the total NO_x reduction. Lessons learned from this demonstration will be applied to others, and will provide further insights on the design parameters that can maximize NO_x reductions and minimize cost.

In the case of cyclone boilers, pilot tests indicate that modification options involving reburning and staged combustion have the potential to reduce NO_x emissions by 50–75%. Reductions of 40–50% are expected in field applications.

Research and demonstration activity with postcombustion NO_x controls uses selective catalytic reduction (SCR). European countries have more than 30,000 MW of coal-fired generating capacity equipped with SCR, while Japan has 6000 MW of SCR on coal-fired boilers. Retrofit installations in Europe and Japan typically achieve NO_x removals of 60–80%, with residual NH_3 emissions of less than 5 ppm (usually 1 to 2 ppm). The capital costs of these retrofits are consistent with EPRI's estimates, which range from $100 to $125/kW. Levelized cost projections for both US and Japanese installations are estimated at 4 to 9 mills/kWh.

The SCR experience abroad leaves several issues unresolved. One key issue is applicability to medium- and high-sulfur coals. Another is the possibility that fly ash will have a lower resale value if it absorbs NH_3. These issues are being studied by EPRI.

Work is also under way in the United States and Europe on selective noncatalytic reduction (SNCR) technologies. Results with urea injection show the potential for 30–50% NO_x reductions, and perhaps up to 75%, with NH_3 emissions below 5–10 ppm. Capital and operating costs are estimated to be $5–15/kW, and less than 3–4 mills/kWh, respectively.

Experience in Europe and Japan confirms earlier assessments that the least-cost approach for meeting stringent NO_x emission limits is to reduce NO_x as far as practical with combustion modifications before adding SCR. This practice would allow the catalyst to operate longer before replacement is required.

ATMOSPHERIC FLUIDIZED-BED COMBUSTION (AFBC)   In the period 1985–1989, approximately 50 circulating atmospheric fluidized-bed units started operating in the United States. These units are used primarily for the cogeneration of steam and electricity. They are operated with only one exception by industrial companies, cogenerators, and independent power producers. The total maximum net power that can be delivered to the US electricity grid by these plants is about 1800 MW. During the past five years 10 bubbling atmospheric pressure fluidized-bed units with a capacity of about 400 MW were in operation. Figure 5 summarizes growth in steam-generating capability for individual boilers for both circulating and bubbling beds.

Major utility-owned fluidized-bed combustion projects are in operation at

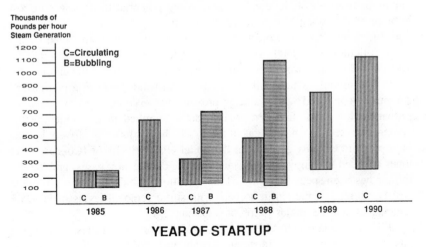

Figure 5   Fluidized-bed combustion single-boiler capacity growth.

four US sites. These are at Northern States Power (Black Dog plant, 130 MW bubbling bed), Colorado-Ute (Nucla plant, 110 MW circulating bed), Montana Dakota Utilities (Heskett plant, 80 MW bubbling bed), and TVA (Shawnee plant, 160 MW bubbling bed). The features of these plants are summarized in Table 8.

A 250 MW circulating fluidized-bed has been selected under Clean Coal II for repowering the A.B. Hopkins generating station of Tallahassee, Florida. This project involves repowering the existing gas/oil-fired Hopkins Unit 2 with a coal-fired circulating fluidized-bed. Bechtel North American Paper Corporation will provide design and construction management. Combustion Engineering, Inc. will provide the boiler, and Westinghouse Electric Corporation will provide the turbine-generator.

In addition to the approximately 500 MW of utility-owned capacity either in operation or startup, 550 MW is either planned or being constructed: the 300 MW of circulating fluid beds at Texas–New Mexico Utilities selected under Clean Coal I and the 250 MW circulating fluidized bed at Southwest Public Service selected under Clean Coal II. The Texas–New Mexico plant is in operation and was put in service smoothly.

PRESSURIZED FLUIDIZED-BED COMBUSTION (PFBC)   PFBC offers a number of potential advantages relative to AFBC systems. Foremost among these is improved efficiency, which results from the use of a gas turbine to extract energy from the pressurized combustion gas. This efficiency gain represents an improvement from AFBC system heat rates of about 10,000 Btu/kWh to 9,000 Btu/kWh for PFBC systems. Other advantages are a lower calcium requirement for sulfur removal as well as the potential to be lower in cost because of the modular nature of the system.

PFBC demonstration projects have been funded under both Clean Coal I and II solicitations by DOE. American Electric Power (AEP) has operated the 70-MW Tidd plant retrofit unit at Brilliant, Ohio. The coal for the unit is a 4% sulfur bituminous, which will be fed as a coal-water paste to a pressurized combustor contained within a large pressure vessel along with cyclones for primary solids separation. Pressurized air is supplied to the boiler by the compressor section of the gas turbine which, in turn, is driven by the high-pressure exhaust gases from the boiler. Parallel sets of two-stage cyclones are used to remove particulates from the combustion gases, and the gas turbine has been specially designed for operation with dusty gases. Dolomite will be utilized as the sulfur sorbent with sulfur removal levels of 90–95% anticipated at Ca/S molar ratios of less than 2.0.

EPRI is working with AEP on a DOE-funded program to test various hot gas cleanup devices on a slip stream containing 1/7 of the product gases. One of these hot gas cleanup systems may be selected for use in the 330 MW

Table 8 Utility fluidized-bed combustion demonstrations

	Northern States Power	Colorado-UTE	Montana-Dakota	TVA
Size (MW)	125	110	80	160
FBC type	Bubbling	Circulating	Bubbling	Bubbling
Feed	Overbed	Overbed	Overbed	Underbed
Supplier	Foster Wheeler	Pyropower/Ahlstrom	Babcock & Wilcox	Combustion Engineering
Scope	Boiler conversion and turbine generator	Add-on boiler	Boiler conversion	Add-on boiler
Coal	Low sulfur subbituminous	Low sulfur, high ash bituminous	Lignite	High sulfur bituminous
Startup	1986	1987	1987	1988
Design heat rate (Btu/ kWh)	10,800	11,300	11,900	9,800

Sporn project funded by DOE under Clean Coal II. That project is scheduled to come on stream in 1995.

SLAGGING COMBUSTOR   Slagging combustors are devices that are planned to (a) reduce NO_x emissions (via combustion stoichiometry control), (b) reduce SO_2 emissions (via sorbent injection), and (c) reject more than 90% of the coal ash as molten slag. Slagging combustors are cylindrical devices in which pulverized coal enters through one end along with tangentially injected combustion air. The high temperatures generated during combustion along with the combustion air vortex create a flowing molten slag layer on the combustion walls while the combustion of coal takes place in suspension. Slag flows down the walls through a collection device and into a slag tap for eventual disposal or commercial use. The development of atmospheric pressure combustors was initiated in the mid-1970s with use of pressurized magnetohydrodynamics (MHD) combustor technology as the basis.

At the present time TRW, TransAlta, CoalTech, and AVCO Research Lab are active in the application of this technology for steam boilers. Their efforts have been aimed at reducing emissions, integration to the boiler island, and equipment durability. TRW has performed large pilot-scale testing and has operated its combustor for more than 6000 hours in a small (35,000 lbs/hr steam) industrial boiler. TRW is participating in a DOE Clean Coal I program to demonstrate its combustor at the 70 MWe Lovett No. 3 boiler of Orange & Rockland utilities. EPRI is one of the project cofunders. To date TransAlta has conducted only pilot-scale tests on combustor emissions control. It has been selected by DOE to demonstrate this technology on a cyclone-fired boiler under the Clean Coal II program. CoalTech is demonstrating its combustor technology on a small (25 MBtu/hr) package (industrial) boiler under DOE's Clean Coal Technology I program. Avco Research has also performed pilot plant testing of its slagging combustor technology.

All of the developers indicate that NO_x emissions are less than 250 ppm (0.3 lbs/MBtu fired). Reported SO_2 removal efficiencies range from 50 to 80% depending on the developer, method of sorbent injection, and combustor operating conditions. Ash removal (in the form of slag) of 80–90% has been reported.

INTEGRATED GASIFICATION COMBINED CYCLE (IGCC)   EPRI's early work identified pressurized oxygen-blown entrained coal gasification coupled with high-efficiency advanced combustion turbines as the most promising candidates for development. Net system heat rates of 9000 Btu/kWh with SO_x emissions of about 10 ppm, NO_x emissions of about 25 ppm, and no particulate emissions are attractive characteristics. Figure 6 is a block flow diagram of an integrated gasification-combined-cycle (IGCC) power plant.

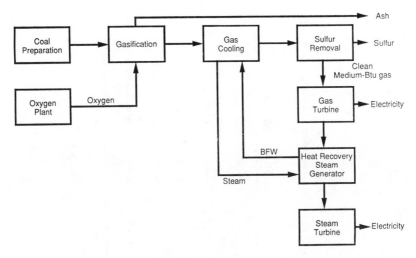

Figure 6 Integrated coal gasification combined-cycle system. Abbreviation used: BFW, boiler feed water.

Over the past 15 years EPRI has supported development work on a number of coal gasification processes. As a result of both experimental work and systems analysis, at least four processes appear to be of commercial significance to the utility industry. These are the Texaco, Dow, British Gas/Lurgi, and the Shell processes. The status of these technologies is provided in Table 9.

The 100 MWe Cool Water Coal Gasification Combined-Cycle Power Plant was based on Texaco Coal Gasification. The Cool Water project brought together Southern California Edison Company, Texaco, Inc., the General Electric Company, Bechtel, and a Japanese consortium led by Tokyo Electric Power Company, and EPRI as principals. ESEERCO and SOHIO Alternate Energy also contributed to the project. This team funded, developed, designed, constructed, and operated the world's first integrated, commercial-scale coal gasification combined-cycle power plant.

Many significant technical and management objectives were accomplished during the course of the Cool Water program. These were in part due to the extensive development program that had been carried out prior to the Cool Water project. For example, EPRI had carried out a large number of engineering and economic evaluations of gasification concepts that resulted in a design with attractive cost and performance characteristics. In parallel, EPRI cosponsored extensive test programs with Texaco, Ruhrkohle, and TVA, in 15 and 150 TPD gasification Texaco-licensed pilot plants. GE carried out combustion test programs on syngas to predict turbine performance and emissions.

Table 9 Status of coal gasification technologies

Technology	Plant throughout (TPD)[a]	Gasifiers	Coal	Plant site	Year of startup	Product
Texaco	1200	1	Low sulfur Utah bituminous	Cool Water Station, California	1984	135 MW electricity
	1500	3	World bituminous, coke	Ube City Japan	1984	H_2 for ammonia
	1200	1	Eastern US high sulfur bituminous	Kingsport, Tennessee	1983	CO/H_2 for chemicals
	800	1	European bituminous	Oberhausen, West Germany	1986	CO/H_2 for chemicals
Dow	1700	1	Subbituminous	Plaquemine, Louisiana	1987	160 MW electricity
BG/Lurgi	500	1	Bituminous	Westfield, Scotland	1984	Experimental data
Shell	2000	1	World coals	Holland	1993	250 MW electricity

[a]TPD = tons of coal per day

From these programs the design bases for the project were developed to minimize technical risk, and the design data were ultimately validated in a five-year program at Cool Water.

The Cool Water test program was completed in January 1989 having, in the judgment of the participants, met or exceeded all of the goals for the project. The project was formally terminated in June 1989 and ownership of the plant was transferred to Southern California Edison. In September 1989 Texaco announced that it was negotiating with Southern California Edison for the purchase of the plant. Texaco plans to gasify a mixture of sewage sludge and coal. If all necessary approvals are obtained and other negotiations completed, the plant, after appropriate modification, will be operational in 1992.

A second IGCC project is under way at Dow Chemical's Louisiana location in a plant with a single gasifier producing sufficient fuel for 160 MW. This project is of particular interest to utilities since it was constructed in phases. Two 100 MW combustion turbines fueled by natural gas were put in service several years before the gasification plant was constructed. As synthesis gas from the coal gasification plant becomes available, an equivalent amount of natural gas is backed out as turbine fuel. This plant started up in April 1987, using subbituminous coal. The plant had exceeded its rated design capacity and achieved acceptable capacity factors by mid-1990. The Public Service of Indiana Co. plans a commercial plant featuring Dow technology.

Presently, more than 10 US utilities are conducting or planning site-specific designs based on the Cool Water and other coal gasification process experience. These utilities are particularly interested in the phased construction approach, since this provides maximum planning flexibility and reduces both the front-end financial burden and overall revenue requirements.

The next major step in IGCC's evolution toward acceptance by the utility industry is the 250 MW project to be built by SEP (a group of Dutch utilities) in the Netherlands. This plant, which will use Shell gasification technology, is closely integrated, and promises very high efficiency. It is scheduled to begin operation in 1993. About 15 IGCC additional projects are planned in Europe. On a worldwide basis about 4000 MW of capacity featuring IGCC systems is planned or under construction.

ELECTRICITY WITHOUT STEAM   Conventional fired boilers have a fundamental thermodynamic limitation that derives from the use of steam as a working fluid in the Rankine cycle. The limitation of Rankine cycles represents a plateau in efficiency of about 40%.

Advanced coal gasification plants, which at this time are conceptual, represent a pathway to high-efficiency electric generation that is possible with an orderly program of development over the next two decades. These advanced generation systems will provide a clean generation system that can produce electricity at a heat rate as low as 6500 Btu/kWh. Such advanced systems incorporate advanced-cycle, high-temperature turbines (2500°F firing temperature), or large-size molten carbonate fuel cells. These new systems may also operate with gasification technology that provides a gaseous fuel containing a high fraction of chemical energy in the form of methane. Figure 7 shows the results of systems analysis performed on new conceptual generating systems. As indicated, two of these systems eliminate the steam cycle completely, and feature molten carbonate fuel cells. By chemically integrating the fuel cell with advanced coal gasification processes (which were operated in pilot scale equipment in the 1970s), overall efficiencies of about 55% are expected. Molten carbonate fuel cells are now being operated only on natural gas at small (100 kW) size. However, commercialization of molten carbonate fuel cells may come about in this decade.

A development program of $100 million–$200 million is required to develop molten carbonate fuel cells. Demonstration of such advanced plants at a large scale is likely to represent an additional investment of about $0.5–1.0 billion. Advanced technology based on coal gasification also has the potential

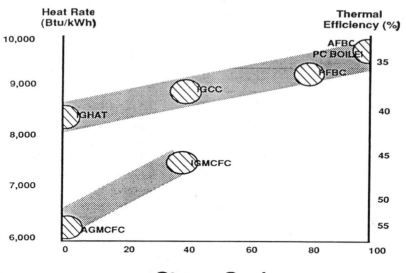

Steam Cycle
Contribution to Gross Power Production (%)

Figure 7 Coal-fired power cycles. Abbreviations used: IGHAT, integrated gasification humid air turbine cycle; AGMCFC, advanced gasification molten carbonate fuel cell; IGCC, integrated gasification combined cycle; AFBC, atmospheric fluidized-bed combustion; PC boiler, pulverized coal boiler; PFBC, pressurized fluidized-bed combustion.

to produce many chemical coproducts, such as methanol and higher alcohols along with electric power.

The Clean Coal Technology development path to highly efficient generation of clean electricity has just started. The realization of this bold challenge is likely to occupy the next generation of scientists and engineers, whose task will be to complete the new systems discussed in this paper.

Annu. Rev. Energy Environ. 1991. 16:25–57

FLUIDIZED-BED COMBUSTION TECHNOLOGY[1]

E. Stratos Tavoulareas

3089 Jennifer Way, San Jose, California 95124

KEY WORDS: power generation, environmental control, clean coal technology.

INTRODUCTION

The need to increase the utilization of coal precipitated by the oil crises of the 1970s, followed by the pressure experienced in the 1980s to minimize the environmental impact of coal combustion, have resulted in the development of various coal-based technologies that have improved environmental performances. These technologies include precombustion coal cleaning methods, postcombustion environmental control technologies, and advanced coal combustion, gasification, and liquefaction technologies. Coal cleaning may involve physical, chemical, or biological methods to reduce the sulfur and ash content of the coal prior to entering the combustion system. Postcombustion

---

[1]Abbreviations used: AFBC, atmospheric fluidized-bed combustion; B&W, Babcock and Wilcox Co.; CaO, calcium oxide; Ca/S, calcium-to-sulfur molar ratio; CE, Combustion Engineering Inc.; CO, carbon monoxide; CURL, Coal Utilization Research Laboratories of Great Britain; DOE, United States Department of Energy; EPA, US Environmental Protection Agency; EPDC, Electric Power Development Corporation of Japan; EPRI, Electric Power Research Institute; ESP, electrostatic precipitator; FBC, fluidized-bed combustion; FERC, Federal Energy Regulatory Commission; FGD, flue gas desulfurization; IEA, International Energy Agency; MonDak, Montana-Dakota Utilities Co.; MW, megawatt electric; NCB, National Coal Board of Great Britain; NO$_x$, nitrogen oxide; NSP, Northern States Power Co.; NSPS, New Source Performance Standards; O&M, operating and maintenance; PC, pulverized-coal; PFBC, pressurized fluidized-bed combustion; ppm, parts per million; PURPA, Public Utilities Regulatory Policy Act; R&D, research and development; SO$_2$, sulfur dioxide; TVA, Tennessee Valley Authority.

1056-3466/91/1022-0025$02.00

environmental control technologies include flue gas desulfurization, catalytic and noncatalytic NO_x control technologies, and processes that simultaneously remove SO_2, NO_x, and particulates from the flue gas.

Of these processes, fluidized-bed combustion deserves particular attention because of its potential to reduce SO_2 and NO_x in the boiler without additional postcombustion equipment. Various FBC design alternatives have already been accepted by the electric utility and industrial market, while others are in the demonstration phase. The technological and market developments related to FBC technology should therefore be of interest to professionals such as engineers, power generation planners, financiers, and regulators.

This chapter describes the evolution of FBC technology, the developmental status of various FBC designs, potential applications and market penetration, and issues associated with the commercialization of FBC technology. The role of government and R&D organizations such as the US Department of Energy, Electric Power Research Institute, and the International Energy Agency in the development and commercialization of FBC technology is also described. Although the article covers both atmospheric and pressurized FBC systems, greater emphasis is given to atmospheric fluidized-bed combustion technology, which is already widely accepted.

HISTORY OF FLUIDIZED-BED COMBUSTION TECHNOLOGY

Fluidized-bed technology dates back to the early 1920s, when F. Winkler invented the first fluidized-bed gas generator in Germany (1, 2). In the 1930s and 1940s, Germany developed the technology and used it in coal gasification and metal refining applications. In the United States, fluidized-bed technology was developed in the 1930s by the petroleum industry to speed the reaction of oil feedstock catalytic cracking, and it has since been established as the primary technology for such applications.

In the early 1960s, the British National Coal Board and the Maoming Petroleum Company in China began research on fluidized-bed technology for the combustion of coal and oil shale, respectively. These developments were driven primarily by the need to burn poor-quality solid fuels.

In the United States, interest in fluidized-bed combustion technology developed in the late 1960s and was accelerated as a result of the oil crises of 1973 and 1979 and the introduction of environmental regulations on gaseous emissions from power-generation facilities. Research focused on the development of a technology that could burn solid fuels and reduce sulfur dioxide and nitrogen oxide emissions without requiring postcombustion treatment of the flue gas by methods such as flue gas desulfurization and selective catalytic reduction.

Also in the 1970s, Lurgi Gesellschaft developed a circulating atmospheric fluidized-bed calciner for the aluminum industry in West Germany, while the same technology was developed in Finland and Sweden for combustion of coal, peat, wood waste, and bark. As a result of the early development efforts, various types of fluidized-bed boilers emerged: bubbling and circulating (see section entitled FLUIDIZED-BED COMBUSTION PROCESS).

In the United States, the initial focus of organizations such as the Environmental Protection Agency, the Department of Energy, and the Electric Power Research Institute, working with private industry, was on the development of bubbling AFBC technology. A 0.5-MW AFBC test facility built in Alexandria, Virginia, with government funding was the first AFBC unit in the United States. Based on the test results from this unit, the first AFBC demonstration unit was designed and operated at the Riversville station of the Monongehela Power Company in West Virginia. Design of the unit began in 1972, and it was operated from 1976 to 1981 with DOE funding. It burned coal and had a 136,000-kg/h (300,000-lb/h) steam-generating capacity.

By the early 1980s, DOE had expanded its AFBC development program to include the construction and operation of three industrial-scale demonstration units:

1. A 45,000-kg/h (100,000-lb/h) AFBC unit burning high-sulfur coal at Georgetown University, Washington, D.C., which has operated commercially since January 1979.
2. A 22,000-kg/h (50,000-lb/h) coal-burning AFBC boiler at the Naval Training Center in Great Lakes, Illinois, which operated commercially from September 1981 to 1984.
3. A 9,000-kg/h (20,000-lb/h) AFBC unit burning anthracite culm in Shamokin, Pennsylvania, which has operated commercially since October 1981.

In parallel with DOE, EPRI focused its R&D efforts on the development of utility-scale AFBC power plants. The first major EPRI project was the construction and operation of a 2-MW fluidized-bed (6' × 6') pilot plant at the Babcock & Wilcox research center in Alliance, Ohio, in 1977. This pilot plant provided data that enabled EPRI and B&W, in conjunction with the Tennessee Valley Authority, to design, build, and operate a 20-MW pilot plant at TVA's Shawnee station in Paducah, Kentucky. This unit was designed to burn high-sulfur Kentucky No. 9 coal and to test alternate coal and limestone feed systems. The TVA 20-MW pilot plant completed more than 20,000 hours of coal-fired operation between 1982 and 1987, and has confirmed such AFBC features as good combustion efficiency, low level of emissions, and fuel flexibility.

Based on experience gained from these facilities, EPRI and a number of electric utilities began a program to demonstrate AFBC technology for utility applications. This program culminated in four large-scale AFBC units, which began operation in 1986 and 1987:

1. Northern States Power's conversion of an existing 100-MW pulverized-coal unit (Black Dog Unit 2) to a 130-MW bubbling AFBC unit.
2. Montana-Dakota Utilities' conversion/upgrade of an existing stoker-fired boiler (Heskett Unit 2) to an 80-MW bubbling AFBC unit firing North Dakota lignite.
3. Colorado-Ute Electric Association's 110-MW circulating AFBC boiler to repower its Nucla generation station.
4. TVA's new 160-MW bubbling AFBC boiler at its Shawnee steam plant near Paducah, Kentucky. The unit provides steam to the existing turbine/balance-of-plant for Unit 10.

Since these AFBC demonstration plants started operations, a large number of AFBC power plants have been commissioned in the United States and overseas. As of mid-1990, a number of AFBC plants are in operation in the 100–160-MW size range, while a number of projects in the design phase involve 250-MW AFBC units. There is also significant activity in Northern Europe, Japan, India, and China in both commercial and R&D projects.

The development of pressurized fluidized-bed combustion technology started in 1969 with the construction and operation of a 6 thermal-megawatt (MWth) PFBC combustor at the British National Coal Board's Coal Utilization Research Laboratories (3–7). Based on the experience gained from this facility, an 80-MWth PFBC facility was built in Grimethorpe, England, in 1980. This facility was established under the auspices of the International Energy Agency, and it was supported by the governments of the United States, Federal Republic of Germany, and the United Kingdom by direct funding of construction and operation and provision of technical expertise. Further development of PFBC technology was undertaken by Asea BB Carbon of Sweden, and more recently by other organizations such as Deutsche Babcock of Germany and Ahlstrom OY of Finland.

Pressurized fluidized-bed combustion technology is also receiving significant attention because of features such as low emissions, high cycle efficiency (more than 40% for a cycle combining both Brayton and subcritical Rankine cycles), and the compactness of the components, which makes it suitable for modular construction and retrofit applications. As of mid-1990, PFBC technology is entering the demonstration phase with three utility-scale projects:

1. American Electric Power Company is retrofitting a PFBC boiler to its existing 70-MW Tidd Unit 1 in Brilliant, Ohio.

2. Stockholm Energi Production AB is building two PFBC modules with combustors and gas turbines and a single steam turbine in Stockholm, Sweden. The output of the plant will be 135 MW (net, electric) and more than 225 MW (thermal).
3. ENDESA (Empressa Nacional de Electricidad S.A.) of Spain is building a 63-MW PFBC unit firing black lignite and utilizing dolomite as a sorbent.

All of these PFBC plants started operating during 1990. A project of a similar size is also planned in Japan by the Electric Power Development Corporation.

THE FLUIDIZED-BED COMBUSTION PROCESS

The key features of FBC combustors that distinguish them from conventional pulverized coal boilers are in situ SO_2 removal and the relatively low gas-side operating temperature (approximately 840°C, 1550°F). Coal and sorbent are injected into the boiler (see Figure 1), while air is blown upwards through the air distributor in the bottom of the boiler. The sorbent and coal ash particles form a dense zone at the bottom of the boiler. When the air flow rate has increased sufficiently, the solid particle mass expands and takes on the hydrodynamic properties of a fluid: thus the term "fluidized bed." As the air velocity increases further, some solid particles are entrained and carried through the top of the boiler.

The calcium carbonate sorbent introduced into the boiler calcines ($CaCO_3$ → CaO + CO_2) and produces porous CaO particles. In a sufficiently oxidizing environment, the CaO reacts with the SO_2, which is generated from the

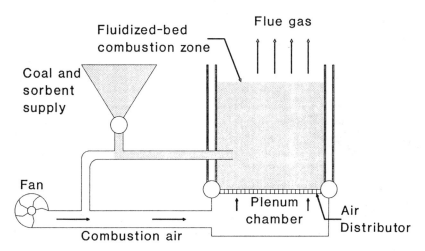

Figure 1 Fluidized-bed combustion process.

sulfur in the coal to form calcium sulfate ($CaO + SO_2 + 1/2\ O_2 \rightarrow CaSO_4$). Under normal conditions, the latter is a benign solid, which is removed from the FBC system as a dry, solid by-product. Thermodynamically, the formation of $CaSO_4$ is maximized at relatively low temperatures (820–870°C, 1500–1600°F). For this reason, the bed temperature of FBC systems is maintained within this narrow range. This low operating temperature is a key factor for most of the advantages of the FBC process, which include (*a*) high SO_2 removal efficiency, (*b*) no slagging of the boiler, (*c*) the ability to handle various fuels without significant performance penalties, and (*d*) low NO_x generation. In addition, the turbulence in the bed results in good combustion and sorbent utilization efficiency even when low-quality fuels are fired.

FBC boilers are usually classified based on their gas-side operating pressure and gas velocity. Thus, boilers operating at near-atmospheric pressure are called atmospheric fluidized-bed combustors, while those operating at 5 to 20 bar are called pressurized fluidized-bed combustors. As the gas velocity increases from 0.3 to 20 m/s (1 to 60 ft/s), the fluidization regime changes (see Figure 2), and the boiler design is adjusted accordingly:

1. fixed FBC at gas velocities of less than 1.2 m/s (4 ft/s),
2. bubbling FBC at velocities of 1.2–3.7 m/s (4–12 ft/s).
3. circulating FBC at velocities of 3.7–9.2 m/s (12–30 ft/s), and
4. transport reactor at velocities of more than 9.2 m/s (30 ft/s).

All four regimes can operate under both atmospheric and pressurized conditions. As of 1990, bubbling and circulating FBC regimes are used for power-generation applications, and they are therefore the focus of this chapter.

AFBC Technology

The development of AFBC boilers has received particular attention because of the potential advantages of the process and the similarities between AFBC power plant components and those used in conventional pulverized-coal technology with respect to fabrication and plant operation. As mentioned in the previous section, two types of AFBC boilers are used for power-generation applications: bubbling and circulating. Similarly to conventional PC plants, both AFBC boiler types operate in a balanced-draft mode, with approximately 20% excess air required for good combustion efficiency.

BUBBLING AFBC BOILERS   In bubbling AFBC systems, air is introduced at the bottom of the boiler (see Figure 3), while coal and sorbent are introduced either through the bottom (underbed feed system) or immediately above the bed (overbed feed system). Although a significant percentage of the coal

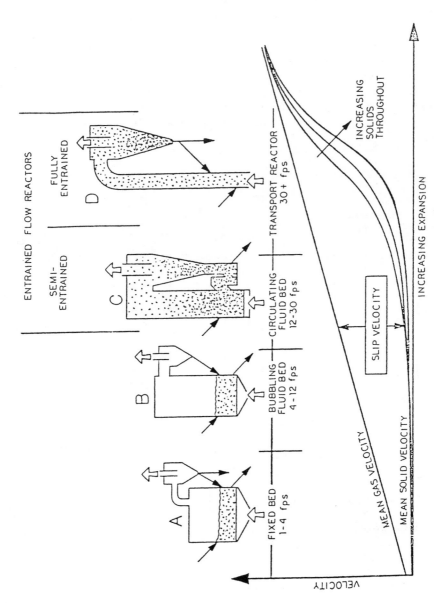

Figure 2 Fluidization regimes (Source: Lurgi Corporation). Abbreviation used: fps, feet per second.

burns upon introduction into the boiler, a large percentage of solids is carried out of the boiler. This material contains some unburned carbon, unreacted sorbent, calcium-sulfur compounds, and coal ash. To increase the efficiency of the AFBC process, a large percentage of these solids is caught in the mechanical dust collector (downstream of the boiler, as shown in Figure 3) and recycled into the boiler. This gives the carbon and sorbent particles a longer residence time in the boiler so as to burn completely and capture sulfur more efficiently.

To maintain the bed temperature within the desirable range, heat-exchangers (evaporators and superheaters) are located in the bed. Additional superheat, reheat, and economizer surface, along with an air heater, are

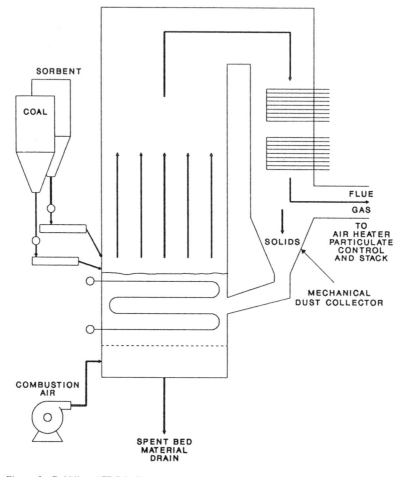

Figure 3   Bubbling AFBC boiler

located downstream of the bed. This part of the boiler is very similar to conventional PC boilers; the few differences are limited to the relatively high dust loading of the gas in AFBC boilers and the higher pressure difference between the gas and air streams at the air heater. This pressure difference has an impact on the design and selection of the air heater. The designer has a choice among a regenerative air heater with special seals to minimize air leakage, tubular heat pipe, or plate-type air heaters. Downstream of the air heater there is either a fabric filter or an electrostatic precipitator for additional particulate removal.

Uniform distribution of both the coal and sorbent feeds is important in bubbling AFBC boilers. The two main differences between underbed and overbed feed systems are:

1. Underbed feeders introduce coal and sorbent through a network of feeder pipes into the bottom of the bed. The coal used in these systems typically has a topsize of less than 0.64 cm (1/4 inch) and a surface moisture of less than 6%. Since current practice requires a coal feed point for every 0.9–1.9 m^2 (10–20 ft^2) of bed area, large boilers may have hundreds of feed pipes, and the associated distribution system tends to be complex and costly. The pneumatic transport blowers required for underbed feed systems entail higher capital cost and auxiliary power requirements than overbed feed systems. Although the sorbent does not have the same distribution requirements as the coal and may be dropped through a few points over the bed, for convenience the sorbent is typically transported pneumatically with the coal. The underbed system, however, yields better coal and sorbent utilization efficiency than the overbed system, especially when firing coal with a low reactivity and containing a high percentage of fines.

2. Overbed feeders, or spreader stokers, broadcast the coal and sorbent over the bed surface. Typical coal size is less than 3.2 cm (1-1/4 inch). They have fewer feed points and the distribution system tends to be simpler and less costly. However, the residence time and mixing of the coal particles and gas in the boiler are less than in a bubbling fluidized-bed, which may result in lower coal and sorbent utilization efficiency and higher operating costs. In particular, if the coal contains a high percentage of fines, both combustion and SO_2 removal efficiency may be affected adversely.

The preferred feeding method largely depends on the coal characteristics. Highly reactive subbituminous coals and lignites can be fed overbed and achieve essentially the same carbon and sorbent efficiency as with underbed feeding. For certain bituminous coals, however, underbed feeding is more efficient. The sorbent utilization efficiency, measured in terms of the calcium-to-sulfur molar ratio needed for a given level of sulfur removal, depends to some degree on the amount of recycled solids, the sorbent reactivity, the operating bed temperature, and the sulfur content of the coal. In addition, the size distribution of the solids and the local mixing between gas and sorbent particles affects the sorbent utilization efficiency. The Ca/S required by bubbling AFBC boilers is slightly higher than that required by circulating AFBC boilers for the same coal, sorbent, and SO_2 removal efficiency.

Load following by decreasing the firing rate throughout the bed is limited to approximately 30% due to constraints on bed temperature. To compensate for this limitation, the bed is typically divided into compartments that can be controlled independently. Thus the boiler load can be reduced by taking bed compartments out of service. Each compartment has separate air and coal/limestone feed systems. This feature results in improved low-load control for bubbling AFBC boilers relative to circulating AFBC boilers, which are not divided into compartments.

CIRCULATING AFBC BOILERS   The circulating AFBC boiler is similar in many respects to the bubbling AFBC system, but its generally smaller sorbent feed particle size and the higher gas velocities in the boiler lead to important physical differences. The bed in a circulating AFBC boiler expands to fill the entire furnace volume (although 90% of the mass may be in the lowest third). The density profile along the height of the furnace changes gradually, and there is no distinct high-density region at the bottom of the furnace. As a result, the heat distribution is more uniform, and no heat-exchangers are needed in the lower part of the furnace (see Figure 4). The difference between the velocities of the gas and the solids (termed the slip velocity), which is larger in circulating than other types of AFBC systems (shown in Figure 2), creates high local turbulence which, in turn, enhances solids mixing and heat and mass transfer. These boilers therefore require an order of magnitude fewer feed points for both the coal and the sorbent.

Since the smaller particles and higher velocities in the circulating AFBC system result in a high flow rate of solids out of the bed (50–100 times higher than in an equivalent bubbling AFBC boiler), virtually all of the material captured in the cyclones is recycled to the combustor. As a result, combustion takes place throughout the furnace volume as the solids "circulate" through the system. In contrast, the bubbling AFBC boiler, which has a definite division between the fluidized bed and the freeboard, burns 80–90% of the fuel within the bed, while the remainder is burned before it exits the boiler. Since the gas velocity in circulating AFBC boilers is higher than in bubbling units, the combustor footprint area is much smaller. However, the required height for circulating AFBC boilers is greater than for bubbling systems.

Another characteristic of the circulating AFBC process is that the combustion temperature must be kept fairly constant throughout the height of the furnace. The main mechanism for controlling the furnace temperature profile

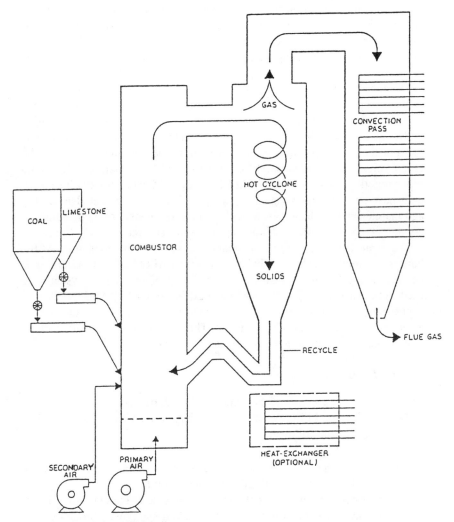

Figure 4   Circulating AFBC boiler

is the introduction of air at different elevations of the boiler, or staged combustion (see secondary air in Figure 4). The introduction of air above the air distributor controls not only combustion in the furnace (the heat release profile), but also the generation of NO_x. Bubbling AFBC boilers may utilize air introduced above the bed (overfire air), but this is usually limited to 15–20% of the total air flow rate. In circulating AFBC boilers, the secondary air flow rate may be as high as 50% of the total air.

Superheat or reheat surfaces are usually located in the upper part of the boiler and the convection pass. Some boiler manufacturers use an external heat-exchanger (shown in Figure 4), which utilizes the heat from the solids caught in the cyclone before they are recycled into the furnace. The presence of the external heat-exchanger may reduce the height of the circulating AFBC boiler.

The location of the cyclone with respect to the heat transfer surface of the convection pass is another difference between bubbling and circulating AFBC systems. The cyclone is located before the convection pass in circulating AFBC units. Therefore, the superheat, reheat, and economizer heating surfaces in the convection pass are not exposed to the high dust loading of the gas. Cyclones in circulating AFBC systems need to be more efficient solid separators than those in bubbling AFBC systems. They are usually lined with both heat- and erosion-resistant refractories. To reduce potential reliability and maintenance problems associated with the refractory, some manufacturers have recently introduced water-cooled solid separators.

The coal feed preparation and handling systems in circulating AFBC boilers are similar to those of bubbling overbed AFBC units. Coal drying is not required since mechanical (as opposed to pneumatic) feed systems are used. The particle size of the sorbent used in circulating AFBC boilers is significantly finer (as low as 150 microns), and less sorbent is required to achieve the same level of sulfur removal. This leads to lower sorbent handling system capacities and generates less ash for disposal.

PFBC Technology

The high gas-side pressure of the combustor (5–20 bar) is the main difference between AFBC and PFBC systems. This high pressure affects the PFBC plant configuration in the following areas:

1. The size of the PFBC boiler is much smaller than either AFBC or pulverized-coal boilers with an equivalent output.
2. The high-temperature and -pressure gas available at the exit of the boiler (see Figure 5) can be utilized to produce additional power through expansion in a gas turbine.
3. The introduction of the gas turbine into the PFBC plant provides the plant designer with the opportunity to configure it as a combined-cycle power plant.

Combustion air enters the PFBC boiler through the air distributor, as with the AFBC boiler, but at an elevated pressure. The bed temperature is maintained between 820 and 920°C (1500–1700°F) for good combustion and SO_2 removal efficiencies. As a result of its improved utilization efficiency, the

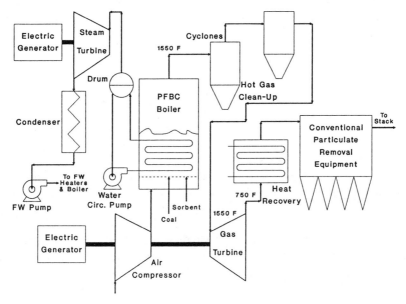

Figure 5 Combined PFBC cycle. Abbreviation used: FW, feed water.

preferred sorbent for PFBC systems is dolomite ($CaCO_3*MgCO_3$) rather than limestone ($CaCO_3$).

Coal and sorbent are fed either in a dry form or as slurries. In the former case, coal ground to 0.32 cm (1/8 inch) or less is transferred pneumatically through a series of pressurized lockhoppers and then injected into the boiler between the air distributor and the in-bed heat-exchangers. The coal needs to be dried to 2–4% total moisture. When slurry is used, it usually contains both the coal and the sorbent along with water.

Similarly to AFBC systems, the gas and a large amount of solids carried downstream of the furnace are cleaned in high-temperature and -pressure devices such as cyclones, electrostatic precipitators, or ceramic filters. Most recent PFBC designs use two stages of cyclones. The need to minimize the dust carryover into the gas turbine requires that these devices be efficient and reliable. For this reason, particular attention is paid to improving existing hot gas cleanup technologies and to developing new ones.

After the gas is cleaned, it expands through a gas turbine, which moves the air compressor (shown in Figure 5) and produces excess electric power (typically, 20–35% of the total power produced). Downstream of the gas turbine, the gas is cooled further through a conventional heat recovery area, cleaned in a conventional fabric filter or electrostatic precipitator, and then discharged into the environment. Steam generated in the PFBC boiler and the heat recovery area is expanded in a steam turbine to produce power. The cycle described here is a combined cycle. However, there are a number of alternate configurations, which are described in the following section.

In addition to the advantages of AFBC technology, PFBC technology has the following further advantages:

1. It is potentially capable of achieving plant heat rates as low as 8000 Btu/kWh as compared to 9500 Btu/kWh for AFBC power plants.
2. It is particularly suitable for repowering applications, i.e. replacement of an existing boiler with a PFBC unit while maintaining the existing steam turbine and the balance of plant.
3. It utilizes very small power plant components (because of the high pressures involved), which make them suitable for shop fabrication and modular construction.

The latter feature alone has the potential to reduce field labor requirements by 30–40% and to make greater use of the trained, highly productive personnel at the fabricating facility. The resulting benefits can be construction schedule reductions of as much as 15% and 10–20% savings in the overall capital costs of the plant.

ALTERNATIVE DESIGN CONFIGURATIONS AND SUPPLIERS OF FBC POWER PLANTS

Alternative AFBC Design Configurations

In the past, bubbling AFBC boilers were offered by a number of US manufacturers: Babcock & Wilcox, Combustion Engineering, Energy Products of Idaho, Foster Wheeler, Keeler/Dorr-Oliver, and licensees of Battelle's Multi-Solid FBC process. However, since 1987, new bubbling AFBC projects have been limited to small plants burning biomass fuels, while boiler vendors are focusing on the development and commercialization of circulating AFBC technology.

Outside the United States, a number of government and private organizations are pursuing the development and commercialization of bubbling AFBC technology. The capability of AFBC technology to burn low-quality fuels, coupled with the moderate environmental requirements of certain countries such as India and China, makes it the technology of choice. In these countries, low-quality fuels (lignite and oil shale, respectively) have been burned successfully in many AFBC power plants. Japan supplements AFBC boilers with selective catalytic devices to meet its strict environmental requirements. Japan has recently decided to proceed with the largest AFBC project in the world, involving conversion of a PC unit to a 350-MW bubbling AFBC boiler.

Circulating AFBC design configurations differ in their use of an external heat-exchanger (shown in Figure 4) and in their location of the superheat and reheat surfaces in the upper part of the furnace or in the external heat-exchanger. Combustion Engineering/Lurgi and Riley Stoker use an external fluidized-bed heat-exchanger. This approach provides more control flexibility and somewhat reduces the boiler height, but it adds to the complexity of the plant. The second approach, employed by Pyroflow boilers (trademark of Ahlstrom OY of Finland), places superheater and/or reheater panels in the upper region of the furnace. This is obviously simpler, but it exposes the high-temperature tubes to a potentially erosive environment and may require a turbine bypass system during startups. In addition, controlling superheat/reheat temperatures at low loads with the latter design may result in increased turbine heat rates.

Ahlstrom and Lurgi are the industry leaders in circulating AFBC technology (see Figure 6). As of 1990, Ahlstrom has more than 40 AFBC power plants in operation worldwide. In North America it markets the technology through Pyropower Corporation, a wholly owned subsidiary. Lurgi has approximately 30 units in operation and under construction. Other circulating AFBC vendors include (a) Gotaverken Energy Systems AB of Sweden, which focuses on the combustion of materials such as peat, wood wastes, and refuse-derived fuels along with coal, (b) Keeler/Dorr-Oliver, with a circulating AFBC configuration similar to Ahlstrom's, (c) Studsvik Energieteknik AB of Sweden, which has licensed its technology to companies such as Babcock & Wilcox of the

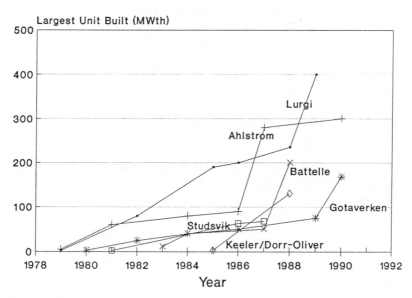

Figure 6 Circulating AFBC technology: development by different vendors. Source: Ref. 9

United States, Babcock-Hitachi of Japan, and Ensaldo of Spain, and (d) Foster Wheeler in the United States.

The Multi-Solid process developed by the Battelle Memorial Institute in Columbus, Ohio, is not strictly a bubbling or a circulating AFBC, but incorporates design features from both technologies. The Multi-Solid AFBC technology has been licensed to Senior Greens, Riley Stoker, and Mitsui Corporation, which have eight plants in operation.

In other recent developments, Deutsche Babcock Werke and Foster Wheeler are independently developing a "hybrid" AFBC boiler that reportedly combines the best features of bubbling and circulating AFBC boilers, such as (a) gas velocities in a range between those of bubbling and circulating AFBC systems, (b) a dense bed in the main combustor, (c) an evaporator/superheater in the upper part of the combustor, (d) a smaller solids circulation rate than that of circulating AFBC systems, (e) use of a smaller, water-cooled cyclone than in circulating AFBC systems, and (f) no external fluidized-bed heat-exchanger.

Alternative PFBC Design Configurations

Similarly to AFBC technology, PFBC power plants utilize both bubbling and circulating FBC boilers. Bubbling PFBC boilers are designed for a gas velocity of approximately 0.6 m/s (2 ft/s). All of the heat transfer surface is in the boiler vessel, and a minimal amount of refractory is used. Circulating PFBC boilers are designed for higher velocities and utilize external heat-exchangers and refractory-lined combustor, cyclone, and gas pipes. All the PFBC demonstration plants under construction and operation as of 1990 employ the bubbling PFBC design. However, development efforts to commercialize the circulating PFBC design have intensified, with particular emphasis on boiler scale-up, erosion reduction, and hot gas cleanup.

With regard to the cycle configuration, the following alternatives are being considered (a) combined PFBC cycle, (b) turbocharged PFBC cycle, and (c) topping PFBC cycle.

In the combined cycle (shown in Figure 5), the air compressor pressurizes the boiler to 5–20 bar. Particulates are removed from the combustion gas in two stages of conventional cyclones. The gas then expands through a turbine, is cooled by a heat recovery boiler, and is further cleaned in a conventional particulate collection device prior to its release into the environment. The gas turbine drives the air compressor and an electric generator; approximately two-thirds of its power is used by the compressor with the remainder producing electricity. The temperature of the gas at the exit of the boiler and the inlet to the gas turbine is approximately 840°C (1550°F). Therefore, gas cleanup occurs at relatively high temperatures and pressures. In addition, the gas turbine is exposed to some dust loading, because final particulate removal occurs downstream of the turbine. Expected net heat rates for the subcritical cycle are approximately 8800 Btu/kWh, while for the supercritical cycle they are approximately 8500 Btu/kWh.

The turbocharged cycle (see Figure 7) differs from the combined cycle in the following respects:

1. The heat recovery boiler has been eliminated downstream of the gas turbine, and the heating surface is placed at the upper part of the PFBC boiler.
2. As a result of this change, the gas exiting the boiler is at a lower temperature (430°C, 800°F), and complete gas cleanup is possible prior to the gas turbine.
3. The gas turbine produces just enough power to drive the air compressor.

All of the electricity produced by the power plant comes from the steam turbine. The heat rate of this cycle is expected to be approximately 600 Btu/kWh higher than that of the corresponding combined cycle. However, the following advantages are claimed for it:

1. The reduced temperature at the exit of the boiler allows utilization of near-conventional hot gas cleanup equipment, which is more reliable and

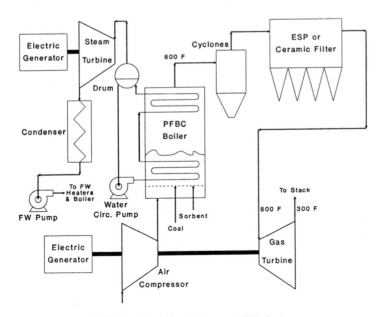

Figure 7 Turbocharged PFBC cycle. Abbreviation used: FW, feed water.

less expensive than the corresponding equipment for the combined PFBC cycle.

2. Particulate loading to the gas turbine is reduced significantly, thus greatly improving its availability.
3. It entails lower overall plant capital costs.
4. It involves reduced technical risks.

The topping PFBC cycle (see Figure 8) is currently under development by the US Department of Energy, the British Coal Research Establishment, and various equipment manufacturers. In this configuration, most of the coal is fed into a carbonizer or mild gasifier. The devolatilized char from the carbonizer is fed into the PFBC boiler, while low-Btu gas produced by the carbonizer is burned in a combuster prior to the gas turbine. Thus the inlet temperature of the gas turbine is raised to more than 1100°C (2000°F), which significantly increases the overall cycle efficiency, achieving a net heat rate as low as 8000 Btu/kWh. The remaining components are similar to those of the combined PFBC cycle. As of 1990, the topping cycle is still under development; R&D efforts are focused on the carbonizer, the gas turbine, and system integration and control.

All three of the PFBC design alternatives described above can incorporate subcritical, once-through subcritical, and once-through supercritical boilers.

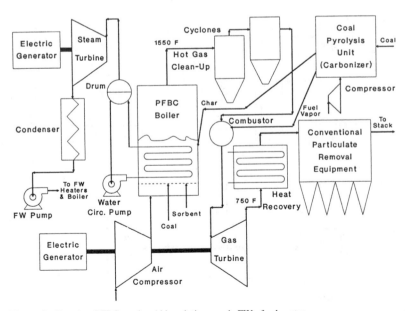

Figure 8   Topping PFBC cycle. Abbreviation used: FW, feed water.

Other cycle configurations have also been introduced, for example, the air-cooled PFBC (Curtiss Wright cycle), which uses air instead of saturated or superheated steam to cool the fluidized-bed. Originally, the technology was developed in the United States. Later on, Snamprogetti of Italy bought the rights to the technology from Curtiss Wright and is continuing its development.

TECHNOLOGY STATUS

As of 1990, AFBC technology has been demonstrated up to the 160-MW size, while PFBC technology is entering the demonstration phase with three 70-MW power plants. AFBC plants up to 250 MW in size (single-combustor units) are offered with commercial performance guarantees. Based on experience with operating AFBC plants worldwide, the following conclusions may be drawn:

1. A significant amount of experience has been generated in designing, building, and operating AFBC power plants of sizes up to 150 MW. The risks assumed by plant owner/operators in this size range are therefore similar to those entailed for conventional power plants.
2. AFBC plants between 150 and 250 MW in size utilizing a single combustor entail additional technical risks with regard to boiler scale-up. However, these risks depend on the specific requirements of the project (e.g. the fuel characteristics and design features of the boiler) and can be mitigated by careful specification, a thorough design review, and supplemental design studies and laboratory testing. As of 1990, many AFBC power plants in this size range are in the design or startup phase and should add to the industry experience with regard to the higher end of the size range.
3. Single-combustor AFBC boilers of more than 250 MW in size entail higher risks and require detailed design studies, test programs, and possibly component development programs.

As of 1990, R&D efforts with respect to AFBC technology are focused on the following areas: (a) scale-up above 200 MW, (b) further improvement of sorbent utilization efficiency, (c) minimization of refractory use in the combustor and cyclone and development of water-cooled cyclones, (d) utilization of noncatalytic postcombustion NO_x control systems to reduce NO_x emission levels to below 100 ppm, and (e) development of utilization applications for solid wastes.

PFBC technology is in the demonstration phase. Three demonstration projects, i.e. American Electric Power's 70-MW Tidd retrofit project in Ohio,

Stockholm Energi Production AB's 135-MW project in Stockholm, Sweden, and ENDESA's 63-MW lignite-firing PFBC in Spain, are currently starting up and should provide valuable information on the pace of PFBC commercialization. All three projects utilize bubbling PFBC technology and are designed and built by Asea BB Carbon of Sweden. In recent years, particular emphasis has been placed on the development of circulating PFBC systems. Various developers are utilizing their experience with circulating AFBC systems or with fluidized-bed catalytic cracking reactors to accelerate the development of this technology. However, most of these projects are still at the pilot-plant scale.

In addition to the PFBC boiler, other plant components and features receiving particular development attention are (a) the coal and sorbent feed systems, (b) hot gas cleanup equipment, (c) the gas turbine, and (d) overall integration and control of the combined-cycle plant.

The topping PFBC cycle is also being developed, mainly in the United States and Great Britain. This cycle is very similar to the combined PFBC cycle and requires the development of only a few components before it can be commercialized. These components are the carbonizer (mild gasifier) and the char-burning PFBC boiler. This technology is also at the pilot-plant scale, although it has the potential to advance rapidly if other PFBC design configurations prove successful.

POTENTIAL APPLICATIONS

Utility Applications

The main application of FBC technology for electric utilities is power generation. This includes both new power plants and retrofit and boiler conversion projects. In the case of new power plants, electric utilities are expected to choose FBC technology for its environmental performance, ability to burn low-quality fuels, and relative economic attractiveness when compared to other power-generation options for the same design requirements. Fuel flexibility, or the ability to switch fuel without significant performance and economic penalties, is another attractive feature of the technology, although provisions need to be made during the design phase to allow for such flexibility.

Retrofit and conversion to FBC may be considered when existing facilities require life extension along with emission reduction. Bubbling AFBC technology is the preferred option for boiler conversions because of its lower height requirements, while PFBC and circulating AFBC are more suitable for retrofit projects (replacement of the existing boiler).

The degree to which US utilities will become involved in retrofit and conversion FBC projects will depend to a large extent on the regulatory environment. If significant emission reduction from existing coal-firing facilities is required, the number of such FBC projects may be very high; however, if only marginal emission reduction is required, other options that are less expensive than FBC, such as low-NO_x burners and fuel switching, may be available to electric utilities.

It is not clear yet how the 1990 Clean Air Act will affect the fluidized-bed market. Even though Clean Coal Technologies, including fluidized-bed combustion, are being encouraged, it appears that selection of the emission reduction technologies adopted by each electric utility will depend on the characteristics of the power-generation system as a whole rather than on the needs of each power-generation unit. Each utility will be required to keep emissions below a certain level, and it will seek the optimum combination of technologies that satisfies the environmental regulations and enhances its competitive position.

In the developed countries, electric utilities are expected to utilize FBC technology with good-quality coals driven mainly by environmental regulations. In less developed areas, such as India, China, Pakistan, Indonesia, and the countries of Eastern Europe, FBC technology is considered more for its ability to burn low-quality fuels such as lignite, oil shale, and biomass, which are abundant in these areas. Small bubbling AFBC power plants may be the dominant design where biomass fuels are abundant and environmental regulations are not strict. In countries with stringent SO_2 and NO_x emission regulations, PFBC and circulating AFBC may predominate.

For electric utilities involved in coal production, another application of FBC technology is the utilization of coal cleaning wastes for power generation. In fact, a recently promoted concept is the integration of a coal cleaning plant with an FBC and a PC unit; the by-products from the coal cleaning plant can be burned in the FBC unit while the clean coal, containing low sulfur and low ash, can be burned in the PC unit without requiring flue gas desulfurization. The ability to change the level of coal cleaning in the cleaning plant and the level of SO_2 removal in the FBC unit provides a great deal of flexibility in optimizing power plant performance while complying with environmental requirements. Because of the limited emission reduction potential (since the sulfur is removed by the FBC unit, which handles only a portion of the total coal flow), this concept is applicable to sites that require moderate levels of emission reduction (30–60% SO_2 removal). However, additional SO_2 can be removed if sorbent injection technologies are employed.

Industrial Applications

AFBC is currently the technology of choice for many industrial facilities. Favorable regulations in the United States, such as the Public Utilities Regulatory Policy Act and Federal Energy Regulatory Commission regulations affecting cogeneration facilities and independent power producers, respectively, have provided an additional incentive to utilize the technology in cogeneration and small power production facilities.

Anthracite culm-firing AFBC facilities are very popular in Pennsylvania, where more than 150 years of anthracite mining has left significant amounts of this fuel above ground. Similar projects (based on coal cleaning wastes) are being developed in Australia and India. Other coal-producing countries, such as Poland and South Africa, are expected to utilize the same technology.

Biomass-firing AFBC power plants are also in operation throughout the world. A number of such power plants using both bubbling and circulating AFBC designs are in operation in California, the northeastern United States, and northern Europe. Many of these facilities are operated by municipalities and used for power generation and central heating. Countries with significant amounts of biomass fuels, such as India, Indonesia, and the Philippines, are also interested in developing and utilizing AFBC technology.

Industrial FBC facilities are not limited to low-quality fuels. A number of coal-firing facilities are also in operation. Most of these facilities are owned and operated by independent power producers and produce electricity along with steam for process applications or enhanced oil recovery.

FBC TECHNOLOGY AS A POWER-GENERATION OPTION

Engineering and economic studies by organizations such as EPRI, DOE, and IEA have concluded that both AFBC and PFBC technologies are competitive in terms of capital and levelized costs with similar PC plants using flue gas desulfurization equipment and are less expensive than integrated-gasification-combined-cycle (IGCC) plants, as shown in Table 1 (8). Table 1 provides cost and performance data for plants of several sizes; the total capital requirements of both AFBC and PFBC units are lower than those of a 300-MW PC plant on a kilowatt basis. However, before drawing general conclusions about the competitiveness of one technology over another, particular attention should be paid to the specifications and design requirements used for the evaluation. For example, an efficiency requirement of 50–70% SO_2 removal from an intermediate-sulfur coal can be met by dry scrubbers or limestone injection, in which case a PC plant may be the technology of choice. Similarly, if 99% SO_2 removal is required, the best option may be an IGCC plant. At an SO_2 removal level of around 90%, FBC technology seems to be more competitive than other power-generation options.

The market and performance of AFBC facilities between 1985 and 1990 have proven that FBC technology has a competitive advantage at least within the size range currently available (50–200 MW). The technology is particular-

Table 1 Alternative power-generation technologies: cost and performance data (end-of-year 1989 dollars)

	PC/ FGD[a]	Circulating AFBC	Combined- cycle PFBC[b]	Turbocharged PFBC[b]	IGCC
Unit size, MW	300	200	340	250	200
Net heat rate, Btu/kWh	9758	7765	8714	9979	9050
Total capital requirements	1496	1603	1508	1464	1886
O&M costs					
fixed, $/kW-yr	30.0	32.4	40.0	30.6	45.7
variable, mills/kWh	2.8	3.1	3.8	2.9	4.3
consumables, mills/kWh	3.2	4.1	2.7	3.6	0.5
by-product (sulfur credit)	—	—	—	—	−1.5

Source: Ref. 8
Plant locations: Eastern and west-central United States.
Fuel: High-sulfur (4%) Illinois No. 6.
Sulfur removal: 90%
[a] Subcritical pulverized coal utilizing wet limestone FGD and producing gypsum as a by-product.
[b] Bubbling PFBC.

ly competitive under conditions when (a) low-quality fuels are available, (b) fuel flexibility is required, (c) moderate to high levels of emission reduction are required, such as 70–95% SO_2 removal and NO_x generation between 100 and 400 ppm, and (d) small modules of up to 200 MW in size and, possibly, phased construction are specified.

With regard to the environmental performance of AFBC technology, there is a perception that it can only achieve a maximum of 90% SO_2 removal. In fact, it has been demonstrated that many AFBC plants can remove as much as 98% of the SO_2. Bubbling AFBC plants can achieve 90% SO_2 removal with a Ca/S ratio of 2.0–3.0. Circulating AFBC plants, on the other hand, have demonstrated 90–95% SO_2 removal efficiencies with moderate amounts of sorbent (Ca/S = 1.5–2.5).

The NO_x generated by most AFBC power plants ranges from 100 to 300 ppm, largely depending on the specific design configuration. The staged-combustion method utilized by circulating AFBC boilers is highly effective and reduces NO_x generation to the 100-ppm level, as indicated in Table 2 (9). Further reduction can be achieved through noncatalytic postcombustion NO_x control systems involving either urea or ammonia injection. Many industrial facilities in California equipped with this type of system have reduced NO_x generation levels to well below 100 ppm.

Regarding the production of wastes, both AFBC and PFBC systems generate larger amounts of solid waste than are generated by PC/FGD and IGCC technologies. However, these wastes are in the form of dry solids, which are benign and relatively easy to handle. In addition, a considerable amount of

Table 2 NO_x emission data from circulating AFBC plants in operation

Unit size (MWth)	Fuel type	No_x (ppm NO at 7% O_2)
65	Bituminous coal	40–260
65	Brown coal	80–120
97	Petroleum coke	90–120
56	Wood waste	110–135
84	Washery rejects	100
226	Bituminous coal	23–40

Source: Ref. 9

effort is being devoted to identifying utilization applications for these materials to obviate the need for disposal.

With regard to the environmental impact of FBC technologies relative to that of other power-generation options, the following observations can be made:

1. Both AFBC and PFBC reduce SO_2 emissions to levels similar to those of PC systems equipped with FGD, but not as low as the levels achieved by IGCC systems. However, NO_x emissions from FBC plants are lower than in comparable PC plants.
2. AFBC generates approximately the same greenhouse gases per megawatt produced as does conventional PC technology. PFBC technology further reduces the generation of CO_2 through better heat rates.
3. Both AFBC and PFBC systems generate larger amounts of solid wastes than do other power-generation options.

THE FBC MARKET

Existing and Planned FBC Installations Worldwide

An informal survey of AFBC installations conducted in 1988 revealed that more than 400 AFBC units with a capacity of at least 9100 kg/h (20,000 lb/h) were either in operation or under construction in North America, Europe, and Asia. This estimate does not include China, which reportedly has more than 2000 small bubbling AFBC units in operation. Most of the bubbling AFBC units in operation were installed during the 1970s and the first half of the 1980s. In recent years, circulating AFBC has become the preferred AFBC design alternative.

More than 120 AFBC boilers, producing a total of 3 gigawatts, are in operation in North America. Two-thirds of these units employ circulating AFBC technology for electric power generation and have a combined capacity of approximately 2500 MW. Bubbling AFBC units have a combined capacity of approximately 570 MW, with many of them being retrofit or conversion projects.

A market study focusing on industrial AFBC units conducted by the US DOE found that sales of these units between 1985 and 1990 have averaged 2.27 million kg/h (5.0 million lb/h) per year (see Figure 9), and this number is expected to triple by the year 2005.

In Europe there are more than 150 AFBC units in operation and under construction, representing close to 3 gigawatts of installed capacity. Seventy percent of this capacity comes from circulating AFBC units. Germany, Finland, Sweden, and France represent approximately 80% of the installed capacity. In Asia, the utilization of AFBC technology is increasing rapidly, with approximately 130 AFBC boilers in operation in Japan, Korea, and India. Many of these boilers are small bubbling AFBC units in India, but 12 of them have a steam-generating capacity of 110,000 kg/h (250,000 lb/h) or more each. The largest AFBC project in the world is a 350-MW bubbling AFBC boiler conversion in Japan, which is in the planning stage. Japan and India utilize primarily bubbling AFBC technology, although they are becoming more interested in circulating AFBC. Korea uses circulating AFBC technology almost exclusively.

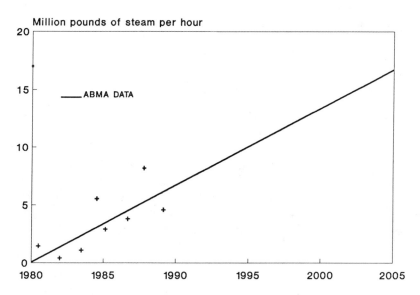

Figure 9 Projection of industrial boiler sales. Abbreviation used: ABMA, American Boiler Manufacturers' Association. Source: Ref. 12

The installed PFBC capacity is mainly represented by the three demonstration plants mentioned previously along with a number of small PFBC pilot plants in Germany, Great Britain, and the United States. Projects in the planning stage include American Electric Power Company's planned 350-MW PFBC project in the United States and a 70-MW demonstration project planned by the Electric Power Development Corporation of Japan.

Potential Market Penetration

The market penetration of FBC technology depends on a number of factors, such as (a) the performance of existing and new FBC power plants in terms of heat rate, SO_2 removal and sorbent utilization efficiencies, NO_x emissions, and reliability, (b) further development of FBC power plant design with possible improvements in performance, reliability, and capital and O&M costs, (c) the rate and size to which the technology is eventually scaled up, (d) the rate of increase in the demand for electricity, (e) the availability and price of oil and natural gas, and (f) the specific environmental regulations imposed on new and existing power-generation facilities.

So far, the performance of large AFBC power plants has been very promising, with heat rate and emissions at expected levels. While some power plants have experienced reliability problems, most of them have been attributed to operator errors and the old equipment utilized in retrofit projects. Some problem areas, such as the coal feed system and the refractory utilized in the boiler and the cyclones, are being addressed through successful design modifications. Sorbent utilization in circulating AFBC boilers is very good (Ca/S = 1.3–2.0), but it is higher in bubbling AFBC systems (Ca/S = 1.8–3.5). In view of the costs of sorbent and of solid waste disposal, significant efforts are being made to further improve sorbent utilization efficiency.

The scalability of FBC technology and the rate of increase in the demand for electricity will also have a significant impact on its acceptance by the market. A slow rate of demand would indicate a smaller average size for new generating units, a scenario which may be favorable to the technology because FBC units are available in smaller units in the 100–200-MW range. However, if the demand for electricity increases significantly and large additions of 500 MW and over are needed, FBC technology may not obtain as large a share of the market. However, even in the latter scenario, it is expected that the multiple-unit concept, in which two or three 150-MW FBC boilers are installed, will be utilized extensively.

Energy prices will also play a significant role in the acceptance of the technology over the short term. Significant increases in the prices of oil and natural gas would incline both electric utilities and industrial users to use coal and technologies such as FBC, PC and IGCC. However, if energy prices remain relatively low and more stringent environmental regulations are imposed, the competitiveness of coal-based technologies will be reduced.

Environmental legislation will have a significant effect on the market penetration of FBC technology. Moderate to high SO_2 removal requirements of 70–95% and NO_x generation limits of 100–300 ppm would be the most favorable scenario for the acceptance of this technology. Conversely, legislation making waste disposal more difficult and costly will penalize FBC technology.

A number of publications have dealt with the topic of AFBC market penetration (10, 11). Most estimates give AFBC technology a market share of up to 20%, based on megawatts of installed capacity, representing up to 50% of coal-fired units (9, 10). The authors' view is that AFBC technology will continue to play a significant role in certain segments of the market, particularly the industrial sector and independent power producers utilizing low-quality fuels and coal. Wide acceptance of FBC technology by the electric utility industry is considered unlikely within the five-year period up to 1996. Following technology development to increase the size of AFBC units to 400 MW in the 1990s, the technology is likely to play a major role in the electric utility and industrial markets after the late 1990s. PFBC technology is expected to follow this trend with a five- to eight-year lag behind AFBC.

High utilization of AFBC technology is expected almost immediately in less developed countries such as India, Indonesia, Pakistan, Turkey, and the countries of Eastern Europe. The availability of low-quality fuels, the preference for relatively small power-generation units, and the similarity between AFBC and PC technologies in terms of their fabrication and operation make AFBC very desirable for these countries. Significant efforts by the donor community to encourage privately funded projects in these countries should further accelerate the utilization of AFBC technology.

ISSUES ASSOCIATED WITH THE COMMERCIALIZATION OF FBC TECHNOLOGY

Technical Issues

In order to further improve the competitiveness of AFBC technology, a number of technical improvements need to be made, including (*a*) development of water-cooled cyclones, (*b*) development of more efficient and compact dust collection devices than the cyclones used currently, (*c*) reduction or elimination of the combustor refractory, (*d*) utilization of materials that reduce the potential for erosion, (*e*) further improvements in the reliability of the coal and sorbent feed systems, (*f*) improvements in the performance of particulate collection devices such as fabric filters and electrostatic precipitators, (g) further improvements in sorbent utilization efficiency, and (h) scaling up of AFBC boilers to the 400-MW range by 1996.

These technological developments are most likely to be made by private industry in response to competitive pressures and the need to comply with changing environmental regulations. However, government agencies and research institutes can act as catalysts to accelerate such developments.

In the case of PFBC technology, the major technical issue is demonstration of its performance, operability, and reliability. Other features that need to be addressed are (a) demonstration of hot gas cleanup technologies, (b) identification of materials to reduce erosion in the boiler, (c) development of instrumentation for speedy diagnosis of malfunctions in the pressurized vessel and the hot gas cleanup system, (d) improvement of gas turbine tolerance to particulates and alkali emissions, (e) improvement of coal and sorbent feed system reliability, and (f) demonstration of the controllability and load following capability of the integrated PFBC system.

Environmental Issues

The environmental performance of FBC systems has proved to be very good with respect to the 1979 US New Source Performance Standards. Further emission reductions could be accomplished by all FBC systems through greater utilization of noncatalytic NO_x reduction devices and by increasing the feed rate of sorbent into the system. One area in which the technology does not compare favorably with other power-generation options is in the volume of solid wastes generated. Even though FBC wastes are benign and relatively easy to handle, their volume is a reason for concern. This problem may be addressed in two ways: (a) by further improving sorbent utilization efficiency to reduce the amount of solid wastes generated, and (b) by developing utilization applications to reduce the amount of solid wastes for disposal.

Regulatory Issues

The regulatory issues that may affect the acceptance of FBC technology are (a) acid rain legislation, (b) global warming-related concerns and potential legislation, (c) legislation affecting the disposal and utilization of power plant wastes, and (d) Federal Energy Regulatory Commission regulations affecting power-generation facilities owned and operated by independent power producers.

The 1990 Clean Air Act Amendments should have a significant impact on the utilization of FBC technology. However, it is not clear yet whether utilities will elect to convert many of their older boilers to AFBC or whether they will add flue gas desulfurization systems to the boilers.

Global warming is still a highly contested issue and no quick scientific or political resolutions are expected. However, it has already been identified as an issue that needs to be considered by utility planners and executives planning power-generation projects and could have a bearing on future decisions on whether to utilize coal or other fuels. In general, however, the availability of coal, together with national security, balance of payments, and unemployment considerations, is expected to sway the fuel decision towards coal in most cases. Therefore, FBC technology is likely to be in competition with other coal-based technologies, such as PC and IGCC, rather than with gas- or oil-based technologies.

With regard to greenhouse gases, AFBC gives rise to roughly the same amount per megawatt generated as PC technology, though more than IGCC. PFBC technology, in particular the combined and topping cycles, has plant heat rates comparable to those of IGCC.

In the United States, the disposal and utilization of FBC wastes is regulated by the Resource Conservation Recovery Act, Subtitle D, covering non-hazardous solid wastes. Although FBC wastes are currently characterized as nonhazardous, local regulations impose design and operating requirements on the disposal site, in terms of requirements for pit liners and monitoring of leachates, etc. Waste disposal may become an issue of more serious concern than it is as of 1990. However, potential developments in the area of waste utilization are likely to have a positive effect on the acceptance of the technology. Many states, in particular the coal-producing states, have already taken action to encourage the utilization of both PC fly ash and FBC wastes. Utilization of FBC wastes is hampered by the variability of its characteristics due to the development stage of the technology and by the lack of overall characterization and utilization information. In the future, it is expected that FBC wastes will have more uniform characteristics, thus increasing the proportion that is utilized and correspondingly reducing the amount of wastes for disposal. High-volume, low-cost applications such as road-base construction seem to be the most promising.

A number of Federal Energy Regulatory Commission regulations and rulings may affect the market penetration of FBC technology in the United States. These regulations would affect life-extension projects, which involve increases of unit output or restoration to original levels and require significant capital investment. Other regulations affecting independent power producers (IPP) may have an impact on the utilization of FBC technology. IPPs have generally utilized AFBC technology more than electric utilities and are expected to continue to do so. They are likely to become even more active if deregulation of the power-generation market continues, and they are expected to utilize FBC technology not only in small facilities firing low-quality fuels but also in larger power plants firing coal.

ROLE OF GOVERNMENT AND R&D ORGANIZATIONS IN THE DEVELOPMENT OF FBC TECHNOLOGY

Organizations such as DOE, EPRI, and IEA have been involved in activities to accelerate the development and commercialization of FBC technology, particularly those requiring substantial capital outlays and representing significant risks for the private sector. The early stages of AFBC technology development in the United States, at the bench and pilot-plant scales, was funded almost exclusively by DOE and EPRI. As the feasibility of the technology was proved, a certain percentage of cofunding was contributed by vendors and architect/engineers, reaching the 50–70% level during the demonstration and commercialization phases.

The demonstration phase of a technology includes the first 2–5 utility-scale projects, and is of particular importance because of its twofold objective to (a) scale up, build, and operate a power plant utilizing a new technology, and (b) monitor plant performance, formulate technical and financial information, and disseminate it to the industry.

The participation of DOE and EPRI in demonstration projects and, more specifically, their contributions to one or both of these objectives, have given rise to debate with regard to (a) whether these organizations should contribute to plant capital expenditures, (b) whether they should participate in more than one demonstration project at a time, (c) whether they should conduct extensive plant performance monitoring and testing programs, and (d) how far the technology transfer plan should go.

Although each technology and project warrants special consideration, the general guidelines followed by such organizations to decide on their level of involvement in the case of demonstration projects are as follows:

1. Financial contributions to capital expenditures usually reflect only the risk associated with the utilization of new processes and plant components, and do not cover the cost of commercially available and proven components.
2. Financial contributions may also be construed as covering part of the higher-than-normal costs of a prototype plant.
3. The main effort of the funding organization (e.g. EPRI or DOE) is focused on documenting all project activities, monitoring plant performance based on comprehensive test plans, and reporting the findings to the industry. Special instrumentation, computer hardware and software, and technical and management expertise may also be provided.

While these guidelines are fairly straightforward, they need to be evaluated on a case-by-case basis. In the past, some projects have been more successful than others because the government or R&D organizations (a) limited themselves to the role of catalysts in forming consortia of the interested market players instead of taking the lead in making design decisions and assuming financial responsibility for the project, (*b*) ensured that the team represented all interested market players, (*c*) gave all promising technical options a chance instead of promoting a single design alternative, (*d*) left critical design decisions for the private-sector participants to make on the basis of technical and economic considerations, (*e*) viewed themselves as technology developers rather than technology advocates, and (*f*) emphasized technology transfer activities.

Technology transfer, in particular, requires a significant amount of effort in terms of planning and executing test programs that yield all the necessary information, such as (*a*) general data on capital and O&M costs, plant reliability, and overall plant performance, (*b*) guidelines for the preparation of design specifications for commercial projects, (*c*) methods and computer software to assist in evaluating commercial bids, (*d*) guidelines for the systematic startup of FBC power plants, (*e*) acceptance testing guidelines, and (*f*) plant operator training support.

The technology transfer media that are commonly used include seminars, publication of technical papers and reports, and development of computer software.

The most important elements in the process of technology transfer include (*a*) the participation of utilities, architect/engineers, and vendors throughout the development of each technology, which makes them aware of its unique features and prepares them for its commercialization and implementation, and (*b*) the mobility of technical experts who follow the technology through the various stages of its development by working first for R&D organizations, then for boiler vendors, and finally for electric utilities, architect/engineers, or consulting organizations.

CONCLUSIONS AND RECOMMENDATIONS

During the 1990s, the application of AFBC technology for power generation has grown from the stage of laboratory experimentation to an installed capacity base of more than three gigawatts in North America and the same overseas. A significant amount of experience has been gained in unit scale-up to 150 MW. A number of power plants in the 100–200-MW size range have been started up between 1987 and 1990 and have successfully met most design specifications. In this size range, some risks associated with the performance of certain AFBC plant components remain, but these can be minimized or eliminated through careful specification and design review. Single-combustor AFBC boilers of more than 200 MW entail higher risks and may need to be accompanied by design studies and test programs to address scale-up issues.

Construction of AFBC power plants continues at a high level and is expected to increase further. Most of the new AFBC power plants in the United States are likely to be built by independent power producers and consortia consisting of deregulated subsidiaries of utilities, private investors, banks, architect/engineers, and equipment suppliers. AFBC technology is expected to be utilized to an even greater extent in less developed countries such as India, Indonesia, Pakistan, China, and eastern European countries.

PFBC technology has entered the demonstration phase with three 70-MW power plants in Europe and the United States. If these projects are successful, commercial activity should increase, with many PFBC power plants in the 70–150-MW size range being built as early as 1995. A number of alternative PFBC cycle configurations are currently under development, but they are not expected to be commercialized before the year 2000.

In the United States, the market penetration of both AFBC and PFBC technologies in the short term will depend to a large extent on the specific environmental regulations that are enacted. Overseas, the acceptance of the technology will vary with the characteristics of the available fuels and the applicable environmental regulations. In power-generation applications, AFBC is expected to play a more significant role than either PFBC or IGCC technologies in the early 1990s because of its good environmental performance and the similarity of its plant design to that of conventional PC plants. Although the introduction of environmental regulations is making electric utilities more process-oriented, in the short term they are likely to prefer systems similar to those they currently use.

During the early 1990s, the focus of R&D efforts is likely to be on the scale-up of FBC units to the 250–400-MW size range, improvement of FBC power plant components such as cyclones and feed systems, further improvement of sorbent utilization, and development of waste utilization applications. The utilization of wastes from FBC plants needs further exploration from both the technical and the regulatory perspectives. Utilization applications should be identified or developed, and regulations encouraging the utilization of power plant wastes should be introduced.

The role of government and R&D organizations in the development and commercialization of FBC technology has been and will continue to be very important. Organizations such as DOE, EPA, and EPRI in the United States and IEA in Europe have acted as catalysts for both the development and commercialization of FBC and other clean coal technologies. Their role is expected to continue to be significant both at the technical and the regulatory levels through support of technological developments, enactment of environmental regulations, and introduction of financial incentives.

Literature Cited

1. Squires, A. 1985. Fluid beds: at last, challenging two entrenched practices. *Science* 230:4732
2. Tavoulareas, S., et al. 1989. *The Current State of AFBC Technology. World Bank publication No. 107.* Nov.
3. Podolski, W. F., et al. 1983. *Pressurized Fluidized-Bed Combustion Technology.* Noyes Data Publications
4. McKenderick, D., Wright, S. J. 1983. The Grimethorpe experimental PFBC facility: an update. *Modern Power Systems.* Oct.
5. Hoy, H. R., Roberts, A. G. 1983. PFBC for combined cycle power generation. *Modern Power Systems.* Oct.
6. Olesen, C. 1983. Full scale PFBC component tests carried out in Sweden. *Modern Power Systems.* Oct.
7. Kinsinger, F. L., McDonald, D. K. 1988. *Combined Cycle using PFBC.*

Paper presented at the Am. Power Conf., Chicago, Ill., April 18–20
8. *Technical Assessment Guide,* Vol. 1, Rev. 6. 1989. *P-6587-L.* Sept. Palo Alto, Calif: Electr. Power Res. Inst.
9. Dry, R. J., La Nauze, R. D. 1990. Combustion in fluidized beds. *Chem. Eng. Prog.* July: pp. 31–47
10. Siegfriedt, W. E., Shibayama, G. 1986. *The Mix of Technologies in the New Power Plants of the 1990s.* Paper presented at the joint ASME/IEEE Power Generation Conf., Portland, Ore., Oct. 19–23
11. *The Role of Repowering in America's Power Generation Future.* 1987. US Dept. Energy. Dec.
12. Smith, D. J. 1990. Non-utility use of fluidized-bed boilers: a growing technology. *Power Eng.* Aug: p. 38. Based on US Dept. Energy's Morgantown Energy Technol. Cent. study

Annu. Rev. Energy Environ. 1991. 16:59–90

CLEAN COAL TECHNOLOGIES AND FUTURE PROSPECTS FOR COAL

Adam Rose

Department of Mineral Economics, The Pennsylvania State University, University Park, Pennsylvania 16802

Thomas Torries and Walter Labys

Department of Mineral Resource Economics, West Virginia University, Morgantown, West Virginia 26506

KEY WORDS: coal economics, research and development, technology adoption, environmental control, government regulation.

I. INTRODUCTION

In analyzing the energy picture of the United States, one is immediately struck by a seeming imbalance with respect to the coal option. It is well known that the energy content of US coal reserves far exceeds that of oil in the Middle East. Yet, despite the concerns over national security and the balance of payments, coal has not displaced oil imports. Even though coal has been highly touted as "the bridge to the future," "America's best chance for energy independence," and "the basis of a major backstop technology" (1–3), gains in coal utilizations have been disappointing, and future prospects are uncertain.

Technological advances have long held the promise of lowering the costs of production, preparation, delivery, and combustion of coal. They have the potential to extend the range of coal utilization, thus opening up entirely new markets. Also, technological advances can help mitigate environmental side-effects associated with the coal cycle, thereby removing obstacles to coal's attractiveness. Despite this potential, however, the direct impact of technology on the coal industry and its customers has been minor in recent decades,

59

1056-3466/91/1022-0059$02.00

especially when compared with technological impacts on such products as telecommunications or chemicals. It remains to be seen whether technology will play a more prominent role than other factors in the future prospects for coal use.

The purpose of this report is to analyze the future potential of coal in the US economy during the next 25 years in light of "clean coal technologies." According to official US Department of Energy (DOE) designations, (4, 5), these technologies pertain only to the beneficiation, transformation, combustion, and postcombustion clean-up stages of the coal cycle; no coal mining or coal transport technologies are included. In general, clean coal technologies offer the prospect of mitigating environmental side-effects of coal utilization, primarily through improved operating efficiencies and lowered costs of air emission controls. If they prove successful, coal users will be able to meet more stringent environmental regulations at little or no additional cost. In assessing the influence of clean coal technologies on coal demand, we focus on the economics of three crucial areas: their development, their deployment, and coal utilization implications of their operation.

II. GOVERNMENT SPONSORSHIP OF COAL TECHNOLOGY RESEARCH AND DEVELOPMENT

Research and development (R&D) in technology related to the coal cycle have been undertaken and sponsored by both government and industry. Periodically, a debate arises over the proper role of each party in this effort. One view is that the marketplace would best dictate allocations for R&D. Private firms reap the gains from technological advance and know what course serves their own interests. Under conditions of perfect competition, these decisions are in the best interests of society as a whole, and therefore any government involvement is seen as interfering with this desirable outcome. Another view is that research is a "public good" from which industry and society as a whole benefit, and that unnecessary duplication will result if companies undertake it in a proprietary manner. Government research centers or private consortia are seen as solutions to this problem, as well as to the large investment requirements or inherent risks in coal-cycle R&D. Another justification for government involvement pertains to the pursuit of broader goals, such as national energy self-sufficiency and a clean environment, which transcend the profitability of individual firms and are not attainable by markets alone.

Billions of dollars have been spent to date on coal-cycle R&D, based usually on reasonable technical principles, but not always on broader strategic planning considerations. The government role in the R&D process has typically been evolutionary in nature. Since the Arab Oil Embargo of 1973–1974, however, there have been three major shifts in direction. The first was Project Independence with its goal of US energy self-sufficiency, which provided a major role for the nation's most plentiful, indigenous energy resource—coal. One aspect of this unsuccessful effort was the push to promote development of coal-based synfuels in the late 1970s. The second major shift came about with the Reagan administration and its laissez-faire posture. This administration, viewing government involvement in the demonstration and commercialization phases as needlessly interfering with the direction of R&D specifically and the marketplace in general, placed its emphasis on more basic research.

The third major shift was legislation aimed at spurring the development of technologies that could simultaneously increase the efficiency of energy technologies and reduce emission of air pollutants. The official "Clean Coal Technology" (CCT) designation arose from the Omnibus Continuing Resolution of October 1984. This resolution called for the removal of $750 million from the Synthetic Fuels Corporation program and initiated a Clean Coal Technology Reserve to provide government assistance to private-sector demonstration projects in this area, with a goal of commercialization by the mid-1990s (hence, its subsequent designation as the Clean Coal Technology Demonstration Program). The resolution also stipulated that the Secretary of Energy was to: (*a*) solicit proposals for these projects from the private sector, (*b*) provide Congress with a comprehensive assessment of the proposals, and (*c*) determine the usefulness of further federal incentives.

The solicitation garnered more than 175 responses in the form of expressions of interest, with a total cost exceeding $8 billion. The responses represented a broad range of technological options scattered throughout 29 states and the District of Columbia (4). The DOE assessment concluded that ". . . new clean coal technologies have the potential to improve the environmental performance and/or economics . . .", but given uncertainties ". . . considerable research and development may be required in order to advance the technology toward readiness for demonstration and evaluation of commercial application" (4; p. 1–4).

As to the issue of appropriate future incentives, DOE argued against them. The agency cited the adequate workings of free market forces and the distortionary effects of subsidies, specifically in the face of the substantial uncertainties involved. Interestingly, DOE also cited its own previous experiences, which it termed "unsuccessful in commercializing new fossil technologies" (4; p. 1–4). Finally, DOE reiterated its intent to continue emphasis on basic R&D, as opposed to commercialization. Congress, however, stipulated that the CCT program could provide a subsidy up to and including 50% of any approved project. At the same time, procedures were established for a repayment of the subsidy from profitable projects.

Based on the initial response, Congress passed PL 99-190 in December 1985, setting aside just under $400 million for demonstration projects. A total of 51 proposals were submitted, of which 9 were originally designated for support.[1]

The original DOE report on CCTs was not well received by Congress. There were criticisms that the assessment had been inadequate and that the policy perspective with regard to promotion of technological advance and implementation was misguided. At the same time, there was growing awareness, stemming from such sources as the Lewis-Davis (8) report and a preliminary National Acid Precipitation Assessment Project (9) report, that pressure was building on the issue of acid rain. If legislation were passed immediately there would have been only three significant proven SO_2 abatement options [flue gas desulfurization (FGD), coal cleaning, and coal switching], all of which were viewed as imposing added costs of operation by industry. On the other hand, most of the CCTs represent improvements in energy transformation efficiency as well as reductions in pollutant emissions. These new technologies thus promise gains that could at least partially offset costs attributable to pollution control.

The United States faces what one DOE official termed a "window of opportunity" for new clean coal technologies because of the aging of existing utility boilers and the forecasted need for significant increase in new capacity by the mid-1990s (10). More than half of the utility boilers currently utilizing coal will be at least 25 years old by then. Annual average growth in electricity demand is projected at 2.5 to 3.0% (11) over the coming decade. One option is to build entirely new facilities, but only at great expense. Another is to modernize, repower, or retrofit existing facilities and thus increase their power output by replacing or enhancing current boiler installations.[2] Such enhancement of existing installations with clean coal technologies might be less expensive than meeting air emission standards with end-of-pipe controls alone, and would definitely be less expensive than building new standard boiler facilities with or without emission controls. More attention thus needs to be given to technologies that are compatible with existing utility configurations.

---

[1]The Clean Coal Technology Program began with substantial industry support (6). Many of the guiding principles of the program are consistent with private industry views as expressed in reports by the National Coal Council (3, 7), an advisory committee to the Secretary of Energy composed primarily of executives of companies involved in the coal cycle, including machinery manufacturers, as well as some labor leaders, public officials, and scientists.

[2]These terms are defined by US DOE (5; p. 30) as follows: *Repowering*—"any technology that replaces a significant portion of the original power plant and reduces atmospheric emissions of pollutants, often while increasing capacity." *Retrofitting*—"the installation on existing facilities devices designed primarily to reduce emissions." *Modernization*—"the upgrading of an aging power plant for the purpose of extending its useful service life in order to postpone the need for new generating capacity."

These factors led to a refinement known as the Innovative Clean Coal Technology Demonstration Program (ICCT). This is a combination of the Department of Interior and Related Agency Appropriations Act of FY 1987 (PL 99-500 and PL 99-591), and the President's $2.5 billion (over five years) "Acid Rain Initiative." In all, the program, as well as estimates of private sector commitments, call for a $6 billion expenditure between 1985 and 1992 to promote CCTs. Approximately $4.9 billion is earmarked for demonstrations; the remainder is for program direction and the Small Business Innovative Research program. Also of interest is the greater role of the states in the CCT funding effort, up from a little over 1.0% during 1980–1985 to almost 5.0% for 1985 onward.

In response to a second round of informational solicitations in late 1986 based on ICCT principles, the Department of Energy received 139 responses (5). Formal proposals submitted in Round II numbered 54, and 16 proposals were approved for federal cofunding (12).

Although $2.5 billion was stipulated in the original legislation to be divided evenly at $500 million per year, Congress appropriated only $575 million for the first two years.[3] Given the slower than planned pace of the various rounds of the program, this was not an inhibiting factor. Overall, the original five-year program has been stretched through 1994, with plans for Round IV under way in 1990.

Round III of the program carried over the emphasis on retrofitting and repowering. Informational solicitations were not requested, but several public meetings were held, including one to encourage participation by western states. A public opportunity notice (PON) received 48 formal proposals, of which 13 were chosen for funding. Cofunding of chosen proposals by private industry was just under 60%, down from about 68% for Rounds I and II (12).

The status of the Clean Coal Technology Demonstration Program is summarized in Table 1. While it has been an identifiable thrust in the direction of desirable technologies and has garnered numerous responses, progress has been slow, especially in light of assessments done at the outset of the Program that called for demonstrations of more than a dozen different types of technologies by the mid-1990s. The Program has been in operation for five years, yet only five projects have reached the operational stage. Also, eight projects have been withdrawn for technical and financial reasons, and four Round I and II projects are still in the "pre-award" stage. It would appear that the approval process is not synchronized with the overall Program target dates. This process involves even more pre-award stages than indicated in Table 1, including: (a) planning and DOE fact-finding, (b) prenegotiations,

---

[3]This is over and above the $250 million of the original $400 million actually committed in 1986–1987. In addition, the federal government has committed $200 million to more basic research on new energy technologies.

Table 1   CCT Project status report (as of 10/25/90)

Phase	Round I	Round II	Round III	Total
Pre-award				
In negotiation	0	2	6	8
At Congress	1	1	2	4
Post-award				
In design/permitting	3	7	4	14
In construction	2	4	1	7
Operating	4	1	0	5
Total in progress	10	15	13	38
Withdrawn	7	1	0	8
Total selected	17	16	13	46

Source: DOE (13).

(c) headquarters panel review, (d) formal negotiations, (e) headquarters panel approval, (f) headquarters executive board approval, (g) DOE Secretary approval, and (h) Congressional approval. Also, the timetable for the Program has been accelerated and decelerated at various times for budgetary and political reasons, thereby increasing uncertainties to the detriment of attracting investors. In any case, CCT Program target dates may be overly optimistic given the many pitfalls associated with demonstrating new technologies.

III. ADOPTION OF NEW COAL TECHNOLOGIES

It is difficult to determine the exact date that a technology is "developed." How these dates are defined, in part, determines the number of technological advances that can be counted in any single time period. By any measure, however, only the following technological innovations had a significant effect (both positive and negative) on coal use in the past decade.

LONGWALL MINING   The number of longwalls in place in the United States has increased from 7 in 1964 (W. J. Merritt, personal communication) to a peak of 112 in 1984 (14). These units greatly increase the tonnage that can be extracted per dollar of capital equipment; and require sizably lower manpower per unit of output. Although longwall mining is not a CCT, this technology evidently offered productivity improvements and was widely adopted where geologic conditions permitted.

COAL BLENDING   Although it is a simple mechanical process that has been around for years, this technique has been used increasingly and has helped fine-tune coal quality to meet specific environmental requirements. It has also helped salvage a market for high-sulfur coal and low-volatile coal previously used for coke in steel making. This "metallurgical coal" suffered from the sharp decline in the steel industry, but more recently has been blended with other types of coal.

FLUE GAS DESULFURIZATION   This is an advanced method of SO_x control mandated for new coal-fired utilities and large industrial boilers by the 1977 Clean Air Act Amendments. This technique has not been adopted voluntarily because of its reported costly and technically problematic operation. By 1988, however, there were scrubber systems in place on more than 64,000 MW of capacity in the electric utility industry (15).

COMBINED-CYCLE GAS TURBINE   This method of burning natural gas or oil first to run a turbine and then using the waste heat to run a conventional steam-boiler has been employed by the electric utility industry for many years. Improvements in this technology plus the abundant quantities of low-priced natural gas and cogeneration possibilities have made this technology even more popular. The immediate effect has been to decrease the demand for coal, although in the future this could change if gasified coal were to become the dominant feedstock.

CONVENTIONAL COAL COMBUSTION   Some technological advances have been made in coal-burning steam-electric generating units that have decreased the cost of producing electricity. These advances increased generating efficiencies by utilizing scale economies and higher operating pressures and temperatures. Even though these technologies were essentially proven since the early 1970s, their use overall declined in the 1980s (16).

Admittedly, many important aspects of the coal cycle do not lend themselves to radically new methods, e.g. the inherent bulk handling aspects of ore removal. On the other hand, where radically new methods were possible, breakthroughs have been very slow in coming, e.g. recent advances in superconductivity that could shift the major transportation link in the coal cycle to the less costly stage of electricity transmission. Also, even when technological advances took place, cost or reliability considerations stifled implementation, as in the case of flue gas desulfurization.

Most other design changes, including those considered to be very promising, have not been transferred from the demonstration phase to extensive commercial use. A major example is Atmospheric Fluidized Bed Combustion (AFBC), which has been a proven technology for over two decades. As of now, there are only about two dozen coal-fired AFBC electricity generating units in the United States, with a combined capacity of about 2000 MW.

Installations planned to come on-line over the next few years would raise this capacity level only to 3500 MW (17), about 1% of the nation's coal-fired generating capacity.

Technological advances that have been available for decades, but have been very slowly adopted, are an enigma. The situation is equally striking if viewed from the perspective of technologies in place today. For example, the boilers in large-scale, coal-fired power plants currently in use are much the same as those brought on-line in the 1920s, despite the fact that hundreds of such plants were built and conditions were ripe for experimentation during the industry's significant growth phase. The increases in energy conversion efficiencies that represented the focus of technological improvement of steam boilers during most of this century leveled off in the 1970s and are likely to have run their course. Still, conventional boilers can be made more attractive from an economic standpoint by some relatively simple improvements. Several of these, such as modular/phased construction and multiple fuel burning capabilities, can reduce capital and operating costs, respectively, by about 20% (11). If the new clean coal technology demonstrations do not result in technically reliable and economically viable outcomes, there will be greater pressure to enhance the steam boiler.

IV. TECHNOLOGY EVALUATION

Coal technologies can be classified according to the stage of the coal cycle as follows:

1. Mining,
2. Preparation,
3. Transportation,
4. Transformation,
5. Combustion,
6. Postcombustion clean-up.

The stages are basically self-explanatory, though some comments are in order with respect to stages 4 and 6. Transformation refers to processes such as coal liquefaction and gasification that "create" a different form of the fuel. Postcombustion clean-up refers to technologies having the singular goal of pollution abatement. In some of the other stages, abatement considerations are subsumed within other activities. Overall, the categories are distinct, though there are minor examples, such as the burning of coal and associated methane in situ, that overlap features of mining and combustion. More important to our subsequent analysis are trade-offs between stages, such as undertaking coal preparation (beneficiation) rather than end-of-pipe pollution abatement. Clean coal technologies "officially" pertain to categories 2, 4, 5, and 6.[4]

In Table 2, we have listed major types of CCTs and their roles in the DOE Demonstration Program. The table indicates that expressions of interest (informational solicitations) have been evenly distributed across the stages of the coal cycle. However, the emphasis on more practical and utility-oriented technologies beginning at the final approval stage of Round I and carrying through from Round II is clearly evident. In effect, no coal transformation (synfuels) projects have been sponsored by the CCT Demonstration Program except for integrated gasification combined-cycle technologies.

It is not our purpose to perform an independent engineering-economic evaluation of CCTs, but rather to review the numerous assessments of individual coal-cycle technologies that have been undertaken recently, primarily by DOE. These have been cross-checked against the scientific literature, which has also been reviewed to fill some gaps. There appears to be a reasonable consensus among these findings and among industry documents such as those circulated by the National Coal Association.[5]

One of the basic assumptions of the CCT programs is that no single technology is capable of meeting all the various energy efficiency and environmental objectives. This is due primarily to technological considerations, but it also emanates from a diversification strategy to reduce risk. Thus, we examine a spectrum of options.

A. Basic Considerations

The economic performance of major CCT options is summarized in Table 2.[6] The DOE analysis indicates that, barring unforeseen technical difficulties,

---

[4]Note that the DOE "Clean Coal Technology" designation is associated only with air pollution, and even more narrowly to combustion-related air pollution or to cleaning methods that can displace or offset such pollution. The designation does not, for example, pertain to particulate release during loading or transporting coal. Also not considered explicitly are water and solid waste considerations, which ironically may become more pronounced with the use of some CCTs.

[5]Limitations of space do not allow us to describe the operating principles of CCTs. The reader is referred to US DOE (5, 12). Excellent summaries indicating some significant differences of opinion are contained in (11, 19). On the other hand, the conclusions reached by the National Coal Council and the National Coal Association are very similar in part because they are based primarily on DOE research (7, 20). For other independent assessments see (17, 21).

[6]A calculation by the authors of costs of several of the CCTs in Table 3 based on a consensus of operating parameters and basic costs from various studies utilizing Congressional Budget Office (see 22) standardized formulas substantiated the DOE estimates.

Table 2 Clean coal technology proposals and accepted projects[a]

Technology	Round I			Round II			Round III	
	IS	PON	AP	IS	PON	AP	PON	AP
Preparation[b]	22	10	2	25	9	1	0	0
Physical cleaning								
Chemical cleaning								
Microbial cleaning								
Transformation								
Underground gasification	4	1	0	0	0	0	0	0
Surface gasification	27	9	0	11	0	0	0	0
Liquefaction	5	3	0	2	0	0	0	0
Combustion								
Fluidized bed	27	9	3	15	13	2	8	1
Gasification combined cycle[c]	—	—	1	—	—	1	3	1
Heat engines[d]	13	0	0	3	0	0	0	0
Alternative fuels	8	0	0	13	0	0	12	2
Fuel cells	8	3	0	1	1	0	0	0
Magnetohydrodynamics	2	0	0	1	0	0	0	0
Advanced combustors	10	4	2	13	5	1	6	1
Repowering[e]	—	—	—	3	5	—	—	—
Postcombustion[b]	33	7	2	49	18	8	13	7
Advanced flue gas desulfurization								
Low-NO_x burner								
Gas reburning/sorbent injection								
Industrial processes	0	3	1	3	2	2	6	1
Other	16	2	0	0	2	0	0	0
Total[f]	175	51	11	139	55	15	48	13

Sources: (4, 5, 12, 18).

[a] Abbreviations used: IS, information solicitation; PON, program opportunity notice (formal proposal); AP, accepted proposal (though not necessarily awarded).

[b] Refers to general category; technologies listed are representative examples.

[c] Integrated gasification combined cycle is subsumed under surface coal gasification in IS and PON stages for Round I and under repowering in IS and PON stages for Round II.

[d] Includes coal-fired gas turbines and diesel engines.

[e] Several other technologies under the Combustion heading are also used in repowering applications.

[f] Totals do not add up to those noted in previous section due to adjustments following initial project awards.

energy conversion efficiency (as compared to a conventional coal-fired power plant) is not adversely affected to a significant degree, and, in a few cases, is actually increased. In fact, the replacement of an old plant with a new plant using any technology will probably result in greater conversion efficiency. The figures in Table 3 refer, in this instance, to comparisons between new CCTs and new conventional installations. The power output of existing plants is not affected significantly either for the conventional and retrofit categories,

Table 3 Technical and economic performance of innovative clean coal technologies

Technology	Efficiency	Power output	Plant life	Incremental cost of electricity[a] (mills/kWh)	Capital cost ($/kW)
Conventional					
Coal cleaning	small increase	no change	slight extension	2–3	additional fuel cost only
Flue gas scrubber	decrease	moderate decrease	no change	9–11	180–200
Retrofit					
Advanced flue gas cleanup	decrease	small decrease	no change	10–12	175–300
Limestone injection burner	decrease	small decrease	no change	5–8	80–110
Slagging combustor	small decrease	small decrease	slight extension	1–2	50–60
Gas reburning	no change	no change	slight extension	depends on gas price	10–20
In-duct sorbent injection	small decrease	small increase	no change	2–4	40–90
Advanced coal cleaning	small increase	no change	slight extension	6–21	additional fuel cost only
Coal slurry	small decrease	small decrease	no change	11–23	20–50
Repowering					
Gasification combined cycle	moderate increase	100–200% increase	moderate extension	0–4	950–1200
Pressurized fluidized bed	slight increase	50–100% increase	moderate extension	4–6	800–1000
Atmospheric fluidized bed	no change	10–15% increase	moderate extension	12–17	700–900

Sources: (5, 23)

[a] Assumes base is conventional coal-fired power plant with only particulate removal devices. For repowering technologies, capital costs are relatively high because the output, i.e. capacity of the plant, is increased.

while the repowering technologies promise very dramatic increases in output. In addition, most of these options promise moderate extension to the useful life of existing facilities.

The effect of the technologies on the cost of electricity is in the range of a 10–20% increase over a conventional coal-fired plant for most alternatives. The range for the capital costs is a bit broader even for the conventional and retrofit alternatives, with several of the options falling far below a 10% increase. The repowering options appear to imply very high additional capital costs, but the numbers need to be adjusted for increased capacity.

Although the incremental cost of utilizing these new technologies is termed an increase in Table 3, the DOE analysis may be misleading. First, the cost of burning alternative fuels (oil and gas) may increase by more than 10–20% above coal burning costs by the mid-1990s. Recall that upward pressure on prices is greater on relatively less abundant resources and those whose prices can be influenced by political events such as the war in the Middle East. Second, it seems more reasonable to use a reference point that includes the full cost of constructing and operating a new steam-boiler facility. That is, it is only fair to include the cost of meeting environmental control requirements. The effect of this can be understood by subtracting the cost for a flue gas desulfurization (FGD) unit from each of the new options. Then, many of the CCTs would provide a definite cost savings, stimulate production of end products, and hence stimulate the demand for coal.[7] In addition to the combustion efficiency and emission control advantages already mentioned, several general design characteristics common to many CCT options represent significant cost-savings. These include modularity, shorter construction time, and fuel flexibility. The first two can save from $150–250/kW of capacity, while the latter can mean as much as a 1¢/kWh energy cost savings (11).

Most of the options promise to make coal use more attractive by virtue of reducing SO_2 emissions substantially; however, NO_x reduction is moderate at best. Aside from the well-known sludge problems of FGD units, no significant waste-disposal problems are associated with the other technologies (5, 24). In light of recent concerns about global climate change, improved combustion efficiency offers some CO_2 reduction (see Ref. 26).

Among the more problematic aspects of evaluating CCTs and their impact on coal demand are projections of tecnnical readiness. Predictions of commercial availability of some major CCTs are presented in Table 4. DOE's first assessment in 1985 projected commercial readiness by the middle of this decade in most cases. However, this key date is moved back two to three

---

[7]It should be noted that the cost of FGD units has been declining, and problems of solid waste disposal are partially overcome by converting sludge to commercial by-products.

Table 4 Commercial availability of clean coal technologies

Technology	1985 estimate "year of commercialization"	1987 estimate "expected availability"	1990 estimate "first commercial plants"
Retrofit			
Advanced flue gas cleanup	1990–1995	1995–2000	1995–1999
Limestone injection multistage burner	1990–1995	1990–1995	—
Advanced combustors	1986–1990	1990–1995	1993–1997
Gas reburning	—	1990–1995	1993–1996
In-duct sorbent injection	—	1990–1995	1991–1995
Chemical & microbiological coal cleaning	1995–2000	1990–1995	1997–2000
Advanced physical coal cleaning	1986–1990	1990–1995	1993–1998
Repowering			
Integrated gasification combined cycle	—	1990–1995	1997–2004
Pressurized fluidized-bed combustion	1995–2000	1990–1995	1997–2003
Atmospheric fluidized-bed combustion	1986–1990	1990–1995	1996–2001

Sources: (4, 5, 23).

years on average by each new DOE assessment (in 1987 and 1990) (compare also Ref. 25). These dates are even more distant for options other than those involving repowering and retrofit; for example, Siegel & Temchin set the date as 2010–2020 for the commercial availability of magnetohydrodynamics (26).

Another way of viewing this situation is to point to the "just around the corner syndrome" that is typical of new energy technology assessments. Most CCT assessments undertaken over the past 30 years, especially those for synfuels, have predicted near-term availability or economic viability; yet we seem to be no closer to the target now than in the late 1950s (27). We should also emphasize the distinction between technical availability and economic viability and note that the DOE studies pertain primarily to the former. The DOE studies are credible in terms of technical progress, but involve a great deal of uncertainty about when or whether the CCTs will be able to compete with conventional systems in the marketplace.

B. Some Major Ongoing Demonstrations

During the evolution of a technology there are usually a number of competing variants. Even though specific technology characteristics are usually unique to individual operations, some general insight can be obtained from an analysis of successful pilot or commercial-scale demonstrations. We turn now to a discussion of examples for IGCC and FBC technologies.

INTEGRATED GASIFICATION COMBINED CYCLE   At the time of this writing (Dec. 1990), demonstration facilities for five types of IGCC technologies are in operation (12, 28, 29). The Cool Water Coal Gasification Project (Daggett, California) is a premier example of an IGCC demonstration operation, resulting from the efforts of a six-party consortium, with the major roles played by Texaco and Southern California Edison. This project, begun in 1984 and recently completed ahead of schedule and within its original budget, is intended to prove the integration of coal gasification and combined-cycle power generation at a commercial scale (120 MW) and with a decided environmental advantage. This also is the case of the Dow Syngas Project in Louisiana (30). Its sulfur removal capability for syngas approached 99% and its total sulfur recovery factor from the coal feedstock is 97% (31). These results pertain to relatively clean coals, but even tests run on high-sulfur midwestern coals indicate sulfur removal in excess of the New Source Performance Standards (NSPS) (Ref. 36).

It is claimed that the Cool Water Project technology, based on the entrained, oxygen-blown Texaco gasifier coupled with a gas turbine and heat-recovery steam generator, is shelf-available (19). In fact, EPRI has for some time been examining IGCC options for member utilities utilizing a variety of fuels (28). Major design parameters and operating standards achieved indicate that the operation holds great promise for widespread application. To date, the project has been without any significant design flaws, has won numerous engineering awards, and according to its backers, has exceeded expectations even with respect to cost-savings (31, 32).[8] Various evaluations of the project (28, 31) indicate that the system's capital and operating costs are both significantly higher than a conventional power plant with SO_2 emission control. However, cost adjustments of first-of-a-kind expenses, scale, and by-product recovery give some experts cause to think that a commercial version of the facility could be operated less expensively than a conventional pulverized coal unit (32). Further, improvements in cost-effectiveness could come from: (a) capital cost savings (e.g. avoidance of capitalizing training and start-up expenses, home office, and special engineering), (b) operations primarily associated with heat rate improvements (e.g. more advanced gas turbines, lower-purity oxygen, improved steam cycle, and overall economies of scale), and (c) general operating and maintenance costs (e.g. less environmental testing and lower ambient temperatures than those found in a desert environment). At coal costs of $1.62/mm Btu, a 360 MW facility is projected by Gluckman et al to have a busbar price of less than 3¢/kWh in 1988 dollars (32). However, we judge this to be optimistic.

The future adoption of IGCC technology depends greatly on industrial use of natural-gas turbines for cogeneration and/or combined-cycle service, which has been growing. Increased use of gas turbines will decrease the demand for coal in the short run. How this technology affects coal demand in the long run depends upon the price of natural gas relative to the cost of gasifying coal. If increased use of natural gas depletes the current US natural gas supply "bubble" and large quantities of Canadian natural gas are not placed on the US market, natural gas prices will increase. If prices increase past the level of gasifying coal, users of gas turbines will install coal gasifiers, thereby increasing the demand for coal at the expense of demand for natural gas. This means that it would be possible for a large number of gas turbines using natural gas, in effect, to switch to coal, though the number and timing of these conversions depend primarily on technology characteristics. The likelihood of this scenario is enhanced because once the gas turbines are installed, decisions to switch to coal are predicated primarily on the additional capital and operating costs of gasifying coal. That is, users' choices would be: (a) continue to burn high-priced natural gas, (b) construct new coal-fired boilers,

---

[8]This does not mean that the experiment is free of operating problems. Also, for a differing view of IGCC economics, see (21).

or (c) add coal gasification plants. Of the three, adding coal gasification plants is the most likely (as long as the gas turbines are not near the end of their productive lives).[9]

FLUIDIZED-BED COMBUSTION   The Tennessee Valley Authority Shawnee Station (Paducah, Kentucky) bubbling-bed variant, which began operation in 1988, is the largest AFBC unit in the United States. It involves the repowering of a 160 MW facility, previously utilizing only a 1.1 million lb/hr capacity pulverized boiler coal, which now generates the steam for a turbine already in place as a baseload unit. The total installed cost is about $190 million. The unit burns high-sulfur bituminous coal and can also be linked to a baghouse for particulate control (34, 35).

There are more than 50 AFBC units with boilers larger than 200,000 lb/hr operating or under construction in the United States (24, 36). AFBCs, either bubbling or circulating, come in a very large number of variants. Experimentation is required with this technology because of its sophistication and its susceptibility to operating problems. In terms of design, significant differences include flue-gas velocity, thermal loading, hot solids separation and reinjection, horsepower, fuel- and limestone-preparation, and use of a heat-exchanger (38). Many of these distinguishing features also involve trade-offs between design parameters to meet alternative objectives (37). A major example is the addition of air to the combustion process, which often leads to a heat loss. The reduction of cost thus comes at the expense of thermal efficiency.

A major assessment of several operating experiences has concluded that "Fluidized-bed boilers have generally lived up to their performance goals" (38). Costs are competitive with conventional combustion technology coupled with pollution control devices, and most units are able to meet NSPS standards. The survey notes that the majority of the problems deal with fuel- and ash-handling equipment in the overall system. Although progress has been made on the major problems of erosion and agglomeration, they are far from being solved (see also 35, 39).

Fluidized-bed boilers theoretically have the ability to burn a wide variety of fuels including municipal and coal-mining wastes. Obviously, greater use of this technology will not increase and may decrease the demand for coal if large quantities of wastes are consumed. However, the many hidden costs of burning such low-grade material, e.g. decreased fuel efficiencies and increased waste disposal, may make burning wastes more costly than burning

[9]Yet despite excellent tests, the Utility Data Institute (33) reported that orders for gas turbines have picked up, but that "Construction of the combined cycle portions is evidently still distant, with none scheduled for at least 10 years. A commitment to the widely publicized coal gasification portion of these facilities is even further away."

coal. Other features of the technology that could hasten its adoption include modularity and the sureness of in-situ combustion pollution control.

V. TECHNOLOGY DEPLOYMENT

Equally as important as the demonstration of technology is its adoption, or deployment. In the case of CCTs, the following factors are likely to slow down the pace. First is the large capital investment, often several hundred million per installation. Premature replacement of an existing facility would take years to recoup and would appear to be unattractive, especially in an era of pursuit of short-run profits. The impact of such an investment on the financial position of an electric utility might even make the investment unattractive in the long run.

Second is the risk factor. Demonstrations at commercial scale may not be considered definitive by many potential CCT customers. The Clean Coal Technology Program defines demonstration as the sustained operation for one year of a facility of significant size, typically identified as 100 MW. However, the experience at one or even several such facilities may not be sufficiently convincing to attract widespread interest. Given the complexity of their operations and the history of technical problems associated with new technologies, utilities and other customers are likely to be cautious. It would seem that the utility industry is even wary of its own research; despite the circulation of findings about the success of FBC and IGCC operations by the Electric Power Research Institute (19, 28, 40), the industry has failed to follow suit.

Third is regulatory agency attitudes, which also have a bearing on the risk factor. Increasingly, state public service commissions have not allowed the costs of nuclear power plant failures to be factored into the rate base. There is the fear that failures of CCTs would meet the same fate. This practice effectively imposes the entirety of the risk of experimenting with new technologies onto utility shareholders, and thus acts as a disincentive to adopting alternatives to conventional power plant configurations. A reversal of this practice would lead to a sharing of this risk by utility customers. Some suggest that this is justified in principle since these customers may benefit from power-generation cost reduction. In practice it is difficult to arrive at a "reasonable" sharing of the risk in light of such great uncertainty. More recently, Congress has shown interest in this issue in the form of the proposed Clean Coal Technology Deployment Act (S879), which designates any CCT project as "prudent" for the purposes of rate-making of the Federal Energy Regulatory Commission (FERC), and encourages the same designation to be given by state public utility commissions (PUCs). Some of these concerns are being addressed by the President's Task Force on Regulatory Relief and the Innovative Control Technology Advisory Panel (41).

Fourth is the uncertainty surrounding, and possibly strategic posturing toward, environmental regulations. A chronic fear by major manufacturers and utilities is that soon after investing in a new environmental control technology, legislation will be passed requiring that some other technology be used. Even with the passage of the Clean Air Act Amendments in late 1990, primarily aimed at acid rain precursors and air toxics, there is mounting pressure for action against the substances thought to contribute to global warming. (CCTs provide only modest mitigation of greenhouse gases, such as CO_2.) At the same time, the recent Clean Air Act Amendments do allow for some delays for utilities planning to install CCTs, but there are doubts about the commercial availability of postcombustion options for Phase 1 of the new law, and whether advanced technologies would be demonstrated by Phase 2 (42). The new Act stimulates the use of CCTs, but there are still a host of other means of compliance, including switching to another fuel and energy efficiency, both of which lower the demand for coal.

Overall, timing is a key issue. Given the history of other technology development programs and the early experience of the CCT program, we reiterate that DOE's projections of availability of most new technologies for commercialization by the mid-1990s are overly optimistic. Moreover, availability is quite distinct from interest and commitment by the private sector. The adoption process for new technologies has typically been characterized by a logistic (S-shaped) time path, indicating a slow reaction to even a definitive demonstration. The large scale of many of these projects, with construction lead-times of 3 to 7 years, delays the process even more. Under normal circumstances, it is thus likely to be at least 20 to 25 years until there is widespread use of these new options. If the federal government wishes to accelerate this process, it may have to take a more active role in technology demonstration and provide greater economic incentives for technology adoption (43).

VI. COAL DEMAND

Factors Influencing Coal Demand

The demand for coal is derived from the demand for outputs it is used to produce (44, 45). Since coal enjoys a variety of applications, its demand depends not only on the general state of the economy, but also on individual end-users (primarily electric utilities, industrial boilers, and steel-makers). The quantity of coal used by electrical utilities has grown considerably over the past 50 years, while the amounts used in transportation, steel making, and other industrial end-uses has declined.

While coal demand generally varies with its own price, the prices of substitute fuels, and the level of economic activity, the latter influence differs according to the particular end-use. For example, future demand for coal by electric utilities depends on the expansion of coal-fired utility capacity. This expansion, in turn depends not only on the growth in the demand for electricity by all users but also the cost of substitute fuels, the limits imposed by coal-burning pollution control requirements, and the ability of technology to meet these requirements. Also, sufficient transportation capacity must be available to move coal to utility sites or bring electric power to consumers. Availability and constraints on the utilization of substitute fuels also affect coal demand. Public concerns remain concerning the reliability of imported oil supplies, the amount of natural gas reserves, and public attitudes toward nuclear power.

Conservation (increases in fuel efficiency) influences total energy and coal demand as well. Household and commercial energy demand has declined because of increased efficiencies in space heating. The fact that the intensity of US energy use (measured in 1000 Btu/$ GNP) is twice as high as those in Japan and West Germany suggests that further reductions in energy consumption though efficiency improvements are possible (46).

The use of coal in the primary metals industry has been declining for several reasons. Not only has the demand for US raw steel declined considerably, but pollution control requirements have checked the growth of the domestic coking industry.

The use of coal in industrial boilers has declined because of substitution by other fuels, and it is not likely to recover soon unless a coal-based modified fuel is developed, such as deep-cleaned or micronized coal (47). Little coal is used in the transportation section, and this is unlikely to change unless new technology is introduced, such as electric vehicles or internal combustion engines able to burn a liquefied product of coal or a coal-water emulsion.

Dynamic adjustments also influence coal demand. When conditions in the coal market change, full adjustment may take many years. When petroleum prices skyrocketed in 1974, coal demand did not increase instantaneously because of the considerable time required to convert the boilers in electricity-generating equipment from fuel oil to coal. Since this conversion process itself was limited, demand lags of a longer-run nature existed until new coal-fired plants could be built. When new coal-burning technology is introduced, it is likely to take several years for coal demand to proceed from the typical slow phase of initial uncertainty to an accelerated pace following dissemination of information on successful operating experience.

Coal demand can also be influenced by government policy. Although the coal market appears relatively unhampered by direct government intervention, indirect influences have a major impact on coal demand. For example, acid rain legislation can influence the quality characteristics of coal mined, which, in turn, can lead to differential regional impacts. The long-standing SO_x controls of new coal-fired plants and the recently passed Clean Air Act Amendments increase the costs of burning coal and thus reduce coal demand. Also of significance is the regulation and deregulation of coal transportation rates and electric power transmission.

Providing a precise measure of the impact of coal-utilizing technology is difficult. The most general interpretation is that coal demand will increase (i.e. its demand curve will shift outward) as new technologies broaden its possibilities for utilization and enhance its attractiveness in current uses. This normally results in the substitution of coal for other fuels or a derived increase resulting from expansion of end-product demand (whose production cost has been lowered). On the other hand, more efficient utilization technology could cause coal use to decrease because less coal input is needed per unit of output.

Baseline Estimates of Future Coal Demand

We now proceed to establish a reference point for our overall assessment of future coal demand. The analysis is based on available coal forecasts. These future scenarios may or may not be based on formal models of coal markets, and most often embody a substantial amount of subjective judgment.

There are numerous forecasts of coal demand by organizations such as the US Energy Information Administration (EIA) (48), the American Gas Association (49), DRI (50), the Gas Research Institute (51) and WEFA (52). Although the assumptions and methodologies underlying these estimates differ, the forecasts do provide a general sense of direction—total coal demand anticipated for the year 2010 ranges between 1195 and 1388 million short tons (Mst), with the EIA forecast being the highest. This indicates that average growth in coal consumption is likely to be in the range of 1.4 to 2.1% per year for the period 1990–2010.

Because the forecasts made by the EIA are close to the predicted values of other forecasts, and because they are widely distributed and generally well-respected, it is of interest to examine the EIA forecasts further.

The EIA forecast of future coal consumption is shown in Table 5. It uses a short ton coal equivalent based on an average heat content of 10,700 to 10,300 Btu/lb of coal. The average heat content used for a specific year depends on the average heat content of the coal mix as forecast by EIA. Electricity consumption figures for the residential, commercial, industrial, and transportation sectors have been adjusted to a 30.5% fuel-efficiency rate in the generation of electricity.

The EIA forecast suggests an increase in energy consumption in all sectors with an overall average increase of 1.5% per year. The electric utility sector shows the greatest growth in energy use and coal consumption while the transportation sector shows the least. Nonutility generation of electricity is

Table 5 Forecasts of US consumption of energy by end-use sector (million short tons 10,600 Btu/lb coal equivalent)

Sector/Energy source	1990	2000	2010
Residential			
Other (including coal)	51	59	68
Electricity[a]	483	611	732
Petroleum & gas	299	276	268
Total	832	945	1069
Commercial			
Other (including coal)	25	22	20
Electricity	443	591	761
Petroleum & gas	156	148	144
Total	624	761	925
Industrial			
Metallurgical coal	52	50	44
Steam coal	80	86	118
Purchased electricity[a]	473	640	823
Other	903	1041	1143
Total	1508	1817	2127
Transportation			
Other (including coal)	43	50	51
Electricity[a]	2	2	3
Petroleum & gas	997	1097	1248
Total	1042	1149	1302
Total actual consumption			
Coal	893	1049	1388
Other	3112	3624	4035
Total	4055	4673	5423
Electric utility generation			
Coal	727	868	1246
Other	575	799	816
Total	1302	1667	2063
Nonutility generation			
Coal	16	27	61
Other	82	139	202
Total	98	166	263

Source: (48) based on most recent data as of 1988.
[a] Gross energy used in purchased electricity includes transmission losses and 30.5% generation efficiency.

forecast to grow at an annual rate of 5–8% per year; while this growth rate is high, quantities are still relatively low through the year 2010.

Table 5 shows the amount of coal or coal equivalent used by each sector. These figures represent an upper bound to the changes in coal demand brought about by particular changes in energy technology for each sector. It is necessary to inspect these potential changes on a sector-by-sector basis because technology adoption will be end-use specific.

RESIDENTIAL   Natural gas and electricity together account for 90% of total energy used in the residential sector. Electricity usage is forecast to increase at a rate of about 2.4% per year to the year 2000 and decline to 1.8% per year by 2010. Natural gas consumption is forecast to decline slightly. If all the forecast increases in electricity consumption were to be derived from the burning of coal, an additional 128 million tons of coal per year would be required by the year 2000 and an additional 121 million tons per year by 2010.

COMMERCIAL   Besides electricity, the major source of fuel for the commercial sector is natural gas. The use of electricity in this sector is forecast to increase by about 2.7% per year. Small decreases are forecast for all other energy forms. If the electricity increase were to be generated entirely by the burning of coal, 148 million more tons per year of coal would be required by the year 2000 and an additional 170 million by 2010.

INDUSTRIAL   Natural gas, purchased electricity, and other industrial fuels, such as industrial and municipal wastes, account for more than 90% of all energy used by the industrial sector. These fuels are used, in part, to produce self-generated electricity. Purchased electricity is forecast to increase by 2.5% per year for the period 1990 to 2000 and 2.1% per year between 2000 and 2010, while other energy sources are forecast to increase by 1.4% per year from 1990 to 2000 and 0.9% per year from 2000 to 2010. If coal were used exclusively to fuel these increases, 167 million more tons per year of coal would be required by year 2000 and an additional 183 million by 2010. In addition, net direct use of steam coal is forecast to increase by only 6 million tons per year by the year 2000 but to increase by an additional 32 million tons per year by 2010. The average yearly increase in coal demand in the industrial sector from 1990 to 2010, as shown in Table 5, is about 31 million tons. The use of natural gas in the industrial sector is expected to decline. This projected decline is particularly critical to the forecast of coal use since coal is likely to be a major substitute for natural gas used in this sector.

TRANSPORTATION   More than 90% of all energy used by the transportation sector is in the form of distillate, jet fuel, and gasoline. Current and forecast use of coal is minor. Any signifiant use of coal by the transportation sector would therefore be achieved only by increased use of electricity (such as electrical railroads or automobiles) or replacement of distillate, gasoline, or residual fuel oil by coal (such as methanol or coal-fired diesel engines). Since the energy use is so large in the transportation sector, even minor percentages of replacement represent significant increases in coal. For every 1 percent of petroleum products replaced by coal, 10 to 12 million more tons per year of coal would be consumed (given the EIA forecast parameters).

ELECTRIC UTILITIES   Electric utilities are the largest consumers of coal and have the second-highest growth rates of coal consumption after nonutility generators. The forecasts given in Table 5 indicate that 141 million tons of additional coal per year will be required by electric utilities by the year 2000 and an additional 378 million tons by 2010. Critical to this estimate are the forecasts of distillate, residual fuel oil, and natural gas used by electric utilities. Forecasts of use of hydro or nuclear power to generate electricity are less subject to error because of the long lead times required for new capacity. In any case, changes in generated power from these are minor compared to those from fossil fuels. These forecasts indicate that noncoal consumption will increase to the equivalent of 224 million tons of coal per year by the year 2000, but will not increase thereafter.

NONUTILITY GENERATION   While generation of electric power by non-utilities is forecast to grow at a rate exceeding 5% per year, total use of energy is small compared to other sectors. Even though coal currently consists of only 20% of total energy use in this sector, coal use is expected to increase. By the year 2010, the total increase in energy use is equivalent to 157 million tons per year of additional coal.

It is worth noting that the transportation sector offers a greater potential for expanded coal use in excess of EIA forecasts than does any other sector. Transportation is currently a major energy user, but not with respect to coal or electricity. The largest total tonnage equivalents shown for year 2010 in Table 5 are electricity for the residential, commercial, and industrial sectors, motor gasoline for the transportation sector, and natural gas for the industrial sector.

Total coal consumption is forecast to increase by 156 million tons per year by the year 2000, of which 141 million would be accounted for by electric utilities. This represents an average total increase of about 15 million tons of coal equivalent per year for the period 1990–2000.

Table 5 represents only one plausible forecast scenario by US EIA. Many other plausible scenarios could also be described resultiing in a wide range of energy and coal use forecasts. An important aspect of understanding prospective coal demands is considering the uncertainty that surrounds them. Uncertainty exists not only because the exact timetable and nature of new technologies are unknown, but also because of episodic events such as oil price shocks and because of cyclical variations in national economic conditions. The impact of this uncertainty falls on three aspects of coal demand forecasting: (a) forecasting total US energy demand, (b) forecasting coal's share of the total energy market, and (c) forecasting the type(s) of coal to be used. To a large degree, overall energy demand must be known before coal's market share can be estimated. In turn, the type of coal demanded and the decision environment (e.g. regulations) can be determined. Recognition of the existence of these uncertainties is key to the planning activities of coal suppliers and users.

Differences in energy forecasts among the various forecasting agencies and organizations cited are mainly due to differences in forecast oil prices and the resulting levels of world economic activity. For example, total coal demand anticipated for the year 2010 ranges from 1195 to 1388 million short tons among the forecasts given. This represents a 16% difference between the high and low forecasts for the 20-year period. The difference is explained largely by a 16% difference in forecast oil prices, which results in the variation of world economic activity and total energy demand.

Substitution of oil for coal is not an important factor in explaining differences in coal demand. Oil substitutes directly for coal at prices around $12 per barrel, which is well below most forecast prices for oil. Nor is coal expected to substitute increasingly for oil. Since coal prices are not forecast to increase significantly (and may well decline) over this time period, subsequent increases in the price of oil will not induce additional substitution of coal for oil.

Another difference among the cited energy forecasts is the assumed proportion of electric power in total energy consumption. A 10–20% variation in coal demand can easily result from the difference between the high and low electricity capacity scenarios for any given forecast.

Any energy forecast must include a range of values because of these uncertainties. The EIA projections of coal consumption range from 6.5% below the base forecast to 4.5% above. The number of plausible scenarios and the range of forecasts reflect the great level of uncertainty that exists in the energy sector. Since total energy demand, oil and gas prices, and the energy regulatory climate cannot be precisely predicted, forecasts of coal use will necessarily encompass a range of plausible values. Still, the high-end and low-end coal forecasts are within 10–15% of the mid-range, even in light of sizeable differences with regard to assumptions about economic growth, oil prices, and environmental regulation. We turn now to examining whether we can expect new technologies to have any greater effect than these other factors.

VII. TECHNOLOGICAL IMPLICATIONS

The impact of new technologies on prospective coal demands can be further elaborated with the use of Figures 1 and 2. In Figure 1, the intersection of supply (S^1) and demand (D^1) represents a "market equilibrium," which simultaneously determines the market price (P^1) and quantity (Q^1). This general

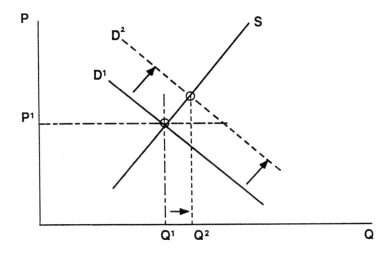

Figure 1   Impact of demand-related technology

characterization would apply both to the market for coal and to the market for any of its end-uses, such as electricity. Technologies that make coal more attractive or broaden its range of application can cause the demand curve for coal to shift upwards. That is, the quantity demanded for a given price, P^1, will increase as technology promotes greater coal utilization. This places upward pressure on the price of coal, which offsets some of the impetus to increase demand, with the new equilibrium output level being represented by Q^2. Other technologies can also affect coal use by causing shifts in the coal supply curve. In the case depicted in the lower market equilibrium diagram (Figure 2), a new technology can diminish the costs of production and hence lower the price of coal from P^2 to P^1. This results in a movement along the demand curve, e.g. the quantity of coal demanded increases from Q^1 to Q^2. Of course, the real world market adjustment in both cases is more complex, but the final impact is qualitatively the same as depicted in the figures.

Five technologies are chosen here to illustrate potential future impacts of CCTs on coal use. These technologies include four options encompassed by the CCT program—Fuel Cells, Pressurized Fluidized Bed Combustion, Advanced Coal Cleaning, and the Limestone Injection Multistage Burner (LIMB). Longwall Mining is included in the examples as a basis of comparison. Some major economic aspects of these technologies are presented in Table 6. Three of the technologies are capable of improving combustion efficiency, while one involves a significant decrease. Two of the technologies affect the output capacity of a power plant, with a pressurized fluidized-bed combustion unit (PFBC) being able to provide as much as 100% enhancement of an existing facility.

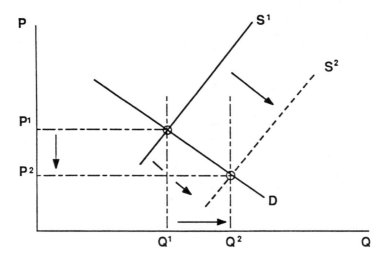

Figure 2   Impact of supply-related technology

Cost estimates are made in comparison with a conventional power plant equipped with a flue gas desulfurization (FGD) unit. This unit adds about 220 kW in capital costs and 10 mills/kWh in operating costs to the conventional plant costs. The Limestone Injection Multistage Burner (LIMB) is less expensive than the basic plant plus FGD and therefore results in an overall capital cost decrease. Two of the technologies, the Fuel Cell and the PFBC, are more expensive than the baseline technology at first pass. However, the latter is able to increase power output, thereby substantially lowering the overall cost of generating electricity on a per unit basis.

The impacts of selected coal-cycle technology advances are presented in Table 7, and summarized below:

Table 6   Economic aspects of selected coal-cycle technology advances

Technology	Energy conversion efficiency	Power output	Incremental cost of electricity[a] (mills/kWh)	Incremental capital cost ($/kW)
Longwall			−5	
Advanced coal cleaning	small increase		−5 to +10	
Pressurized FBC	slight increase	40–50%	−4 to −6	600–800
Fuel cells	small increase		+15	1600
LIMB	moderate decrease	small decrease	−5 to +5	−100

Source: Adapted from (53, 54).
[a] Negative numbers imply cost savings in comparison with a conventional power plant supplemented by a flue gas desulfurization FGD unit.

Table 7 Economic impacts of selected coal-cycle technology advances

Technology	Effect on coal supply curve	Effect on electricity supply curve[a]	Effect on coal demand curve	Overall change in coal use
Longwall	small downward	small downward	no change	slight increase
Advanced coal cleaning	small upward	slight downward	slight downward	slight increase
Pressurized FBC	no change	small downward	moderate upward	moderate increase
Fuel cells	no change	moderate upward	slight downward	no change
LIMB	no change	slight downward	small upward	small increase

[a] Comparison with power plant and FGD unit combination.

LONGWALL MINING   This technology reduces the cost of extracting coal, thus shifting the coal supply curve downward (rightward). The result is a less than proportional but still downward movement in coal prices that should lower the cost of electricity generation. To the extent that coal demand is price sensitive, this makes coal more attractive as a utility fuel, all other things being equal. However, since coal is already an inexpensive fuel relative to its substitutes, the increase in demand due to this substitution effect is likely to be small. In addition to this substitution effect, there is also an output effect stemming from the increased demand for electicity as a result of its lower price, though this is also not very price sensitive. The overall (net) result is likely to be only a slight increase in coal use.

ADVANCED COAL CLEANING   This alternative causes a small upward shift in the delivered coal supply curve. The cost of advanced coal cleaning is comparable to that of an FGD unit, so there is no change in the electricity supply curve on that basis. However, there is a slight downward shift in the electricity supply curve because of the increase in combustion efficiency associated with this alternative. There would also be a slight downward shift on coal demand from the increase in combustion efficiency. The overall effect is likely to be only a slight increase in coal use.

PRESSURIZED FLUIDIZED-BED COMBUSTION   This technology would lead to a slight increase in combustion efficiency and also relatively lower costs than FGD units. The lower cost per kilowatt-hour will shift the supply curve for electric power generation downward. (The lower cost is attributable primarily to the increased power output of an existing facility able to accommodate a PFBC unit, a feature that overcomes the large incremental capital investment in this alternative.) The increase in electric power capacity of an existing facility is likely to have a significant effect on the demand for coal. This technology could lead to the largest increase in coal use of any of the alternatives listed in Table 7; however, it would still be only a moderate amount.

FUEL CELLS   The development of fuel cells does not have an effect on the coal supply curve. Because of its high cost, however, the implementation of this technology in a power plant would mean an upward shift in the electricity supply curve. Coal demand would shift downward slightly due to the increased combustion efficiency, which decreases coal input per unit of output. No other significant effect on electricity demand is likely as a result of the development of this technology. Also, we do not anticipate widespread use of fuel cells into the beginning of the next century given its relatively early stage of current development and relatively high cost.

LIMB   This technology would have no effect on the coal supply curve. The decrease in combustion efficiency associated with it would to some extent offset its cost savings, so there would be very slight if any change in the electricity supply curve. On the demand side, there would be a small power output decrease, but this would be offset by the decrease in combustion efficiency leading to an increase in coal input per unit of electricity output. At the same time, this technology is capable of generating widespread interest and use. Therefore, it is not unreasonable to project a potential upward demand shift for this technology. The overall effect on coal use, however, is still likely to be small.

VIII. CONCLUSION

The coming two decades hold some promise of increased coal use, but new Clean Coal Technologies are not likely to add much to the upswing. The major contribution of CCTs may be in helping to maintain coal's market share in the face of new environmental regulations that could otherwise induce a shift from coal to other fuels.

How much of a potential increase in coal use will be realized depends upon the five major factors affecting CCTs: (a) strictness of the compliance schedule for the new Clean Air Act Amendments, (b) the need for new power generation capacity and the ability of utilities to extend plant life, (c) the ability of electric utility and non-utility producers to use innovative technology in meeting future electricity demands, (d) the price of natural gas and other fuels, and (e) various institutional and regulatory factors.

Phase 2 compliance deadlines under the new Clean Air Act Amendments are 1997 with a variance until 2003 for utilities planning to install CCTs. Thus, planning must be undertaken by 1997, which is likely to be before most CCTs have been satisfactorily demonstrated. The problem is that before 1997 utilities may install FGD units or switch to other fuels, precluding the adoption of CCTs when they do become available.

Most forecasters expect electricity use to increase, probably at a greater rate than experienced in the early 1980s. Some utilities have begun to adopt CCTs, either through their own construction or by increasing purchases from nonutility generators. However, until recently there was more momentum for options that would make less use of coal, such as cogeneration by non-utilities. The price of natural gas is forecast to increase in the 1990s, but this forecast is highly uncertain. Regulatory reform has begun but faces a long road, since so much of it has to traverse through state public service commissions.

Electric utilities do not have, at the present time, great incentives to invest heavily in new CCTs because of the risk of failure and uncertainty about future regulations. Utilities are more inclined to extend existing plant life than to invest heavily in new CCTs or new generation capacity. Deployment of innovative technologies, such as CCT, is encouraged by the participation of nonutility generators, but this group is likely to remain only a small part of the electric power generation industry for the next 20 years.

The many factors affecting coal demand are difficult to forecast and are inexorably intertwined. For these reasons, we acknowledge the uncertainty surrounding our analysis. Most of the emerging technologies examined are very complex and their success cannot be guaranteed. Moreover, their adoption by coal producers and users is very difficult to predict. Finally, the demand for coal is highly susceptible to external forces, such as international and domestic politics, climatic conditions, and unforeseen technological developments.

Although the theme of this paper has been the future prospects for coal use, it should be kept in mind that this not an end in itself, but prized because of its broader ramifications. These include:

1. Gains in employment, especially in depressed regions of the United States.
2. Enhanced energy security associated with a renewal of interest in tapping our major domestic fossil fuel.
3. Improved balance of payments from lessening the dependency on foreign oil.
4. Increases in tax revenues, especially at the state and local levels.

Of course, there are more direct ways of achieving some of these goals, but there is no doubt that increased coal use will help promote them.

If previous experience is any indication, considerable inertia on the part of key decision-makers works against innovative technologies. The broader net benefits to society from increased coal use may far outstrip the private gains to companies in the coal cycle: however, the marketplace will not be able to bring about the optimal outcome. No doubt these factors are the basis for federal, and, increasingly, state government willingness to share the risk (cost) of innovations. But the present 50% cost-share arrangement is an arbitrary level, and does not extend to the deployment phase.

More formal analysis of the costs and benefits of coal use, in addition to more knowledge of the R&D and adoption process, are needed before policy goals can be formulated in this area. Only then can the adequacy of existing market institutions, regulations, and policy inducements be evaluated, and additional remedies, if needed, be formulated on a sound basis.

ACKNOWLEDGMENTS

The authors were ably assisted by Shih-Mo Lin, Stephen Shaw, and Amit Mor. This review is based heavily on a study funded by the U.S. Office of

Technology Assessment (OTA). Supplementary funding for subsequent work was provided by The West Virginia University Energy and Water Resources Center and The Pennsylvania State University Mineral Research Institute. The authors wish to thank Steve Plotkin of OTA for his helpful suggestions and comments during the initial phase of this research. We are grateful to George Sall and Lowell Miller of DOE, who supplied us with numerous reports relating to Clean Coal Technologies. Very useful comments were received from John Byam and other staff of the Morgantown Energy Technology Center and from William Poundstone. Of course, the authors are solely responsible for the final contents. Note also that this paper was written before the passage of the most recent Clean Air Act Amendments in late 1990.

Literature Cited

1. Landsberg, H., Arrow, K. J., Bator, F. M., Dam, K. W., Fri, R. W. et al. *Energy: The Next Twenty Years,* Cambridge: Ballinger. 628 pp.
2. Wilson, C. 1980. *Coal: Bridge to the Future.* Ballinger. 247 pp.
3. National Coal Council. 1990. *The Long-Range Role of Coal in the Future Energy Strategy of the United States.* Washington, DC. 156 pp.
4. US Dept. Energy. 1985. *Report to Congress on Emerging Clean Coal Technologies,* Washington, DC
5. US Dept. Energy. 1987. *Second Report to Congress on Emerging Clean Coal Technologies Capable of Retrofitting, Repowering and Modernizing Existing Facilities,* Washington, DC. 282 pp.
6. Wooten, J. M. 1986. Technology for new coal markets. *Economics of Internationally Traded Minerals,* ed. W. R. Bush. Soc. Min. Eng.
7. National Coal Council. 1986. *Clean Coal Technology.* Washington, DC. 47 pp.
8. Lewis, D., Davis, W. 1986. Joint Report of the *Special Envoys on Acid Rain.* Washington, DC: US Dept. Energy
9. National Acid Precipitation Assessment Program (NAPAP). 1987. *Interim Assessment: The Causes and Effects of Acidic Deposition.* 750 pp.
10. Wampler, J. A. 1987. *Testimony to the Committee on Energy and Natural Resources.* US Senate, May 18, Washington, DC
11. Electr. Power Res. Inst. 1987. How advanced options stack up. *EPRI J.* 12:4–13
12. US Dept. Energy. 1990. *Clean Coal Technology Demonstration Program: Annual Report to Congress.* Washington, DC

13. US Dept. Energy. 1990. Project status report (mimeo). Office of Fossil Energy, Washington, DC
14. Sprouls, M. W. 1990. Longwall census. 1990. *Coal* 95:36–47
15. Rubin, E. S. 1989. Implications of future environmental regulation of coal-based electric power. *Annu. Rev. Energy* 14:19–45
16. Joskow, P. L., Rose, N. L. 1985. The effects of technological change, experience, and environmental regulation on the construction cost of coal-burning generating units. *Rand J. Econ.* 16:1–27
17. Wolk, R., McDaniel, J. 1990. *Electricity without steam: advanced technology for the new century.* Presented at Assoc. Edison Illuminating Cos. Power Generation Meet, Feb. 8, 1990. Orlando, Fla.
18. US Dept. Energy. 1990. *Comprehensive Report to Congress: Proposals Received in Response to the Clean Coal Technology III Program Opportunity Notice.* Washington, DC
19. Electr. Power Res. Inst. 1988. Coal technologies for a new age. *EPRI J.* 13:4–17
20. National Coal Association (NCA). 1986. *Clean Coal Technology: Securing Our Energy Future.* Washington, DC
21. Blair, P. 1987. New electric power technologies. *Energy Sys. Pol.* 10:189–217
22. Congressional Budget Office. 1989. *The Role of Technology and Conservation in Controlling Acid Rain.* Washington, DC: CBO. 46 pp.
23. US Dept. Energy. 1990. *Clean Coal Technology: The New Coal Era.* Washington, DC
24. US Dept. Energy. 1989. *Clean Coal*

Technology Demonstration Program: Final Programmatic Environmental Impact Statement. Washington, DC

25. Torrens, I. M. 1990. Developing Clean Coal Technologies. *Environment* 32:11–14, 28–33

26. Siegel, J., Temchin, J. 1991. Role of clean coal technology in electric power generating in the 21st century. In *Energy and the Environment in the 21st Century,* ed. J. Tester. Cambridge: MIT Press

27. Griffin, J., Steele, H. 1986. *Energy Economics and Policy.* Orlando: Academic. 2nd ed.

28. Electr. Power Res. Inst. 1990. Cool water coal gasification program: commercial-scale demonstration of IGCC technology completed. Technical Brief, Palo Alto, Calif.

29. Preston, W. E. 1990. *The Texaco coal gasification process: a means to produce clean, cost competitive power from coal.* Am. Inst. Chem. Eng. 1990 Spring Natl. Meet., April

30. Electr. Power Res. Inst. 1989. Shell coal gasification process—an update. Technical Brief, Palo Alto, Calif.

31. Clark, W. N., Shorter, V. R. 1987. Cool water coal gasification: a mid-term performance assessment. *Min. Eng.* 2:97–102

32. Gluckman, M. J., Spencer, D. F., Watts, D.H., Shorter, V. R. 1988. A reconciliation of the cool water IGCC plant performance and cost data with equivalent projections for a mature 360 MW commercial IGCC facility. *Proc. 7th Annu. EPRI Contractor's Conf. Coal Gasification. 1988.* Palo Alto, Calif.: Electr. Power Res. Inst.

33. Utility Data Inst. 1987. *Projected Capital Costs, U.S. Steam-Electric Plants.* Washington, DC

34. Yeager, K. E., Baruch, S. B. 1987. Environmental issues affecting coal technology: a perspective on U.S. trends. *Annu. Rev. Energy* 12:471–502

35. Smith, D. J. 1990. FBC: Is it the answer for new coal-fired capacity? *Power Engineering.* 94:27–30

36. Electr. Power Res. Inst. 1990. *Colorado-Ute 110-MW AFBC plant: test program.* Technical Brief. Palo Alto, Calif.

37. Sens, P. F., Wilkinson, J. K., ed. 1988. *Fluidized Bed Combustor Design, Construction and Operation.* Essex, England: Elsevier Applied Science Publishers. 149 pp.

38. Makansi, J., Schwieger, R. 1987. Fluidized-bed boilers. *Power* 131:S1–S16

39. Moll, J., Simbeck, D., Wilhelm, D. 1989. *Track record for fluidized bed boilers.* 4th Annu. Cogeneration and Independent Power Market Conf., New Orleans, La., April, 1989

40. Electr. Power Res. Inst. 1988. *Issues and responses: 1988.* Palo Alto, Calif.

41. US Dept. Energy. 1989. *Report to the secretary of energy concerning commercialization incentives.* Washington, DC

42. Torrens, I. M. 1990. *Electric utility response to new acid rain and air toxic legislation in the United States.* Paper presented at the UGB Congr. Tech. Inf. Exposition, Power Plants 1990, Sept. 18–21, 1990. Essen, Federal Republic of Germany

43. Kahn, E., Stoft, S. 1989. *Designing a Performance Based Cost-Sharing System for the Clean Coal Technology Program.* Berkeley: Univ. Calif.

44. Labys, W. C., Sharokh, F., Liebenthal, A. 1979. An econometric simulation model of the U.S. market for steam coal. *Energy Econ.* 1:19–26.

45. Gordon, R. 1987. *World Coal.* Cambridge: Cambridge Univ. Press. 145 pp.

46. Chandler, W. U., Geller, H. S., Ledbetter, M. R. 1988. *Energy Efficiency: A New Agenda.* Am. Counc. for an Energy Efficient Econ., Washington, DC. 75 pp.

47. National Coal Council. 1990. *Industrial Use of Coal and Clean Coal Technology.* Washington, DC, 101 pp.

48. US Energy Inf. Admin. (EIA). 1990. *Annual Energy Outlook 1990: Long-Term Projections.* Washington, DC

49. Am. Gas Assoc. (AGA). 1989. *Total Energy Resource Analysis Model. The 1989 AGA-TERA Base Case.* Alexandria, Va.

50. DRI/McGraw-Hill. 1989. *Energy Rev.* Autumn

51. Gas Res. Inst. (GRI). 1989. *GRI Baseline Projections of U.S. Energy Supply and Demand.* Washington, DC

52. The WEFA Group. 1989. *Long-Term Economic Outlook.* 3rd Quarter 1989, Bala Cynwyd, Pa.

53. US Dept. Energy. 1985. *Supplemental Report to Congress on Emerging Clean Coal Technologies,* August. Washington, DC

54. Grayson, R. L. 1988. Longwall mining cost differentials. Personal communication, West Virginia Univ.

Annu. Rev. Energy Environ. 1991. 16:91–121

CREATING THE FUTURE: INTEGRATED RESOURCE PLANNING FOR ELECTRIC UTILITIES

Eric Hirst

Oak Ridge National Laboratory, Oak Ridge, Tennessee 37831

Charles Goldman

Lawrence Berkeley Laboratory, Berkeley, California 94720

KEY WORDS: competition, demand-side management, electricity, environment, public utility
              commission.

1. INTRODUCTION

Integrated resource planning (IRP) helps utilities and state regulatory commissions assess consistently a variety of demand and supply resources to meet customer energy-service needs cost-effectively. Key characteristics of this planning paradigm include: (*a*) explicit consideration of energy-efficiency and load-management programs as alternatives to some power plants, (*b*) consideration of environmental factors as well as direct economic costs, (*c*) public participation, and (*d*) analysis of the uncertainties and risks posed by different resource portfolios and by external factors.

IRP differs from traditional utility planning in several ways, including the types of resources acquired, the owners of the resources, the organizations involved in planning, and the criteria for resource selection (Table 1). References 1–5 discuss IRP and its development.

This paper reviews recent progress in IRP and identifies the need for

91

Table 1 Differences between traditional planning and integrated resource planning

Traditional planning	Integrated resource planning
Focus on utility-owned central-station power plants	Diversity of resources, including utility-owned plants, purchases from other organizations, conservation and load-management programs, transmission and distribution improvements, and pricing
Planning internal to utility, primarily in system planning and financial planning departments	Planning spread among several departments within utility and often involves customers, public utility commission staff, and nonutility energy experts
All resources owned by utility	Some resources owned by other utilities, by small power producers, by independent power producers, and by customers
Resources selected primarily to minimize electricity prices and maintain system reliability	Diverse resource-selection criteria, including electricity prices, revenue requirements, energy-service costs, utility financial condition, risk reduction, fuel and technology diversity, environmental quality, and economic development

additional work. Key IRP issues facing utilities and public utility commissions (PUCs), discussed in this paper, include:

1. Provision of financial incentives to utilities for successful implementation of integrated resource plans, especially acquisition of demand-side management (DSM) resources;
2. Incorporation of environmental factors in IRP;
3. Bidding for demand and supply resources;
4. Treatment of DSM programs as capacity and energy resources;
5. Development of guidelines for preparation and review of utility resource plans; and
6. Increased efforts by the US Department of Energy (DOE) to promote IRP.

Many other planning issues are important, but are not discussed in this paper. Such issues include alternative ways to organize planning within utilities; the role of collaboration and other forms of nonutility involvement in planning (6, 7); the relationships among competition, deregulation, and utility planning; treatment of electricity pricing as a resource; fuel switching (primarily between electricity and gas); treatment of uncertainty in utility planning and decision making (8, 9); the appropriate economic tests for utility DSM programs; ways to measure the performance of DSM programs (10); development and use of improved data and planning models; and transfer of information among utilities and commissions. Many of these topics are covered in (11).

2. REWARDING UTILITIES FOR EFFECTIVE IRP IMPLEMENTATION

One feature that distinguishes IRP from traditional utility planning is the explicit inclusion of demand-side programs as utility resources. Unfortunately, traditional regulation discourages utility DSM investments. This conflict between the interests of customers and shareholders occurs because "each kWh a utility sells . . . adds to earnings [and] each kWh saved or replaced with an energy efficiency measure . . . reduces utility profits" (12). Disincentives to utility investments in DSM include:

1. Failure to recover all program costs,
2. "Lost revenues" caused by DSM programs that reduce electricity use and result in utility under-recovery of allowed fixed costs between rate cases,
3. Concerns that DSM investments, which are generally not put in rate base, will ultimately reduce the utility's rate base and earnings.

PUCs and utilities have developed various approaches to overcome these disincentives (Table 2). For example, at least nine states allow DSM investments to be included in rate base.

Table 2 Regulatory incentives for utility DSM incentives

State	Rate-basing	Lost revenue adjustment	Decoupling profits from sales	Higher rate of return	Bounty	Shared savings
CA	X		X	X		X
CT	X			X		
ID	X			X		
IA	X					
KS				X		
MA	X	X			X	
MT	X			X		
NH		X				X
NY		X	X	X		X
NC				X		
OK	X					
OR	X	X				
RI						X
TX	X					
VT		X				
WA				X		
WI				X	X	

Several states adjust for DSM-induced revenue losses. These adjustments ensure that utilities collect from customers the net revenue that they would have gained from employing their generating resources had the DSM program not reduced electricity sales. A related option is to decouple utility profits from sales. This approach is used in several states, most notably in California where the PUC uses an Electric Revenue Adjustment Mechanism (ERAM) and in New York where the Commission (13) approved a "Revenue Decoupling Method" for one utility. In both states, these regulatory mechanisms guarantee that utility earnings are independent of the amount of sales achieved. ERAM has been used in California since 1982 and accounts for the over- or under-collection of authorized base revenues (essentially all nonfuel costs) caused by discrepancies between actual and forecast sales of electricity (14). Utilities use balancing accounts for over- (or under-) collection of revenues. These revenues are then returned to (or collected from) customers the following year through an adjustment to the price of electricity. This mechanism breaks the link between sales and profits, thus eliminating a major disincentive to utility DSM programs.

PUCs in several states have also approved various types of financial incentives to utility shareholders for exemplary delivery of DSM services (Table 2). In addition, investigations on incentive proposals are being conducted by about 10 other PUCs (15). These bonus incentives can be grouped into three broad groups: increased rate of return, bounty, and shared savings (12).

The simplest approach is to adjust the utility's allowed rate of return for a specified accomplishment, such as achieving target levels of energy savings or DSM spending. The adjusted rate of return is applied either to the utility's DSM investment or to its total rate base. For example, Washington allows utilities to earn a 2% higher return on equity for conservation expenditures than for other investments.

The bounty approach is to pay utilities for specified achievements. For example, the PUC in Massachusetts recently approved an incentive for Massachusetts Electric that provides "a fixed payment for each kW and kWh saved that is verified through an after-the-fact evaluation and monitoring system" (16). The major advantage of such a method is its administrative simplicity. However, the utility has no incentive to minimize DSM program costs and its bonus does not depend directly on the benefits provided by its DSM programs.

Shared-savings mechanisms are probably the most popular approach. Under this approach, the utility keeps a fraction (typically 10–20%) of the net benefit provided by its DSM programs. The net benefit is the difference between the total benefits and program costs. Total benefits are typically defined as the amount of energy saved by the program multiplied by the

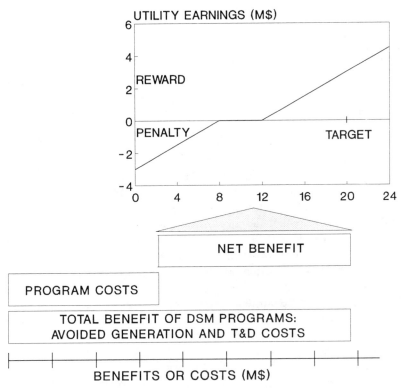

Figure 1 Schematic showing the mechanics of a shared-savings mechanism to reward utility shareholders for implementation of cost-effective DSM programs.

avoided energy cost plus the amount of demand reduction multiplied by the avoided capacity cost. [These benefits reflect the utility's reductions in capital costs (caused by the DSM-program savings, which allow the utility to defer construction of new plants) and in operating costs.] Program costs include administrative costs and financial incentives to customers. This type of mechanism encourages utilities to run ambitious DSM programs, provides a continuing incentive to control costs, and represents a reward for value received. Versions of such incentives are now in place in California, New York, Rhode Island, and New Hampshire (Table 2).

Figure 1 shows how such an incentive mechanism might work. The bars at the bottom of the figure show the relationships among total benefits, program costs, and the net benefit. The top part shows how the net benefit might be shared between utility shareholders and customers. In this example, shareholders earn an additional $3 million if the utility achieves its target net benefit of $20 million. If the utility's DSM programs are more effective than expected, earnings increase (but are typically capped at a prespecified upper limit). On the other hand, shareholders pay a penalty (i.e. earnings decrease) if the utility is unable to achieve 40% of the target. And shareholders receive no benefit if the net benefit falls in the "deadband" range of 40–60% of the target. This approach ensures that utility shareholders face both risk and reward.

Shared-savings incentives may not be appropriate for all types of DSM programs. Some DSM programs are offered primarily for equity reasons (e.g. direct assistance to low-income customers) and may not offer significant net resource savings (17). Shared-savings incentives are best suited for DSM programs that provide least-cost resource options and involve installation of energy-efficient hardware as opposed to behavioral changes.

Incentives can affect a utility's overall planning process. For many utilities, DSM-program planning, design, and evaluation have not been especially important activities. With incentives, utility earnings depend in part on the savings from DSM programs. Thus, both utilities and PUCs are emphasizing evaluation because evaluations identify the energy and demand savings provided by DSM programs. The visibility of DSM programs is increased because they affect the utility's earnings (so top management wants to know the results of these programs). And regulators are requiring better measurement of the actual benefits of DSM programs before utilities receive incentive payments. Thus, incentive mechanisms can help close the loop among DSM-program planning, implementation, and evaluation.

In summary, providing incentives to utilities will not solve all the problems associated with planning and resource acquisition. But they can be a critical element of a successful IRP process by aligning the utility's financial interest with aggressive pursuit of a least-cost energy strategy.

Because states differ in ratemaking practices (e.g. historic vs future test years and fuel-adjustment clauses) and utilities differ in structure and corporate culture, it is unlikely that any one mechanism will be suitable everywhere. Figure 2 shows projected DSM expenditures and annual earnings from incentive mechanisms for five utilities. These DSM incentives differ in structure, size, and riskiness to the utility. Pacific Gas & Electric's incentive is quite attractive (37% of total DSM expenditures), but the company will not earn any incentive unless it exceeds minimum performance targets that are quite high. In contrast, Southern California Edison's incentive is lower (11% of total expenditures) but is structured so that the utility has a greater chance of earning additional money even if customer participation is lower than expected.

The size of bonuses for implementing exemplary DSM programs needs to be evaluated in the context of spending levels and resource value of these programs, the risk/reward relationship facing the utility, the impact on cus-

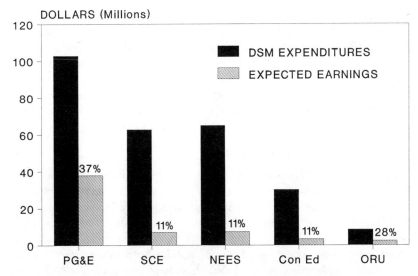

Figure 2   Planned DSM-program expenditures and the associated financial incentive for five utilities, in California (Pacific Gas & Electric and Southern California Edison), New England (New England Electric System), and New York (Consolidated Edison and Orange & Rockland Utilities).

tomer rates, and the impact on the utility's overall return on equity. Superior incentive mechanisms are likely to be state- and utility-specific and evolve out of negotiations among key parties. A variety of incentive mechanisms are being tested in many states and the effects of these schemes should be carefully evaluated. The credibility and future viability of incentive mechanisms hinges on the ability to measure energy savings and load reductions carefully and accurately (10), because these measurements determine the payments to utilities. Finally, development of incentive mechanisms is in one sense a large-scale experiment, one which is being implemented in an ad hoc fashion.

3. INCORPORATING ENVIRONMENTAL FACTORS IN UTILITY PLANNING

Electricity production imposes significant burdens on the environment. In the United States, roughly two-thirds of the SO_2 emissions and one-third of the NO_x emissions, both of which cause acid deposition with its accompanying harmful effects on trees, lakes, and man-made structures, come from power plants. Electric utilities account for about one-third of US emissions of CO_2, the principal greenhouse gas (18). In addition, electricity production causes water pollution, solid waste, and land-use problems. These effects of electricity production and transmission are externalities, defined as any cost not reflected in the price paid by electricity consumers (19). Existing federal and state regulations internalize some of the environmental costs associated with electricity production (e.g. regulations that limit emissions), and it is likely that this approach to managing environmental externalities will be expanded in the future. For example, the 1990 Clean Air Act Amendments mandate significant reductions in SO_2 and NO_x emissions by utilities during the 1990s, and federal legislation may ultimately be enacted to reduce CO_2 emissions.

State PUC Regulatory Initiatives

Regulators and utilities have long recognized environmental concerns about generation and transmission. For example, decisions to locate power plants in rural areas or place distribution lines underground reflect such concerns. However, the scope of regulatory practices has broadened significantly as many PUCs are now focusing on policies that address residual emissions, which include air, water, solid waste, and land-use externalities that are present after applicable state and federal laws have been met (20). Moreover, some PUCs are including environmental costs explicitly in utility resource planning and selection. PUCs are responding to increasing concerns regarding the effects of global warming and acid rain and public opposition to construction of power plants and transmission lines because of adverse land-use, visual, and water-quality impacts.

In roughly one-third of US states, PUCs now have procedures for considering environmental externalities in resource planning and acquisition (21). In six of these states, PUCs require utilities to consider environmental costs in resource planning but do not specify the methods to be used. Typically, utilities confine their efforts to characterizing and describing environmental impacts of various resource options, and adopt a qualitative approach.

Other states (e.g. California, Massachusetts, New Jersey, New York, Oregon, Vermont, and Wisconsin) adopted orders that establish quantitative measures of environmental impacts. As examples,

1. The Wisconsin Public Service Commission (PSC) (22) requires that, in resource planning, utilities use a "noncombustion" credit of 15% for nonfossil supply and demand-side resources because of reduced pollution.
2. The Vermont PSC ruled that utilities should discount DSM-resource costs by 10% and increase supply-side resource costs by 5% in resource planning to capture costs not already included in the monetized prices of supply sources.
3. The New York Public Service Commission (23) decided to assign the most environmentally disruptive resource (a coal plant built to New Source Performance Standards) an environmental cost of 1.4¢/kWh, based on

estimated costs of controlling air emissions. All other resources are assigned some fraction of that total, depending on their environmental score.

4. The Massachusetts Department of Public Utilities (24) concluded that environmental externalities should be assigned monetary values using implied valuation methods (i.e. cost of control estimates as a proxy for environmental damages in the absence of comprehensive damage cost estimates) and specified externality values to be used by all utilities in resource planning and bidding processes (Table 3). The adopted values for various emissions (in $/ton) are high and are likely to have a significant negative effect on various resource options (e.g. coal-fired technology).

Wisconsin and Vermont use a percentage adder that either increases the cost of supply resources or decreases the cost of DSM resources in utility planning. Differentiation among supply technologies is limited (e.g. coal vs gas-fired projects) and focuses on generic technical characterizations rather than individual projects. A more detailed evaluation of the environmental impacts of specific projects occurs in a separate regulatory process or is

Table 3 Environmental externality values adopted by the Massachusetts Department of Public Utilities (24)

	Cost (1989-$/ton)
Nitrogen oxides	
Ambient air quality	6500
Greenhouse	0
Total	6500
Sulfur oxides	1500
Volatile organic compounds	5300
Total suspended particulates	4000
Carbon monoxide	
Ambient air quality	820
Greenhouse	50
Total	870
Carbon dioxide	22
Methane	220
Nitrous oxide	3960

conducted by other state agencies. In contrast, the New York and Massachusetts approaches, which affect both planning and resource acquisition, differentiate among technologies based on individual project characteristics (e.g. expected emissions and project size).

Options for Incorporating Environmental Externalities

Table 4 lists the major approaches used to incorporate environmental externalities in utility resource planning and selection, and in system operation. Initially, utilities will often characterize and describe environmental effects of various generation options without trying to value these impacts. Resource options are described in terms of their environmental attributes (e.g. emission types and rates, water and land use). This type of work is a useful starting point and typically forms the basis for a more detailed analysis of externalities.

Ranking and weighting procedures estimate the attractiveness of resource options using relative environmental damages. Ranking and weighting schemes vary widely in sophistication as well as technical basis, and could rely heavily on subjective judgments, or conversely on a synthesis of information drawn from a detailed characterization of environmental effects of resource options.

New England Electric Systems (NEES) developed a rating and weighting approach, which it uses in its long-range planning process and proposes to use in competitive bidding (25). Based on a survey of experts, NEES assigns various weights to environmental factors: global climate change (11%), acid rain (15%), land use (10%), water use (16%), emissions to air (17%), ozone (18%), and other factors (15%). Actual emission rates are then multiplied by the environmental factor as well as a percentage weight (e.g. the 15% weight

Table 4 Alternative approaches to incorporating environmental externalities

	Resource planning	Acquisition and selection	System operation
Characterization and description of environmental effects	X		
Ranking and weighting	X	X	
Assigning monetary values			
Implied valuation (cost of control)	X	X	
Direct assessment and valuation of damages	X	X	
Resource set-asides		X	
Full-cost dispatch			X
Emission quantity targets	X	X	X

for acid rain comprises 40% for SO_2 and 60% for NO_x) to develop a total environmental score for a resource option. NEES then assigns the highest rated project (i.e. the most environmentally degrading) a cost adder of 15%; the costs of other projects are increased based on the ratio of their score to the highest score. This method is easy to understand, but its lack of scientific basis is troubling. Therefore, this approach may be useful primarily as an interim method.

Two general approaches have been used to assign monetary values to environmental externalities. The first approach involves direct assessment and valuation of actual environmental damages imposed on society by different generating technologies. Costs to society are estimated by tracing impacts through each step of the fuel cycle (i.e. emissions, transport of pollutants, and the effects of these pollutants on plants, animals, people, etc). The extent of each effect that arises from an externality is estimated and a value is assigned to that effect. For example, SO_2 emissions can be linked to lost forest products, damage to buildings, and human respiratory problems. Ottinger et al (18) used this approach to calculate "starting point" values for various generating technologies (Table 5). Values for new coal- and gas-fired technologies were 3.2¢/kWh and 1.0¢/kWh respectively; estimated values for existing coal-fired plants were much higher (5.8¢/kWh).

Direct assessment and valuation of environmental damages is the preferred approach conceptually; however, technical and methodological issues are complex and data limitations are severe. For example, valuation is not

Table 5 Monetary estimates of environmental costs (¢/kWh)

	Direct damage cost valuation	Implicit valuation
Coal-fired plant,		
New source pollution standards	3.2	1.3
Combustion turbine		
with #2 oil	2.5	0.8
with natural gas	1.3	0.5
Gas-fired combined cycle	1.0	0.5
Nuclear plant	2.9	—
Biomass	0 to 0.7	—
Wind	0 to 0.1	—

Sources: Koomey (60), based on Ottinger et al (18) and the Consolidated Edison bidding system in New York.

possible for some environmental-resource damages, dose-response relationships are uncertain, valuing intangible costs (e.g. recreation facilities and
endangered species of wildlife) is difficult, valuing human mortality is controversial, and much of the direct-costing research is area-specific and may
not be transferable to other regions (26).

The second approach, called "implied valuation," relies on the cost of
control (or mitigation) of the pollutants emitted by the generating technology
to estimate the value of pollution reduction (19). The rationale for this
approach is that the cost of controls provides an estimate of the price that
society is willing to pay to reduce the pollutant. This approach has several
limitations (e.g. it cannot directly be applied to pollutants such as CO_2 that are
not now regulated), but its principal disadvantage is that pollution-control
costs bear little relation to the actual damages imposed by power plant
emissions. Nevertheless, PUCs that want to assign monetary values to externalities often favor this approach (e.g. Massachusetts, New York, and
California).

Approaches that rely on "set-asides" for specified technologies in resource
selection processes have also been proposed by planning agencies in California. Regulators would specify the preferred mix of new generating resources
to be acquired by the utility, which would best account for environmental
externalities. For example, a utility might conduct two competitive resource
procurements (one that would include all types of projects while the other
would be restricted to non–fossil fuel–based projects); the size of each
resource block would be determined by the regulatory process (27). Proponents argue that this approach is a reasonable interim response given the
uncertainties associated with assigning monetary values to externalities.
However, "set-aside" schemes are often criticized by classical economists
who argue that predetermined capacity limits for specific technology types
produce inefficient economic outcomes.

Limits on emission levels or targets for quantities of pollutants (e.g. lower
and maintain CO_2 emissions at 1985 levels) is another method that has been
proposed. Proponents of the environmental constraint approach argue that it
avoids the complexity of direct damage assessment methods, and allows
environmental quality improvements to be traded off explicitly against increased costs to the utility and its ratepayers (20). This approach is a particularly attractive way to address CO_2 emissions because of the inadequacy of
"implied valuation" techniques (i.e. it is now not possible to define a cost of
abatement option for CO_2) and the global consequences and uncertainties of
greenhouse gases.

The previous approaches focus primarily on long-range resource planning
and selection. Bernow et al (28) suggested that a utility's short-run operational decisions should also reflect environmental externalities, principally

because of the enormous levels of pollution produced at existing generation facilities. They proposed that utilities dispatch their power plants on the basis of "full cost" dispatch, which includes both direct costs (fuel and variable operation & maintenance) as well as environmental externality costs. This approach is quite controversial and, if adopted, could cause sharp increases in electricity prices.

Emerging Issues

Many utilities argue that it is inappropriate for PUCs to place significant additional costs that result from environmental externalities only on electric utility consumers, and not on consumers of other fuels. Thus, electric utilities and others are likely to raise questions about the role of PUCs in addressing environmental externalities versus the roles of federal and state government agencies directly responsible for environmental quality (e.g. the US Environmental Protection Agency).

Increased attention to environmental concerns may provide an important impetus for public policy makers and PUCs to broaden the boundaries of IRP. For example, PUCs may ask gas and electric utilities to compare the social costs and benefits of providing energy services (e.g. water heating or cooking) using gas directly vs through gas-fired electric generation. Future public policy concerns about the environmental effects of energy technologies may force significant changes in the demands for electricity and gas. For example, the policies of local air quality boards that limit vehicle emissions (and, for example, encourage electric cars) or national legislation affecting greenhouse gas emissions could affect future electricity and gas uses.

4. NEW APPROACHES TO ACQUIRING ELECTRIC POWER RESOURCES

Nonutility power production has emerged as a major source of new generating capacity, principally because of the 1978 Public Utility Regulatory Policies Act (PURPA). Cogenerators and small power producers built nearly 15,000 MW of nonutility capacity during the 1980s, although there were significant regional variations in nonutility capacity additions. Under PURPA, PUCs are responsible for pricing arrangements under which electricity is purchased from Qualifying Facilities (QFs) at the utility's avoided cost. Avoided costs were determined administratively and some states, which sought to encourage QF suppliers, offered long-term contracts based on forecasts of avoided costs. In several states, the response by private producers was much greater than expected, partly because avoided-cost forecasts turned out to be high given events in world oil and gas markets. Some utilities also claimed that the obligation-to-purchase provisions of PURPA and the open-ended nature of standard-offer contracts introduced substantial uncertainty about how much power would ultimately be developed. Thus, PURPA fundamentally altered the market position of private producers and stimulated the development of competitive forces in electricity generation. But PURPA was not an unqualified success, because the supplier response created major planning and operational problems for some utilities.

Many utilities and PUCs are using competitive resource procurements (CRPs) as one way to obtain supply and DSM resources. Since the first CRP was issued by Central Maine Power in 1984, utilities and PUCs in 27 states have adopted or are developing competitive bidding systems (29). Thus far, capacity offered by private producers has often been 5–15 times greater than the utility's requirements (see Figure 3 for results from the most active states). However, some utilities have found that a significant fraction of bids do not meet the requirements specified in the bid package and are therefore not considered seriously. For example, Central Maine Power received bids for more than 2300 MW of generating capacity in response to a 1989 solicitation; only about 1000 MW remained as realistic options after CMP's initial review of the bids. Figure 4 summarizes the types of technologies and fuels proposed by bidders as well as the distribution among winning bidders from utility

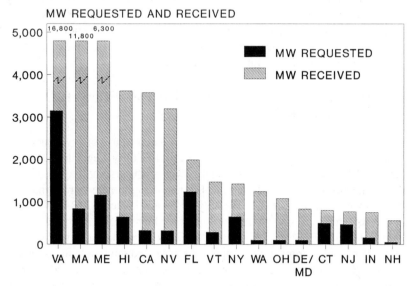

Figure 3 As of December 1989, electric utilities had received almost 1100 bids in response to their competitive resource procurements (29). Although utilities requested a total of 10,100 MW, the bids totaled 56,000 MW.

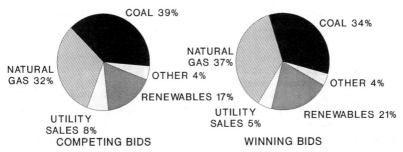

Figure 4   Coal and natural gas dominate the bids received and accepted by electric utilities (29).

CRPs as of December 1989. Gas- and coal-fired projects dominate, accounting for about 70% of the winning projects, while various renewable resources provided about 20% of the capacity of the projects selected by utilities.

The popularity of competitive bidding is strongly linked to the failures of power plant construction during the 1970s and 1980s (e.g. massive cost overruns, significant disallowances, rate shocks) as well as problems associated with implementing PURPA. CRPs are attractive to private producers because the purchasing utility offers long-term contracts, which they need to get financing on attractive terms. For the utility, competitive acquisition allows it to ration contracts for nonutility resources in an efficient manner. Moreover, these contracts usually transfer to private developers some of the risks associated with project siting and permitting, construction cost overruns, operating problems and outages, and environmental impacts. In addition, a competitive process can reduce the burden of estimating avoided cost by providing a market benchmark to determine value.

Despite its theoretical virtues, there are formidable practical problems involved with developing competitive procurement programs. Traditional utility planning requires trade-offs among financial, operating, and environmental features of resource alternatives. Competitive bidding requires the utility to address these issues through arms-length contracting. To assess bids, a utility must account for and value the multiple attributes of projects. This unbundling and explicit valuation of attributes is a new phenomenon in resource planning. Typically, utility bidding systems differentiate projects on pricing terms, operating characteristics, project status and viability (e.g. likelihood of successful development), and in some cases environmental impacts. Determining the economic value of these nonprice factors is probably the most difficult problem that utilities confront in designing bidding systems (30).

Two design features are particularly critical for utilities as they develop CRPs: (*a*) the method used to assess or score proposals, specifically the extent to which the utility discloses assessment criteria and the weight assigned to each feature before bid preparation; and (b) incorporation of DSM options into bidding schemes.

Bid Evaluation Criteria

Utilities take two general approaches to the bid solicitation and evaluation process. In the first approach, the utility develops an explicit scoring system that clearly states the assessment criteria and weights for various features. Bidders self-score their projects, assigning points in various categories (e.g. price, level of development, dispatchability) based on project characteristics. This self-scoring approach can be considered an "open" system and is used in Massachusetts, New Jersey, and New York.

A principal advantage of self-scoring systems is their transparency. Regulators can easily audit the utility's project rankings and there should be little controversy over the utility's selection of the winning bids. Some PUCs favor self-scoring because it allows them to shape utility planning decisions early in the resource-acquisition process rather than through after-the-fact prudence review of contracts. However, some utilities are concerned that self-scoring denies them the flexibility needed to select the optimal mix of projects. Another potential disadvantage of self-scoring systems is that they assume that projects can be evaluated independent of their interactions with other projects. When the utility's resource need is small compared to the existing system, the independence assumption is reasonable; however it becomes increasingly untenable for resource procurements that are large relative to the existing utility system.

In the second approach, utilities reveal bid evaluation and project selection criteria in qualitative terms only, providing only general guidance about its preferences (30). Bidders submit detailed proposals, which provide the basis for the utility's evaluation and ranking of projects. In this approach, the utility retains more discretion to select the optimal mix of projects as well as flexibility to negotiate with bidders in light of all offers received. This approach can be considered "closed" because the utility has information about the evaluation process that is not available to bidders. Prominent examples of this approach include procurements issued by Virginia Power, Florida Power and Light, and Public Service of Indiana.

The closed approach acknowledges the inherent complexity in optimizing resource selection because the value of proposed projects is multidimensional and uncertain, particularly over long times. Theoretically, this approach allows the utility to select the most efficient mix of bids, because it explicitly recognizes the interactive effects among individual projects and their effects on the utility system. In one sense, closed systems reflect the fact that the private power market is highly competitive and currently a buyer's market.

Because private suppliers are abundant, utilities prefer not to provide pre-specified bid evaluation criteria and their associated weights to bidders, but rather disclose only broad planning objectives (31).

Some utilities have experimented with hybrid approaches that combine elements of self-scoring and closed systems. For example, Central Maine Power includes elements of self-scoring, although the utility retains substantial flexibility to select attractive projects for further negotiation. Niagara Mohawk uses a self-scoring system for initial screening and then negotiates with bidders in the initial award group. The Massachusetts Department of Public Utilities (24) recently proposed a similar approach. This hybrid approach represents an attractive option that could successfully balance the utility's need for flexibility and discretion with the need to ensure fairness. Bid evaluation methods are an evolving art rather than a science and we expect continued experimentation with information requirements and risk-sharing between utilities and private power producers.

Bidding for DSM Resources

Another key issue that arises in CRPs is the types of resources and entities to include [e.g. QFs, independently owned generation facilities, energy service companies (ESCOs), large commercial and industrial customers, as well as the sponsoring utility]. Among these entities and resource options, the appropriateness of including "saved kWh and kW" provided by ESCOs or individual customers has been the subject of vigorous debate (32–35). Much of this debate has focused on theoretical issues related to integrating DSM and supply resources in the same "all-source" bidding process and the proper pricing of demand-side resources (36).

Including demand-side options in a utility's bidding solicitation raises interesting design and implementation issues (37). These include ways to measure the expected energy and demand savings and whether the ceiling price for DSM bids should be based on avoided cost or on the difference between avoided cost and average revenues (to reflect lost revenues). Some of these issues are not unique to DSM bidding and arise in utility DSM programs also. Determining the relationship between utility-sponsored DSM programs and DSM bidding efforts is likely to be a critical issue in the future because an increasing number of state PUCs are encouraging utilities to develop comprehensive DSM programs, stimulated by the offer of financial incentives for utility shareholders (38).

Table 6 summarizes results from utilities that include DSM options in their bidding approach, including the MW offered by bidders as well as the MW selected by the utility. In addition, results are shown for recent supply-side procurements conducted by New England Electric and Boston Edison, along with results from their DSM performance contracting programs involving

Table 6 Supply and DSM resources (MW) in utility bidding programs

Utility	Amount of resource requested	Supply projects		DSM projects	
		Proposed	Winning	Proposed	Winning
Central Maine Power	100	666	0	36	17
Central Maine Power[a]	150–300	2338	NA[b]	30	NA
Orange & Rockland	100–150	1395	181	29	18
Public Service Electric & Gas	200	654	210	47	47
Jersey Central	270	712	235	56	26
Puget Power	100	1251	127	28	10
PSI[c]	550	1800	640	78	15
Niagara Mohawk	350	7115	405	162	36
Separate auctions					
New England Electric	200	4279	204	NP[d]	35
Boston Edison	200	2800	200	NP	35

[a] This row represents a different CRP than does the top row
[b] NA = not announced
[c] Public Service Company of Indiana
[d] NP = not applicable.

ESCOs. Typically, there have been 5 to 20 DSM bids submitted by ESCOs and individual customers. The DSM bidders have a stronger likelihood of winning (40–80%) than do supply-side projects. The amount of DSM savings proposed by winning bidders, while significant (10–47 MW over a 3–5-year period), represents a small part of a utility's overall DSM program for the same time (5–20%). In addition, the amount of DSM savings proposed by bidders typically represents about 20–25% of the total amount requested by the utility. Initial results reflect current infrastructure limitations in the ESCO industry as well as ESCO caution about the risks of guaranteeing the energy savings and their limited experience with DSM bidding.

In summary, experience with bidding for DSM resources is limited. Initial experience suggests that DSM bidding may have a small role in a utility's overall DSM strategy but may not be appropriate for all market segments. For example, it is difficult to imagine DSM bids for the new construction market. The relative immaturity of the ESCO industry is in marked contrast with the strength of private power producers. In practice, this means that the quantities offered under DSM bidding programs will be small, and will not reflect the full market potential of DSM. Most utilities are skeptical about DSM bidding and prefer other ways to deliver DSM programs.

Future Prospects

Because of increased load growth and retirement of existing units, US electric utilities will need about 100,000 MW of new generating capacity during the next 10–15 years. Utilities and their regulators must decide on the proportions of utility-owned and nonutility-owned generation. The expectation that the private power market will supply a significant fraction of new capacity hinges on the success of competitive resource procurement processes. Success will be measured by the extent to which winning bidders develop working projects, continued benefits to ratepayers from competition (e.g. lower costs of power with adequate reliability), willingness of private suppliers to offer operational flexibility, and the perceived fairness of the process to all parties (e.g. the extent to which utilities refrain from potential abuse of market power by giving preference to unregulated subsidiaries or affiliates).

5. TREATING DSM PROGRAMS AS RESOURCES

Program Potential and Performance

A recent study estimates that, as of 1990, utility DSM programs cut annual electricity use and peak demand by 1% and 4%, respectively (39). The study's business-as-usual forecast of the effects of such programs shows reductions in annual electricity use of 3% in 2000 and 6% in 2010 and reductions in peak demand of 7% and 10%. Oak Ridge National Laboratory conducted a survey of 24 utilities to determine the present and likely future effects of their DSM programs. These utilities account for one-third of US electricity sales. Results showed a planned contribution of DSM programs to incremental energy resources (from 1990 to 2000) of 16% and a contribution to incremental capacity resources of 28% (40). These numbers show that DSM programs are likely to account for an increasing share of total electric-utility energy and capacity resources.

Neither of these estimates takes into account the likely increases in such programs because of (a) growing public concern about environmental quality; (b) a combination of pressure and the promise of financial incentives from state regulatory commissions; and (c) the increasing difficulties associated with the siting, licensing, and construction of new power plants and transmission lines.

In part, the effects of utility DSM programs are likely to increase because the potential for cost-effective savings remains large (41). Both the ambitious estimates from the Rocky Mountain Institute and the cautious estimates from the Electric Power Research Institute, presented in Figure 5, show large opportunities to reduce electricity consumption. For example, the Electric Power Research Institute analysis shows that it is technically feasible to cut electricity use by 24 to 44% by the year 2000 in addition to the 9% already included in utility forecasts. Similar studies have been conducted at the state or regional level for the Pacific Northwest, California, Michigan, and New York. The New York study, for example, identified cost-effective efficiency improvements amounting to one-third of 1986 use (42).

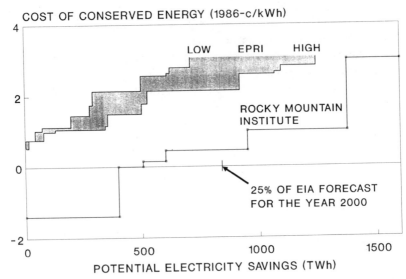

Figure 5 Conservation supply curves for US electricity use developed by the Electric Power Research Institute and by the Rocky Mountain Institute.

Also, the effects of utility DSM programs are likely to increase because utilities are gaining valuable experience in running such programs. Nadel's (43) review of such programs shows typical and "best" results obtained to date (Table 7). Typical utility programs aimed at commercial and industrial customers reach less than 5% of the eligible customers, cut annual electricity use per participant by less than 10%, and cost less than 4¢/kWh-saved. The most successful programs do considerably better. Nadel suggests several ways to increase participation and energy savings. For example, customer response to energy audits alone is generally quite limited. Followup services, such as arranging for contractor installation of measures and financial incentives, greatly increase customer propensity to adopt measures recommended in the audits. Wisconsin Electric's estimate of the peak-demand savings caused by its DSM programs almost doubled (from about 250 MW to 450 MW) between 1987 and 1989 (44). The company's experience in running its Smart Money Energy Program, begun in 1987, surely played a major role in the increase.

DSM Uncertainties

Uncertainties associated with DSM programs fall into two classes, neither of which has been adequately incorporated into utility resource planning. The small unit sizes and short lead times of DSM programs reduce the uncertain-

Table 7 Summary of results from electric-utility DSM programs for commercial and industrial customers (43)

Program type	Cumulative participation (%)		Savings (%)		Utility cost (¢/kWh)
	Average	Best	Average	Best	
Audits	1–4	60–90	4–5	6–8	0.9
Lighting rebate	<1–3	10–25	3	—	1.0
HVAC[a] rebate	<1	10	11	—	2.9
Motor rebate	<1	15	5	—	0.6
Industrial	0–3	5–9	—	—	0.8
New construction	—	—	—	30	3.0

[a] Heating, ventilation, and air conditioning

ties that utilities face. The Northwest Power Planning Council (45) accounts for these benefits of DSM programs in its analysis of alternative resources. The Council's resource-planning model treats future load growth as a stochastic variable; resources are acquired by the model to meet medium load growth and options are acquired to meet high load growth. Ford & Geinzer (46) examined the effects of energy-efficiency standards for new buildings on the uncertainties associated with future load growth. Because the energy savings caused by such standards is positively related to economic and load growth, such standards reduce utility uncertainty.

In addition to these uncertainty-reduction benefits, DSM programs create new uncertainties for utilities. These uncertainties concern likely future participation in programs and their associated energy and demand savings. New England Electric (47) conducted probabilistic assessments of all the resources, both supply and demand, in its integrated resource plan. Experts from different parts of the company assessed the likelihood that different projects would achieve their planned outputs at different times in the future. The purpose of this analysis was "to provide an estimate of how certain [New England Electric System] can be that a given resource plan will meet future needs." This analysis showed that the company's DSM programs had an 80% chance of reducing peak demand by at least 400 MW in 1995 and a 50% probability of cutting demand by at least 580 MW that year.

Finally, utilities in New England, New York, Wisconsin, California, and the Pacific Northwest are rapidly expanding their DSM programs, with budgets expected to double or triple within a few years. These utilities are counting on DSM programs to defer construction of new power plants to a much greater extent than was true in the past. Uncertainties exist about the availability of skilled people to plan, design, and implement these programs and the materials and high-efficiency equipment these programs promote.

Information Needs

From the perspective of resource planning (rather than program design and implementation), the greatest need is for additional data on the costs and performance of DSM programs. Utilities need data in three areas: baseline data on the efficiency and utilization patterns of end-use systems in the existing stock of equipment, buildings, and factories; data on the costs and performance of energy-efficiency and load-management technologies; and data on customer participation in and costs of different types of utility DSM programs.

Much baseline data exists on some customer groups, building types, and end-uses, especially for the residential sector. Information is most limited for the industrial sector, especially related to process uses. Data on the performance of individual technologies is abundant, but widely scattered and often collected in inconsistent ways.

However, the information on actual DSM programs (the third of the three topics) is especially limited and crucial. Additional information is needed on program participation, energy savings, and costs. For example, program participation, the ratio of the number of participants to the number of eligible customers, is defined in various ways. Ambiguities arise over annual vs cumulative rates, consistent definitions of participants and nonparticipants, and differences among retrofit, replacement, and new-construction markets.

DSM-program costs sometimes include customer costs and sometimes do not. Sometimes costs refer only to the direct costs of the devices and their installation. Other times, costs include the administrative activities (e.g. staff training, marketing, quality control, program evaluation, and corporate overhead) as well as direct costs.

Utilities usually base their estimates of program energy savings primarily on engineering analysis rather than on field measurements of actual electricity use. Substantial evidence shows that such engineering calculations typically yield estimates of electricity savings higher than actually achieved.

Field measurements can include monthly electricity bills, special short-term (e.g. 30-days) metering of circuits in a facility affected by a DSM program, time-of-use load-research data, or end-use load-research data. These data can be used to compute total electricity savings and net savings. Total savings is the reduction in electricity use experienced by customers that participate in a program. Net savings is the portion of total savings that can be directly attributed to the program. Thus, net savings is the difference between total savings and the savings that participants would have achieved on their own had the utility program not existed.

A few organizations have, in recent years, developed reporting formats and definitions for utility DSM programs. The data collection instrument developed by the Northeast Region Demand-Side Management Data Exchange (48) is probably the most comprehensive.

Additional work is needed if, as seems likely, the role of utility DSM programs will continue to grow. Agreement must be reached on the appropriate cost-effectiveness test(s) to use in assessing the benefits and costs of DSM programs. And utilities and PUCs need more information on the actual costs and energy savings of DSM programs. Such field data will reduce the perception on the part of some utilities that DSM programs are "risky" because they depend on the (often unseen) actions of millions of customers.

6. GUIDELINES FOR PREPARATION OF UTILITY PLANS

Many electric utilities periodically prepare long-term resource plans, often in response to requirements from their PUC. These plans inform regulators and customers about the utility's analysis of alternative ways to meet future energy-service needs and the utility's preferred mix of resources to meet those needs. The plan is an opportunity for the utility to share its vision of the future with the public and to explain its plan to implement this vision.

In principle, utility plans should be assessed on the basis of the utility's resource-acquisition activities. But IRP is so new that insufficient implementation has as yet resulted from these plans. Currently, utility plans can be assessed only on the basis of their planning reports. This section discusses guidelines for long-term resource plans, based on the written reports (7, 49). The purpose of these guidelines is to help PUC staff who review utility plans and utility staff who prepare such plans. Several utility plans contain many of the positive features in the guidelines (45, 50–55).

The "goodness" of a plan can be judged by at least four criteria (Table 8):

1. The clarity with which the contents of the plan, the procedures used to produce it, and the expected outcomes are presented;
2. The technical competence (including the computer models and supporting data and analysis) with which the plan was produced;
3. The adequacy and detail of the short-term action plan; and
4. The extent to which the interests of various stakeholders are addressed.

Report Clarity

The primary purpose of a utility's IRP report is to help utility executives decide (and PUC commissioners review) which resources to acquire, in what amounts, and when. Thus, the report must be useful both within and outside the utility. The utility's plan should be well-written and appropriately illus-

Table 8 Checklist for a good integrated resource plan (49)

Clarity of plan—adequately inform various groups about future electricity resource needs, resource alternatives, and the utility's preferred strategy

- Clear writing style
- Comprehensible to different groups
- Presentation of critical issues facing utility, its preferred plan, the basis for its selection, and key decisions to be made

Technical competence of plan—positively affect utility decisions on, and regulatory approval of, resource acquisitions

- Comprehensive and multiple load forecasts
- Thorough consideration of demand-side options and programs
- Thorough consideration of supply options
- Consistent integration of demand and supply options
- Thoughtful uncertainty analyses
- Full explanation of preferred plan and its close competitors
- Use of appropriate time horizons

Adequacy of short-term action plan—provide enough information to document utility's commitment to acquire resources in long-term plan, and to collect and analyze additional data to improve planning process

Fairness of plan—provide information so that different interests can assess the plan from their own perspectives

- Adequate participation in plan development and review by various stakeholders
- Sufficient detail in report on effects of different plans

trated with tables and figures. The report should discuss the goals of the utility's planning process, explain the process used to produce the plan, present load forecasts (both peak and annual energy), compare existing resources with future loads to identify the need for additional resources, document the demand and supply resources considered, describe alternative resource portfolios, show the preferred long-term resource plan, and present the short-term actions to be taken in line with the long-term plan. Important decision points should be identified, and the use of monitoring procedures to provide input for those decisions should be explained. The most significant effects of choosing among the available options (e.g. capital and operating costs, resource availability, and environmental effects) should be discussed. The report should also describe the data and analytical methods used to develop the plan.

Technical Competence

Typically, computer models are used for a variety of functions in developing a plan, such as load-forecasting; screening, selection, and analysis of demand and supply resources; and calculations of production costs, revenue requirements, electricity prices, and financial parameters. These models are used to analyze a wide range of plausible futures and available resources in developing the utility's preferred resource portfolio. The basic structure of the models used, the assumptions upon which they are based, and the inputs utilized should be explained.

The technical competence of a utility's IRP is reflected most critically in the ways that the demand and supply resources are presented as an integrated package. The analytical process used to integrate these different resources should be discussed. The criteria used to assess different combinations of resources (e.g. revenue requirements, annual capital costs, average prices, reserve margin, and emissions of pollutants) should be clearly stated.

Results for different combinations of supply and demand resources should be shown explicitly. It is not enough to treat demand as a subtraction from the load forecast and then do subsequent analysis with supply options only. Subtracting DSM-program effects from the forecast and using the resultant "net" forecast for resource planning eliminates DSM programs from all integrating analysis. This approach makes it difficult to assess alternative combinations of DSM programs and supply resources and the uncertainties, risks, and risk-reduction benefits of DSM programs (e.g. small unit size and short lead time).

Demand-side resources must be treated in a fashion that is both substantively and analytically consistent with the treatment of supply resources; demand and supply resources must compete head to head. The plan must show how the process truly integrates key parts of the company: load forecasting, DSM resources, supply resources, finances, rates. And the important feedbacks among these components (especially between rates and future loads) should be shown.

A thorough analysis of a variety of plausible future conditions and the options available to deal with them is essential. This analysis should consider uncertainties about the external environment (e.g. economic growth and fossil-fuel prices) and about the costs and performance of different resources. The analysis should show how utility decisions are affected by different assumptions about these factors and show the effects of these uncertainties and decisions on customer and utility costs. The assumptions must be varied in ways that are internally consistent and plausible. Differences among resources in unit size, construction time, capital cost, and operating performance should be considered for how they affect the uncertainties faced by utilities. Finally, the links between the results of these multiple runs and the utility's resource-acquisition decisions must be demonstrated.

Action Plan

The action plan, in many ways the "bottom line" of the utility's plan, must be consistent with the long-term resource plan. This is necessary to ensure that what is presented as appropriate for the long haul is implemented, and implemented in an efficient manner. The action plan also should be specific and detailed. The reader should be able to judge the utility's commitment to different actions from this short-term plan. Specific tasks should be identified for the next two to three years, along with milestones and budgets. For example, the action plan should show the number of expected participants and the expected reductions in annual energy use, summer peak, and winter peak for each DSM program. The action plan also should discuss the data and analysis activities, such as model development, data collection, and updated resource assessments, needed to prepare for the next integrated resource plan.

Equity

A final criterion by which a plan can be judged is the effect of its recommended actions on various interested parties. Because the interests of stakeholders differ, the ways in which they will be affected by short- and long-term costs, power availability, and other results of utility actions will likewise differ.

Without the involvement of customers and various interest groups a plan may ignore community needs. Accordingly, the plan should show that the utility sought ideas and advice from its customers and others in developing the plan. Energy experts from a state university, state energy office, PUC, environmental groups, and organizations representing industrial customers could be consulted as the plan is being developed. For example, utilities in New England are working closely with the Conservation Law Foundation to design, implement, and evaluate DSM programs (6).

Additional work is needed to refine the guidelines discussed here and to ensure that they are helpful to utilities and PUCs. In particular, PUCs should articulate better the reasons they want utilities to prepare such plans and how they will use the plans in their deliberations. This articulation should avoid the "data list or cookbook approach" and focus on the purposes of the planning report. In the long-run, the success of IRP should not be measured by assessing utility reports. Rather, the level and stability of energy-service costs, the degree of environmental protection, and the extent to which consensus is achieved on utility resource acquisitions will be the important criteria.

7. FEDERAL ROLES TO PROMOTE INTEGRATED RESOURCE PLANNING

Because electricity production consumes almost 40% of the primary energy used in the United States, electricity must be a major part of national energy policy. In addition, public concerns about environmental quality, economic productivity, and national security suggest a larger role for the federal government in working with utilities to expand their planning and to carry out DSM programs. The Department of Energy can influence utility planning and resource acquisition in several ways. These methods include Federal Energy Regulatory Commission regulation of wholesale contracts, technical assistance from DOE's Integrated Resource Planning Program, collection of data on utility DSM programs by DOE, and expansion of the DSM programs run by the federal Power Marketing Agencies. Only the last topic is discussed here; see Ref. 56 for examination of the other topics.

The federal Power Marketing Agencies (PMAs, part of DOE) and Tennessee Valley Authority (TVA) (an independent federal corporation) account for one-tenth of the electricity consumed in the United States. Traditionally, TVA and the Bonneville Power Administration (the largest PMA) have operated large DSM programs, which saved energy for their customers and served as examples for other utilities. Unfortunately, short-term budget considerations during the mid-1980s forced reductions in these programs at both agencies. Indeed, TVA canceled all its conservation programs in 1989. Bonneville, on the other hand, plans to increase its conservation budgets over the next several years. To be specific, Bonneville (57) plans to spend $440 million to acquire an additional 1200 GWh/year in conservation between 1990 and 1993, after building up to 2100 GWh/year of conservation through 1988, equivalent to almost 3% of total Bonneville sales that year.

The Western Area Power Administration (58), the second largest PMA, runs a small Conservation and Renewable Energy Program for its 800 utilities. Western's service area encompasses 15 western states, a region of the country which, except for California, has little integrated resource planning and few DSM programs. As a consequence, Western's program could have large "technology transfer" effects throughout the region.

Western's current program emphasizes flexibility for its customer utilities, primarily because these utilities differ substantially in size, mix of customers, dependence on other producers for their electricity supplies, and load/resource balance. Because of this diversity, Western imposes few requirements on its customers, operates a decentralized program from its five regional offices, and offers extensive technical assistance. Depending on the size and type of utility, it is required to conduct from one to five "activities," chosen from a list of approved projects developed by Western.

The major price paid for the program's simplicity and flexibility is the lack of justification for utility selection of conservation projects. Because Western requires no cost-effectiveness analysis, it is not clear whether or how individual DSM programs fit into any overall resource plan. In addition, there is little documentation of program benefits and costs in the present system.

Fortunately, Western (59) is revising the program, with a final rule expected to be issued in mid-1991. Western could set aside a fraction of its low-cost federal hydropower for assignment to those utilities that run especially effective conservation programs. Access to additional low-cost power would be a very effective way to encourage utilities to pursue aggressively cost-effective DSM programs. Alternatively, Western could purchase energy savings and load reductions achieved by its customer utilities.

New federal legislation could require the federal power authorities to expand their DSM programs and to consider explicitly environmental and social factors in their benefit/cost analyses of all resource alternatives. Such legislation would be a logical extension of the 1980 Pacific Northwest Electric Power Planning and Conservation Act (P.L. 96-501), which explicitly made conservation the electricity resource of choice. Even without new legislation, the PMAs could expand their DSM programs, as the Western and Southwestern Power Administrations already are.

8. CONCLUSIONS

A growing share (now almost 40%) of the primary energy consumed in the United States flows through electric utilities. Therefore, the economic and environmental effects of utility actions are enormous. Integrated resource planning is a new way for utilities to meet the energy-service needs of their customers and is widely used by utilities and PUCs along the east and west coasts and in the upper midwest. In this review, we discussed guidelines for preparation and review of IRP plans based on the approaches used by the most advanced utilities and commissions and highlighted the key issues associated with use of DSM programs as resources. We also examined three topics that are particularly important as utilities move from a period of excess capacity to one of resource need: new approaches (i.e. competitive procurements) for acquiring energy and capacity resources, ways to include environmental costs in resource planning and acquisition, and regulatory changes to overcome disincentives to utility DSM programs. Utilities and PUCs that successfully address these issues will be well positioned to acquire substantial amounts of cost-effective DSM resources and to harness the private-power market. These utilities and their customers will benefit from reasonably priced electric-energy services produced in an environmentally benign fashion. Because IRP is a comprehensive and open process, its implementation is likely to result in fewer controversies over utility resource-acquisition decisions.

Much work is needed to convert the potential benefits of IRP into reality. Perhaps the greatest need is to transfer information, ideas, and enthusiasm from those utilities and PUCs active in IRP to those less active. It is also important to push the frontiers of resource planning in all the topics discussed here.

Literature Cited

1. Cavanagh, R. C. 1986. Least-cost planning imperatives for electric utilities and their regulators. *Harvard Environ. Law Rev.* 10(2): 299–344
2. Gellings, C. W., Chamberlin, J. H., Clinton, J. M. 1987. *Moving Toward Integrated Resource Planning: Understanding the Theory and Practice of Least-Cost Planning and Demand-Side Management. EPRI EM-5065.* Palo Alto, Calif: Electr. Power Res. Inst.
3. Hirst, E. 1988. Meeting future electricity needs. *Forum Appl. Res. Public Policy* 3(3)
4. Hirst, E. 1988. Integrated resource planning: the role of regulatory commissions. *Public Util. Fortn.* 122(6):34–42
5. Krause, F., Eto, J. 1988. *Least-Cost Utility Planning, A Handbook for Public Utility Commissioners.* Vol. 2, *The Demand-Side: Conceptual and Methodological Issues.* Lawrence Berkeley Lab. Natl. Assoc. Regul. Util. Comm.
6. Ellis, W. B. 1989. The collaborative process in utility resource planning. *Public Util. Fortn.* 123(13):9–12
7. Schweitzer, M., Yourstone, E., Hirst, E. 1990. *Key Issues in Electric Utility Integrated Resource Planning: Findings from a Nationwide Study. ORNL/CON-300.* Oak Ridge, Tenn: Oak Ridge Natl. Lab.
8. Hobbs, B. F., Maheshwari, P. 1990. A decision analysis of the effect of uncertainty upon electric utility planning. *Energy* 15(9):785–802
9. Hirst, E., Schweitzer, M. 1990. Electric-utility resource planning and decision-making: the importance of uncertainty. *Risk Anal.* 10(1):137–46
10. Argonne Natl. Lab. 1989. *Energy Conservation Program Evaluation: Conservation and Resource Management. Proc. 1989 Conf., Argonne, Ill.*
11. Hirst, E., Destribats, A., eds. 1990.

Integrated resource planning. *Proc. ACEEE 1990 Summer Study on Energy Efficiency in Buildings,* Vol. 5. Washington, DC: Am. Counc. Energy-Efficient Econ.
12. Moskovitz, D. 1989. *Profits & Progress Through Least-Cost Planning.* Washington, DC: Natl. Assoc. Regul. Util. Comm.
13. NY Public Service Comm. 1989. *Opinion and Order Approving Demand-Side Management Rate Incentives and Establishing Further Proceedings. Opinion No. 89-29,* Albany, NY
14. Marnay, C., Comnes, G. A. 1990. *Ratemaking for Conservation; The California ERAM Experience.* Berkeley, Calif: Lawrence Berkeley Lab.
15. Reid, M., Chamberlin, J. 1990. Financial incentives for DSM programs: a review and analysis of three mechanisms. See Ref. 11, Vol. 5, pp. 157–66
16. Mass. Dept. Public Util. 1990. *Investigation by the Department on its Own Motion as to the Propriety of Rates and Charges . . . by Massachusetts Electric Company. DPU 89-194 and DPU 89-195.* Boston, Mass.
17. Schultz, D., Eto, J. 1990. *Designing Shared Savings Incentive Programs for Energy Efficiency: Balancing Carrots and Sticks. LBL-29503.* Berkeley, Calif: Lawrence Berkeley Lab.
18. Ottinger, R. L., Wooley, D. R., Robinson, N. A., Hodas, D. R., Babb, S. E. 1990. *Environmental Costs of Electricity.* NY: Pace Univ., Oceana Publications
19. Chernick, P., Caverhill, E. 1989. *The Valuation of Externalities from Energy Production, Delivery, and Use.* Prepared for Boston Gas Co.
20. Violette, D., Lang, C., Hanser, P. 1990. Framework for evaluating environmental externalities in resource planning: a state regulatory perspective. In *Proc. Natl. Conf. Environ. Ex-*

ternalities, pp. 63–84. Washington, DC: Natl. Assoc. Regul. Util. Comm.

21. Cohen, S. D., Eto, J. H., Goldman, C. A., Beldock, J., Crandall, G. 1990. *A Survey of State PUC Activities to Incorporate Environmental Externalities into Electric Utility Planning and Regulation. LBL-28616.* Berkeley, Calif: Lawrence Berkeley Lab.

22. Wis. Public Service Comm. 1989. *Findings of Fact, Conclusion of Law and Order. 05-EP-5.* Madison, Wis.

23. NY Public Service Comm. 1989. *Opinion and Order Establishing Guidelines for Bidding Program. Case 88-E241. Proc. Motion of the Comm.* as to the guidelines for bidding to meet future electric capacity needs of Orange and Rockland Utilities, Inc. *Decision 89-7*

24. Mass. Dept. Public Util. 1990. *Investigation by the Department on its own Motion into Proposed Rules to Implement Integrated Resource Management Practices for Electric Companies in the Commonwealth. DPU 89-239.* Boston, Mass.

25. Destribats, A. F., Hutchinson, M., Stout, T. M., White, D. S. 1990. Environmental costs and resource planning consequences: New England Electric's rating and weighting approach. See Ref. 20, pp. 85–99

26. Chernick, P., Caverhill, E. 1990. Monetizing externalities in utility regulations: the role of control costs. See Ref. 20, pp. 41–62

27. Dasovich, J., Morse, D. 1990. Incorporating environmental externalities in LCUP and selection: the California experience. See Ref. 20, pp. 410–12

28. Bernow, S., Biewald, B., Marron, D. 1990. Full cost economic dispatch: recognizing environmental externalities in electric utility system operation. See Ref. 20, pp. 151–72

29. Wellford, T., Robertson, H. 1990. *Bidding for Power: The Emergence of Competitive Bidding in Electric Generation.* Washington, DC: Natl. Independent Energy Producers

30. Kahn, E., Goldman, C., Stoft, S., Berman, D. 1989. *Evaluation Methods in Competitive Bidding for Electric Power. LBL-26924.* Berkeley, Calif: Lawrence Berkeley Lab.

31. Kahn, E., Stoft, S., Marnay, C., Berman, D. 1990. *Contracts for Dispatchable Power: Economic Implications for the Competitive Bidding Market. LBL-29447.* Berkeley, Calif: Lawrence Berkeley Lab.

32. Cavanagh, R. C. 1988. *The Role of Conservation Resources in Competitive Bidding Systems for Electricity Supply.* Testimony before the Subcommittee on Energy and Power, US House of Representatives, Washington, DC

33. Cicchetti, C. J., Hogan, W. 1989. Including unbundled demand-side options in electric utility bidding programs. *Public Util. Fortn.* 123(12)

34. Joskow, P. L. 1988. Testimony before the Subcommittee on Energy and Power, US House of Representatives, Washington, DC

35. Joskow, P. L. 1990. Understanding the "unbundled" utility conservation bidding proposal. *Public Util. Fortn.*

36. Goldman, C., Hirst, E. 1989. Key issues in developing demand-side bidding programs. In *Demand-Side Management: Partnerships in Planning for the Next Decade. Electr. Counc. New England Natl. Conf. Util. DSM Programs. EPRI CU-6598.* Palo Alto, Calif: Electr. Power Res. Inst.

37. Goldman, C., Wolcott, D. 1990. Demand-side bidding: assessing current experience. See Ref. 11, Vol. 8, pp. 53–68

38. Raab, J. 1990. Is there room for integrated demand-side bidding. See Ref. 11, Vol. 8, pp. 223–30

39. Faruqui, A., Seiden, K., Benjamin, R., Chamberlin, J. H., Braithwait, S. D. 1990. *Impact of Demand-Side Management on Future Customer Electricity Demand: An Update. EPRI CU-6953.* Prepared for the Electr. Power Res. Inst., Palo Alto, Calif., and the Edison Electric Inst., Washington, DC

40. Hill, L., Schweitzer, M., Hirst, E. 1991. *Integrating Demand-Side Management Programs into the Resource Plans of U.S. Electric Utilities. ORNL/CON-311.* Oak Ridge, Tenn: Oak Ridge Natl. Lab.

41. Fickett, A. P., Gellings, C. W., Lovins, A. B. 1990. Efficient use of electricity. *Sci. Am.* 263(3):64–74

42. Miller, P. M., Eto, J. H., Geller, H. S. 1989. *The Potential for Electricity Conservation in New York State. Report 89-12.* Prepared for the NY State Energy Res. Dev. Auth., Albany, NY

43. Nadel, S. 1990. Electric utility conservation programs: a review of the lessons taught by a decade of program experience. See Ref. 11, Vol. 8, pp. 179–206

44. Carlson, D. G. 1990. *Direct Testimony on behalf of Wisconsin Electric Power Company before the Public Service Commission of Wisconsin. Docket No. 6630-UR-104.* Milwaukee, Wis.

45. Northwest Power Plan. Counc. 1986. *Northwest Conservation and Electric Power Plan.* Portland, Ore.
46. Ford, A., Geinzer, J. 1988. *The Impact of Performance Standards on the Uncertainty of the Pacific Northwest Electric System.* Prepared for Off. Conserv., Bonneville Power Admin., Portland, Ore.
47. New England Electr. 1989. *Conservation and Load Management Annual Report.* Westborough, Mass.
48. Northeast Region Demand-Side Manage. Data Exch. 1989. *NORDAX: A Regional Demand-Side Management Database, Final Report. No. 07-89-32.* Washington, DC: Edison Electr. Inst.
49. Hirst, E., Schweitzer, M., Yourstone, E., Eto, J. H. 1990. *Assessing Integrated Resource Plans Prepared by Electric Utilities. ORNL/CON-298.* Oak Ridge, Tenn: Oak Ridge Natl. Lab.
50. Carolina Power & Light Co. 1989. *CP&L Exhibits 1, 2 and 3, Before the NC Util. Comm. Docket No. E-100, Sub 58.* Raleigh, N.C.
51. Green Mountain Power Corp. 1989. *Integrated Resource Plan.* South Burlington, Vt.
52. Northwest Power Plan. Counc. 1989. *1989 Supplement to the 1986 Northwest Conservation and Electric Power Plan.* Vol. II. Portland, Ore.
53. Pacific Power & Light Co. and Utah Power & Light Co. 1989. *Planning for Stable Growth,* Portland, Ore.
54. Puget Sound Power & Light Co. 1989. *Demand and Resource Evaluation: Securing Future Opportunities, 1990–1991.* Bellevue, Wash.
55. Seattle City Light. 1989. *Strategic Corporate Plan 1990–1991.* Seattle, Wash.
56. Hirst, E. 1989. *Electric-Utility Energy-Efficiency and Load-Management Programs: Resources for the 1990s. ORNL/CON-285,* Oak Ridge, Tenn: Oak Ridge Natl. Lab.
57. Bonneville Power Admin. 1990. *1990 Resource Program. DOE/BP-1405.* Portland, Ore.
58. Western Area Power Admin. 1986. *Conservation & Renewable Energy Customer Sourcebook.* Golden, Colo.
59. Western Area Power Admin. 1990. Request for public comments on the Western Area Power Administration's conservation and renewable energy guidelines and acceptance criteria. *Fed. Regist.* 55(17):2574–77. Golden, Colo.
60. Koomey, J. 1990. *Comparative Analysis of Monetary Estimates of External Environmental Costs Associated with Combustion of Fossil Fuels. LBL-28313.* Berkeley, Calif: Lawrence Berkeley Lab.

Annu. Rev. Energy Environ. 1991. 16:123–44

SOVIET ENERGY DILEMMA AND PROSPECTS

Yuri Sinyak

International Institute for Applied Systems Analysis, A-2361 Laxenburg, Austria

KEY WORDS: energy intensity, material intensity, Soviet energy scenarios, Soviet economic restructuring, global climate change.

0. INTRODUCTION

The USSR is the largest energy producer and the second largest energy consumer in the world. Its share in global energy demand reached 17% in 1988. The USSR's higher energy intensity than in other industrialized countries can be partially justified by such factors as the climate, poor infrastructure, and vast territory; however, the low energy efficiency and high per capita energy consumption are also explained by an inefficient economic structure, a high level of material intensity, along with enormous wastes and losses, a backward technological level, and a large stock of obsolete, old equipment used in all sectors of the national economy, especially in the energy sector. Presently, the energy savings potential is approximately equal to half of domestic energy consumption. Improvement in energy efficiency at all levels of the national economy is the primary goal of the national energy policy for the next two decades.

Though short-term energy prospects for the 1990s are quite uncertain, some fundamental ideas concerning long-term energy projections have emerged. Several long-term scenarios of energy development, among them a stabilization of energy consumption after 2000–2005 and a 20% reduction of CO_2 emissions by 2030, are investigated in this paper. Expected improvements in the Soviet energy system will require changes in energy management, including reduction of centralized planning, decentralization and privatization of energy-producing enterprises, energy price reforms, reshaping of investment patterns, and persistence in conversion from military to civilian production.

123

1056-3466/91/1022-0123$02.00

1. CRITICAL FEATURES OF THE SOVIET ENERGY SYSTEM

These are difficult times for the Soviet Union, and its energy system, always considered a backbone of economic and social progress, is no exception. The nation suffers heavily today from severe political, economic, human, financial, and technological problems, and recently the signs of an energy crisis have appeared. It would have sounded strange several years ago, but now the reality is that this country, which has huge energy resources, feels painful energy shortages. To explain this predicament, we must first analyze specific features of the Soviet energy system and then review the main weaknesses in its past development.

The following main features of the Soviet energy system have influenced supply and demand over a long period in the past and continue to play a decisive role:

1. vast domestic energy resources,
2. a long-distance energy transportation system from eastern to western parts of the country,
3. severe climatic conditions.

The amounts of domestic energy resources (notably coal, natural gas, and hydropower) are vast enough to sustain the country's economic development in the foreseeable future through domestic energy production and a relatively high level of energy export.[1]

Large territory and uneven energy resource distribution make the problem of long-distance energy transportation particularly important. Despite the fact that over the past 10–20 years virtually all the new highly energy-intensive industries have been developed in Siberia, the European part of the country's energy requirements have grown, while its regional cheap energy resources have become scarce (virtually all coal of the European USSR is produced in mines, mainly deep ones, and from thin beds.) Thus, ever-increasing amounts of fuel and energy need to be moved from the eastern region to the European region. Bulk energy transportation increased manyfold over a short period (1970–1985); from 125 million tons of coal equivalent (tce) to 975 million tce per year. More than half of the eastern resources are in the form of coal and natural gas, the transportation of which over a distance of 2.5–4 thousand km

---

[1]The last statement does not necessarily apply to the long run and is highly criticized now within the Soviet Union. However, in the short and medium term, the export of energy resources (mainly crude oil and natural gas) is considered a major source of hard currency needed to continue perestroika's changes in the USSR and to pay the rapidly increasing foreign debt.

is extremely expensive. As a result, the European USSR has been and will remain a zone of expensive fossil fuels. This situation justifies the high priority placed on the search for efficient ways to transport large amounts of fuel and electricity over large distances.

The climatic conditions in most of the USSR differ considerably from those in most of western Europe and the United States: long severe winters and moderate summer temperatures. This climate results in a larger share of low-temperature processes (notably space heating) in energy demand. The number of space heating degree-days in the USSR is almost double those in the majority of other developed countries. Two-thirds of the Soviet population live under harsh climatic conditions compared with only 16% in the United States and less than 5% in Western Europe. The climate is one of the leading factors underlying the high level of energy consumption in the USSR and at the same time is the source of a large potential for energy-efficiency improvement.

The economic difficulties of the past several years exaggerate the impacts of these objective factors. The crisis has touched almost all economic sectors. The deficit of the state budget reached more than 90 billion rubles in 1989 and the internal state debt increased to 460 billion rubles by the middle of 1990 (for comparison, the national income was 620 billion rubles in 1988). Inflation has jumped to an unprecedented 8–11% per year. The supply of consumer goods has diminished dramatically.[2] Industrial production suffers from stagnation and decline—industrial output decreased by 1.2% in the first half of 1990 compared with the same period of the previous year. Lastly, the gross national product (GNP) fell by 2% in the first half of 1990 (1–4).

On the backdrop of the economic crisis, the energy supply situation has become more and more complicated from year to year. During the past 3–4 years fossil fuel supply disruptions have become regular. Production of the most efficient fuels (crude oil and natural gas) has encountered large difficulties arising from a steady decline in investments in the energy sector because of the higher priorities of social and environmental problems and increased domestic political tensions. At the beginning of the 1990–1991 winter season, 28 large cities in the Russian republic alone faced a 20% shortage of heat supply. Out of 650 thousand km of heat supply mains, about 50% needed replacement, but the actual rate of replacement was expected to be very low—not more than 10 thousand km per year—only one-third the needed replacement rate. Half of the existing main lines are completely obsolete: each km of mains has on the average one breakdown per year (5).

---

[2]According to some estimates, there were fewer than 20 commodities available in free sale in the 1990 spring-summer season from more than one thousand items included in a statistical observation poll.

Electricity supply is also in a critical situation: generating equipment is practically obsolete, and no reserve capacities exist. The sector suffers from a lack of capital investments (in the 1960s the share of the electricity sector in total industrial investments was about 11–12%; this share fell to only 6% in the 1980s) (6). All existing nuclear reactors are operating with a decreased load factor, which increases their safety, but also reduces their capacities, adding to the overall shortage of thermal power plant capacity. Four nuclear power plants were planned to start operation in 1990, but none has been put into operation because of strong public opposition (7). Opposition to energy facility construction, especially nuclear plants, has become an effective tool in the political struggle for power in the Soviet Union. By the beginning of 1990, the public had demanded a stop to construction or a shut down of about 60 new power plants with a total installed capacity of around 160 million kW (6); many of these protests succeeded in halting or delaying the construction of modern power plants, including nearly all new nuclear power plants, and old, inefficient power plants have had to continue to be used. The situation may even worsen in several years because no reserve capacities exist.

National tensions in Soviet republics (Middle Asia, Caucasian Region, Moldova) between Russians and the native population also have very strong negative impacts on the energy supply. As a result of the tensions, many Russians, who are usually the most qualified workers in those regions, may leave the areas, and a decline in electricity generation and in the reliability of the electricity supply would follow.

Much public anxiety is connected with the current ecological situation in the USSR. The following examples illustrate the negative impacts that energy systems are having on the environment in the USSR. Pollutants released into the atmosphere by these systems amount to 100–120 million tons of chemical substances per year, including 20 million tons of fly ash and soot, 40 million tons of carbon dioxide, 30 million tons of sulfur dioxide, 20 million tons of hydrocarbons, and 10 million tons of nitrogen oxides. About 40 million tons of pollutants (mainly hydrocarbons and NO_x) are exhausted by automobiles and, with the high share of automobiles in large cities, this exhaust creates especially dangerous ground-level concentrations of toxic pollutants. Hundreds of cities have pollutant concentrations higher than national standards permit (8). Thermal power plants release 2 million tons per year of salts and other chemical components and produce 2–3 cubic kilometers of waste water (part of which contains oil products), requiring 10–20 times more fresh water to dilute concentrations of hazardous substances to standard levels. The construction of hydroelectric stations was accompanied by the loss of millions of hectares of productive agricultural lands, unrecoverable damage to fisheries, degraded biological health of many rivers, and a changed landscape in vast regions. Fossil fuel production causes extensive damage to the environment [e.g. in the northern part of the Tjumen region, the total cumulative damage to natural ecosystems from crude oil and natural gas production and transportation over the past 20 years equals more than 450 billion rubles (9); in regions with coal surface mining such as Kuzbass or Ekubastuz, millions of hectares of agricultural land have become unusable]. Concerns over the safety of nuclear power have reached a critical level, which makes further development doubtful in many regions of the USSR (as well as in other countries). According to some findings, the economic damages of environmental pollution caused by energy systems are several times greater than fuel market prices (in large cities it is even five- to tenfold). These costs are rarely taken into account when energy policies or technologies are developed.

For many years the USSR's administrative system of planning has resulted in conflict between long-range prospects, often based on assessments of cost-effective energy solutions, and short-term management, which demanded the fulfillment of plans by whatever means. This planning takes place under a double energy price system: *state prices* of fuel and energy based on average cost calculations,[3] which are used for everyday transactions and payments, and *marginal energy prices,* derived by optimizing the state/regional energy balances and comprising the full cost of energy production, processing, transportation, and distribution, plus differential rent reflecting the usefulness of different energy sources.

The latter price system has been recommended for assessing new energy-related technologies. However, because the prices arising from the two systems differ by a factor of at least two, use of the marginal system makes new technologies unable to compete against established ones. This results in many ineffective, short-sighted solutions to energy system problems.

In sum, the Soviet energy sector is experiencing many problems with a variety of causes, adding to the country's serious economic difficulties. The way out of these problems depends on the rate and depth of the economic restructuring. The next section describes the energy sector's main weak points, which require immediate attention and action.

2. MAIN SOURCES OF SOVIET ENERGY SUPPLY DIFFICULTIES

In the seven decades of Soviet history, many factors have contributed to the low efficiency and high environmental burden of the energy system—several severe wars, the USSR's superpower role after World War II, the armaments

---

[3]According to Marxist theory, only labor produces wealth. Therefore, until recently Soviet state prices did not account for the cost of capital (except that connected with amortization). This situation produced strong discrepancies and wrong conclusions on the attractiveness of different resources.

race, worldwide support to Communist and anti-Western regimes, etc. Over the past 20 years Soviet conservation policy was less effective than those of some Western countries (Table 1). Since the early 1970s, energy/GNP in the USSR had decreased only 16% by 1985, as compared to 21% on the average for the International Energy Agency (IEA) countries, 31% for Japan, and 23% for the United States. So modest a reduction in the USSR energy/GNP ratio can be explained as follows:

1. Abundance of domestic fossil-fuel resources and a secure energy supply resulted in late awareness of the importance of energy efficiency.
2. Over the years, the policy of "economic independence" from the West has resulted in slow structural changes, with continuously increasing shares of energy-intensive sectors and industries;
3. The administrative economic system together with low energy prices did not stimulate energy conservation and end-use efficiency;
4. The existing planning practice, in which the energy ministries play a leading role in shaping the economy and resource distributions, concentrated on expanding energy-production sectors instead of saving energy for end-users;
5. Slow rates of technology replacement resulted in an ever-increasing share of obsolete equipment with low energy efficiency and weak environmental standards.

I consider now only the most important of the above factors: unfavorable economic structure, immense material losses and wastes, and backward technological level of equipment (16).

For a long time the misguided concept of the preferable development of

Table 1 Energy/economy intensities for different countries, tons of oil equivalent (toe)/10^3 US$ (1985)

Country	1985	Percent reduction compared to	
		1973	1979
USSR	0.74–1.05	14	3
Some IEA[a] countries			
USA	0.45	23	16
Japan	0.15	31	22
FRG	0.27	18	14
France	0.22	—	—
Italy	0.32	21	13
Average of IEA countries	0.45	21	15

[a] International Energy Agency.
Sources: (26–28)

heavy industries ("production of the means of production") was the main principle of economic growth in socialist countries.[4] Moreover, in the 1970s the economy was oriented towards complete self-sufficiency from domestic production ("economic isolation"). In practice such a policy caused an exaggerated accumulation in the national product (sometimes reaching 40% compared to 15–20% in Western economies) and hyperbolic growth of certain basic materials industries. In fact, the share of highly energy-intensive industries remains very high: only 25% of the national industrial output is connected with the production of low energy-intensive consumer goods; the remaining 75% are heavy-industry commodities, which are characterized by very high energy intensity.[5] During the 1980s, heavy industries received almost 90% of total industrial investments. This factor alone explains the high energy intensity of the Soviet economy.

The military complex, supported by the external political goals connected with the "superpower syndrome," plays a major role in the overdevelopment of heavy industries. Without any real return to the economy, the military sector absorbs a large quantity of material and intellectual resources of higher quality than average for the economy. At present military expenses account for about 30% of GNP compared to less than 6% for the United States (10). There are strong indications that this factor has until now restrained positive changes in the economy as well as in the society as a whole.

Therefore, the first steps that should be taken in order to reduce the energy intensity of the Soviet economy are the profound restructuring of the economy and much more involvement in international trade.

High energy intensity, however, is not the only disease of the Soviet economic system. During the past few years, the Soviet economy has also had a very high level of material intensity, which decreased from 1975 to 1987 by only 5%, less than for Western countries.[6] Unfortunately, lessons of the oil

---

[4]This principle was considered one of the primary laws of the socialist political economy, and means that in the normal production process the growth of capital goods production must exceed on the average the growth of consumer goods output. Thus, it was a "ruling principle" that the share of heavy industries and capital construction in the gross national product should always increase while the share of consumer goods should decline.

[5]For example, the energy/value added ratio in Soviet industry was, at the beginning of the 1980s: heavy industries—iron and steel 17 tce/10^3 rubles, fuel and energy 8, chemicals 8, building materials 7; and less energy-intensive industries—machinery 1.3, textile 0.8, food production 3. During the past 20 years the share of heavy industries in the total industrial output increased from 60% to more than 70%.

[6]For comparison, in the European Community the iron and steel capacities had the following reductions: cast iron, 13% (from 137 million tons in 1975 to 119 million tons in 1988); steel, 12% (from 190 to 167 million tons); hot-rolled stock, 14% (from 143 to 123 million tons). The share of continuous steel casting at the same time rose from 15.3% of the total output to 76% compared to only 17% in the USSR. The reduction of oil-refinery output in Western Europe reached 30% (from 20.8 in 1976 to 14.4 million barrels per day in 1985). These measures account for most of the material intensity decline during the past 15–20 years.

Table 2 Iron and steel losses in metal processing (Ref. 24)

	1970	1980	1985
Total steel production (10^6 tons)	116	148	155
Percent of waste in iron and steel production	29	28	28
Steel consumption in machinery (10^6 tons)	59.9	82.2	89.1
Percentage of losses during metal processing	22	22	21
Of which percentage of losses in metal shaving	48	44	44

crises of the 1970s have not been taken into account in the USSR. As a result, the Soviet economy remains technologically backward and wasteful of materials. An example of Soviet industry's high material losses is provided by the ferrous-metal industry (Table 2). The share of metal losses in steel production is very high (28%) and practically did not decline during the past few years. Very high losses (almost one fifth) also occur in metal processing, nearly one half of which are due to shaving. Both values indicate the low technological levels in metal production and treatment, which indirectly reflect the high levels of lost energy. The situation is no better in other industries.

According to recent estimates made by the USSR State Committee of Science and Technology and the USSR Academy of Sciences (11), annual industrial wastes amount to 450 billion rubles (about half of total industrial production). Losses in agriculture amount to 20–40%, i.e. another 50–100 billion rubles (for example, the total number of combines that remain unrepaired in Soviet agriculture is more than 70 times greater than the US annual production of combines (14). Bad production management causes a 30% loss of labor time valued at 85 billion rubles. In total, losses in the national economy are comparable today with the gross national product. In 1988 the losses (excluding the value of ecological damage) reached 38–40% of GNP (11).

This is only one side of the equation, however. The overdevelopment of heavy industries results in the production of a large number of goods that are "useless" for the purposes of public consumption and serve only the expansion of heavy industries. According to assessments by Abel Aganbegian, the value of these "useless" products reaches 25% of industrial production, which is equal to 8–9% of GNP (12).[7]

The high energy intensity in the Soviet economy may also be explained by the low efficiency and quality of technologies and equipment in use. Only 16% of the total machinery stock (including energy conversion and end use) is on par with world standards and one fourth to one third is even below national standards, which are often far behind world standards. Especially poor are the

[7]Other Soviet economists consider these numbers as underestimated (for example, M. Lemishev attests that the level of "useless" production in the USSR reaches 60–70%.)

iron and steel and chemical-production industries, which are the most energy-intensive. More than 40% of the machinery and processes are 10 years old or older. Use of such old equipment during the past 20–25 years was 21% in the iron and steel industry and more than 16% in the energy sector and chemical production. Even more serious is the fact that more than one fifth of the machinery used in the iron and steel industry and in chemicals production is totally obsolete. All of these factors play decisive roles in explaining the low energy efficiency of the Soviet energy system. The situation will change very little over the next few years, because such change would require high investments in modernization and restructuring of the Soviet economy. Such investments are presently lacking because of low past GNP growth and the current priority given to resolution of social problems, which have recently reached critical dimensions.

The situation is even worse with outdated and obsolete energy equipment (Table 3). In total, the share of modern equipment in the energy sector does not exceed 10%, while the share of outdated and obsolete equipment is more than 50%. This situation does not allow rapid improvements in energy efficiency and is the reason for the high energy/GNP ratio in the USSR.

Analyzing all these factors leads one to the conclusion that the Soviet economy has a huge energy-saving potential. The Institute of Energy Research of the USSR Academy of Sciences (13) evaluated the scope of this

Table 3 Energy equipment characteristics (as of April 1, 1986) (%) (Ref. 24)

	Best world standard	National standard	Out-of-date and obsolete
Energy equipment			
Steam boilers	10.5	52.9	36.6
Waste-heat boilers	5.7	48.5	45.8
Steam turbines	2.6	36.6	60.8
Gas turbines	5.2	76.3	18.5
Hydro turbines	2.0	21.5	76.5
Compensating capacities	11.7	42.3	46.0
Diesel engines (excluding auto)	6.7	52.3	41.0
Electrothermal equipment			
Electric arc furnaces	4.9	38.6	56.5
Metal treating furnaces	7.8	47.4	44.8
Induction melting and heating equipment	9.8	51.1	39.1
High-frequency industrial equipment	9.6	59.9	30.5
Electric welding equipment			
Open-arc welding machines	12.2	50.0	37.8
Contact welding machines	7.1	44.7	48.2
Special welding machines	9.8	50.7	39.5

Table 4 USSR energy saving potential for the next 15–20 years, million tce

Fossil fuel	400 (36%)
Electricity	185 (16%)
Heat (centralized supply)	155 (14%)
Motor fuel	190 (17%)
Raw materials	195 (17%)
Total	1125 (100%)
Economic restructuring	225 (19%)
Technology improvements	470 (42%)
Energy efficiency	430 (39%)

potential at about half of the current energy demand. Table 4 illustrates the main areas and the range of expected saving potential in the USSR. The prospects would be even brighter (especially in the area of economic restructuring) if demilitarization of the Soviet economy were undertaken, which now looks much more realistic than it did several years ago. Rough estimates show that approximately 40% of this potential (400–450 million tce) can be saved with measures having a payback period of less than two years (Figure 1). In total, the saving potential with a payback period less than five years is equal to roughly 60% (600–700 million tce). If we assume that the average capital investments in the Soviet energy sector at the current energy construction pattern are equal to 250 rubles per tce,[8] then 50% of energy savings could be achieved with less capital investment than required for equivalent expansion of energy production. Unfortunately, an energy-saving policy could not be realized on a broad scale in the past because of the strong influence of energy-producing ministries that were less interested in the national interest than in expanding their own domains. Now, with the start of economic reforms, there is hope that it will be possible to utilize this potential to reduce energy consumption by improving both energy and economic efficiency.

3. SOVIET ENERGY PROSPECTS IN VIEW OF CURRENT GLOBAL PROBLEMS

Until recently, many still believed that intensive development of the energy sector would provide the basis for further economic growth in the USSR.

---

[8]According to recent energy statistics (25), primary energy production in the Soviet Union increased from 1976 to 2424 million tce over 1980–1988. At the same time, capital investments in the energy sector were equal to 198.2 billion rubles. When we assume that the capacity replacement rate was about 2.5% per year, we obtain that gross capacity growth reached 790–795 million tce. Thus, the average capital investments in primary energy production can be roughly estimated at around 250 rubles/tce.

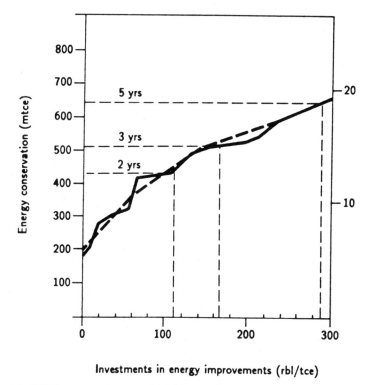

Figure 1   USSR energy saving potential [author's estimates with the use of data from (23)].

However, the economic crisis has changed such beliefs. Growth of primary energy production stopped even earlier than that of other sectors of the national economy. In 1989 primary energy production decreased by 0.5–1% from the 1988 level (1). In 1990, the production level is expected to be about 2% lower than in 1988 [recalculated from data in (14)]. Natural gas production is the only industry within the energy sector that retains the possibility of significant long-term growth, though this growth rate cannot compensate for decreasing production of other forms of energy. However, the total decrease of industrial output, especially in the basic material sectors, is likely to improve the Soviet energy balance over the next few years.

Negative rates of economic development will evidently remain for the near future. The situation in the coal industry (after numerous social and political conflicts) is becoming more stable, and the current decline in coal mining will be halted because of improvements in worker productivity and opening of new mines in the eastern part of the country. On the other hand, the crisis in the oil industry has a long-term character and huge efforts will be required over some years to stabilize the declining crude oil production. The situation in the electricity sector is also very serious and uncertain, as electricity generation faces a high probability of decline in the near future. Therefore, in the near-term a further decline in primary energy production as a whole is anticipated.[9]

However, the economic reforms will produce positive results sooner or later,[10] and through the 1990s it is possible that the Soviet economy and its energy system may slowly come out of the presently critical situation.

Although Soviet energy policy for the 1990s has been formulated without special emphasis on the issue of climatic change, this important factor cannot be ignored for the long-term aspects of energy systems development.

The energy sector is one of the main sources of anthropogenic CO_2 emission (and partly of CH_4); therefore, in order to mitigate global climate impacts from these sources, it is essential to improve the structure of industrial production and energy utilization, in particular. In addition, the global nature of the greenhouse effect requires treatment of the problem at an international level. Within the USSR, the solution of the problem is hampered by a lack of agreement among scientists on the scope and consequences of global climate change.

Opinions vary widely among Soviet experts studying global climatic processes. Prof. M. Budyko is well known in the USSR and abroad for his optimistic view of the problem.[11] He was one of the first, at the beginning of the 1960s, who called attention to global warming. He claimed, in 1971, at the Conference on Physical Climatology that the global atmosphere was accumulating increasing quantities of CO_2 released from fossil fuel consumption, and predicted that with a high probability, there will be a 1°C increase of the global average temperature by the beginning of the next century, and 3°C by the middle of the century. However, he does not see any catastrophe in these changes; in fact, he even recommends speeding up the process of global warming to reach, as soon as possible, a new climatic period similar to the Pliocene Era, when the global climate was more favorable in many aspects

---

[9]However, the most recently published report on Soviet energy prospects (15), which is more conservative in its estimates of energy consumption and production growth than it was in the past, still predicts further expansion of the energy sector until 2000 that will hardly be achievable in view of today's difficulties and tensions in the country.

[10]This paper is not the place to discuss all the aspects of economic reforms in the USSR (some of which are still in the embryonic state), but it is worth mentioning a few main directions: increasing the share of consumer goods in the GNP, developing residential and conmmercial sectors and passenger transportation, reducing military expenditures, expanding foreign trade, price reform, and demonopolization.

[11]Prof. M. Budyko is Head of the Climate Department in the Institute of Hydrometeorology in Leningrad, USSR. His views on global warming have been published in two books (17, 18) as well as in his recent paper (19) and interview (20).

than it is nowadays.[12] He explains recent increases in extreme meteorological phenomena such as tornados, hurricanes, and droughts as resulting from instability of climatic processes during the transition period from one climatic era to another. He states that it could take 20–30 years to achieve a new stable level of global climatic processes and restore the average annual number of abnormalities to thier usually observed level. He claims that the territories in higher latitudes (Canada, Alaska, Siberia) will then have more favorable conditions, but the most benefits of global warming will be received by Africa, in which he expects the Sahara to become a savanna. These global changes require preparation, but all steps to reduce CO_2 emissions and to delay global warming should be weighed with the cost-benefit approach.

The position of Prof. G. Golitzin is more anxious.[13] He essentially agrees with Budyko's projection of the average temperature increase during the next century, but believes that the temperature is neither the sole nor the most important climatic factor affecting humanity. Global warming in combination with higher CO_2 atmospheric concentration could be followed with higher growth of practically all types of plants. However, precipitation is far more important, especially with respect to agriculture. For the major portions of territories located in the lower latitudes the situation with precipitation will be unfavorable (the climate of the Middle East and Africa would become dry, causing a gradual increase in desert areas[14]). However, for middle and higher latitudes he predicts improvement of climatic conditions (increase of vegetation growth periods, decrease in the number of frosts). The most drastic changes are expected in the regions with so-called risky agriculture: in the north where the present number of frosts is large, and in the south where the same is true for droughts. On the whole, Golitzin, as well as Budyko, points out that the temperature increase would be favorable for the Soviet Union, since it would lengthen the growing season and intensify precipitation. At the same time, they both claim that a rise in sea level will not be a major problem. Golitzin is less optimistic than Budyko regarding the expected consequences of global warming, mainly because of insufficient data on the impact of human activity on the biosphere and nature, and particularly the atmosphere and ocean. Golitzin is convinced that humans are capable of coping with global warming and neutralizing negative trends because of their gradual onset. However, he insists that humanity should take control of the present

[12]Unfortunately, Prof. Budyko does not explain how he intends the speeded up warming to be achieved: by increasing CO_2 emissions by the further growth of fossil fuel burning, or by freezing nuclear and renewable energy programs? Beyond doubt, this is one of the weakest points in his "optimistic" theory.

[13]His opinions on climatic changes are expressed in two articles (21, 22).

[14]This example of exactly opposite opinions expressed by two leading Soviet scientists on the direction of climatic changes in Africa demonstrates how fragile and unreliable the theories of climatic changes still are.

situation, by restricting consumption of fossil fuels associated with the emission of pollutants that disturb the natural equilibrium.[15]

In conclusion, leading Soviet scientists agree on future temperature changes and their origin, but hold different views on the possible outcomes of these phenomena. The social consequences of global warming are hard to predict: they may have a positive, as well as a negative, character depending on the territories and types of biological species affected. The level of warming will probably be less important than the speed with which it will come, in view of the low adaptation capability of many species to rapid changes in temperature and precipitation. In the event of rapid changes, the costs of preventing or softening negative consequences could substantially increase.

The uncertainty about the consequences of global warming and the time scale of its possible appearance make this problem (at any rate until now) of less priority for the USSR and other Eastern European countries than the short-term economic and political problems with which they are confronted. Of course, there is not a complete unawareness of the problem—several publications on the subject have appeared in scientific and popular magazines and journals—but the global warming issue remains in a distant second place in comparison with the acute national problems that must be solved in the very near-term future.

Nonetheless, in 1988 the USSR Academy of Sciences launched a large research program on ecology and the biosphere, part of which focuses on climatic changes and global warming. A special subprogram is directed at the following energy-ecology-climate problems: finding the linkages between pollutants produced by the energy sector and human health and biosphere degradation, and evaluating damages caused by these interactions; elaboration of new, more effective measures of pollution abatement; energy systems risk assessments; and making energy-climate projections. The program will contribute to the development of a long-term national energy policy aimed at lower environmental degradation and higher economic efficiency, taking into account the broad scope of externalities that compose the social cost of energy resources. In 1989 the USSR government also began to treat the global warming issue seriously, by making a special decision to prepare the state program to provide sustainable long-term economic development under climate-change conditions.

Recently, the first projections of Soviet energy development to 2030 were issued (23). Several scenarios for social and economic development with

---

[15]Both Soviet scientists consider the ozone depletion problem more dangerous at the moment. It is much less investigated and does not seem to have such a time inertia as the CO_2 problem, and it might result in sharp changes in UV radiation within a very short period of time without giving human beings and biota the chance to adjust to new life conditions.

different rates of economic restructure and efficiency improvements are considered. According to these scenarios, Soviet energy demand, which is about 2000 million tce in 1990, may reach 2500–3600 million tce in 2030, depending on the scope and extension of measures implemented and economic activity over the period under consideration. A lower level of primary energy demand is considered as a means of delaying global warming, but this scenario would require a high level of investments in economic restructuring and energy-efficiency improvements plus a considerable expansion of international cooperation. These all depend heavily on the success of economic and political reforms in the Soviet Union.

The International Institute for Applied Systems Analysis (IIASA), in its new study on global energy and climate change, uses a different approach and analyzes the following three scenarios (29):

1. *Moderate Scenario* with some improvements in economic restructuring and energy conservation.
2. *Enhanced Scenario* with large improvements and the aim of stabilizing primary energy consumption after 2005–2010.
3. *20% CO_2 Reduction Scenario* focussed on ways to decrease CO_2 emissions compared to 1990 within the framework of the stabilization concept.

These scenarios represent three cases: first, a moderate, "dynamic-as-usual" scenario, with moderate improvements in economic and energy systems, and two "exotic" scenarios—primary energy stabilization and enforced CO_2 reduction after 2005–2010. The latter pathways could result in too heavy an impact on the national economic system at the first stage of their realization, which would lower economic growth rates until 2005–2010 compared to the moderate scenario (dynamic-as-usual). Less energy-intensive economics, however, are assumed to improve economic progress over the long term. There is abundant proof of this concept; however, success is highly dependent on the conditions of the nation to which the concept is applied.

Of course, one should understand that this is one of the first attempts to project long-term Soviet energy futures that emphasize the CO_2 problem. Therefore, the projections are quite vulnerable because the economic system is not yet fully determined, and there is no sign of a stochastic approach to the problem. The following steps should fill the gaps and produce more consistent and detailed projections.

During elaboration of scenarios, special attention is paid to final energy assessments for different energy-consuming sectors (material production—industry and agriculture, transportation, residential and commercial sectors, and non-energy use of fossil fuels as a feedstock). The estimates take into account social and economic goals, as well as improvements in energy efficiency by final energy consumers, and ecological and safety constraints

Table 5 USSR energy supply and demand for 1990 (estimates) in millions of tce

Activity	Solids	Liquids	Gaseous	Renewables	Nuclear	Electricity	Heat	Total
Primary energy production	468	880	1025	76	75			2524
Balance of trade	−30	−255	−145					−430
Domestic consumption	438	625	880	76	75	−5		2089
Electricity generation	−168	−120	−315	−76	−75	212	211	−331
Other energy sectors	−15	−35	−102			−58	−33	−243
Materials production	155	168	287			101	134	845
Transportation		172	11			13	2	198
Residential/commercial	90	40	115			35	42	322
Non-energy use	10	90	50					150
Final energy	255	470	463			149	178	1515

Source: (23)

that have arisen recently in the USSR as a result of public opposition to particular energy technologies, especially nuclear power.

In evaluating primary energy production levels, estimates of primary energy reserves and resources are used as well as state-of-the-art energy production systems. Also included are difficulties that could affect future energy production and of course, comparative economics of energy producttion and utilization.

Tables 5–8 summarize the final and primary energy demand evaluations for 1990 and the three scenarios for 2030. As the tables show, the scenarios differ significantly in the levels of energy consumption and its structure. Under the Moderate Scenario, primary energy consumption is expected to grow by a factor of 1.2 as compared to 2005 and more than 1.5 compared to 1990. In contrast, under the other two scenarios—Enhanced and 20% CO_2 Reduction— energy consumption nearly stabilizes after 2005–2010.[16]

As a result of increased electrification and remarkable improvements in the efficiency of electricity generation, the final/primary energy ratio is projected to decline steadily: from 72% in 1990 to 70% in 2005 and further to 67% (Moderate and Enhanced Scenarios) and even 57% (20% CO_2 Reduction Scenario), meaning increased energy conversion losses in the system. The structure of final energy consumption will also change dramatically, especially in the "exotic" scenarios.

---

[16]Soon after 2005 a weak growth of primary energy consumption is assumed, followed by ever diminishing rates that will become negative after 2010–2015 and result in about the same level of energy consumption around 2030 as is projected for 2005.

Table 6 Moderate scenario, 2030, million tce

Activity	Solids	Liquids	Gaseous	Renewables	Nuclear	Electricity	Heat	Total
Domestic consumption	780	660	1400	200	250	−30		3260
Electricity generation	−485	−70	−520	−150	−250	455	330	−690
Other energy sectors	−15	−45	−170			−100	−45	−375
Materials production	210	75	350	20		200	205	1060
Transportation		265	30			35		330
Residential/commercial	50	30	220	30		90	80	500
Non-energy use	20	175	110					305
Final energy	280	545	710	50		325	285	2195

Analysis of these scenarios shows the feasibility, in principle, of stabilizing primary energy demand after 2005–2010. However, even with enforced and rapid changes in the energy system, this hardly seems achievable within the existing economic system. It would require active stimulation of the process by immediate changes in the capital investments pattern in the direction of heavy energy savings, technological and structural improvements in industrial and agricultural production, and the development of nonfossil energy options, which are not possible within the existing economic system.

In the Soviet economy electrification has always been considered a symbol of technological and social progress, and this seems valid for the future as well. Thus, electricity generation may continue to increase when primary energy consumption is already stabilized or even reduced, and the concept of accelerated growth of electricity generation compared to primary energy is dominant in the projections under consideration. Electricity generation is expected to reach about 2500 TWh in 2005 (growth by a factor of 1.4 over 15 years) and increase further to 3500–5000 TWh by 2030. Such a program would require increasing net installed generating capacities by 160 GWe by 2005 and by an additional 370–815 GWe by 2030. The electricity sector's share of total domestic primary energy consumption will increase from 27% in 1990 to 30% in 2005 and further to 33% (Moderate Scenario), 36% (Enhanced Scenario), and more than 50% (20% CO_2 Reduction Scenario) by 2030.[17]

The technological structure of the electricity sector will probably change

---

[17]The changes in final energy consumption are even more impressive: the electricity share rises from 10% in 1990 to 12% by 2005 (only 2% absolute growth over 15 years) and to 15–35% by 2030.

Table 7 Enhanced scenario, 2030, million tce

Activity	Solids	Liquids	Gaseous	Renewables	Nuclear	Electricity	Heat	Total
Domestic consumption	480	520	1300	150	180	−30		2600
Electricity generation	−330	−50	−570	−100	−180	430	240	−560
Other energy sectors	−10	−30	−150			−75	−35	−300
Materials production	105	55	280	20		185	130	775
Transportation		230	30			40		300
Residential/commercial	30	20	160	30		100	75	415
Non-energy use	5	135	110					250
Final energy	140	440	580	50		325	205	1740

substantially, especially if it is necessary to implement a strategy of greenhouse gas abatement. In this case the share of thermal power plants in electricity generation should be reduced from the current 73% to 50% in 2030, and the share of new and renewable electrictiy sources should increase from 13% in 1990 to 24% (almost 50% of installed capacity). The role played by nuclear energy is likely to be more modest than has been projected in the past, reaching 20–25% of total capacity in 2030.

If the scenarios involving significant reduction of CO_2 emission rates are to be followed, significant changes in the electricity generation system will be required and these must start immediately. Otherwise it will be technically

Table 8 20% CO_2 reduction scenario, 2030, million tce

Activity	Solids	Liquids	Gaseous	Renewables	Nuclear	Electricity	Heat	Total
Domestic consumption	150	400	1265	460	355	−30		2600
Electricity generation	−90	−20	−950	−300	−325	615	250	−820
Other energy sectors	−10	−30	−150			−75	−35	−300
Materials production	40	25	45	35	10	225	140	520
Transportation		175		55		70		300
Residential/commercial	5	15	60	40		215	75	410
Non-energy use	5	135	60	30	20			250
Final energy	50	350	165	160	30	510	215	1480

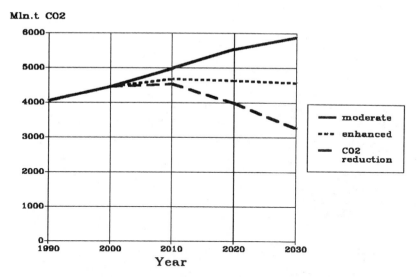

Figure 2 CO_2 releases (millions of tons) by the energy sector: Moderate Scenario, Enhanced Conservation Scenario, and 20% CO_2 Reduction Scenario.

impossible to introduce the required changes within the very limited time left before undesirable levels of global warming become inevitable.

Figure 2 presents CO_2 projections for the three scenarios. Obviously, energy supply and consumption has a high inertia before 2000–2005 since the CO_2 released into the atmosphere will continue to increase by 10–20% until 2005. However, in the first quarter of the next century changes in the national energy policy could give some positive results, but as stressed above, strong governmental and societal support of the necessary transformations will be required. The electricity sector is now the second largest CO_2 emitter in the USSR, accounting for about one third of total emissions. By 2030, however, this sector will take the lead with 40–55% of emissions. Therefore, measures of CO_2 reduction will be most significant in this sector.

It is worth mentioning that the general line of the USSR's national energy policy, of which IIASA's scenarios are marginal cases, even today is in accordance with the CO_2 reduction goals because of its orientation towards energy conservation, wider use of natural gas, nuclear energy, and renewable sources. Reductions of CO_2 emissions per tce of primary energy consumed take place under all three marginal scenarios: from 1.95 t CO_2 in 1990 to 1.85 t in 2005 and further to 1.75–1.8 t by 2030 under the Moderate and Enhanced Scenarios, or to 1.25 t by 2030 under the 20% CO_2 Reduction Scenario.

For their realization the three IIASA scenarios require different efforts. If we measure these efforts by the quantity of capital investment in the energy

Table 9   Approximate capital investments in energy sector (1990 to 2030) in billions of rubles

Energy sector	Moderate scenario	Enhanced scenario	20% CO_2 reduction scenario
coal	125	95	65
oil production	650	590	540
oil refining	75	70	60
gas production	280	280	275
gas transport	700	700	685
electricity, total	875	1065	2330
fossil	145	160	150
renewables	275	410	1190
nuclear	165	135	215
electricity transport	290	350	775
energy conservation	595	1150	925
Total	3300	3940	4880
% of the national income	6.3	6.9	9.3

Source: (29)

[a] Assumed fuel costs: coal = 100 rbl/tce; oil = 210 rbl/tce (300 rbl/t); oil refining = 25 rbl/tce (30 rbl/t); natural gas = 60 rbl/tce (70 rbl/1000 m^3); natural gas transportation and distribution = 250% of natural gas production; thermal power plants = 250 rbl/kWe (life time = 20 years); renewables = 1000 rbl/kWe (life time = 10 years); nuclear = 1000 rbl/kWe (life time = 20 years); electricity lines = 50% of electricity generation; energy conservation—Moderate Scenario = 150 rbl/tce, Enhanced and 20% CO_2 Reduction Scenarios = 200 rbl/tce.

sector during 1990–2030 (which of course is a simplification), then we can compare the scenarios from the viewpoint of the economic system and its constraints. Table 9 contains estimates of the capital investment required, showing total investments required during 1990–2030 of 6.5% of the cumulative national income produced within this time period for the Moderate Scenario, 6.9% for the Enhanced Scenario, and 9–10% for the Reduced CO_2 Scenario, a level that seems unreachable. However, it should be noted that even a lower level of capital investment looks high compared to a typical 5–6% for developed countries, Consequently, it is to be expected that energy system development in the USSR will encounter serious financial difficulties. Implementing this policy of reduction will be even more difficult when one takes into account the enormous changes required in public opinion.

In the end, it is necessary to point out that in spite of the importance of seeing the problem of global warming at national levels, it is a global problem requiring a global approach to its solution. The following aspects of the problem need serious additional investigation:

1. It would be extremely desirable to evaluate and compare the rates of global climate changes arising from technological greenhouse gases emission and deforestation. The research should determine the steadiness of basic biochemical cycles, disturbed by economic activities.

2. Global climate change, if a reality, would inevitably cause a reconstruc-

tion of the world economy. Additional data must be obtained on the capacity-
of the world economy to adapt to such rapid changes. The necessary investments and their probabilities should be evaluated as well.

3. Because global changes would touch everybody, prevention of
catastropic impacts on human beings requires shared charges, responsibilities,
and efforts. This implies that the political and legal aspects of environmental
regulation must be given more attention. Particularly, the acceptability or
nonacceptability of different international options influencing the energy
policy of the USSR and other countries should be central to the investigation.
A sustainable future implies ability to solve international conflicts and find the
roots of arguments. The problems of environmental regulation have to be
analyzed as an important element of global security.

4. CONCLUSIONS

In conclusion, the USSR energy system should radically improve its energy
efficiency and reduce CO_2 emissions after 2000. These will require:

1. a shared public opinion of the importance of reducing CO_2 emissions,
2. sufficient finance to start the reconstruction of the energy system immediately,
3. "open doors" for technological transfers from the West to the USSR,
   especially those of energy savings and new safe and environmentally
   benign technologies,
4. some lead time before global warming becomes significant (the sooner
   society understands the necessity of change, the less expensive the consequences will be in all aspects).

The general economic reforms required to implement an effective energy
strategy are the following:

1. demonopolization of large energy-supply systems and privatization at the
   end-use level, and introduction of a market economy concept replacing
   centralization in the energy sector;
2. changes in the energy-pricing system that orient it toward marginal costs
   and to more allowance for social costs instead of average costs;
3. changes in investment policy with greater emphasis on energy savings and
   conservation rather than on energy production;
4. conversion of the military industries with wider use of their productive
   capacities and intellectual potential for the production of modern energy-
   saving equipment instead of armament production oriented toward foreign
   markets;
5. conditions for the expansion of Western capital investments in the USSR
   and creation of a favorable environment for their profitability;

6. R&D collaboration with Western partners in high-priority areas such as energy savings, power-plant modernizations, research and development on joint European electricity and natural gas supply grids, access to research, student and specialist exchanges;
7. sufficient time to assess and improve the new energy policy.

Literature Cited

1. USSR economy in 1989. 1990. *Izvestia.* Jan. 29*
2. Gaidar, E. 1990. About good intentions. *Pravda.* July 27*
3. Big fall in E Europe output. 1990. *Financial Times.* July 24
4. USSR economy in the 6 months. 1990. *Izvestia.* July 29*
5. *Izvestia.* Sept. 9, 1990*
6. Semenov, A. 1990. USSR energy prospects. *Energy Construction* 1:5–8*
7. *Izvestia.* Oct. 4, 1990*
8. *USSR State Environmental Program for 1991–1995 and up to 2005.* 1990. Suppl. to *Economics and Life,* No. 41, Oct.*
9. Katasonov, V. 1990. Tyumen's apokalipsys. *Green World* 4:4*
10. *Izvestia.* July 4, 1990*
11. *Izvestia.* January 4, 1990*
12. *Pravda.* August 2, 1990*
13. Vigdorchicky A., Achmedov, E., Djemileva, R., Tchulkov, A., Tchupiatov, V. 1988. Energy savings: methodology and prospects. *Energy and Transport* 4:38–49*
14. Gaidar, E., Shatalin, S. 1989. *Economic Reform: Causes, Directions, Problems.* Economica. Moscow. 55 pp.*
15. Comm. Alternative Energy Scenarios, USSR Acad. Sci. 1990. *USSR Alternative Energy Scenarios.* Moscow. 77 pp.*
16. Sinyak, Y. 1990. *Energy Efficiency and Prospects for the USSR and Eastern Europe.* Cent. Res. Inst. Electr. Power Industry Japan–Int. Inst. Appl. Syst. Anal. *EY9001.* Dec.
17. Budyko, M. 1977. *Global Ecology.* Moscow: Mysl*
18. Budyko, M. 1980. *Climate in the Past and Future.* Leningrad: Gudrometeoizdat.*
19. Paper presented to World Climate Congr., Nov. 1988, Hamburg
20. Zurueck ins paradies? 1990. Interview to *Der Spiegel.* Jan. (In German)
21. Golitzin, G. 1986. Climate and economic priorities. *Kommunist.* No. 6, pp. 97–105.*
22. Golitzin, G. 1989. We are facing climatic change. *Priroda i Czelovek.* No. 8, pp. 28–29.*
23. Makarov, A., Bashmakov, I. 1989. *The Soviet Union: a Strategy of Energy Development with Minimum Emission of Greenhouse Gases.* Presented at Workshop on Greenhouse Gas Control Strategies, Budapest, Hungary, 6–9 Dec. 1989
24. Goscomstat SSSR. 1988. *USSR Materials and Technology Supply.* Moscow*
25. Goscomstat SSSR. 1989. *USSR Economy in 1988. Statistical Yearbook.* Moscow. 766 pp.*
26. Int. Energy Agency. 1987. *Secretariat Report*
27. *BP Statistics.* June 1990
28. *Statistical Abstracts of the United States,* p. 822. 1989
29. Sinyak, Y. 1991. USSR: Energy efficiency and prospects. *Energy* 16(5): 791–815

*(In Russian)

Annu. Rev. Energy Environ. 1991. 16:145–78

REGIONAL ENERGY DEVELOPMENT IN SOUTHERN AFRICA: GREAT POTENTIAL, GREAT CONSTRAINTS

P. Raskin and M. Lazarus

Stockholm Environment Institute—Boston Center, Tellus Institute, 89 Broad Street, Boston, Massachusetts 02110

KEY WORDS: energy and development, SADCC, regional cooperation, energy planning.

A DECADE OF REGIONAL COOPERATION

Ten years ago, nine countries in southern Africa launched a grand experiment in international cooperation. They formed the Southern African Development Coordination Conference (SADCC), an organization dedicated to promoting development at the national level through coordination at the regional level.

SADCC was formally established in April of 1980. The original members were Angola, Botswana, Lesotho, Malawi, Mozambique, Swaziland, Tanzania, Zambia, and Zimbabwe. Following the political settlement that led to the withdrawal of South Africa and popular elections in 1990, Namibia will become the official 10th member of SADCC, once administrative procedures are completed.

SADCC's founding declaration etablished four objectives: 1. reduction of economic dependence, particularly, but not only, on the Republic of South Africa; 2. the forging of links to create a genuine and equitable regional integration; 3. the mobilization of resources to promote the implementation of national, interstate, and regional policies; and 4. concerted action to secure international cooperation within the framework of a strategy for economic liberation (1).

The first objective suggests a political mission that complemented

145

1056-3466/91/1022-0145$02.00

SADCC's basic goals of mutual social and economic development. National ideological differences notwithstanding, member states resolved to combat the system of apartheid in neighboring South Africa. The political and development objectives of SADCC were viewed as strongly coupled. More self-reliant regional development would gradually reduce dependence of so-called "front line states" on South Africa, the economic giant of the region, thus allowing more effective political pressure.

Organizational Structure

To pursue these ambitious objectives, SADCC established a sector-based structure for planning and action, illustrated in Figure 1. Sector programs were distributed among member countries, with Angola, the sole oil producer, assuming responsibility for energy. The Technical and Administrative Unit (TAU) of the SADCC Energy Sector was established in Luanda, Angola in 1982. To coordinate overall SADCC program implementation, SADCC set up an Executive Secretariat in Gaborone, Botswana, and a cycle of annual government meetings.

Efforts to build an effective regional energy program would have faced significant political, economic, and institutional obstacles under the best of circumstances. In the 1980s, the region was beset by international destabilization, civil war, and socio-economic hardships. A review of SADCC's first decade of regional energy coordination must be situated within this historical context.

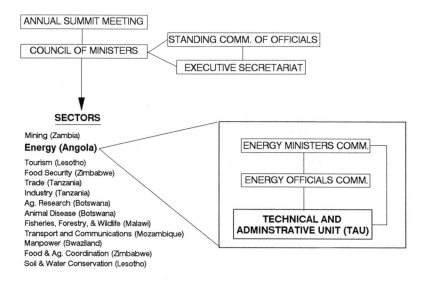

Figure 1   SADCC organizational structure. Based on (7).

The Promise of Energy Coordination

Viewed in the aggregate, the region enjoys an abundant endowment of primary energy resources—petroleum, coal, hydropower, biomass, and solar. Yet, as the scene is surveyed in more detailed resolution, severe imbalances come into focus.

At the national level, key resources are lacking. All countries but one are totally dependent on imported oil. National electric systems vary from over-capacity to undercapacity. Coal is unevenly distributed and developed. Significant exploitation of the region's renewable resources remains a distant possibility.

At the local level, the abundance of aggregate national wood resources is irrelevant. Households and small producers throughout the region face advancing wood shortages. Driven by land use changes and by excessive woodfuel collection for rural users and rapidly growing cities, local degradation of woody biomass is a serious problem for all member states. More than a woodfuel problem, the deterioration in traditional biomass production systems threatens rural ecosystems, social patterns, and livelihoods.

This article addresses the fundamental question: to what degree can regional coordination contribute to energy development in member countries? The following sections review the energy situation in the SADCC region, appraise the first decade of regional efforts, and suggest directions for the future.

SADCC IN THE 1980s

SADCC countries gained independence over the past 30 years. The colonial legacy left economies that remain highly dependent on raw material export and vulnerable to fluctuations in world commodity markets. During the 1980s, economies suffered as the terms of trade deteriorated (2, 3). For example, Zambia's export earnings fell two-thirds from 1976 to 1985 as the real price of copper fell 60%.

The need to build more resilience into economies through diversifying the industrial base and establishing reliable regional markets shapes SADCC's longer-term objectives. However, the regional transport infrastructure, established during the colonial era, was configured for trade with Europe, rather than for commerce among SADCC nations. Appropriately, SADCC has made the transport sector its highest priority.

Figure 2 illustrates the region's geographic situation. The six landlocked SADCC countries depend on a few supply routes from Mozambique, Tanzania, and the Republic of South Africa. The important Benguela railway to Angola has been out of service throughout the 1980s. Malawi, Zimbabwe, and Zambia depend on the Mozambique ports of Nacala, Beira, and Maputo, with trade also passing through South Africa. Lesotho, Swaziland, and Botswana are currently supplied almost entirely from South Africa.

Figure 2   SADCC countries, capitals, and major rail routes.

Vital SADCC transport network and energy facilities (refinery, oil depots, transmission lines) were major targets of the destabilization activities of the 1980s. Trade and transport were disrupted, particularly supplies of oil, coal, and electricity. The flow of goods through South Africa increased, undermining a major SADCC objective.

The Regional Economy

Compared with Europe, SADCC has a greater total land area (5.86 million square kilometers), with less than 20% the population (77 million) and less than one-half percent the GDP (20 billion US$) (4–6). With an average income of under $300 per capita, the SADCC region is one of the world's poorest. The regional economy is predominantly agricultural, with about 80% of the population rural. Agriculture accounts for about one-quarter of SADCC's measured GDP (7). Manufacturing capacity is unevenly distributed, concentrated most heavily in Zimbabwe (8). Mining accounted for about half of SADCC's export earnings in 1986 (9). The informal economy is still significant though not well represented in GDP figures.

Although it increased in the 1980s, intra-SADCC trade represents only about four percent of the total trade volumes of SADCC countries (7).

Intra-SADCC trade has been dominated by a few countries, most notably Zimbabwe. The only trading partners exceeding $50 million are Zambia-Zimbabwe and Botswana-Zimbabwe. There is potential for significantly more trade within SADCC, in particular, as we shall see, in energy commodities. However, impediments to increased SADCC trade—payments required in scarce foreign exchange, limited transport system, and underdeveloped markets—must be overcome.

In general, economic development in the SADCC region during the 1980s was sluggish.[1] The costs of destabilization were extremely high, particularly in Angola and Mozambique. The cumulative lost GDP from 1980 to 1988 has been estimated at $62 billion, more than three times the annual SADCC GDP (7). Declining commodity prices, food crises, and chronic foreign exchange shortages compounded SADCC's economic problems. The future will be much brighter if political transformations continue within South Africa, conflicts in Angola and Mozambique are resolved, and sound national and regional development strategies are advanced.

Energy Trends

From 1980 to 1988, per capita consumption of modern[2] fuels fell 12%, as population grew 29% and total consumption only 13%. Modern energy use trends are illustrated in Figure 3. From 1980 to 1984, total modern fuel consumption actually decreased 4%, whereas from 1984 to 1988 consumption increased 18%. Throughout the 1980s, coal use, excluding power generation, declined.

SADCC energy use grew more slowly during the 1980s than in most other developing regions. In Latin America and Asia, total modern fuel consumption increased 44% and 22%, respectively, compared with the 13% rise in SADCC (10, 11). The relatively slow growth can be traced to sluggish economies, restricted availability of modern fuels, and limited substitution of electricity for coal and oil products. There is no evidence that improvements in energy efficiency, noted in other countries during this period (8), contributed significantly to the SADCC trend.

As shown in Table 1, biomass fuels dominate the 1988 regional energy mix. Imprecise biomass demand data (4, 12), the lack of time-series data, and the highly localized nature of surveys make it difficult to track trends in biomass use. However, it is clear that a significant substitution of modern for biomass fuels—the so-called energy transition—has not occurred in SADCC in the 1980s. As per capita modern fuel consumption declined, there even are

---

[1]With a per capita income four times the regional average, Botswana has been a notable exception.

[2]Throughout this article, fuels are divided into the categories of biomass (fuelwood, charcoal, and agricultural residues) and modern (all other fuels).

Table 1 Population, GDP, and energy statistics, 1988

Country	Population (millions)	1987 GDP (million US$)	Total final energy use[a] (1000 toe)	Share of final energy use[a]						
				Firewood[b]	Charcoal[b,c]	Other biomass[b,c]	Electricity	Coal	Petroleum products	Ethanol
Angola	9.4	4,900	2,533	37.9%	18.1%	0.0%	2.4%	0.0%	41.6%	0.0%
Botswana	1.2	1,552	700	37.6%	0.0%	0.0%	8.7%	22.0%	31.7%	0.0%
Lesotho	1.7	373	614	52.8%	0.0%	23.6%	2.1%	3.6%	17.9%	0.0%
Malawi	8.3	1,248	3,346	92.1%	0.0%	1.8%	1.2%	1.0%	3.8%	0.1%
Mozambique	14.9	1,570	6,957	89.4%	5.4%	0.0%	0.7%	0.2%	4.4%	0.0%
Swaziland	0.7	542	666	17.0%	0.0%	41.9%	7.8%	17.6%	15.8%	0.0%
Tanzania	23.2	3,002	14,115	94.0%	0.0%	0.0%	0.7%	0.3%	5.1%	0.0%
Zambia	7.5	2,209	3,972	48.6%	15.0%	0.0%	13.6%	7.7%	15.1%	0.0%
Zimbabwe	8.9	4,377	5,365	42.4%	0.0%	0.0%	13.4%	28.2%	15.7%	0.3%
SADCC	75.8	19,773	38,268	74.3%	3.7%	1.3%	4.3%	5.7%	10.7%	0.0%

Source: (4)

[a] The data differ with some country-level data (53) and international statistics (10, 11).

[b] Biomass shares are rough estimates, and are inconsistent with other sources, including a recent TAU report (12), in which the total biomass shares are: Angola-Botswana (48%), Lesotho (47%), Malawi (94%), Mozambique (82%), Swaziland (35%), Tanzania (92%), Zambia (54%), and Zimbabwe (47%).

[c] Not reflected here is the extensive use of charcoal in Tanzania, and the use of other biomass in most SADCC countries. In Tanzania, charcoal accounts for approximately 7%, and other biomass 6%, of total final consumption (53).

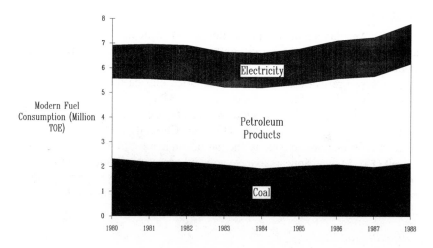

Figure 3 Energy trends in the SADCC region. Sources: (4, 52)

indications of "backward fuel switching" in recent years from modern fuels to woodfuels in urban areas and from woodfuel to agricultural residues in rural areas (12).

PETROLEUM

Oil has played a critical role as a source of energy for all modernizing countries in this century. With the notable exception of Angola, the SADCC countries depend exclusively on imports for their petroleum, a major drain on scarce foreign exchange. Securing reliable sources of oil at reasonable cost is a precondition for steady development of these fragile economies, and a major issue on the regional energy agenda.

SADCC ends its first decade no nearer to this goal than it began it. This was a period of mercurial international oil prices, with the price in both 1980 and in late 1990 approaching $40 per barrel as the second and third oil shocks rippled from the Mideast. During the mid-1980s, prices fell to one third of that value. The period of soft oil prices reduced the sense of urgency for establishing a comprehensive oil strategy. Meanwhile, regional political and military conflict disrupted supply routes and destroyed infrastructure; development faltered, and capital investments were severely limited.

Several factors may permit more effective regional oil sector development in the future. These include the current relaxation of civil and international political and military destabilization, the prod of higher international oil prices, and the institutional preparations for joint action begun during SADCC's first decade.

Resources

In 1988, Angolan crude oil production stood at about 23 million tonnes against reserves of about 273 million tonnes (see Table 2), for a reserve-to-production ratio of only 12 years. Only 6% of the crude was processed in Angola's own refineries; the rest was exported outside the SADCC region, mostly to the United States.

Regional oil and gas exploration to date has been minimal and based on inadequate geological information (13, 14). Currently, active hydrocarbon exploration is ongoing in Angola and has begun in most of the other countries as well (15). Malawi appears to have good oil prospects. While exploration of promising sites should continue to be promoted as a top priority, prudent strategic planning must proceed on the assumption that significant new discoveries will not be forthcoming in the near term.

Natural Gas

Natural gas has an important niche in areas near production fields. Natural gas resources have been established in Angola, Tanzania, and Mozambique (see Table 2). In Tanzania, about half of reserves have been committed to an ammonia/urea fertilizer plant, due to begin operation in 1990. As for the remainder, a pipeline to Dar es Salaam for power generation is being considered (16). In both Mozambique and Tanzania, the establishment of a fleet of compressed natural gas vehicles has been discusssed (17, 18).

Petroleum Products

Petroleum product balances for 1988 are displayed in Table 3. As in almost all developing countries (19), the transportation sector dominates oil consump-

Table 2 Fossil fuel reserves and production, 1988

	Coal (million tonnes)		Crude oil (million tonnes)		Natural gas (billion cubic meters	
	Reserves[a]	Production	Reserves[b]	Production	Reserves[b]	Production
Angola	0	0.00	273	22.28	54	0
Botswana	2425	0.48	0	0.00	0	0
Lesotho	0	0.00	0	0.00	0	0
Malawi	3	0.00	0	0.00	0	0
Mozambique	2422	0.08	0	0.00	65	0
Swaziland	775	0.17	0	0.00	0	0
Tanzania	1898	0.02	0	0.00	116	0
Zambia	69	0.52	0	0.00	0	0
Zimbabwe	1187	5.07	0	0.00	0	0
SADCC	8779	6.34	273	22.28	235	0

Source: (4)

[a] Includes proven and additional probable reserves.

[b] Proven reserves only. Estimates vary widely.

Table 3 Petroleum product balances, 1988 (1000 toe)

	SADCC	Angola	Botswana	Lesotho	Malawi	Mozambique	Swaziland	Tanzania	Zambia	Zimbabwe
Refinery output	2550	1373	0	0	0	0	0	550	627	0
+Import	2258	239	219	96	127	341	105	307	0	824
−Export	−640	−513	0	0	0	−7	0	−110	−10	0
−Electricity generation	−136	−29	−1	0	−1	−18	0	−36	−23	−28
+Stocks & errors	−28	−21	2	13	0	−12	0	4	−13	−1
Final consumption	4004	1049	220	109	126	304	105	715	581	795
Agriculture	219	39	11	5	7	24	0	48	13	72
Mining	215	47	25	0	0	0	0	0	143	0
Industry	443	89	1	4	6	30	13	180	69	51
Transport	2455	658	167	56	90	163	86	340	283	612
Other	307	143	6	0	17	24	0	40	50	27
Household	365	73	10	44	6	63	6	107	23	33

Source: (4)

Table 4 SADCC refinery characteristics

	Angola	Mozambique	Tanzania	Zambia	Zimbabwe
Location	Luanda	Maputo	Dar es Salaam	Ndola	Mutare
Year of startup	1958	1960	1966	1973	1965
Year of shutdown	—	1984	—	—	1966
Maximum feed[a]	1650	740	750	1100	920

Source: (4)
[a] 1000 tonnes crude/year

tion, accounting for 61% of the SADCC total. With most of the future growth in oil likely to occur in this sector, transport strategies are paramount in curbing increased oil import requirements, an issue we return to below. Though Zimbabwe and Malawi have active programs for blending ethanol with gasoline, further substitution strategies are not promising in the short term.[3]

Refineries

Of the five refineries in the region, only three are currently operating, as shown in Table 4. These refineries provide less than half of the region's petroleum product requirements. Based on simple hydroskimming technologies, the refineries have product mixes poorly matched to regional product requirements. As illustrated in Table 5, the region produces an excess of fuel oil, and a deficit of petrol and diesel oil. Substantial quantities of fuel oil are exported, while refined middle distillates are imported, even in oil-rich Angola.

The 1980s were unfavorable to local refining. Inefficient refineries, outputs not well matched to market demands, and soft international petroleum product prices made it economically preferable in most cases to import petroleum products rather than to import and refine crude oil. For this reason, Mozambique shut down its refinery in Maputo in 1984 and Zimbabwe has left its refinery closed since it became independent (7).

Supply Routes

As noted earlier, the weakened transport network severely restricted petroleum trade throughout the 1980s. Hostilities in Mozambique disturbed the natural supply routes through its ports throughout the 1980s. Less direct and more expensive alternate routes were used to the detriment of inland countries. As of 1990, recent political settlements in Angola and Mozambique promise the possibility of improving oil delivery arrangements in the region.

---

[3]Currently, ethanol is blended with gasoline at 13–20% in Malawi and Zimbabwe. Swaziland, which also produces large amounts of molasses for sugar mills, has concluded that ethanol production does not appear economic for the time being.

Table 5 Consumption and refinery output of petroleum
products in SADCC, 1988 (1000 tonnes)

Product	Output	Demand	Surplus or Deficit
LPG	34	63	−29
Gasoline	347	763	−416
Kerosene	100	249	−149
Jet fuel	264	499	−235
Diesel	847	1740	−893
Fuel oil	970	347	623

Source: (4)

South Africa remains a critical petroleum supply conduit for Lesotho, Botswana, and Swaziland, and an important emergency backup source for Malawi and Zimbabwe (20). Dependency on its petroleum supply lines gives South Africa important political leverage, a situation contrary to SADCC's goals, potentially compromising SADCC's ability to place anti-apartheid pressure on its neighbor to the south.

Directions for Petroleum Policy

In considering the prospects for the development of SADCC's petroleum sector, three types of actions can be distinguished: upgrade of the current infrastructure, mid-term institutional initiatives, and long-term strategic possibilities for greater regional integration.

UPGRADING THE EXISTING SYSTEM   The existing transport infrastructure requires upgrading, repair, and diversification. SADCC has initiated several projects, and more are under consideration. The Beira and Northern (via Tanzania) corridors have been given priority by SADCC, and repair, strengthening, and extension of other corridors are under way or under active discussion (21).[4] In addition, four of the region's refineries can be rehabilitated and upgraded for about $120 million to improve the match between refinery output and regional demand (20).

An increase in the strategic petroleum storage facilities would help to cushion the region against supply interruptions. It has been proposed that these facilities should achieve 90-day strategic stocks in Botswana, Malawi, Swaziland, and Zimbabwe and 120-day stocks in Lesotho (21).

[4]These include Mozambique (Maputo)–Swaziland, Mozambique (Maputo)–Zimbabwe, Mozambique (Nacala)–Malawi, Tanzania (Dar-es-Salaam)–Zambia, and Angola (Lobito)–Zambia. Emergency repair for the Tanzania–Zambia oil pipeline has begun, but lacks funding required for further improvements (7). The Beira–Mutare pipeline is being extended to Harare and sized so that Botswana and Malawi could also recieve supplies.

INSTITUTIONAL INITIATIVES   International cooperative arrangements could pool expertise, resources, and marketing leverage to the benefit of members. Perhaps the major achievement of SADCC to date is to set the stage for the rapid deployment of these mutually strengthening arrangements. Possibilities include a joint regional exploration program to improve the geological data base, to better evaluate petroleum potential, and to provide a stronger foundation for negotiating with international oil companies (15, 22). A regional petroleum training center that has begun operations in Sumbe, Angola (15) could be expanded to provide training and assistance on legal and contractual issues, administrative systems for petroleum operations, and financial control. Human resources could be developed through well-structured training courses, job exchanges, establishing a manual of common practices, and training missions outside the region (23).

SADCC could also benefit from a program for the joint procurement of petroleum products (21). A joint procurement program could take advantage of economies of scale in both the bulk purchase and transportation of petroleum.[5] A new or existing agency could be mandated to negotiate on behalf of member states and to coordinate financing and allocation (21).

STRATEGIC OPTIONS   Integration of Angola into the SADCC petroleum supply and trade system remains the key long-term oil policy question, one posed at the very inception of the SADCC Energy Sector (24). Angola's unique oil abundance within SADCC stands in contrast to the region's lack of international trade and to the complete import dependence of other SADCC countries.

A recent study found that Angola-refined oil was competitive with Mideast oil, that refineries could easily be adjusted to accept Angolan light crude, but that increased use of Angolan products did not offer significant economic advantages (20). Nonetheless, SADCC trade in Angolan oil could play an important role in catalyzing regional development. Cross-sectoral trade or barter arrangements could exchange Angolan oil for commodities from other member countries. Such arrangements could strengthen local economies and substantially build regional oil and overall economic independence. A thorough investigation would require that SADCC look beyond its sectoral planning structure.

Despite Angola's expressed willingness to supply crude and refined oil products to member states based on favorable reciprocal trade agreements, more serious consideration has been undermined by several factors. Most

[5]For most sub-Saharan African countries, the landed cost of petroleum products and crude oil reflects the small volume of transactions and lack of competition in procurement. According to a recent World Bank proposal, multi-country purchasing, taking advantage of international competitive bidding procedures and larger vessels, could save $1–4 per barrel.

notably, 15 years of Angolan war and insecurity have created widespread perception of risk. At the same time, the 1980s were an era of secure and abundant international oil. Lastly, SADCC cooperation and capabilities are only beginning to reach the level required for such a major effort at regional integration.

SADCC as a region is self-sufficient in oil. At the same time, all but Angola face major problems of foreign exchange drains for imports and supply insecurity. After 10 years, SADCC has no more oil security than at the beginning of the decade. However, the basis has been set for strengthening institutional cooperation and, finally, taking a hard look at how a major effort at a regional petroleum policy, potentially involving Angolan oil, could help be an engine for mutual reinforcement of the separate national economies. Meanwhile, oil dependence can be moderated through better regional coordination, including demand management activities, as addressed below.

COAL

SADCC countries have ample coal resources for meeting current require-ments, for expanding future coal use, and for further developing export markets. As with other SADCC resources, the production of coal is sharply constrained by weak infrastructure, scarce capital, and limited human re-sources.

As an alternative to oil and wood, coal can help address two major energy issues in the region. Coal can substitute for most processes, with transporta-tion a notable exception. Indeed, coal stoves have been introduced in Mozam-bique as a substitute for charcoal, coal briquetting is under serious considera-tion, and new coal-fired electric plants have been proposed (25, 26).

At the same time, the environmental consequences of coal utilization must be managed very carefully. Coal mining can have significant local impacts on land and water systems, and coal burning can be a major source of air pollution in households and local areas, and, with increasing concern about acid rain and climate change, at regional and global levels, as well. Since coal is likely to play a significant role in near-term development strategies for SADCC, effective efforts to achieve high levels of end-use efficiency and pollution control are needed, as well as a regional coal strategy informed by environmental considerations.

Resources

Seven of the nine SADCC countries produce at least some coal. The reserve figures shown in Table 2 reflect official SADCC statistics. However, there is considerable uncertainty; a recent assessment by the U.N. Economic Com-mission for Africa reported coal reserves of 14,504 million tonnes (27), much higher than the 8779 million tonnes reported in Table 2. Coal production in Africa is dominated by the Republic of South Africa, which accounts for about 90% of total continental reserves. Of the remaining reserves, more than 90% are found in the SADCC region (27).

Current coal production remains a small fraction of either reserve figure. Despite the large potential, SADCC coal production dropped during the 1980s, nowhere more dramatically than in Mozambique. Ideally situated for increased production due to high-quality coking coals and port facilities, Mozambique suffered heavily from the destabilization. Production decreased from 500,000 tonnes in 1981 to 83,000 tonnes in 1988 (27, 4).

As shown in Table 6, Zimbabwe dominates coal use in the region, accounting for 75% of total use for electricity generation, coking, and final consumption. The strong coal program in Zimbabwe is rooted in Rhodesian government policies of the 1970s to foster economic self-sufficiency in order to blunt the international embargo against the minority regime. Zimbabwe's success in substituting coal for use in agriculture and industry may offer useful lessons for other SADCC countries (27).

Trade

Intra-SADCC trade, at 70,000 tonnes, accounts for a small fraction of total trade and consumption (4). With its superior transport infrastructure, South Africa's exports to Botswana, Lesotho, Mozambique, and Swaziland, and Zimbabwe (for ferrochrome smelting) exceed SADCC's internal coal trade. Swaziland is the primary SADCC coal exporter to the international market at about 160,000 tonnes in 1988; at the same time, due to coal quality incompatibility, it imports most of its domestically used coal from South Africa (29). In recent years, Zimbabwe has increased its exports outside SADCC, reaching 58,000 tonnes in 1988, and Zambia has begun to export as well.

As with oil and other commodities, weak transport infrastructure places a bottleneck on intraregional and international coal trade. For such trade to flourish, transport development is essential. While the rehabilitation of the railway lines currently under way will provide sufficient capacity to handle trade growth, locomotives and rolling stock require repair and replacement (30).

Institutions

Effective development of SADCC coal resources will require coordination among three SADCC institutions—SATCC, the Mining Sector, and TAU. SATCC, the Southern Africa Transportation and Communication Commission based in Mozambique, manages the development of railways. The Mining Sector, based in Zambia, handles projects related to coal extraction and processing. Finally, TAU is responsible for projects addressing coal use.

Table 6 Coal balances, 1988 (1000 tonnes)

	SADCC	Angola	Botswana	Lesotho	Malawi	Mozambique	Swaziland	Tanzania	Zambia	Zimbabwe
Production	6506	0	613	0	37	83	167	17	524	5065
+Import	412	0	32	32	16	110	184	38	0	0
−Export	−291	0	0	0	0	−15	−160	0	−41	−75
−Electricity generation	−3075	0	−430	0	0	−86	−21	0	0	−2538
−Coke ovens	−255	0	0	0	0	0	0	0	0	−255
+Stocks, errors, & losses	−107	0	9	0	−6	−73	0	0	−37	0
Final consumption	3190	0	224	32	47	19	170	55	446	2197
Agriculture	377	0	0	0	0	0	0	0	0	377
Mining	437	0	185	0	0	0	4	0	178	70
Industry	1656	0	10	0	47	11	93	55	268	1172
Transport	375	0	0	0	0	8	33	0	0	334
Other	278	0	27	0	0	0	7	0	0	244
Household	67	0	2	32	0	0	33	0	0	0

Source: (4)

Currently, TAU has no active coal projects. The Mining Sector, meanwhile, has several active projects, including a regional coal analysis laboratory in Malawi, and national projects to promote coal development in Malawi, Mozambique (including coke production), and Swaziland (displacing imports with domestic production).

Overlapping agency responsibilities can hinder development. In recognition of this problem, TAU is seeking a senior advisor responsible for coordination of the coal sector. In 1990, with no such advisor at TAU, all plans for developing projects essentially were on hold (31).

Given the institutional constraints and the destabilization activities, it is not surprising that coal development in the 1980s was not significantly influenced by SADCC initiatives. Nonetheless, some regional activities show promise for going beyond what is possible at the national level. With sufficient expertise in place, SADCC could provide needed coordination on regional international transport links; effective policies for switching to coal in electric, manufacturing, rural-industrial, and even household sectors; marketing strategies; intraregional barter and trade agreements; and staff training. SADCC could also help to pool and share information on coal resources and technology, and to evaluate alternative institutional arrangements (e.g. joint government-private ventures and operations, etc).

In its recent draft strategy for development and increased utilization of coal in the SADCC region, TAU has suggested some of these ideas, including the establishment of an institute for coal-related research and development (31). However, human resources remain the critical constraint in proceeding towards an effective coal development strategy that combines integrated, detailed, and coordinated coal supply and demand goals with workable policies, regulations, and investment plans to implement them.

ELECTRICITY

As a region, SADCC has far more installed electric generating capacity than currently required, with vast additional resources for future development. However, there are significant national and local problems. Several countries are dependent on south African imports. Others rely on fossil-fuel generation, while inexpensive hydropower is available in neighboring countries. Many areas experience inadequate service due to limitations of transmission and distribution networks. Electrification in rural areas is still the exception.

Resources

Electricity balances are presented in Table 7. Hydro accounts for two-thirds of the electricity produced in SADCC. Two countries—Zambia and Zimbabwe—account for nearly 80% of current electricity production, driven by large mining and industrial sector requirements, respectively. Zimbabwe and Botswana are the only SADCC countries with major coal-fired electric plants; small plants in Swaziland and Mozambique also use coal. The use of oil for electricity generation in SADCC is restricted to peaking plants and isolated diesel generators.

Installed capacity information by country and generating type is collected in Table 8. Peak loads of the member countries are less than half of the installed capacity. Nevertheless, without new capacity, Zimbabwe, Malawi, and Botswana are projected to face power shortages in the early to mid 1990s, and Lesotho, Mozambique, and Swaziland will rely heavily on imports (32).

More than 2000 MW of the SADCC total installed capacity of 7000 MW comes from a single hydro facility, Cahora Bassa, in northern Mozambique. Conceived by the Portuguese authorities prior to independence, Cahora Bassa was built for export to South Africa via a direct DC line. A major target of anti-government sabotage, the line was destroyed in 1981. Cahora Bassa output has been negligible since that time. Nevertheless, if security can be established and export markets developed, a 1250 MW extension could be added at the site.

SADCC has abundant hydro resources, with an identified potential of nearly 46,000 MW, or 215,970 GWh in firm energy. Of this potential, only 5,200 MW, providing 31,000 GWh in firm energy, is now in place (exceeding current SADCC electricity requirements of 22,800 GWh).

Regional Coordination

The degree of coordination of the national electric systems will be a measure of the scope for coordination of the region. As development philosophies cover a spectrum from autarky to regionalism, electric system strategies may range from national self-sufficiency to regional power pooling, with many intermediate possibilities.

Strong grid interconnections in SADCC, as with other regions, can promote system reliability and help meet growing electricity demands at lowest possible cost. Load diversity among interconnected systems reduces overall capacity requirements. Combined systems are less vulnerable to power plant forced outages since surplus capacity from elsewhere can substitute. Systems jointly dispatched on a least-cost basis reduce total operating costs. If construction plans are coordinated, the region could take advantage of the best available siting opportunities and economies of scale in construction. In the hydro-rich SADCC region, there are the further attractions of utilizing excess capacity from large dams built on small systems, such as in Zambia and Mozambique, and in the conjunctive use of separate hydro systems.

However, the vastness of the SADCC region, spanning 3000 km from the Atlantic to the Indian Ocean and 3500 km from Lake Victoria to southern

Table 7 Electricity balances, 1988 (GWh)

	SADCC	Angola	Botswana	Lesotho	Malawi	Mozambique	Swaziland	Tanzania	Zambia	Zimbabwe
Generation (hydro)	13,881	758	0	0	581	165	211	1,203	8,300	2,663
Generation (oil/coal)	6,989	84	804	0	2	169	200	127	20	5,583
+Imports	2,261	0	59	174	0	341	255	0	86	1,346
−Exports	−1,330	0	0	0	0	0	0	0	−1,330	0
−Losses	−2,542	−121	−138	−23	−111	−107	−52	−228	−668	−1,094
Final consumption	19,259	721	725	151	472	568	614	1,102	6,408	8,498
Agriculture	1,234	20	0	0	58	0	103	93	175	785
Mining	6,624	0	400	0	0	0	35	0	4,687	1,502
Industry	5,924	195	103	26	209	268	346	199	494	4,084
Transport	6	0	0	0	0	0	0	0	6	0
Other	2,718	310	165	75	119	105	43	499	637	765
Household	2,754	197	57	50	85	195	86	311	410	1,363

Source: (4)

Table 8 Electric generating capacity, 1988 (MW)

	SADCC	Angola[a]	Botswana	Lesotho	Malawi	Mozambique	Swaziland	Tanzania	Zambia	Zimbabwe
Hydro	5,220	287	0	2	145	2,142	41	330	1,608	666
Coal-fired	1,554	0	212	0	0	47	0	0	0	1,295
Oil-fired	368	176	7	2	23	86	10	60	0	5
Total capacity	7,142	463	219	4	168	2,275	50	390	1,608	1,966
Peak load	3,160	151	138	49	100	102	87	209	919	1,406
Reserve margin	126%	207%	59%	-92%	68%	2,139%	-43%	87%	75%	40%
Hydro potential	45,970	17,680	0	360	660	12,840	110	5,590	5,810	2,920

Sources: (4), (35).
[a] Peak load is an estimate for 1987 from (32).

Botswana, is an impediment to comprehensive grid interconnection. Fortunately, the major power facilities and much of the electricity demand lies within a concentrated area centered around Zimbabwe, including parts of Zambia, Botswana, Mozambique, Malawi, and Swaziland. Zambia already exports power to Zimbabwe from its Kariba hydro station over a 2000 MW capacity line, the only large intra-SADCC transmission link. An existing skeleton of 330 kV transmission lines in this area forms the basis for the future strengthening of a regional grid. An important element in regional planning is the possibility of capturing some of the low-cost hydropower available at the 2075 MW Cahora Bassa station. Mozambique supports the concept of an intra-SADCC redirection of Cahora Bassa power if satisfactory pricing and contractual arrangements can be established (33).

The economic advantages of regional integration are limited by the competing objective of national power system self-sufficiency. This factor has contributed to difficulties in arranging the sale of excess Zambian hydropower to capacity-short Zimbabwe (another major factor is the Zambian requirement for payment in foreign exchange). The emergence of a political climate in which strict control over national electric systems is significantly relaxed in favor of international arrangements will be a long and difficult process. The preconditions are mutual trust, clear mutual benefits, sound plans, and successful experiences in joint ventures.

SADCC Initiatives

In practice, SADCC will need to build toward power pooling slowly. First, pair-wise interconnections must be strengthened and bulk power interchanges increased. A stated priority within SADCC is the establishment of principles for cross-border tariff setting (34). Then, member systems can engage in formal contingency planning to improve system reliability and reduce vulnerability to extreme events, for example, loss of key units, dry years, and unexpected jumps in demand. Later, more coordinated operations and planning can be considered.

During its first 10 years, SADCC commissioned a study of regional electric interconnections, proposed several interconnection projects, and implemented two of them (32, 34). Proposed cross-border interconnections fall into two broad categories: larger lines to connect neighboring grids and smaller ones to supply isolated areas from a neighboring system. The two completed SADCC projects include one of each type. A 340-km, 220-kV transmission line completed in 1990 links the Zimbabwe to Botswana electric systems, permitting Botswana to displace South African imports with lower-cost power from Zambia and Zimbabwe (32). A 66-kV line completed in 1988 enables power from Northern Botswana to supply an isolated region in Zambia. Bilateral projects outside the SADCC framework have linked other isolated areas. The feasibility of several other, larger interconnections has also been studied (34).

Additional regional projects need to be actively explored. In a region lacking trained professionals, spare parts, and equipment, member-country needs can be met efficiently by joint training programs, common maintenance procedures (35), industrial workshops, and service units. These institutional efforts would provide a foundation for greater regional coordination of national power systems.

SADCC's first 10 years have yet to spawn significant regional development of the electric sector. However, a stronger regional perspective may be emerging. The World Bank and TAU have developed an important proposal for more thorough investigation of regional interconnection possibilities. Particularly promising is the process begun by the newly formed SADCC Electricity Sub-Committee, which brings electric utility representatives at the operational level together to discuss and pursue joint opportunities.

In the long term, SADCC's success in regional energy coordination can be gauged by the degree to which it has stimulated the evolution of a more optimal regional electricity system. While concrete accomplishments are modest to date, SADCC can capitalize now on a legacy of greater mutual trust that it forged during its first decade.

BIOMASS

With 95% of SADCC households relying primarily on biomass fuels to meet their energy needs, the sustainability of biomass resources is arguably the most important and immediate energy issue affecting human well-being in the region. The socio-economic and environmental implications of biomass collection and use are complex. The increasing scarcity of biomass resources in many parts of SADCC could be an important obstacle to both rural and urban development.

Biomass Consumption

Biomass dominates the energy accounts of the region. Some 80% of the region's final energy consumption comes from woodfuel, charcoal, and agricultural residues.[6] The range among countries in biomass use is significant, as shown in Figure 4. This figure illustrates a negative relationship between biomass use and per capita GDP among SADCC countries, which suggests that economic development in SADCC indeed would lead to a transition towards modern fuels. Decreased woodfuel availability alone, on the other hand, does not appear to be capable of forcing such a transition. In Tanzania, Malawi, and Mozambqiue, the biomass share tops 90%, despite many regions of scarcity (4, 12).

Regionally, charcoal accounts for only approximately five percent of final

---

[6]Ethanol derived from sugar or molasses is included as a new and renewable resource, not a biomass-derived source in TAU's accounts.

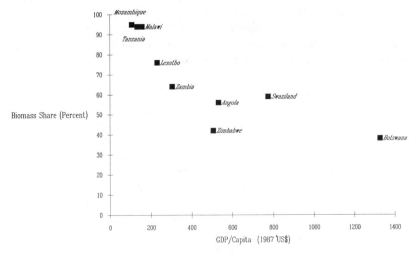

Figure 4 Biomass share of total final energy use vs economic development. Source: (4). Per capita GDP data are for 1987; biomass share data are for 1988.

woodfuel consumption. Nonetheless charcoal production places considerable pressure on wood resources. Traditional earthen kilns consume about eight tonnes of wood to produce one tonne of charcoal. Demand for charcoal, used extensively in urban areas in several SADCC countries, will grow with the rapid urbanization now proceeding in the region.

Although would shortages have led to an increase in the use of agricultural residues in some countries, data on residue use are poor. To date, there has been notably little study on their availability, use, and effects on soil fertility (12).

Almost 90% of wood consumption is in the household sector, mostly as energy but also for building poles. On purely quantitative grounds, then, the regional energy story is about trees for households. While rural households account for more than 80% of biomass fuel use, urban households and small-scale rural industries also depend heavily on biomass fuels. Most urban households in Mozambique, Tanzania, and Zambia cook with woodfuels (12). Small-scale rural industries—brick making, fish curing, tobacco drying, beer making, etc—are heavily woodfuel consumers. In Malawi, a country experiencing scarcity throughout its central and southern districts, tobacco drying accounts for one-quarter of total woodfuel use (12).

Wood Shortages

Comprehensive and reliable data on woody biomass supplies are not available. Nevertheless, field surveys and experience indicate that all SADCC countries are experiencing growing wood shortages (12, 24, 36–38). Wood deficits are caused largely by the expansion of agricultural production into lands previously available for rural wood supply (12, 38,39). At the same time, growing demands stress the contracting resource base.

As trees become more scarce, rural households may respond in a number of ways. Wood gatherers, mostly women, may travel further, spend cash, if available, for newly commodified woodfuel products, simply make do with less, or use agricultural residues, with possible degradation of soils. Wood shortages tend to undermine both socio-economic development and rural environments.

Scope of the Problem

Throughout the region, there is a lack of institutional coordination, data, resources, and effective intervention strategies to address this complex problem (25, 12, 40). At the national level, only Malawi and Tanzania appear to have the institutional capability to develop and implement major programs for tackling biomass/woodfuel issues (12). At the regional level, TAU has only recently begun to address these issues with the urgency they deserve.

Indeed, the social, economic, scientific, and institutional dimensions of the wood energy problem in Africa have only recently been clarified (38, 39, 41, 42). The problem is complex, specific to local physical and cultural conditions, and intimately bound to broader issues of development. In rural contexts, woodfuel is largely collected on farms, not forests. Trees are often multi-purpose crops. In addition to woodfuel, they provide fodder, timber, food, shade, and erosion control. Effective interventions must understand the way that woodfuel patterns are imbedded in the matrix of rural production, and are affected by local conditions, gender differences (men control tree planting, women collect wood), and land access rights.

These considerations suggest two important lessons for policy that underscore the vexing nature of the problem. First, solutions must be tailored to local conditions, must involve farm families who can best unterstand the problem, and must have a heavy "bottom up" component to complement centralized programs. Second, the rural woodfuel problem is inherently multi-sectoral, involving energy, forestry, agriculture, land use, environment, and rural development issues. Therefore, to address the problem, ministries and agencies must coordinate their planning and efforts.

The urban woodfuel problem involves another set of complex issues. The basic elements of an urban strategy are fuel switching and more efficient use of wood at the end use. These transitions will occur only if fuel prices are judiciously set, alternative fuels are securely available, and end users can afford modern-fuel cook stoves. However, setting and enforcing prices in dispersed wood and charcoal markets present immense practical difficulties, and urban poverty is a counterweight to any solutions involving additional capital investment. The urban woodfuel problem must be addressed in the context of an integrated strategy for energy development.

Regional Opportunities

The local specificity of rural woodfuel problems and effective solutions suggests that SADCC, as a regional organization, has limited scope. Nevertheless, SADCC can play a significant, albeit indirect, institution-strengthening function for its members. Recent SADCC workshops, seminars, and surveys on biomass energy are signs that TAU has begun to assume this role.

Just as national energy departments need to coordinate with other agencies that address aspects of the multi-sectoral biomass problem, TAU will need to harmonize its emerging biomass activities with other SADCC sectors. In particular, the SADCC Forestry Sector, based in Malawi, recently has launched several relevant projects, including an agroforestry research network and a regional tree seed center (12).

A Five-Year Woodfuel Implementation Strategy was approved by the Energy Ministers in September 1987 for 1988–1992. TAU is directed to foster national woody biomass programs by providing consulting, training, and information dissemination, and to catalyze regional research on woodfuel supply and demand and fuel switching. These are ambitious objectives. To achieve them, TAU must improve its own organizational capacity to design, implement, and fund a concrete biomass fuel action plan.

As of 1990, the specific woodfuel programs and projects that TAU has advanced for urgent implementation—covering the broad areas of Policy, Planning, and Coordination; Manpower Development; and Intensification of Mass Awareness/Public Participation—have yet to receive required donor funding. Additional projects with significant promise have been proposed recently, including Energy in Small Scale Industry and a regional Biomass Assessment, building on earlier studies and emphasizing development of local expertise in both remote-sensing and ground-based analysis techniques.

To date, insufficient attention has been given to biomass both within SADCC and among donors. TAU encountered significant difficulties in setting up its Woodfuel section, even though the official SADCC Woodfuel Strategy gives the highest priority to the development of this section (12). Still, the SADCC framework can significantly advance national efforts to address woodfuel problems, by creating a center of excellence that could, first and foremost, provide training, and also develop assessment methods, prepare materials for extension efforts, disseminate information, and conduct appropriate research.

NEW AND RENEWABLE SOURCES OF ENERGY

New and renewable sources of energy (NRSE) are insignificant in the current energy budgets of member countries. Regional initiatives can complement, support, or stimulate national efforts to develop alternatives to traditional fuels.

The Potential

While data are fragmentary, NRSE potential in SADCC countries appears to be significant[7] (43, 44). Solar radiation averages 4 kWh/m^2, sufficient for most photovoltaic and thermal-solar devices. Abundant crop and animal wastes can be used in biogas digesters for agriculture, commercial, and household applications. Wind resources are suitable for water pumping, and possibly for small-scale power generation. Zimbabwe and Malawi already produce substantial quantities of ethanol (38 million and 12 million liters per year, respectively), and four other countries have excess molasses that could be used as an ethanol plant feedstock.

The promise of NRSE for developing countries is well known. Greater reliance on indigenous resources would mean decreased fuel imports and increased employment resulting from local production and assembly of NRSE equipment, and the emergence of a distribution and maintenance infrastructure. SADCC countries can build on their fledgling NRSE industries, which include limited capacity for the production of solar water heaters, photovoltaic modules, biogas lamps and burners, and wind pumps.

NRSE applications have relatively benign environmental repercussions. NRSE also contributes to energy security by substituting dispersed small-scale for centralized technologies, thus reducing the dangers of catastrophic losses of key installations. As a set of small-scale, dispersed, and flexible applications, NRSE can be blended gradually into appropriate niches.

The Constraints

Despite its impressive promise, NRSE activity has diminished in SADCC countries during the past decade. The oil price drop undermined, at least temporarily, the political will for alternative energy development. Energy institutions, such as electric utilities, have been skeptical of and ill-prepared for NRSE, preferring the familiarity of centrally engineered, maintained, and financed facilities to the perceived risks of dispersed and smaller-scale technologies.

Additional impediments include high import duties and sales taxes, foreign-exchange restrictions, and inadequate distribution, installation, and

---

[7]As used in SADCC, NRSE includes solar, wind, small-scale hydro, and nontraditional fuels such as ethanol and biogas. It excludes large-scale hydro and traditional biomass fuels.

maintenance services. Local NRSE markets are below threshold levels for economies of scale in manufacture and for establishing an adequate service infrastructure (44, 45). The economic advantages of NRSE depend on life-cycle cost comparisons in which relatively high initial costs are amortized against avoided fuel expenses over the life of the equipment. Creative financing arrangements are required to overcome the "first-cost hurdle," especially with cash-poor consumers.

SADCC Activities

In 1989, TAU adopted a bold framework consisting of five programs for regional NRSE activity (45). The Technology Development and Diffusion program would address impediments to greater market penetration of NRSE technologies, including the evaluation of foreign exchange, taxation, and financing treatment in each country; the design of credit schemes for NRSE products; and the establishment of regional standards and testing. The Canadian-funded Rural NRSE Pilot Programme, the only major regional NRSE project to date, will identify the most cost-effective alternatives to conventional energy sources in rural and peri-urban areas.

The Production Support program would assist NRSE production facilities, including advising on production equipment and operation, marketing, and raw material procurement. The Training Program envisions seminars for development and energy planners on NRSE; courses for engineers, extension personnel, and entrepreneurs on NRSE technology design, implementation, and production; on-the-job training for industrial workers; and training and materials for educators to promote awareness of NRSE potential among young people.

The Information Dissemination program would include a regional directory of NRSE-related organizations, suppliers, consultants, training opportunities, and international activities. A data bank on solar, wind, and biomass resources and other data also is envisioned. Finally, a regional public awareness campaign would publicize the benefits of NRSE technologies.

The NRSE Institution Strengthening program would build a regional and national capacity to achieve this ambitious strategy. For TAU, this requires a core of qualified and dedicated staff under capable supervision, as well as an adequate library, computer facilities, and other equipment. The regional NRSE team would need to work with high-caliber national counterparts and have a clear institutional mandate.

CONSERVATION

Conservation is arguably the most attractive energy resource in the region today. There are clear indications that substantial opportunities exist to conserve energy through better housekeeping and more efficient equipment in all sectors. Moreover, the growing SADCC economies have an opportunity to place efficiency at the foundation of their energy strategy. By avoiding the wasteful energy practices of previously industrialized countries, they could reap huge economic and environmental benefits. This will require carefully designed and comprehensive programs to promote cost-effective conservation and efficiency. Regional coordination may assist national conservation efforts, but a role for SADCC has yet to be clearly defined (15).

SADCC Activities

While several member countries have initiated some conservation activities, (e.g. cook-stove programs, energy audits and educational campaigns), as of 1990, TAU has implemented only one regional conservation project (46). The Canadian-funded Industrial Energy Conservation Pilot Project has conducted 30 audits and identified substantial opportunities for energy saving. Much of the industrial plant stock in the region predates independence and more recent energy-efficiency features.

Audit results suggest the potential for average savings of 15–20% across various industries, with even higher savings for coal and petroleum (30–50%). Big savings could be achieved through "no-cost" improvements in maintenance and production control procedures, and "low-cost" investments (e.g. insulation of steam pipes and hot water tanks, lighting conversions, control equipment, condensate recycling, and heat-exchangers) (47).

Although many of the low-cost savings were quickly implemented at audited plants, significant technical and management barriers to widespread adoption of efficiency improvements remain, particularly those requiring longer payback periods, financing, and foreign exchange. The project is offering training seminars to build recognition of energy conservation cost savings and technical expertise.

Transport

The transport sector accounts for 60% of petroleum consumption in the region. The importance of energy conservation in transportation began to be recognized in a 1989 seminar on Energy and Transport sponsored jointly by the energy and transportation agencies of SADCC (48). Several broad strategic directions were noted, including improved vehicle maintenance and efficiency, alternative fuels, road maintenance, public transport, and greater technology transfer and spare parts availability.

Compared to that of industrialized countries, transport stock is aging and poorly maintained. The potential of programs to increase the efficiency of the existing equipment is correspondingly greater. At the same time, regional policies to maximize the efficiency or new vehicle stock should be actively considered. Such policies could include import fees for vehicles, increasing steeply with decreasing vehicle efficiency or engine size for a given vehicle class. Regional coordination on conservation-oriented import policies could provide a larger and more coherent market for more efficient equipment to the advantage of all member countries.

Regional Conservation Center

Regional conservation efforts are only beginning. The first step is building the expertise within the energy sector at TAU. Then, a more ambitious approach may be warranted. The industrial energy project, which is based in Zimbabwe, could be expanded into a regional energy conservation center. The center would provide information, training, and consultative services to member countries.

The center could also assist member countries wishing to reform their energy pricing policies. For example, the relationship between fuel prices and conservation needs to be thoroughly examined. Fuel price subsidies (e.g. for diesel fuel) distort decision-making by consumers and managers, impede conservation, and induce unintended trade consequences. Also, pricing and financing mechanisms can be developed for treating demand-side investments by electric utilities as direct alternatives to capacity expansion.

Joint regional initiatives to develop model efficiency standards, building codes, import policies, demonstration projects and so on, would benefit all members by pooling expertise and avoiding duplication By fostering an organized internal conservation market in the region, the center would stimulate the emergence of local conservation entrepreneurs and energy-efficient equipment production. National programs are needed for designing and implementing an effective blend of conservation measures. Regional assistance activities can help achieve this exceedingly difficult yet vitally important task.

APPRAISING THE FIRST DECADE

The gap between regional resource abundance, and national and local shortages sets the context for SADCC's energy-sector activities. The promising scope for regional energy initiatives was understood from the start:

> . . . an optimal scenario for coordinated regional energy development is likely to look far different from what would emerge from a national focus alone. What appears cost-effective in a limited national context may not be the best alternative when viewed from a regional perspective. Coordinated planning could move the region toward greater energy self-sufficiency, avoid duplication of effort, achieve economies, promote joint initiatives, and channel scarce development resources toward the areas and sectors of greatest need. (49)

Against this possibility, SADCC's regional energy achievements to date have been modest. Present national energy patterns are not significantly different than they would have been in the absence of SADCC. Initiating effective regional action under conditions of underdevelopment is always daunting. For SADCC, inherent obstacles have been compounded by South Africa's destabilization program, civil wars, and economic stagnation.

Foundation for the Future

SADCC's first-decade energy achievements must be evaluated against a standard that is mindful of these obstacles. The important question is whether a process has begun that can provide the basis for significant and concrete achievements in the future.

The most important success is that TAU has fostered an atmosphere for more effective regional energy cooperation in the years to come (50). The cycle of annual meetings and subsectoral seminars has advanced significantly the awareness of key regional actors of the importance of regional energy issues. More prosaically, but crucial in the long run, the process has begun to build the interpersonal trust necessary for cooperative working relationships.

At the same time, TAU has developed projects and promoted them with SADCC's "international cooperating partners," the bilateral and multilateral assistance agencies that convene at Annual Consultative Conferences to consider SADCC's portfolio of proposed projects. Most projects still lack funding, many are essentially national in character, and more than 85% of funding is for electricity projects (15). Nevertheless, by early 1990, $155 million in funding had been secured. This process has provided TAU with valuable experience in obtaining and directing funding to priority regional projects in the future.

Institutional Constraints

The challenge now is to convert the achievements of the first 10 years into programs that can materially improve energy development in the region. Important obstacles will have to be overcome.

First, human resources must be developed. As of 1989, only 5 of 47 TAU staff had achieved bachelor level degrees and only a few had relevant outside experience, none in international organizations (50). The organization itself is the training ground for the regional energy development cadre. With this fragile skill base, the departure of key professional staff—a frequent problem in African energy departments—becomes a rolling organizational crisis.

Second, the practical difficulties of operating from Angola are disabling. The Angolan infrastructure, beset by 15 years of civil war, has been unreliable, communications poor, and working and living conditions demoralizing. Also, Angola, as a one of two Portuguese-speaking countries of SADCC (Mozambique is the other) has had to build its translation and English-speaking capabilities. All of this is costly in time and resources. Before anything else, internal administrative procedures, staff accountability, and the conditions of work must improve.

Perhaps the most difficult obstacle facing the energy sector is the hesitance of member states to engage in regional efforts. National energy institutions have often been reluctant to direct their limited staff resources toward providing TAU with the data and other necessary elements for effective coordination and planning. From the beginning, there has been a tension between SADCC's broad goals and its organizational practice. The call for strong regional development planning was clear from the establishment of SADCC and its energy sector (51). Yet SADCC was designed as a coordinating, not a decision-making body. A decentralized approach to organizational structure was adopted, complex formal treaties and new bureaucracies were avoided, and national concerns for control in matters affecting national interests were respected assiduously. Not surprisingly, the energy sector has not provided, and has not been called on to provide, bold and dynamic leadership in advancing a regional energy agenda.

TAU's minimalist interpretation of the mandate for regional coordination was both necessary and appropriate. It was necessary because of the severe operational constraints during the 1980s and practical difficulties in establishing a functioning organization. It was appropriate to allow the member states gradually to gain a better appreciation of the mutual benefits of cooperation and to build reciprocal confidence.

Regional Energy Planning

TAU can capitalize on this experience—and the apparent relaxation of tensions in Southern Africa, generally, and in Angola, specifically—to build a more active role for itself in the future. The degree to which this can be achieved will depend on the development of strong internal leadership and a willingness of the energy sector to begin to look comprehensively at alternative regional futures and strategies for achieving a desirable one.

Regional energy planning during the first decade was severely hampered by the devotion of very limited staff resources at too early a stage to modelling exercises, some of which were unnecessarily complex and unworkable. This negative experience has led TAU to defer integrated energy planning efforts indefinitely. Planning activities at TAU have turned more narrowly towards cost-benefit analyses of proposed projects. Once sufficient resources are in place, reconsideration of comprehensive, integrated energy planning will be essential.

If TAU is to pursue an active planning role, it will need to build comprehensive and quantitative understanding of the regional energy picture, synthesize national energy information into this regional picture, and explore the trade-offs among alternative regional programs, projects, and policies.

Such efforts, while challenging, can yield substantial rewards, helping SADCC to set its own energy agenda with outside agencies and consultants. This activity will require furthering the development of staff capacity, establishment of appropriate methods, and active coordination with national energy planning efforts.

Environment

The SADCC Energy Sector currently has neither formal responsibility nor staff resources for addressing the environmental aspects of energy development. There is a growing appreciation of the importance of including environmental concerns at the early stages of energy development, coming from within SADCC and from outside assistance agencies. Zimbabwe recently established the region's first Ministry of the Environment, and is requiring environmental assessments of proposed hydropower development on the Zambezi River. Increasingly, donor agencies also are requiring assessments of proposed energy projects. Furthermore, international negotiations on controlling climate change have put developing-country energy strategies high on the agenda of global policy issues.

As the call for environmentally sensitive energy choices mounts, SADCC will need to establish the expertise, methods, and policies for sustainable energy development. The environmental and public health risks of alternative energy strategies, so-called externalities, will need to be internalized into analyses of the relative costs of alternative energy options. TAU can consider developing model SADCC guidelines and methods for environmental assessments of proposed energy projects. To address the complex environmental issues related to regional energy development, TAU might need to establish a new environmental section to complement its existing structure.

Uncertainty to the South

SADCC's fortunes are coupled to the outcome of the political struggle within South Africa. Most favorable to SADCC would be an internal settlement leading to a democratic South Africa. Destabilization would end and new opportunities for regional cooperation would open. SADCC would then face the dilemma of whether formally to include South Africa or to maintain an independent regional organization as a counterweight to its economic influence (7). Another possibility is that the democracy movement in South Africa is contained. The antagonism between the minority government of South Africa and the SADCC states would persist, with the likely continuation of destabilization and sabotage in the region. An intermediate possibility is a prolonged period of internal chaos within South Africa with uncertain effects on SADCC. Clearly, international action to hasten democracy in South Africa also provides powerful, albeit indirect, assistance to SADCC.

Crossroads

As SADCC enters its second decade, regional energy development in Southern Africa is at a crossroads. Along one path, SADCC continues to provide limited services to nationally based energy development. Along the other, SADCC evolves into a forceful agency for advancing regional strategies beneficial to all members. The latter direction requires a forward leap in in-house skills, international trust, and organizational leadership. Yet the potential benefits are so great for the well-being of the people of South Africa that the bold path toward genuine regional energy coordination beckons.

ACKNOWLEDGMENTS

We thank the Swedish International Development Agency for supporting our work with the SADCC Energy Sector over the past decade. We extend our appreciation to the staff at TAU, in particular to J. T. Carvalho Simoes, Regional Coordinator, and Antonio M. M. Pinto, Head of Planning, and also to the Norwegian, Canadian, and Belgian consultants to TAU. We are indebted to Victoria Johnson for collecting, organizing, and reviewing documents used in preparing this chapter.

Literature Cited

1. Nsekela, A. J., ed. 1981. *Southern Africa: Towards Economic Liberation*. London: Collings
2. Bhagavan M. R. 1985. *The Energy Sector in SADCC Countries: Policies, Priorities and Options in the Context of the African Crisis. Research Report No. 74*. Upsala: Scandinavian Inst. African Stud.
3. Mongula, B. S., Ng'andwe, C. 1987. Limits to development in Southern Africa: energy, transport and communications in SADCC countries. In *SADCC Prospects for Disengagement and Development in Southern Africa*, ed. S. Amin, D. Chitala, I. Mandaza, pp. 85–108. London: Zed
4. SADCC Energy Sector Tech. Adm. Unit (SADCC/TAU). 1990. *SADCC Energy Statistics—1988*. Draft. Luanda, Angola
5. World Resour. Inst. 1990. *World Resources 1990–91*. New York/Oxford: Oxford Univ. Press
6. Tebicke, H. L. 1989. Energy use in transport modes in the SADCC region—a preliminary study. In *Proc. Energy and Transport, SADCC Energy Sector and SATCC Semin., Maputo, Mozambique*
7. Hanlon, J. 1989. *SADCC in the 1990s—Development on the Frontline. Special Report No. 1158*. London: Econ. Intell. Unit. 159 pp.
8. Sathaye, J., Ghirardi, A., Schipper, L. 1987. Energy demand in developing countries: a sectoral analysis of recent trends. *Annu. Rev. Energy* 12:253–81
9. Peet, R. 1984. *Manufacturing Industry and Economic Development in the SADCC Countries*, Vol. 5. Stockholm: Energy, Environ. Dev. Africa, Beijer Inst. 119 pp.
10. Int. Energy Agency. 1990. *World Energy Statistics and Balances: 1985–1988*. Paris: OECD/IEA
11. Int. Energy Agency. 1989. *World Energy Statistics and Balances: 1971–1987*. Paris: OECD/IEA
12. SADCC/TAU. 1989. Further development of the woodfuel strategy. Annex 2 in *Proc. Semin. Woodfuel in the SADCC Region, Arusha, Tanzania*
13. SADCC/TAU. 1990. *Energy Sector Working Group—TAU*. Lusaka, Zambia
14. Marques, M. 1986. Mozambique presentation. In *Proc. Oil and Gas Exploration in the SADCC Region—SADCC Energy Sector Semin*. Arusha, Tanzania: SADCC/TAU
15. SADCC/TAU. 1989. *Programme Doc. 1990 Annu. Consult. Conf.*
16. Mwalyego, Y. S., Moshi, W. A. E. K. 1987. Songo Songo gas: appraisal, drilling and plans for use. *SADCC Energy, Quarterly 1987—Vol. V, No. 13:12–13*
17. Saintes, P. 1990. Energy to fuel trans-

port, transport to distribute energy. *SADCC Energy*, Quarterly 1990—Vol. VIII, No. 20:43–45

18. dos Santos, A. 1987. Considerable gas reserves make ammonia production feasible. *SADCC Energy*, Quarterly 1987—Vol. V, No. 15:15–16

19. Imran, M., Barnes, R. 1990. *Energy Demand in the Developing Countries. World Bank Staff Commodity Working Paper, No. 23.* Washington, DC

20. BEICIP. 1988. *CCE/SADCC Study of SADCC Oil Supply Strategy—Final Report.* Exec. Summ.

21. SADCC/TAU. 1988. *Conclusions and Recommendations of the Harare Workshop on the Draft Final Report on SADCC Oil Supply Strategy.* Arusha, Tanzania

22. SADCC/TAU. 1987. *Regional Cooperation in Petroleum Exploration Activities.* Luanda, Angola

23. Bandlien, E. H. 1986. Regional cooperation in petroleum exploration activities in SADCC. See Ref. 14, pp. 495–511

24. Simoes, J. T. C., ed. 1984. *SADCC: Energy and Development to the Year 2000,* Vol. 2. Stockholm: Energy, Environ. Dev. Africa. SADCC Energy Sector/Beijer Inst. 184 pp.

25. SADCC/TAU. 1990. *National Surveys of Biomass/Woodfuel Activities.* Luanda, Angola

26. Kaoma, J. 1988. Coal briquettes domestic fuel. In *Proc. SADCC Energy Sector Coal Semin.,* Gaborone, Botswana

27. Moshi, F. S. 1988. Prospects for development and increased utilization of coal resources in SADCC countries. See Ref. 26

28. Deleted in proof

29. Myeni, P. S. 1990. Residential and industrial demand management to promote use of indigenous energy resources: the case of Swazi coal. In *Proc. Specialized Sub-Committee on Energy Planning.* Nhlangano, Swaziland: SADCC/TAU

30. Makela, P. 1988. Bulk handling and transport facilities in SADCC. See Ref. 26

31. SADCC/TAU. 1990. *Proc. Technical Sub-Committee on Petroleum and Coal, Lobatse, Botswana*

32. SADCC/TAU. 1988. *Coordinated SADCC Interconnector Studies, Summary report.* Prepared by Shawnigan/Lavalin (Can. Int. Dev. Agency), Luanda, Angola

33. Electr. Mocambique. 1990. Planning cross border power sharing. In *Proc. Specialized Sub-Committee on Energy Planning.* Nhlangano, Swaziland: SADCC/TAU

34. SADCC/TAU. 1990. *Proc. Electr. Sub-Committee Meet., Harare, Zimbabwe*

35. SADCC/TAU. 1988. *Maintenance of Mechanical Equipment for Power Stations in the SADCC Region*—Vol. 1, SADCC project 3.0.3, prepared by SWECO

36. O'Keefe, P., Munslow, B., eds. 1984. *Energy and Development in Southern Africa, SADCC Country Studies,* Part I, Vol. 3. Stockholm: Energy, Environ. Dev. Africa, Beijer Inst. 193 pp.

37. O'Keefe, P., Munslow, B., eds. 1984. *Energy and Development in Southern Africa, SADCC Country Studies,* Part II, Vol. 4, Stockholm: Energy, Environ. Dev. Africa, Beijer Inst. 227 pp.

38. SADCC/TAU. 1987. *Wood Energy Development: A Study of the SADCC Region.* ETC Found. on behalf of the SADCC Energy Sector, Netherlands

39. Leach, G., Mearns, R. 1988. *Beyond the Woodfuel Crisis—People, Land and Trees in Africa.* London: Earthscan

40. O'Keefe, P., Soussan, J., Munslow, B., Spence, D. 1989. Wood energy in eastern and southern Africa. *Annu. Rev. Energy* 14:445–68

41. O'Keefe, P., Raskin, P., Bernow, S. 1984. *Energy and Development in Kenya: Opportunities and Constraints,* Vol. 1. Stockholm: Energy, Environ. Dev. Africa, Beijer Inst. 190 pp.

42. Munslow, B., Katerere, Y., Ferf, A., O'Keefe, P. 1988. *The Fuelwood Trap: A Study of the SADCC Region.* London: Earthscan

43. Maya, R. S. 1989. New and renewable sources of energy in SADCC: situation review. In *Proc. Seminar On NRSE in the SADCC Region, Arusha, Tanzania*

44. SADCC/TAU. 1990. *New and Renewable Sources of Energy in the SADCC Region. A Summary of Resources and Activities, and a Directory of Organizations, Manufacturers and Distributors.* Luanda, Angola

45. SADCC/TAU. 1990. Strategy and programmes to promote activities in NRSE in the SADCC region. In *Proc. Specialized Sub-Committee on Woodfuel/NRSE, Maseru, Lesotho*

46. SADCC/TAU. 1990. *Complete Documentation of Energy Projects Available for Funding.* Lusaka, Zambia

47. Maunder, Britnell, Inc. 1990. CIDA/SADCC Industrial Energy Conservation Pilot Project: Mid Term Review. Draft Report. Prepared for the Canadian Int. Dev. Agency

48. In *Proc. Energy and Transport.* See Ref. 6, Working Group Recommendations

49. Raskin, P. 1982. *Issues in SADCC Energy Planning: Usage Patterns, Resource Potential, and Regional Possibilities.* Discussion Document for the First SADCC Energy Semin. Harare, Zimbabwe

50. Group of Independent Consultants (GICON) A/S. 1989. *Project Review: Norwegian Development Co-operation Programme to TAU, Energy Sector, SADCC.* Report to the Norwegian Agency for Int. Dev.

51. SADCC/TAU. 1986. *What is TAU?,* Luanda, Angola

52. SADCC/TAU. 1987. *SADCC Energy Situation.* Garborone, Botswana

53. Mrindoko, B. J. 1990. The application of computerized energy planning in Tanzania using the LEAP model. In *Proc. Specialized Sub-Committee Energy Planning.* SADCC/TAU. Nhlangano, Swaziland

Annu. Rev. Energy Environ. 1991. 16:179–203

ENERGY-INTENSITY TRENDS IN BRAZIL

Howard S. Geller

American Council for an Energy-Efficient Economy, Washington, DC 20036

David Zylbersztajn

Institute of Electrotechnology and Energy, University of São Paulo, São Paulo, Brazil

KEY WORDS: Energy use, energy and developing countries, energy policy, energy trends.

I. INTRODUCTION

Energy has had a substantial impact on the Brazilian economy during the past 20 years. Oil accounted for more than 40% of Brazil's total import bill, which was equivalent to nearly 50% of export revenues during the early 1980s (1). A negative balance of payments due in part to costly oil imports led to mounting foreign debt, which in turn severely limited economic growth and investment during the 1980s. At the beginning of the 1990s, the foreign debt stands at about $115 billion or one-third of Brazil's annual gross domestic product (GDP), and threatens to continue to drain economic resources from the country.

Investment in the energy sector also rose dramatically during the 1970s and early 1980s. At their peak, capital-intensive energy supply projects absorbed 5% of total GDP and 24% of total investment (2). Large investments in energy supply enabled industries and services to expand, but they increased foreign debt and limited spending for other purposes. Large investments in energy supply are of concern considering Brazil's "social debt"—the lack of adequate housing, health care, educational facilities, and other social services. For example, there is a shortage of at least 10 million housing units (around 30% of families are inadequately housed), illiteracy levels exceed 40% in some poorer states, 78% of children do not complete primary school, and total public spending on health care, nutrition, water, and sanitation is less than 3% of GDP (3, 4).

179

Industrialized countries, with their mature economies and high saturation of energy-intensive goods and services, have demonstrated that economic growth can be decoupled from energy growth. However, this decoupling is more difficult to achieve in developing countries where the industrial base is growing, transport and electrical services are expanding, etc. In most developing countries, growth in energy consumption has equalled or exceeded growth in economic output during the past 20 years. A superficial examination of the data places Brazil in this category (see Table 1). Population increased 40%, GDP increased 89% according to official figures, and energy consumption increased 88% during 1973–1988 (5). As discussed below, however, national energy intensity declined significantly during this period if logical adjustments to the data are made.

Energy use per unit of GDP is a simplistic measure of energy intensity. Various structural and technological factors influence this aggregate statistic. In Brazil, it appears that energy intensity has declined in a number of specific areas in recent years. On the other hand, economic development policies have caused energy intensity to rise in other areas. Given the critical role that energy plays in the economy, it is useful to analyze the overall trends and the factors affecting them.

This paper reviews national and sectoral energy-intensity trends since 1973. The reasons for changes in energy intensity are examined, including both technical and political-economic factors. Since energy use and intensity are influenced by the type of energy source, energy supply trends are reviewed as well. In the conclusion of the paper, suggestions are provided concerning how Brazil can increase energy productivity and lessen adverse impacts caused by growth in energy use in the future.

Table 1 Population, GDP, and energy consumption in 1973 and 1988

Parameter	1973	1988
Population	103.0	144.4
GDP (billion 1987 dollars)		
Official figures	154.6	292.7
Purchasing power correction[a]	330.6	679.0
Primary energy use[b]		
Official figures	3.80	7.15
Hydropower correction[c]	3.33	5.40

[a] Based on the adjustment of GDP values to reflect purchasing power parity with the dollar. The value for 1988 is an estimate since adjustment factors are only available through 1985 (see footnote 2 to text).
[b] Excludes energy sources used for industrial feedstocks.
[c] Based on the direct energy content of hydroelectricity.

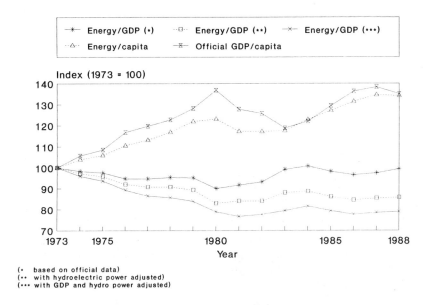

Figure 1   National energy-intensity trends

II. NATIONAL ENERGY-INTENSITY TRENDS

According to the 1989 National Energy Balance, national energy intensity (primary energy consumption per unit of GDP) declined about 1% between 1973 and 1988 (see Figure 1 and Table 2) (5). National energy intensity declined 10% during 1973–1980, a period when per capita GDP rose about 37%. Energy intensity increased, however, during the recession of 1981–1983—per capita energy use was relatively inelastic while per capita economic output sharply declined. Energy use then grew in tandem with GDP during the 1983–1988 economic recovery. The potential for reducing energy intensity when economic output is rapidly increasing (and vice versa) is consistent with trends prior to 1973. For example, during the "economic miracle" of 1968–1973 when GDP rose about 11%/yr on average, energy use rose only 65% as fast as GDP (6).

Energy-intensity values derived from the National Energy Balance are misleading for two reasons. First, all electricity is considered as if it were generated from oil in a thermal power plant using a coefficient of 13.1 GJ per MWh (i.e. assuming a 28% generating efficiency).[1] In reality, about 95% of

[1]In the Brazilian National Energy Balance, it is assumed that a tonne of oil equivalent contains 45.2 GJ of energy. In studies of other countries (e.g. by the International Energy Agency), a tonne of oil equivalent contains 41.8 GJ.

Table 2   National energy-intensity trends[a]

Year	National energy balance[b] (GJ/1000 $)		Hydropower adjustment[c] (GJ/1000 $)		Hydropower and GDP adjustment[d] (GJ/1000 $)	
1973	24.6		21.5		10.1	
1978	23.5	(− 5%)	19.5	(− 9%)	8.6	(−14%)
1983	24.4	(− 1%)	18.9	(−12%)	8.0	(−21%)
1988	24.5	(− 1%)	18.4	(−14%)	8.0	(−21%)

[a] Primary energy consumption per unit of gross domestic product. GDP is given in 1987 dollars. Values in parentheses are the changes relative to 1973.
[b] Hydroelectricity counted in terms of the equivalent fuel input to a thermal power plant; official GDP figures.
[c] Based on the direct energy content of hydroelectric supplies.
[d] Based on the direct energy content of hydroelectric supplies and adjustment of GDP values to reflect purchasing power parity with the dollar. Value for 1988 is an estimate.

electricity consumed in Brazil is hydropower, which has a direct energy content of 3.6 GJ per MWh generated. With hydropower counted this way, primary energy use equalled 5.40 EJ in 1988, rather than 7.16 EJ as reported in the National Energy Balance.

Figure 1 and Table 2 include the trends in primary energy intensity with hydropower valued in terms of its direct energy content. With this adjustment, national energy intensity declined 14% during 1973–1988. All of the net improvement occurred during the 1970s. The decline is related to the growing importance of electricity (i.e. hydropower) in the overall energy mix. Consumption of electricity increased by nearly 260%, while consumption of all other forms of energy increased only 50% during this 15-year period. For reference, population increased 2.3%/yr on average during this period, reaching 144 million by 1988.

The second problem with energy-intensity values derived from the National Energy Balance is that GDP is given in constant US dollars using one particular exchange rate. However, the official GDP trends do not reflect "purchasing power parity" with the dollar, i.e. they do not use exchange rates that maintain equal buying value over time. Researchers have calculated adjusted GDP levels for various countries maintaining purchasing power parity with the dollar.[2] With this adjustment, Brazil's GDP increased by a factor of 1.85 during 1973–1985, rather than by a factor of 1.70 as the official data indicate.

[2] The adjusted GDP values were taken from a database compiled by R. Summers and A. Heston of the University of Pennsylvania. The data are available only through 1985. Values for 1986–1988 are estimates based on the increases in GDP reported by the government of Brazil. The Summers and Heston data were provided by the Worldwatch Institute, Washington, DC, Sept. 1990.

With both the hydropower and the GDP adjustments, primary energy intensity declined by about 21% during 1973–1988. Once again, all of the improvement occurred during the 1970s. National energy intensity measured in this manner remained nearly constant during 1980–1988 (see Figure 1). In addition, these adjustments lead to a 60–70% reduction in the absolute magnitude of primary energy intensity. Most of this difference is due to a substantial increase in GDP in any particular year when it is adjusted to reflect purchasing power parity.

In the remainder of this paper, primary energy consumption and energy intensities are presented and discussed with both the hydropower and purchasing power parity adjustments taken into account. Most tables contain both unadjusted and adjusted data, but the text refers only to the latter.

III. ENERGY SUPPLY TRENDS

Tables 3 and 4 show primary energy supply in 1973 and 1988 disaggregated by sector and energy source. As of 1988, the industrial and transport sectors together account for about 65% of overall energy use. Energy supply in 1988 was relatively diversified with four sources (petroleum, wood, sugar cane, and electricity) each providing 15% or more of the total. Major uncertainties in the energy supply data are discussed in Section VI.

Table 5 shows the changes in the composition of primary energy supply during 1973–1988. It is estimated that the fraction of total energy supply provided by wood dropped from 36% in 1973 to 15% in 1988. At the same time, the fraction of consumption provided by sugar cane products (i.e.

Table 3 Primary energy consumption (PJ) by source and sector, 1973[a]

| Source | Sector | | | | | | |
	Residential	Service	Industrial	Transport	Energy	Agriculture	Total
Wood[b]	830	9	171	—	—	195	1206
Charcoal[b]	25	2	71	—	—	1	98
Sugar cane products	—	—	179	7	6	—	192
Electricity[c]	52	59	139	3	12	2	267
Petroleum products	100	19	407	830	88	45	1489
Mineral coal	—	—	59	—	—	—	59
Natural gas	—	—	1	—	1	—	2
Other	—	—	11	—	4	—	15
Total	1006	88	1038	842	111	243	3328

[a] Based on the direct energy content of hydroelectric supplies, not on the values in the National Energy Balance.
[b] Data on wood and charcoal are estimates contained in the National Energy Balance.
[c] The fraction of electricity provided by hydropower was 93% in 1973.

Table 4 Primary energy consumption (PJ) by source and sector, 1988[a]

| Source | Sector | | | | | | Total |
	Residential	Service	Industrial	Transport	Energy	Agriculture	
Wood[b]	426	6	260	—	—	114	806
Charcoal[b]	33	2	259	—	—	—	296
Sugar cane products	—	—	218	265	315	—	798
Electricity[c]	182	170	502	5	22	28	909
Petroleum products	216	34	353	1101	155	154	2013
Mineral coal	—	—	342	—	—	—	342
Natural gas	—	—	64	—	38	—	102
Other	—	—	122	—	15	—	137
Total	857	213	2120	1371	544	296	5403

[a] Based on the direct energy content of hydroelectric supplies, not on the values in the National Energy Balance.
[b] Data on wood and charcoal are estimates contained in the National Energy Balance.
[c] The fraction of electricity provided by hydropower was 96% in 1988.

ethanol and bagasse) and hydropower increased from 13% to 30%. Thus the total share provided by renewable sources remained nearly constant between 1973 and 1988 at approximately half of total energy use.[3] This renewables fraction is very high considering that Brazil is a highly urbanized, wealthier developing nation.

Petroleum is still the primary energy source in Brazil, although its relative importance declined significantly during the past decade. Absolute consumption of gasoline declined by 46% and absolute consumption of industrial fuel oil declined by 10% during 1973–1988. Consumption of diesel, liquified petroleum gas (LPG), and nafta increased, on the other hand. Petrobras (the state-owned oil company) was not able to adjust refinery outputs in order to keep up with the changes in product requirements. Thus about 35% of gasoline production was exported and 20% of LPG consumption was imported in recent years (5).

Dependence on net energy imports (imports minus exports) increased from 32% of total energy consumption in 1973 to a high of 37% in 1979. Import dependence then declined to 16% in 1985, but rose again during the late 1980s, reaching 22% in 1988 (5). Expanding production and use of sugar cane products, hydropower, and domestic petroleum contributed to the overall increase in self-sufficiency during this period, as did reductions in energy intensity in certain areas. The upturn in import dependence in the late 1980s was caused mainly by stagnating domestic petroleum production.

[3]If hydropower is counted based on the oil required to generate an equivalent amount of electricity (i.e. following the convention in the National Energy Balance), then hydropower contributed 36% and all renewable sources contributed 63% of total energy use in 1988, while hydropower contributed 18% and all renewable sources contributed 58% in 1973.

Table 5 Primary energy supply trends by source[a]

Source	Fraction of annual energy consumption (%)			
	1973	1978	1983	1988
Wood[b]	36	25	20	15
Charcoal[b]	3	3	5	5
Sugar cane products	6	7	13	15
Hydroelectricity	7	9	13	15
Other renewable	0	0	1	1
Total renewable	(52)	(44)	(52)	(51)
Petroleum products	46	51	41	39
Mineral coal	2	3	5	6
Other nonrenewable	1	2	2	4
Total nonrenewable	(49)	(56)	(48)	(49)
Total energy (PJ)	3328	4167	4384	5403

[a] Based on the direct energy content of hydroelectric supplies, not on the values in the National Energy Balance.
[b] Data on wood and charcoal are estimates contained in the National Energy Balance.

While the fraction of energy supplied by fossil fuels remained roughly constant, total energy consumption increased substantially during 1973–1988. Thus, carbon emissions (the principal cause of the greenhouse effect and global warming) increased as well. From fossil fuels only, carbon emissions increased by an average of 4.1% per year, from about 37 million tonnes in 1973 to 67 million tonnes in 1988. If wood and charcoal produced without reforestation (i.e. biofuels unsustainably harvested) are included along with fossil fuels, then carbon emissions grew from around 64 million tonnes in 1973 to 95 million tonnes in 1988.[4] In recent years, energy use was a much smaller source of carbon emissions than was deforestation related to land clearing in Brazil. For example, carbon emissions from deforestation were estimated to equal about 340 million tonnes in 1987 (7a).

IV. SECTORAL ENERGY INTENSITY TRENDS

A. *Residential Sector*

In 1988, the residential sector accounted for 16% of primary energy consumption. During 1973–1988, with hydropower counted directly, total residential energy consumption declined 15% and residential energy use per capita fell

[4]Wood and charcoal in theory are renewable energy sources. In reality, replanting and sustainable harvesting are not widely practiced. For example, about 78% of the charcoal produced in Brazil is from native forests, although the amount of charcoal produced from forest plantations is rising (7).

Table 6 Residential sector energy-intensity trends[a]

Year	Population (millions)	National energy balance[b] (GJ/capita)	Hydropower adjustment[c] (GJ/capita)
1973	103.0	10.67	9.76
1978	115.9	9.54 (−11%)	8.14 (−17%)
1983	129.8	8.59 (−19%)	6.60 (−32%)
1988	144.4	8.36 (−22%)	5.92 (−39%)

[a] Primary energy consumption per unit of gross domestic product. GDP is given in 1987 dollars. Values in parentheses are the changes relative to 1973.
[b] Hydroelectricity counted in terms of the equivalent fuel input to a thermal power plant; official GDP figures.
[c] Based on the direct energy content of hydroelectric supplies.

39% (see Table 6). This reduction was caused primarily by the shift from wood, a very inefficient source of energy as used in households, to electricity and bottled gas—modern and more efficient energy sources. Wood accounted for 82% of household energy use in 1973 but only 50% of the total in 1988. Electricity and LPG increased their combined share from 12% to 45% of the residential total during this period.

Even though residential energy use declined, energy services greatly expanded. About 80% of households were electrified in 1985 compared to 47% in 1970, and about 62% of households possessed refrigerators in 1985 compared to only about 26% in 1970 (8). Similarly, the use of pipeline or bottled gas for cooking increased from 43% of households in 1970 to 78% of households in 1985. Moreover, the shift to higher-quality fuel sources is estimated to have increased the average efficiency of energy utilization in the residential sector from 11% in 1970 to 26% in 1987.[5]

The average household in Brazil consumed 1260 kWh in 1988 (1500 kWh considering only electrified households). However, electricity use is directly related to household income, with the wealthiest income class consuming about six times as much electricity as the poorest income class (8). Nearly 50% of electrified households consume less than 1200 kWh per year, but these same households account for only about 16% of total residential electricity use.

B. Services Sector

The services sector (defined as the commercial and public sectors) accounted for only about 4% of national energy use in 1988 with hydropower directly counted. Electricity dominates in this sector, representing 76% of total energy

[5] Utilization efficiency is the useful energy obtained per unit of energy consumed (See Ref. 5, p. 19).

Table 7   Service sector energy-intensity trends[a]

Year	National energy balance[b] (GJ/1000 $)		Hydropower adjustment[c] (GJ/1000 $)		Hydropower and GDP adjustment[d] (GJ/1000 $)	
1973	2.82		1.29		0.60	
1978	3.05	(+ 8%)	1.30	(+ 1%)	0.57	(− 4%)
1983	3.80	(+34%)	1.48	(+15%)	0.62	(+ 4%)
1988	3.84	(+36%)	1.51	(+18%)	0.65	(+ 8%)

[a] Primary consumption in the commercial and public sectors per unit of gross domestic product produced by these sectors. GDP is given in 1987 dollars. Values in parentheses are the changes relative to 1973.
[b] Hydroelectricity counted in terms of the equivalent fuel input to a thermal power plant; official GDP figures.
[c] Based on the direct energy content of hydroelectric supplies.
[d] Based on the direct energy content of hydroelectric supplies and adjustment of GDP values to reflect purchasing power parity with the dollar. Value for 1988 is an estimate.

demand as of 1988. With the hydro adjustment alone, energy consumption per unit of services GDP[6] rose 18% during 1973–1988 (see Table 7). With both the hydropower and GDP adjustments, the overall rise in energy intensity during 1973–1988 was limited to about 8%.

Energy intensity rose in spite of substitution away from wood and charcoal, which accounted for 5% of sectoral energy use in 1988 compared to 11% in 1973. Rising energy intensity is attributed to growth in energy services such as air conditioning and uniform artificial illumination in modern commercial buildings. A survey of electricity use in commercial buildings in metropolitan São Paulo shows that relatively new, large office buildings and shopping centers typically consume more than 200 kWh/yr per square meter of floor area. In contrast, older, medium-sized establishments typically consume 100 kWh/yr per square meter or less, and small establishments often use less than 30 kWh/yr per square meter (9). There are no efficiency requirements for new buildings in Brazil. The general trend has been to construct glass and concrete "boxes" that rely less on natural lighting, task lighting, and natural ventilation than do older buildings.

There is considerable electricity waste in commercial buildings in Brazil. For example, one study of four large office buildings occupied by the national telecommunications company found that three of the four buildings were heavily overlit. The study estimated that total electricity use in the three wasteful buildings could be reduced by 22–31% simply by removing unnecessary light fixtures and reducing the operating hours of lighting and air

[6] The fraction of GDP attributed to the services sector increased from 48% in 1973 to 52% in 1988.

conditioning systems (10). Use of energy-efficient lamps, ballasts, fixtures, and controls could provide even greater savings.

C. Transportation Sector

Transportation accounted for 25% of national energy use in 1988. During 1973–1988, transport energy use per unit of total GDP fell 21% with economic output adjusted to maintain purchasing power parity with the dollar (see Table 8). Furthermore, transport services as measured by passenger-kilometers of personal travel and tonne-kilometers of freight transport increased faster than GDP during this period. Thus the direct fuel intensity of transport fell even more than 21%. Reductions in transport energy intensity were caused by vehicle efficiency improvements, shifts to more efficient fuels, and shifts to more efficient transport modes.

Personal travel over roads (i.e. use of cars, taxis, and buses) accounts for about 94% of total personal travel in terms of passenger-kilometers as estimated by the National Transportation Planning Department (11). Brazil had about 70 automobiles per 1000 inhabitants in 1985, slightly less than the world average and about 20% of the level in Western Europe. Table 9 shows the fuel use, vehicle fleet, and fuel-intensity trends for personal transport as best as can be determined (1985 is the last year for which data are available). Energy use refers to automotive gasoline, ethanol, and an estimate of the diesel fuel used by buses. According to these estimates, fuel use per vehicle declined approximately 47% and fuel use per passenger-kilometer declined approximately 54% during 1973–1985. However, the passenger-kilometer figures should be viewed as approximate since they are estimated by federal officials. In spite of this uncertainty, it is likely that the energy intensity of personal travel declined substantially.

Table 8   Transportation energy-intensity trends[a]

Year	National energy balance[b] (GJ/1000 $)		GDP adjustment[c] (GJ/1000 $)	
1973	5.47		2.56	
1978	5.29	(− 3%)	2.35	(− 8%)
1983	4.88	(−11%)	2.07	(−19%)
1988	4.70	(−14%)	2.03	(−21%)

[a] Transport energy consumption per unit of gross domestic product. GDP is given in 1987 dollars. Values in parentheses are the changes relative to 1973.
[b] Hydroelectricity counted in terms of the equivalent fuel input to a thermal power plant; official GDP figures.
[c] Based on the adjustment of GDP values to reflect parity purchasing power with the dollar. Value for 1988 is an estimate.

Table 9 Fuel use, vehicle, and energy-intensity trends for passenger travel[a]

	Fuel use[b] (PJ)			Vehicles[c]	Passenger-kilometers[c]	Fuel intensity	
Year	Gasoline	Ethanol	Diesel	Total	(10^6)	(10^9)	(GJ per vehicle)	(GJ per 10^9 P-km)
1973	348	7	56	411	3.63	161.9	113	2.54
1978	368	35	100	504	6.84	314.7	74	1.60
1983	272	118	108	498	8.89	461.4	56	1.08
1985	245	186	120	551	9.66	499.4	57	1.10

[a] Based on travel in automobiles, taxis, and buses.
[b] It is assumed that trucks accounted for 25% of gasoline use in 1973, 20% in 1978, 10% in 1983, and 8% in 1985. Also, it is assumed that buses accounted for 22% of road diesel use in all years.
[c] Based on vehicle registrations and estimates of vehicle use made by the National Transportation Planning Department.
Sources: (5, 11, 17).

The expansion of ethanol fuel use during this period contributed to rising energy efficiency. Ethanol is about 43% more efficient (in terms of vehicle-kilometers per GJ of fuel) than pure gasoline when the ethanol is used in a gasoline-ethanol blend. Ethanol is about 20% more efficient when used in its pure form.[7] In effect, the shift from gasoline to higher-quality ethanol enabled automobile manufacturers to improve the efficiency of cars sold in Brazil. Manufacturers were able to increase the compression ratio of auto engines once ethanol was introduced on a large scale, owing to the higher octane number of ethanol compared to gasoline. One study estimates that the average fuel economy of new gasoline cars[8] increased about 50% between 1973 and 1983 (13).

By 1988, ethanol surpassed gasoline in terms of contribution to transport energy use in Brazil (5). More than 80% of the ethanol was used in a pure form and the remainder was mixed with gasoline. Also, gasoline accounted for only 12% of total petroleum products consumed in 1988.

Dynomometer tests of the fuel economy of all new car models began in Brazil in 1979. Figure 2 shows the trends in the sales-weighted average fuel economy (km per liter) of new ethanol and gasoline cars during 1983–1989. Ethanol cars represented about 90% of cars sold in Brazil during this period. Average fuel economies increased about 10% during 1983–1989, with most of the gains during the early years. Because of the higher calorific value of

[7] A liter of ethanol has about 30% less calorific value than a liter of gasoline. Use of ethanol fuel in light vehicles in Brazil has shown that one liter of ethanol can replace one liter of gasoline in a blend of about 80% gasoline, 20% ethanol. When pure ethanol is used, about 1.25 liters of ethanol replace one liter of gasoline (12).
[8] Gasoline cars refer to cars that use either pure gasoline or a gasoline-ethanol blend since these fuels are interchangeable. Beginning around 1980, vehicle manufacturers began producing both gasoline and ethanol cars. The latter can only burn pure ethanol.

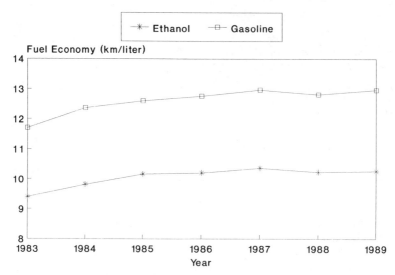

Figure 2   Average fuel economy of new cars

gasoline relative to ethanol, the fuel economy (in terms of km/l) of gasoline cars is about 25% greater than that of ethanol cars.

The fuel intensity of freight transport declined approximately 35% during 1973–1985 (see Table 10). One factor that caused this reduction was the shift to more efficient transport modes. The fraction of total tonne-kilometers moved by trucks declined from about 69% in 1973 to 54% in 1985, the fraction moved by train increased from 18% to 24%, and the fraction moved by boat increased from 10% to 18%. Compared to road transport, cargo ships consume about 9% as much energy per tonne-kilometer and trains consume about 18% as much energy per tonne-kilometer on average (14).

The shift from gasoline- to diesel-fueled trucks also reduced overall fuel intensity since trucks with diesel engines are more fuel-efficient than trucks that burn gasoline. The fraction of diesel-fueled trucks increased from only about 50% of the truck fleet in 1973 to more than 85% of the fleet by 1985 (11). In addition, the fraction of semi-heavy and heavy trucks increased from about 15% of the truck fleet in 1973 to 28% of the truck fleet by 1985. At full capacity, heavier trucks are less energy intensive than light or medium trucks in terms of energy use per tonne-kilometer.

D. *Industrial Sector*

The industrial sector (excluding energy production) accounted for 39% of total energy use in Brazil in 1988. The overall energy intensity of the industrial sector rose substantially during the past 15 years. With the hydro-

Table 10 Fuel use, quantity transported, and energy-intensity trends for freight transport[a]

Year	Fuel use[b] (PJ)					Freight carried[c] (billion tonne-km)	Fuel intensity (GJ per 10^9 T-km)
	Gasoline	Diesel	Fuel Oil	Electricity	Total		
1973	116	231	25	2	374	170.0	2.20
1978	92	399	47	3	541	286.8	1.88
1983	30	439	51	4	524	342.1	1.53
1985	21	485	67	4	577	406.1	1.42

[a] Based on freight transport by truck, train, and boat; excluding air and pipeline transport, which accounted for about 4% of total freight transport in recent years.
[b] It is assumed that trucks accounted for 25% of gasoline use in 1973, 20% in 1978, 10% in 1983, and 8% in 1985. Also, it is assumed that all forms of freight transport accounted for 78% of road diesel use in all years and that 90% of diesel, fuel oil, and electricity use by trains and boats is for freight transport. Electricity is based on the direct energy content of hydropower. The small amounts of coal and wood used for freight transport are excluded.
[c] Based on data collected by the National Transportation Planning Department.
Sources: (5, 11, 14, 17).

power adjustment alone, industrial energy use per dollar of industrial output[9] increased 25% during 1973–1988 (see Table 11). With both the GDP and the hydropower adjustments, the overall increase in industrial energy intensity was about 15%.

Rising industrial energy intensity was caused primarily by structural shifts, not by declining energy efficiencies. Rapid growth in the production of energy-intensive metals (i.e. aluminum, iron and steel, and steel alloys) strongly influenced energy-intensity trends in this sector. Based on physical output, production of aluminum, raw steel, and steel alloys increased 450%, 245%, and 430%, respectively, during 1973–1988. For comparison, total industrial production (based on monetary value) increased 63% during this period (5). In 1988, the primary metals industries were more than four times more energy intensive (in terms of GJ per dollar of output) than the industrial sector as a whole (5). Also, the average energy intensity within the primary metals industries more than doubled between 1973 and 1988 due to strong growth by the most energy-intensive segments of the industry (i.e. production of primary aluminum and steel alloys). Because of these trends, the increase in production of primary metals between 1973 and 1988 contributed 50% of the growth in total industrial energy use but only 10% of the growth in industrial product.

Much of the growth in metals production was destined for export. If export of primary metals had not occurred (i.e. if production equalled domestic

[9] The fraction of GDP attributed to the industrial sector declined from 38% in 1973 to 33% in 1988.

Table 11 Industrial sector energy-intensity trends[a]

Year	National energy balance[b] (GJ/1000 \$)	Hydropower adjustment[c] (GJ/1000 \$)	Hydropower and GDP adjustment[d] (GJ/1000 \$)
1973	22.0	17.8	8.3
1978	24.7 (+12%)	18.9 (+ 7%	8.4 (+ 1%)
1983	29.8 (+35%)	21.1 (+19%)	8.9 (+ 7%)
1988	32.3 (+47%)	22.1 (+25%)	9.5 (+15%)

[a] Industrial energy consumption per unit of gross domestic product provided by the industrial sector (excluding energy production). GDP is given in 1987 dollars. Values in parentheses are the changes relative to 1973.

[b] Hydroelectricity counted in terms of the equivalent fuel input to a thermal power plant; official GDP figures.

[c] Based on the direct energy content of hydroelectric supplies.

[d] Based on the direct energy content of hydroelectric supplies and adjustment of GDP values to reflect purchasing power parity with the dollar. Value for 1988 is an estimate.

consumption as of 1988), industrial energy consumption would have been 21% lower and national energy consumption about 8% lower (15, 16). However, export of primary metals generated about \$4.7 billion in revenue in 1988, representing 14% of export revenue, nearly 5% of industrial output, and about 1.5% of gross domestic product. Thus without primary metals export, industrial energy intensity would have been 17% lower and national energy intensity 7% lower than the actual values in 1988.

Trends in energy consumption per tonne of output within individual industries show that substantial efficiency improvements occurred in a number of important areas (see Table 12). Average energy intensity declined 28%, 26%, and 17% in the cement, nonferrous metals, and paper and pulp industries, respectively (5). Industrial energy efficiency improvements resulted from waste reduction, installation of new, more efficient capacity, process modifications, efficiency gains in some cases due to increasing electrification, and retrofits such as installation of heat recovery equipment or controls. The technological factors affecting the trends in individual industries are briefly discussed below.

1. CEMENT Cement manufacturers increased the fraction of cement produced using the dry process from about 20% to 80% between 1970 and 1987 (17). In Brazil as in other countries, the dry process requires less energy per tonne compared to the older wet process. Also, the average size of cement-producing furnaces increased and the market share for cements to which active ingredients were added increased (18). Both of these technical changes reduced average energy intensity as well.

Table 12   Energy-intensity trends within specific industries

Industry	National energy balance[a] (GJ/tonne)			Hydro adjustment[b] (GJ/tonne)		
	1973	1988	Change	1973	1988	Change
Cement	6.9	5.2	(−24%)	5.9	4.2	(−28%)
Iron and steel	31.6	30.8	(− 3%)	27.9	25.9	(− 7%)
Steel alloys	95.0	116.4	(+23%)	48.1	56.7	(+18%)
Nonferrous metals	310.8	236.0	(−24%)	134.1	98.9	(−26%)
Chemicals	12.8	19.4	(+51%)	9.1	12.3	(+35%)
Textile	40.2	55.3	(+38%)	27.3	28.6	(+ 5%)
Paper and pulp	30.0	27.0	(−10%)	23.5	19.5	(−17%)

[a] Hydroelectricity counted in terms of the equivalent fuel input to a thermal power plant.
[b] Based on the direct energy content of hydroelectric supplies.

2. IRON AND STEEL   The overall energy intensity of the iron and steel industry declined about 7% during 1973–1988. However, a reduction of around 20% was registered between the year of maximum energy intensity (1975) and 1988. This improvement resulted from modernization of older capacity as well as introduction of more efficient new capacity. In particular, inefficient open hearth furnaces represented less than 1% of steel production in 1988 compared to 36% of total production in 1973 (19). Also, the use of energy-efficient continuous casting increased from 12% to 49% of total production during 1973–1988. However, use of purchased scrap (i.e. recycling) declined as a fraction of total steel production during this period, due to the growth of steel exports.

3. STEEL ALLOYS   Production of steel alloys (e.g. silicon steel), an electricity-intensive process, increased more than 400% during 1973–1988 (5). Furthermore, the average energy intensity of the steel alloys industry increased about 18% during this time period. Increasing average energy intensity was due to very rapid growth in the most energy-intensive segments of the industry. Alloys that required more than 8.0 MWh/tonne grew from 15% of total production in 1976 to 36% of total production in 1986 (20). However, electric arc furnaces used to produce steel alloys in Brazil are modern and are equivalent in efficiency to those used in industrialized countries (20).

4. NONFERROUS METALS   The nonferrous metals industry is dominated by aluminum production, which increased by a factor of three between 1981 and 1987 alone (15, 16). The energy efficiency of aluminum production in Brazil is relatively high because much of the capacity is new and large-scale. It is reported that newer production facilities consume around 15.5 MWh/tonne, about 8% less than the world average (21). The significant reduction in

average energy intensity in this sector during 1973–1988 appears to be due to improvements in the average efficiency of aluminum production as well as changes in product mix within the sector.

5. CHEMICALS   The overall energy intensity of the chemicals industry increased substantially during 1973–1988. Again this was primarily due to structural factors, i.e. high growth in more energy-intensive segments of the industry. In particular, production of chlorine and caustic soda increased about 345% and production of ammonia-based fertilizers increased nearly 400% during 1973–1988. For comparison, production of all chemicals only increased about 90% during this time period. Chlor-soda production is around three times as energy intensive and ammonia-based fertilizers are nearly four times as energy intensive as chemicals production as a whole.

Nonetheless, efficiency improvements did occur within the chemicals industry. For example, one large producer of chlorine and soda reports reducing its average energy intensity 20% during 1979–1988 (22). Also, the energy intensity of ammonia production using natural gas as the feedstock was 36 GJ/tonne as of 1983, about 15% less than the average energy intensity of ammonia production in North America, according to Brazilian researchers (23).

6. PAPER AND PULP   Rapid expansion and the installation of new, efficient paper mills contributed to the reduction in the energy intensity of the paper and pulp industry during 1973–1988. Also, energy conservation measures such as better insulation, better process control, and heat recovery boilers were installed in existing mills, and recycling expanded to the point where used paper accounted for 30% of total fiber consumption as of 1988 (24). In addition, the use of waste materials (e.g. black liquor) increased from 17% to 37% of energy consumption during this 15-year period (5).

V. ROLE OF POLICIES

A wide range of energy and economic policies influenced the energy-intensity trends discussed above. Some of these policies led to reductions in energy intensity, others increased energy intensity.

A. Policies that Decreased Energy Intensity

1. PROMOTION OF MODERN ENERGY SOURCES   The rapid growth in electricity and bottled gas use in households was encouraged by various government policies. First, the homes of the poor were connected to the electricity grid in rural as well as urban areas. Second, the prices of electricity and LPG were subsidized for residential customers, and the prices were

substantially reduced in real terms during the past 20 years (see Figure 3). In addition, households that use small amounts of electricity pay much less per kWh than high-consumption households. As indicated above, the substitution of modern energy sources for wood enabled energy services to expand greatly without increasing total energy use in the residential sector.

2. THE NATIONAL ETHANOL FUEL PROGRAM (PROALCOOL)   In 1975, the federal government created the ethanol fuel program in order to reduce oil imports and to support the sugar industry. Ethanol producers were given highly subsidized financing until the early 1980s and a guaranteed market for authorized ethanol production. In the first phase of the program, gasoline was mixed with up to 22% anhydrous ethanol. In the second phase of the program, hydrated ethanol was produced and used in cars modified to operate solely on this fuel. The relative prices of ethanol and the gasoline-ethanol blend were maintained so that it was less costly to operate an ethanol car. Also, ethanol cars had a lower sales tax. Production and use of ethanol rapidly increased, enabling Brazil to avoid about $10 billion in oil imports during 1975–1989. As mentioned above, increasing use of ethanol fuel led directly to vehicle efficiency improvements.

3. RELATIVE PRICING OF LIQUID FUELS   In addition to pricing pure ethanol and gasoline in order to encourage ethanol use, the government substantially increased the gasoline-to-diesel price ratio during the 1970s (see Figure 3).

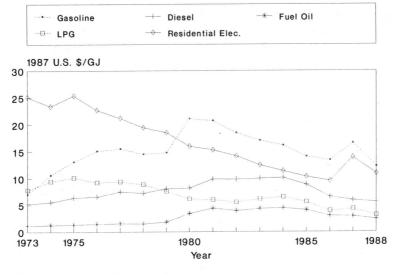

Figure 3   Average retail energy prices

This made it cost-effective to convert trucks and buses from gasoline to diesel engines. The latter are more energy-efficient than the former. Also, virtually all heavy trucks and about one-third of light-duty trucks sold beginning in the late 1970s had diesel engines (25).

4. ENERGY CONSERVATION POLICIES AND PROGRAMS   Specific energy conservation programs were not begun until a National Energy Commission was created at the time of the second oil price shock. In the early 1980s, a program was undertaken to reduce oil consumption by industries. This effort included company-specific supply quotas for fuel oil, protocols between the government and major industry associations that committed companies to reduce oil consumption, and low-interest financing for conservation and oil substitution projects. The effort was quite successful—the cement, paper, and steel industries reduced their consumption of petroleum products by 95%, 70%, and 50%, respectively, between 1980 and 1985 (5). Much of this savings resulted from fuel switching, but waste reduction and process efficiency improvements also occurred (e.g. in the cement and paper industries).

As part of the program for economizing fuel (PECO), a voluntary protocol between the government and automobile producers was signed that called for a 20% improvement in the average fuel economy of new cars between 1983 and 1987. Only one car manufacturer (Ford) achieved this goal; the other three major manufacturers realized overall improvements of only 3–8%. There were no penalties for failing to achieve the 20% efficiency improvement.

At the end of 1985, a national electricity conservation program (PROCEL) was started. This program supports technology development and demonstration, education, promotion, direct installation of efficiency measures, legislation, and incentives. With support from PROCEL, private companies started marketing a wide range of electricity conservation measures in the late 1980s. It is estimated that PROCEL reduced national electricity use in 1989 by approximately 0.5–1.2% (26). Much larger savings targets have been established for 1990–2010.

B. Policies that Increased Energy Intensity

1. ECONOMIC DEVELOPMENT STRATEGIES DURING THE 1970s AND 1980s   The rapid economic growth that occurred in Brazil during the late 1960s and early 1970s was dominated by production of durable goods such as cars and appliances. Intermediate goods such as metals, chemicals, and petroleum were still heavily imported. Domestic aluminum production, for example, equalled about 60% of consumption during this period (see Figure 4). Steel production did not keep up with consumption and the balance of

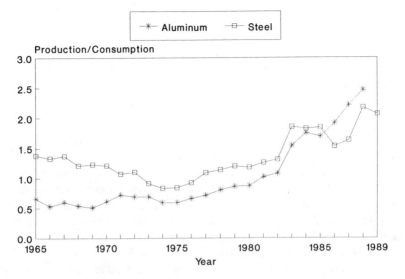

Figure 4   Self-sufficiency in basic metals

trade in this product became negative starting around 1973. Rapid growth of basic materials imports led to balance of payments problems and a growing foreign debt even prior to the 1973 oil price shock.

In 1974, the federal government decided to give greater priority to production of energy-intensive intermediate goods (e.g. steel, aluminum, and chemicals). A national development plan spelled out this intention, and a variety of fiscal and tax incentives were provided to private and state-owned companies. It is estimated that the financial subsidies provided by the federal government to export industries were equivalent to about 60% of the value of exports during 1969–1988 (27). This overall development strategy eventually produced large trade surpluses, but it also increased energy consumption and the energy intensity of the economy during the 1980s.

The government particularly encouraged production and export of minerals and primary metals. Roads were constructed in the Amazon basin, mineral surveys were carried out, mining and mineral processing companies received financial and technical assistance, and mineral-rich lands were leased. Also, a large hydroelectric power plant (the Turcurui plant) was constructed in the north region during 1976–1985. Most of the electricity from this plant is provided to aluminum smelters at highly subsidized prices (28, 29). These policies significantly increased metals production and export (see Figure 4). The fraction of aluminum, raw steel, and steel alloys that were exported reached 59%, 57%, and 43%, respectively, in 1988 (15, 16).

The second oil price shock and corresponding jump in international interest rates triggered the debt crisis of the early 1980s. Also, inflation rose sharply in 1979–1980. The federal government responded to these problems by tightening monetary policy, devaluing the currency, and adopting other policies that contributed to the recession of 1981–1983. As shown in Figure 1 and discussed above, overall energy intensity increased in the early 1980s when GDP declined.

2. PRICING POLICIES DURING THE 1980s   With inflation racing out of control, energy prices were frequently used by the government as an instrument for fighting inflation. The prices of gasoline, diesel fuel, LPG, and electricity for residential consumers declined substantially in real terms during 1980–1988 (see Figure 3). Besides discouraging energy conservation, these price reductions created a financial crisis for the energy companies (particularly state-owned electric utilities). In the mid-1980s, revenues from electric tariffs were insufficient to cover utility operating expenses and debt service requirements, let alone finance new investments (30).

Energy pricing also was used to promote substitution among energy sources. As mentioned above, the relative prices of liquid fuels helped to decrease energy intensity in the transport sector and subsidized LPG helped to reduce wood use in the residential sector. In addition, energy prices were used to promote use of electricity and substitution of electricity for liquid fuels. A special electricity tariff was offered to industrial customers who installed electric boilers and furnaces during 1981–1986 in order to make use of excess hydroelectric generating capacity (the electrotermia program). By 1986, this program accounted for 13% of industrial electricity use (31). The discounts to all industrial customers were stopped at the end of 1986 when surplus power was no longer available, but many industries continued using new electrothermal equipment. Promoting the adoption of electric boilers and furnaces reduced final energy use if hydropower is counted based on its direct energy content. But it increased national energy use and energy intensity if all electricity is counted based on the fuel input to a thermal power plant.

3. INSTITUTIONAL ARRANGEMENTS IN THE ENERGY SECTOR   Large government-owned institutions dominate the energy sector in Brazil. Petrobras maintains a monopoly on petroleum exploration, production, and transport, while Electrobras, its subsidiaries, and large state-owned utilities dominate electricity generation and supply. Thus energy planning and decision making is highly centralized, serving the interests of these companies and their contractors. The overall energy efficiency and productivity of the country is not always of primary concern.

In the electricity sector, utilities have taken a strong interest in constructing hydroelectric plants. Other efficient sources of power, such as industrial cogeneration, have been neglected and discouraged. Substantial cogeneration potential exists in the alcohol, chemicals, metals, and paper industries at potentially attractive electricity prices (32). But cogeneration facilities produced less than 3% of total electricity supplies as of 1986 (17). All of this power was consumed on-site. In effect, utilities have prevented industries from providing power to the grid by not offering reasonable prices and by requiring burdensome interconnection procedures (32).

4. PRIORITIES IN THE TRANSPORT SECTOR   Transportation policies in Brazil have emphasized road transport and use of automobiles. For example, the rate of growth of the automobile fleet was about twice the rate of growth of the bus fleet during 1973–1985 (11). Urban buses and other mass transit systems are heavily overloaded and service is deteriorating in many cities (33). This situation is related to low fares and limited government funding of mass transport systems. Thus, even though energy use per passenger-kilometer declined substantially during 1973–1985, further gains would have been possible had public transport not deteriorated.

Regarding transport modes, road construction has received the largest portion of federal transportation funds. With a few exceptions, the train and shipping systems are plagued with problems such as delays in loading and unloading due to inadequate transfer facilities or handling procedures (14). Only a small portion of transport funding (3% during 1986–1989) has gone into improving the productivity of existing transport systems. As a result, the fraction of freight transported inefficiently by trucks is relatively high. Moreover, it is predicted that the fraction of freight carried by trucks will increase during 1990–2000 unless other transport modes are improved (14).

5. LACK OF MINIMUM EFFICIENCY REQUIREMENTS   Brazil has not adopted minimum efficiency requirements for new buildings, appliances, cars, or other major energy-using equipment. Experience in industrialized countries has shown that such requirements can significantly reduce energy use in a cost-effective manner. For example, the average fuel efficiency of new cars produced in the United States doubled between 1973 and 1985 due in large part to minimum efficiency standards (34). The lack of minimum efficiency requirements in Brazil has allowed wasteful practices such as overlighting of commercial buildings and production of inefficient cars, buildings, motors, appliances, and other equipment to continue on a wide scale.

VI. UNCERTAINTIES IN THE DATA

We have used the best data available in preparing this report. While we believe that the results are reasonably accurate and conclusions valid, we must acknowledge uncertainties in a number of areas.

First, the data on wood and charcoal use in the National Energy Balance are estimates. The data were derived from questions about fuel use in the national census conducted in 1970 and 1980, as well as more frequent industrial surveys and residential surveys in particular locales. In the case of wood, there is uncertainty in both the amount consumed and average calorific value.

Second, data on the amount of passenger vehicle use (e.g. passenger-kilometers) are estimates made by the federal transportation planning department. Vehicle stock data, on the other hand, are reasonably accurate since they are based on mandatory vehicle registrations. Also, freight transport data should be reasonably accurate since they are compiled from shipping records and reports. Another uncertainty in the transport sector concerns the allocation of gasoline and diesel use between personal and freight transport. Based on values reported by the National Energy Commission in particular years and general trends, estimates were made for the entire 1973–1985 time period.

Third, detailed information is lacking concerning the reasons for changes in overall energy intensity in some heterogeneous industries (e.g. nonferrous metals or chemicals). This makes it difficult to determine the relative importance of structural and technological factors when evaluating the overall change in energy intensity within these industries.

A fourth area of uncertainty pertains to GDP. Methods of counting GDP have changed over time as have the inflation indices used to convert to constant value within Brazil. In addition, the informal sector (i.e. economic activity that is not reported to or recognized by the federal government) is excluded from official GDP figures while it contributes to national energy consumption. It is estimated that GDP would rise on the order of 35% if the informal sector were included (35). Furthermore, the relative magnitude of the informal sector (i.e. output generated by the informal sector divided by official GDP) appears to be growing over time.

VI. CONCLUSION

Determining whether or not national energy intensity declined in Brazil depends on how energy use and GDP are counted. If hydroelectricity is counted based on the energy input to a thermal power plant and official GDP figures are used, than there was virtually no change in overall energy intensity between 1973 and 1988. However, if hydroelectricity is based on its direct energy content and GDP figures are adjusted to reflect purchasing power parity with the dollar, than overall energy intensity fell about 21% between 1973 and 1988.

Large reductions in energy intensity occurred in the residential and transportation sectors during 1973–1988. Energy intensity also declined in the cement, nonferrous metals, paper, and steel industries. The main causes for these reductions appear to be: (*a*) rapid growth in electricity and bottled gas use coupled with a drop in wood use in residences (i.e. a shift to energy sources that are used more efficiently), (*b*) shifts to more efficient fuels, vehicles, and transport modes in the transportation sector, and (*c*) modernization and process changes in industry. Also, energy intensity declined when economic output rapidly expanded, and vice versa.

The overall energy intensity of the industrial sector increased substantially during 1973–1988. This was due in part to rapid growth in the production of energy-intensive intermediate goods such as aluminum, steel, and chemicals. Also, structural shifts favoring more energy-intensive products occurred within the steel alloy and chemical industries. Rapid growth in the export of primary metals had a particularly strong impact on energy use and intensity— industrial energy use would have been 21% lower and national energy intensity 7% lower in 1988 if primary metals export had not occurred. Energy-intensive industries and metals export were encouraged by the government in order to improve the balance of trade and to service the foreign debt.

Energy-pricing policies had mixed impacts on energy-intensity trends during this period. Subsidized and declining prices for electricity and LPG accelerated the shift to modern fuels in the residential sector. Also, the relative prices of liquid fuels encouraged shifts to more efficient fuel types. However, falling energy prices during the 1980s probably reduced interest in energy efficiency and conservation. Specific conservation policies and programs appear to have had a modest impact on overall energy intensities during 1973–1988.

Public officials in Brazil can take a number of steps to increase energy efficiency and reduce energy intensity in the future. First, certain energy conservation programs that were started during the 1980s (e.g. PECO and PROCEL) could be continued and strengthened in order to increase their impact. Second, new policies could be adopted that would increase energy efficiency. In particular, minimum efficiency requirements for buildings, cars, motors, lighting products, and appliances could result in significant energy savings (26). Lowering import duties and sales taxes on energy-efficient automobiles, microprocessor-based control systems, and other efficient equipment could reduce energy intensity, as could increasing taxes and duties on inefficient equipment. These types of policies are being developed and advocated by a new comprehensive energy rationalization program established by the federal government in 1990.[10]

Institutional arrangements in the energy sector also could be improved from the perspective of raising energy productivity. Enabling private power pro-

---

[10]Private communication with José Roberto Moreira, Secretary-General, National Program for Rationalizing the Production and Use of Energy, São Paulo, June 1990.

ducers to compete fairly with public utilities, directing utility planners to take a "least-cost energy services" perspective, and providing utilities with financial incentives for reducing the cost of energy services could help to stimulate industrial cogeneration as well as more efficient electricity use. To accomplish these objectives, the laws governing electric utilities need to be changed.

Structural factors will continue to have a strong effect on national energy intensity. In the current political environment, the export of energy-intensive basic materials such as aluminum, steel alloys, and cellulose is expected to increase during the 1990s (36). Continued promotion and subsidy of energy-intensive basic materials industries should be reexamined in light of the costs to Brazilian society for supplying energy and other services, the capital intensity and relatively small number of jobs created, and the adverse environmental impacts. Likewise, efforts could be made to derive more value-added from metals and other energy-intensive materials (e.g. increase fabrication of metallic goods and reduce export of primary metals). In the transportation sector, improving the efficiency, availability, and quality of both public transport and nonroad freight transport could further increase energy efficiency and provide other benefits.

While progress in reducing energy intensity has occurred in some areas, inefficient energy use is still widespread. A study of the potential for increasing the efficiency of electricity use as well as shifting away from electricity-intensive materials production concluded that projected electricity use in 2010 could be reduced by 30% (26). In the transportation sector, the energy intensity of passenger transport could be further reduced on the order of 50% if public transport is improved and the efficiency of the vehicle fleet is upgraded. By taking advantage of these opportunities for raising energy productivity, the energy sector can support rather than hinder economic and social development in the future.

ACKNOWLEDGMENTS

Support for this work was provided by the University of São Paulo and the US Environmental Protection Agency. The authors thank Andrea Ketoff, Alan Poole, and Robert Socolow for providing comments on earlier drafts.

Literature Cited

1. Inst. Brasil. Geogr. Estat. 1989. *Anuario Estatistico do Brasil*. Rio de Janeiro, Brazil (In Portuguese)
2. Alvim, C. F. 1989. Investimentos energeticos. *An. I Congr. Brasil. Planaj. Energ.* Campinas, Brazil: Univ. Est. Campinas (In Portuguese)
3. United Nations Dev. Program. 1990. *Human Development Report 1990*. New York: Oxford Univ. Press
4. World Bank. 1988. *Brazil—Public*

Spending on Social Programs; Issues and Options. Report No. 7086-BR. Vol. 1. Washington, DC
5. Minist. Minas Energ. 1990. *Balanco Energetico Nacional 1989*. Brasilia, Brazil (In Portuguese)
6. Furtado, A. 1990. As grandes opcoes da politica energetica Brasileira: o setor industrial de 80 a 85. *Rev. Brasil. Energ.* 1(2)
7. Brito, J. O. 1990. Carvao vegetal no

Brasil. *São Paulo Energ*. 64:27–31 (In Portuguese)
7a. Goldemberg, J. 1990. Energy and environmental policies in developed and developing countries. *Proc. Conf. Energ. Environ. 21st Century*. Cambridge, Mass: Mass. Inst. Technol.
8. Jannuzzi, G. de M. 1989. Residential energy demand in Brazil by income classes. *Energy Policy* 17(3):254–63
9. Jorge Wilheim Consult. Assoc. 1988. *Consumo de Energia nos Setores de Comercio e Servico*. São Paulo, Brazil (In Portuguese)
10. Lomardo, L. L. B. 1989. Comparacao de eficiencia energetica predial. *An. Simp. Nac. Conserv. Energ. Edif*. São Paulo: Esc. Politec., Univ. São Paulo (In Portuguese)
11. Empres. Brasil. Planaj. Transportes (GEIPOT). various years. *Anuario Estatistico dos Transportes*. Brasilia, Brazil (In Portuguese)
12. Penna, J. C. 1990. Chega de inverdades sobre o Proalcool. *Folha de São Paulo*, March 11, p. B-9 (In Portuguese)
13. Alvim, C. F. 1988. *Evolucao do Consumo de Gasolina e Alcool em Funcao da Composicao da Frota*. Brasilia: Comis. Nac. Energ. (In Portuguese)
14. Banco Nac. Desenvolv. Econ. Soc. 1987. *Perspectives do Setor de Transporte Interno de Carga*. Rio de Janeiro, Brazil (In Portuguese)
15. Ramos, F. 1989. *A Conservacao de Energia Eletrica e a Politica de Exportacao de Metais Basicos*. São Paulo, Brazil: Eletropaulo (In Portuguese)
16. Ramos, F. 1989. *São Paulo Energ*. 55:3–12 (In Portuguese)
17. Subgrupo de Ref. Basicas. 1989. *Relatorio Final*. Brasilia: Com. Nac. Energ. (In Portuguese)
18. Tolmasquim, M. T. 1988. Energy demand and adaptation to the oil shocks—Brazil 1973–1985. Luxembourg: *Proc. 10th Int. Assoc. Energ. Econ. Conf*.
19. Inst. Brasil. Sider. various years. *Anuario Estatistico da Industria Siderurgica Brasileira*. Rio de Janeiro, Brazil (In Portuguese)
20. Com. Nac. Energ. 1988. *Industria de Intenso Consumo de Energia: Ferroligas*. Brasilia (In Portuguese)
21. Com. Nac. Energ. 1989. *Estudo Sobre Energointensivos*. Brasilia (In Portuguese)

22. Programa em dez anos recompensa Carbocloro. 1989. *Bol. PROCEL* 25:3. Rio de Janeiro: Electrobras (In Portuguese)
23. Inst. Pesq. Tecnol. 1985. *Conservacao de Energia na Industria de Fertilizantes*. São Paulo, Brazil (In Portuguese)
24. Inst. Pesq. Tecnol. 1985. *Conservacao de Energia na Industria de Celulose e Papel*. São Paulo, Brazil (In Portuguese)
25. de Carvalho, A. V. 1990. Evolution of energy inputs and rationalization efforts in the Brazilian transportation sector. *Rev. Energ*. 14(1). Quito: Org. Lat. Energ.
26. Geller, H. S. 1991. *Efficient Electricity Use: A Development Strategy for Brazil*. Washington, DC: Am. Counc. Energy-Efficient Econ.
27. IPEA/IPLAN. 1989. *Para a Decada de 90: Prioridades e Perspectivas de Politicas Publicas, Vol. 2—Setor Externo*. Brasilia (In Portuguese)
28. Machado, R. de C. 1988. *A Industria do Aluminio Neste Final de Seculo*. Ouro Preto, Brazil: Edicao Fund. Gorceix (In Portuguese)
29. Brasiliense, R. 1990. Energia da empresa vem de Tucurui. *J. Brasil*. Rio de Janeiro, Sept. 18 (In Portuguese)
30. Rodrigues, A. P., Hermann, J. 1989. Condicoes economico-financeiras do setor eletrico no Brasil. *São Paulo Energ*. 59:19–38 (In Portuguese)
31. Pagy, A. 1989. *Panorama da Eletrotermia*. Rio de Janeiro: PROCEL, Eletrobras (In Portuguese)
32. *An. I Simp. Brasil. Co-Geracao Energ. Ind*. 1989. Campinas, Brazil: Univ. Est. Campinas (In Portuguese)
33. Poole, A. D., Pacheco, R. S., de Mello, M. A. C. 1990. *Urban and Collective Transport Policy in Developing Countries: A Case Study of Brazil*. Ottawa: Int. Dev. Res. Cent.
34. Schipper, L., Howarth, R. B., Geller, H. S. 1990. United States energy use from 1973 to 1987: the impacts of improved efficiency. *Annu. Rev. Energy* 15:455–504
35. World Bank. 1990. *World Development Report 1990*. Oxford: Oxford Univ. Press
36. Dept. Estud. Mercado. 1990. *Modelo Setorial Macroeconomico*. Rio de Janeiro: Eletrobras (In Portuguese)

Annu. Rev. Energy Environ. 1991. 16:205–33

ENERGY AND ENVIRONMENTAL POLICY IN CHINA[1]

*Robert D. Perlack and Milton Russell**

Oak Ridge National Laboratory, Oak Ridge, Tennessee 37831

*with Zhongmin Shen**

KEY WORDS: coal, environmental degradation, efficiency, legislation, price reform and planning.

INTRODUCTION

China is the largest developing country in the world. Chinese energy and environmental developments, policies, and trends are of world significance both because of their large direct impacts and because of their demonstration effect on the rest of the developing world. The picture in China over the past few decades is a hopeful one. Per capita income is up, energy production has supported growth even though energy, especially electricity, remains a constraint on development, additions to oil and gas reserves have kept up with production, and prospective energy output appears adequate over the intermediate term to support continued rises in living standards. The environmental consequences of the past 40 years of development have been severe—measured in ecological conditions such as deforestation and health risks such as air pollution—but the past decade has shown substantial interest in implementation of policies that will bring these effects under control. No adequate summary statistics are available, but casual observation suggests

*University of Tennessee, Knoxville, Tennessee 37996

that, as a whole, environmental conditions in China are no longer getting worse. Yet, substantial improvement remains in the future.[2]

Chinese policies since 1978 (and perhaps before) have emphasized the programmatic benefits of use of incentives and market-like forces to achieve national goals. Use of this approach has always been moderated by other policies, including commitment to a peculiar Chinese form of socialization. The scope for use of market forces (and of the authority given the rule of law) has widened and narrowed over the past decade, but until the reaction against the demonstrations in 1989, it was generally a matter of two steps toward the market and the rule of law and one step back. Whether 1989 represents a turning point in this process is not yet clear. What can be observed, however, is that the official rhetoric has consistently been that the openness to foreign trade and investment will be maintained, and that economic reform—making prices conform more closely to costs and incomes to production—remains a prime goal. The heady pace of change in this direction was first checked in 1988 in the face of an inflation so severe that it was beginning to have destabilizing effects, especially in urban areas. The events of 1989 created massive uncertainty about which direction change should go, and uncertainty about how fast.[3]

The purpose of this paper is to provide an overview of the Chinese situation as it relates to energy and the environment. Given the dominance of coal as an energy source and the environmental consequences of its production and use, more attention will be devoted to this fuel than to others. As noted below, however, China has a distinctly modern sector superimposed on a traditional

[2]This paper was written mostly using Chinese sources in the open literature. Most of this material was in English, but some was in Chinese, translated or utilized by Zhongmin Shen. Even sources such as the United Nations and the World Bank rely on Chinese government data. No material used in the preparation of this article came from materials restricted by the Chinese Government to use by Chinese citizens. The nature of these sources and of conditions in China are such that independent verification is often impossible. Further, as is always the case, government "goals" and "expectations" are often more wish than reality. In the same vein, as the saying goes, "it's a long way from Beijing to the village," meaning that policies, however well articulated and devoutly supported, may not be implemented, certainly as intended. The reader is cautioned that data, plans, and policies must always be interpreted in this light. The authors did have the opportunity to observe conditions in China for a substantial time (a total of a little less than six months), and over different geographical areas (11 provinces) and time periods (1984–1990), and to hold discussions (interpreted) on these topics with officials, academics, managers and workers, and farmers. This experience provides some field grounding, but by no means allows for the in-depth coverage that could be wished. The reader should be alert to these limitations of this paper in interpreting its conclusions.

[3]This paper is written (winter, 1990–1991) at a time when the outcome of these changes is unclear, but reflects the view (not necessarily the prediction) that forces that have been in motion and the policies that they sponsored will continue in effect and become more influential over actual behavior in the future.

sector that depends on noncommercial or traditional fuels. The intriguing aspect of China's economy is that nationwide trends are all but impossible to detect because of the differences between the commercial and the traditional sectors. Averages are particularly meaningless when it comes to energy and environmental conditions in China.

Primary energy production in China's modern sector has grown rapidly since the founding of the People's Republic in 1949. Today, China is the third largest consumer of primary energy in the world after the United States and the Soviet Union with total primary production of commercial energy of more than one billion tons of coal equivalent (tce) (1).[4] In 1986, China surpassed the United States as the world's largest producer and consumer of raw coal. Crude oil production rose from negligible amounts in 1949 to a present output of more than 2.7 million barrels per day (2). Hydroelectric production has increased 12-fold since the surge in economic growth began in the late 1960s. Total electricity generation from all sources in China is about 545 TWh and is growing annually by more than 13%. These statistics for the modern sector do not include the traditional biomass fuels—the main source of energy for more than 75% of the Chinese population—as these are not a part of the commercial economy. These traditional biomass fuels are used for cooking and space heating and provide some 80% of total household energy, the equivalent of 330 million tce (2).

Not surprisingly, energy production, conversion, and use are major sources of pollution in China. Coal mining and oil production disturb the land and pollute the water with silt, acid, and salt. Direct combustion of coal releases smoke, particulates, and heavy metals to the atmosphere and emits sulfur dioxide (SO_2), and is the major cause of atmospheric pollution in China. All of the pollutants from coal combustion threaten public health, reduce visibility, increase the dirt in cities, and cause acid rain that harms ecological systems and materials. Petroleum use causes ozone pollution, which damages health and crops. The lead added to gasoline causes mental retardation and emotional and development problems in children. All fossil-fuel burning produces carbon dioxide (CO_2), a greenhouse gas associated with prospective climate change. Currently, China is the third largest country source of greenhouse gas emissions (3). Environmental degradation from energy production has also been severe in China. The construction of large hydroelectric dams has destroyed valuable crop lands and drowned natural ecological systems. The construction of large coal-fired thermal power plants has raised concerns about the availability of water resources and the quality of these resources after use. The intensive use of biomass in rural areas has caused massive deforestation and has led to widespread soil erosion and siltation with its concomitant impacts on agricultural productivity and watersheds.

[4]This total is an unofficial estimate for 1990.

The expectation over the next decade and beyond is for continued growth in use of all sources of energy. China's economic goal is to double its present GNP by the year 2000. Meeting this goal, according to plans, will require annual production of 1.4 billion tons of coal, 200 million tons of petroleum, 30 billion cubic meters of natural gas, and 1200 TWh of electricity (4–6). The prospects for meeting petroleum and natural gas goals are less certain than meeting those for coal; declines in production from old oil fields are likely, and it will be a difficult task to get the output from new fields to market over the next decade. Successful exploration holds the key beyond 2000. Natural gas production, now only two percent of gross energy supply, is expected to double. Coal-fired and hydroelectric facilities coming on-line in the 1990s will add significantly to generation capacity. The prospects are less certain for reducing unit energy demand with the implementation of conservation policies and technical improvements to modernize outmoded heavy industry, and price reform to provide better resource allocation. The availability of sufficient capital to finance energy and environmental protection projects remains in doubt.

It is generally conceded that coal will continue its dominance of China's energy system for the foreseeable future; even under the most optimistic conservation and efficiency scenarios it is unlikely that coal will provide much less than the current three-quarters of total commercial energy even by 2025 (5, 7). Moreover, any possible reductions in domestic coal consumption could be offset by increases in coal exports. The prospects of China's doubling or even tripling coal production and increasing its energy consumption to levels now equal to those in the United States could have ominous environmental consequences both regionally and globally. How China might continue to receive the benefits of energy use while minimizing the negative environmental effects in major urban areas, the rural countryside, as well as globally is the focus of the remainder of this paper.

ENERGY PRODUCTION, USE, AND THE ENVIRONMENT

The Modern Energy Sector

Energy consumption to fuel China's modern sector has grown rapidly since 1949, averaging nearly 10% annually. As shown in Figure 1, this growth has been largely driven by coal. Coal's share of primary energy consumption declined from more than 90% in the early 1950s to a low under 70% in 1976. Following world oil price increases in the late 1970s and recognizing its comparative production potentials for coal and oil, China pursued a domestic coal-for-oil substitution policy. This policy helped to free oil for the export

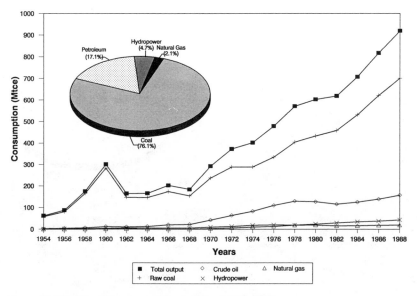

Figure 1 Primary energy consumption in China (1954–1988)

market (8, 9).[5] In addition, China adopted a number of policies and reforms to encourage local governments and collectives to develop and operate small coal mines. One such reform was to allow operators of coal mines to sell any output produced above the state-mandated quota at higher "market or negotiated" prices. In addition, China began investing in modernizing and mechanizing its state-owned mines, which has had the effect of increasing productivity.[6] Today, China's dependence on coal—which accounts for more than three-quarters of its commercial energy—is preeminent among nations. For the two largest producers and consumers of coal, the United States and the Soviet Union, coal accounts for less than 25 and 22% of total primary energy consumption, respectively (10).

As shown in Figure 2, about 58% of primary energy consumption is accounted for by the industrial sector. The urban residential sector uses about a quarter of final consumption. The agriculture, transportation, and the combined commerce and public service sectors consume relatively small amounts—about 7, 5.5, and 3.1%, respectively (11). In comparison with other developing countries, China consumes more energy in the industrial sector (58% versus an average of 50% for all developing countries) and much less in transportation (5.5% versus 22%) owing to a lack of private transport

[5]In the early 1980s, oil exports were providing about 25% of China's export earnings. By 1986, however, oil export earnings had fallen by 50% (8).

[6]Between 1981 and 1988, the share of mechanized mines increased from 40 to 58% and productivity increased by about 25% (2).

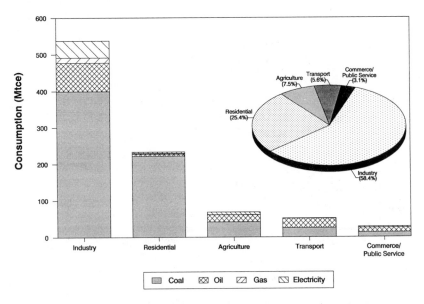

Figure 2   Commercial energy consumption by primary energy and by sectors in China (1988)

(12). For comparison, in the United States, industry accounted for 31%, transport 35%, agriculture 1%, and commercial and residential sectors 30% (13).

In the end-use sectors, coal provides about 72% of the energy in the industrial sector (including power), 61% in agriculture, 94% in the residential sector (including cooking and space heating), and 45% in the combined commerce and public service sector (2). About 49% of the energy used in the transport sector (nearly all rail) comes from coal. Examination of energy consumption by fuel type shows that nearly 90% of all coal used in China is accounted for by the industrial (57%) and residential sectors (32%). Specifically, about 675 million tons are consumed in the industrial sector (including 250 million tons for thermal power generation) and about 210 million tons for residential use (2). The remainder is used in coking (70 million tons) and transport (23 million tons).

As expected, energy-related environmental impacts largely involve the production and combustion of coal. In coal production there are mining wastes, effluent from coal washing, runoff, and acid mine drainage. Water resources, which are relatively scarce in China when measured in per capita terms and compared to other countries, are particularly vulnerable to energy production (14)[7]. The planned increase in the construction of thermal power

[7]Renewable water resources in China are estimated at 2470 cubic meters per capita compared with 7690 worldwide (15).

plants, especially in the north, may exacerbate the problems of water supply and diminish water quality.

Unlike in other major countries, coal use in China is not confined to thermal power generation. In China, most of the coal is used in inefficient, small- and medium-scale industrial boilers, in building boilers to provide residential heat, and in coal stoves for cooking. In Beijing alone there are more than 6000 heating boilers and one million household coal-burning stoves (16). This heavy and dispersed coal consumption has produced severe urban air pollution in the form of SO_2, suspended particulates, and acid rain. The problem results not only from inefficient combustion at all stages of use, but also from poor matching of coal type with end-use and the general absence of postcombustion pollution control. Moreover, there is little coal desulfurization other than what is done in washing or adding binders to briquettes. In terms of impact on public health (17), suspended particulate matter is the most significant pollutant, because several kinds of toxic metals and organic compounds are adsorbed on the airborne particles. SO_2 emissions are also a serious problem, as they act synergistically with suspended particulates and are precursors to acid rain. The impacts of suspended particulates and SO_2 are confined mainly to the large and medium-size cities with severity peaking during the fall, winter, and spring heating months (17). NO_x emissions are increasing in China with increasing vehicular traffic, but at present are confined to main streets in major cities (17).

Total annual emissions of SO_2 and suspended particulates in China are approximately 15.6 million tons and 14.0 million tons, respectively (18). The most recent estimates of average daily SO_2 concentrations indicate a level 105 $\mu g/m^3$ over all cities in China with concentrations slightly higher in southern cities (119 $\mu g/m^3$) and lower in northern cities (93 $\mu g/m^3$). The average daily concentrations of total suspended particulates were about 526 $\mu g/m^3$ in northern cities, about 318 $\mu g/m^3$ in southern cities, and about 432 $\mu g/m^3$ over all of China. The World Health Organization (WHO) has established guidelines of 100–150 $\mu g/m^3$ of air for SO_2 and 150–230 $\mu g/m^3$ of air for total suspended particulates. According to the Global Environment Monitoring System (GEMS), the average (mean) values for five major cities in China were 66 days that exceeded SO_2 guidelines and 204 days that exceeded particulate guidelines per year (19). Moreover, the average daily concentrations greatly exceed WHO particulate guidelines in both summer and winter and exceed SO_2 guidelines in winter (17). However, the trend in total emissions for both SO_2 and suspended particulates is down (19).

The problem of acid rain tends to be confined to southern areas and to the middle and lower reaches of the Yangtze (16). It is estimated that the acidity of rain is below pH 5 over an area of 1.3 million square kilometers (20). The alkaline soils of the north tend to act as a natural buffer. The economic losses from all sources of pollution in China are estimated to be about 2% of GDP (20).

Because of China's high reliance on coal combustion huge amounts of carbon dioxide (CO_2) are released. In 1986, China accounted for about 10% of global carbon emissions from all fossil fuels, and 20% from coal. In total, China accounts for about 7% of all world greenhouse gas emissions (15). China's fossil carbon emissions are less than half of the world per capita average of 1.1 tons and considerably less than the 4.9 tons per capita of the United States (21). However, China emits five to six times as much fossil carbon per unit of economic activity as industrialized countries (21). Of particular concern is the growth rate of China's fossil carbon emissions, which were estimated to be increasing at 5.5% annually in the 1980s as compared with a rate of less than 2% worldwide (21). However, it is unlikely that the high growth will be sustained in the 1990s. An increase in coal production from about 1 billion tons in 1990 to 1.4 billion tons in 2000 would increase coal-based carbon emissions from 500 million to about 700 million tons—an annual increase of less than 4%.[8]

The Rural Energy Sector

Households, agriculture, and township industries in rural areas account for nearly half of China's total energy consumption (2).[9] As shown in Figure 3, rural households account for about two-thirds of total rural energy use of which about 80% is traditional biomass fuels (22). Firewood and straw each account for about half of total biomass use. A very small fraction of household energy comes from biogas. Households consume slightly more than half of the rural commercial energy, which is composed of approximately 70% (175 million tce) coal. About a quarter of total rural energy consumption or slightly more than half of the rural commercial energy goes to township enterprises. In 1987, just over 100 TWh of electricity were consumed in rural areas (2).

China's rural areas produce a significant proportion of the commercial energy they consume. Locally operated coal mines produced about 40% of the total national coal production (23). Moreover, rural areas produced about 25% of their electric power, primarily from small hydroelectric facilities and small coal-fired thermal plants. On a per capita basis, rural energy consumption is only one-fourth of that in urban areas. Lower energy use is due both to structural reasons and to insufficient supplies and allocation policies. For

---

[8]This estimate assumes 0.714 tce per ton of raw coal, 29.0 GJ per tce, and 24.1 kg of carbon per tce.

[9]This estimate assumes 253 million tce out of a total country primary energy consumption of 859 million tce in 1987. Added to this estimate was 330 million tce of biomass energy, which is not assigned a monetary value.

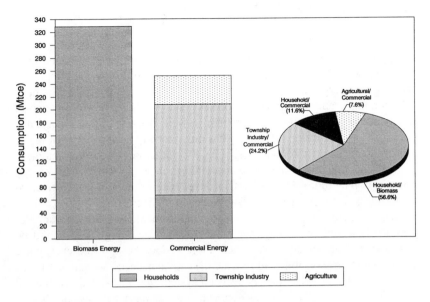

Figure 3 Rural energy consumption in China (1987)

example, 70% of rural households have access to electricity; however, supply
is severely constrained relative to urban areas and per capita consumption is
only 13 kWh per year (23).

The use of biomass has been a source of much deforestation in China. To
be sure, deforestation has also resulted from land clearing for agriculture and
grazing, excessive timber production, and failed large-scale industrial movements, such as the Great Leap Forward of the 1950s. Regardless of the
specific causes, deforestation has resulted in erosion, and, in some cases,
lower agricultural productivity, local climatic changes, loss of plant and
animal life, and increased incidence of flooding. During the 1980s, China was
very active in reforestation. Of the approximately 125 million hectares of
forest land in China, about 25% or 31 million hectares have been planted,
including those set aside for fuelwood (18). In 1990 it was reported that just
over 5 million hectares had been planted (24). If these estimates are correct,
China stands out as the world's largest reforester. If the 31 million planted
hectares sequester 4 tons of carbon each year then China is currently offsetting
about a quarter of the carbon that is released during the combustion of coal.[10]

[10]This estimate assumes an annual growth rate of 8 dry tons with half of the dry weight being
carbon. Of course, the planted trees will continue to sequester carbon as long as they are growing.
In the longer term, mature trees will have to be harvested and permanently stored and the area
replanted with new seedlings if trees are to continue to sequester and offset fossil-based carbon
emissions.

THE ENERGY SECTOR AND POLICY

Energy Inefficiency

Energy-use efficiency was not seen to be an important issue in China until growing industrialization and rising incomes in the 1980s placed greater demands on energy supply than could be met. More recently, the problem of pollution has been given greater emphasis. Energy efficiency was recognized as a means to solve the pollution problem as well. Because of this history, it is not surprising that energy efficiency has only recently been a matter of major concern to decision makers. Energy supply facilities are very capital-intensive and require long lead times to construct—much longer times than energy consumption facilities. Once shortages developed in China, it was difficult to reallocate investment fast enough to keep up with growing demand. There was no automatic system to make mutual adjustments so that, for example, peak loads would be managed or the most valuable energy uses would be served first. This situation was made more serious because the energy-use sectors had no responsibility to provide the capital needed and were separated from energy-supply planning. A decision system that was suited to China in its earlier, slower phase of development did not meet the new needs that arose.

This problem was made worse by the fact that there was little opportunity for energy prices to guide production and use. A flexible pricing system would have provided the needed incentives for rapid adjustment. Energy prices, especially coal prices, were not allowed to rise to cover the increases in costs and to take more advantage of the opportunity to sell oil and coal abroad for foreign exchange when world prices rose in the 1970s.[11] Under these conditions it is not surprising that shortages appeared in the 1980s and that, relative to other products in China, energy was underpriced and therefore overused, underproduced, and underexported. Low energy prices became a greater source of inefficiency with the reform of the economy, the rise of the responsibility system, and the increase in the level of living of many Chinese.

Because of the pricing practice, industrial users consumed energy as if its value to the country were much lower than it really was. Processes were designed, machinery and appliances built, and buildings constructed that use more energy than justified considering its real value in other uses. Low prices also caused people to operate those facilities in ways that use more energy than they would if managers took account of energy's true value. Wasteful production of excess energy, and wasteful idleness of factories and other facilities when sufficient energy was not available, have resulted.

The rationalization of energy prices began in 1979 with an increase in the

---

[11]To be sure, China substituted coal for oil and promoted oil exports to generate foreign exchange.

price of state-produced coal. In the early 1980s, the price of electricity was raised in areas of short supply (Northeast, North, and East China). In 1985, a two-tier price system for energy resources based on state-controlled prices and negotiated market prices went into effect.[12] The policy became to distribute some energy (coal) at the previous low price as an allocation, but then to allow firms and individuals to purchase more on the free market. This policy changed again in 1989 when more coal from the China National Coal Corporation was allocated by quota than previously and the price for coal beyond the quota was restricted to 2.5 times the price for quota coal (25).

This two-tier pricing system has given producers the incentive to produce more because all above quota that they bring to market is sold at a much higher price and profit, and it has given consumers incentives to save energy because the energy they avoid using is more expensive than before. Still, problems remain. Energy producers do not receive nearly enough funds from user energy bills to cover their costs of production, much less to invest to increase supplies and improve productivity. Producer deficits are made up through subsidies, and some investment funds must come from the central authority, which has limited funds available and many other demands to meet. Further, because quotas are difficult to adjust with changing conditions, energy is often not directed to its best uses.

The basic problem with the two-tier coal pricing system is that it creates market distortions. That is, better-quality coal (higher heating value, lower sulfur, etc) is sometimes sold in the allocated market at a lower price than poorer-quality coal in the free market (26). The environmental consequences of this policy are best illustrated in the urban household coal market. In some cities high-quality, low-volatile (smokeless) anthracite coal destined for the allocated urban residential market may get diverted to industrial uses at higher prices. The environmental implications are obvious in terms of air emissions, aesthetics, and public health.

Despite the shortcomings of these price reforms in the 1980s, the energy intensity of China's economy dropped by more than 30%. However, the country is still a very inefficient user of energy.[13] Popular comparison measures suggest that energy use is one-fourth to one-half as efficient as in major developed countries and slightly worse than in other major centrally planned economies. For example, China's energy intensity in the mid-1980s was about 5 times that of Japan, 2.5 times that of the United States, 4 times that of Western Europe, and about 1.8 times that of coal-intensive India (13, 14).

---

[12]Further, the price of locally produced coal was released from state control. The price of coal from state mines was also increased by 41%.

[13]According to Granzer, measures to conserve energy so far have mostly been administrative in nature (e.g. penalties for consumption above established quotas) and a result of better operation (27).

Interpretation of gross efficiency and intensity data must be made carefully, however, before any specific conclusions are drawn. First, energy use is strongly affected by the proportions of each industry in a country's total production. A country with a high proportion of heavy industry relative to light industry will have higher energy use per capita and by value of output. Second, a developing country will require a large amount of energy to build dams, railroads, factories, and housing. Energy to create this physical capital was expended decades earlier in other countries further along the development path, and does not appear as a current use. Third, energy use will depend, properly, on the relative cost of different inputs.[14] Finally, the quality of any cross-country comparisons must acknowledge the role of the traditional sector and the fact that exchange rates may not measure accurately true levels of living or output. In sum, the wisdom about the efficiency of energy use in China or in any other country must be gained from examination of the conditions in that country, not from measures imported from abroad.

Nonetheless, it appears that energy use in China is much less efficient than it could be, and that improvements would raise the level of living and greatly reduce pollution. For example, in electricity generation, the average heat rate for thermal power plants in China is about 11,630 kJ/kWh (2), whereas steam plants in the United States, Japan, and Europe are more efficient with heat rates of 9,500–11,600 kJ/kWh (12). Line losses are reported to be about 8.2%, which compare favorably with developed-country standards. The most extensive example of energy inefficiency in China no doubt lies in the 400,000 small- and medium-size industrial boilers and the relatively large number of dispersed building boilers and coal stoves used in the urban residential sector. The industrial boilers consume an estimated 425 million tons of coal annually, and the residential heating boilers and coal stoves consume an estimated 210 million tons (2).[15] Typical thermal efficiencies for industrial boilers are 55–60% as compared to 75–80% in developed countries (14). Improving the efficiency of industrial boilers, say from 60 to 75%, could save 100 million tons each year. A similar level of improvement in the urban residential sector could save an additional 50 million tons. Of course, any reductions in coal use would have a proportional effect in reducing coal-related emissions and improving urban air quality. The high coal use and emissions from these boilers are not only attributable to the inherent inefficiency of small-scale units and lack of pollution control equipment, but also to the wrong type and poor quality of coal used for the application.

[14]For example, coal in the ground and the labor to dig it is cheap in China relative to the additional capital required to use it more efficiently. Fuel-inefficient machinery will be scrapped in a country where capital is abundant and foreign exchange no problem, while it may properly be kept in operation in other countries where the conditions are different.

[15]Specific data apparently do not exist on the number of dispersed building boilers, coal stoves, fuels (anthracite, briquettes), and the percentages used in cooking and space heating.

Energy Shortages and Resources

The reforms in the coal industry in the mid-1980s greatly increased supplies. However, shortages of coal still exist owing to a combination of inadequate production and inadequate rail capacity to move coal from mines to major consumption points. The lack of sufficient coal supplies compounds problems in the power sector, which has inadequate capacity to meet full demand in most locations virtually all the time.[16] When peak loads occur, forced outages can idle up to one-third of industrial capacity, and brownouts are common among residential and commercial users. Although electricity generation is growing very rapidly, the country's supply is short by some 18–20 GW of capacity and 90–100 TWh for annual generation (6). Moreover, one-third of its industry is claimed to be unable to operate at full capacity. The shortages of electricity are estimated to cause annual economic losses of 200 billion yuan ($42 billion) (28). Further, there is no spare capacity to accommodate periodic maintenance and repair needs, which adds to the instability of supply. Though substantial new capacity has been built, growing demand has required that antiquated and less efficient thermal units be maintained on-line. The problem of inadequate production and distribution also contributes to shortages of certain coal grades and qualities for specific end-use applications, such as anthracite (smokeless coal) for the urban residential market.

The Chinese economy also suffers from shortages of petroleum products. Exports of oil have been to a large extent done at the short-term expense of greater domestic consumption (9). The overemphasis on production drilling at the expense of exploration is partly responsible for the lack of real oil production opportunities and prospects (14). Currently, oil reserves are sufficient, but new fields need to be developed rapidly to offset impending declines at mature fields (30). This problem is compounded by questions about Chinese oil exploration, development, and production technology, which are antiquated and much behind Western standards. Even if major new petroleum reserves were developed, China faces inadequate distribution facilities and a stressed and outmoded refinery industry. The petroleum refinery industry was planned and built when the Chinese market was largely diesel and fuel-oil centered, and even the low-technology refinery mix could produce a surplus of low-octane gasoline to be exported. The growth of van and automobile transport and of farm mechanization over the past decade has increased the demand for high-octane gasoline, which has sometimes been in very short supply.[17] Refineries have resorted to heavy use of tetraethyl lead as

[16]In 1988 it was reported that thermal power plants in four major networks received about 93% of their allotted quota of coal, forcing them to reduce generation (29).

[17]In 1988 there were nearly 5 million motor vehicles in China; this number is expected to reach 13 million by the turn of the century (31).

an additive, which has serious implications for public health, as demonstrated by research on the effects of lead on children.

As noted earlier, there are critical shortages of both traditional biomass fuels and modern sources of energy in the rural areas of China. The labor required in the collection of fuelwood has prompted many of the rural poor to use inferior fuels, such as crop residues, and to cut back on basic energy needs. The use of crop residues sometimes depletes fertility and harms soil conditions. The much-heralded farm and village biogas production program was never as successful as expected. Today, biogas represents only a very small fraction of rural energy consumption despite the claim of millions of units in operation (22). The technology has improved, however, and the number of active installations now appears to be growing after a decade of decline. These facilities still require unpleasant and arduous labor, however, and the rising prosperity of the agricultural sector in the provinces where they are most practical has led to resistance to their use. The rapid growth in rural economic activity over the past decade and the shift from traditional to modern energy has been significant. This growth will continue to place demands on local governmental units to develop small coal mines, small-scale hydropower projects, and other resources that have previously gone un-exploited.

China ranks with the United States and the Soviet Union in terms of energy reserves, but the spatial distribution of reserves is a major problem in China. Coal reserves are very large and are of relatively good quality.[18] Although coal is found in all provinces, the deposits in the South are much inferior in quantity, quality, and exploitability (14). Coal is a bulky commodity and it is not surprising that its movement is the largest component of all transportation requirements in China—notwithstanding the substantial use of local coal.

The most important coal-production areas in terms of size and quality are concentrated in the Provinces of Shaanxi, Shanxi, Anhui, and Shandong and western Inner Mongolia, some 600–1000 kilometers from the industrial region running from Beijing through Tianjin to Qinghuangdao, the major coal export port. The distances are greater to other coastal industrial centers. The necessity of transporting coal from north to south or west to east long distances to the major points of consumption has overloaded the rail system, resulting in much unmet demand. In many cases, inferior local resources are used because the high transport costs do not allow higher-quality coals to compete in local markets.

Use of mine-mouth generation is starting, but transmission of electricity through long-distance grids is rudimentary in China. Major expansions of

---

[18]China has about 15% of the world's economically recoverable reserves (14, 15). About 86% of China's coal reserves are the rank of bituminous or anthracite, whereas only 14% of China's coal reserves are lignite.

transportation and transmission networks have occurred in the last five years, but these have not offset previous inadequacies, much less kept up with growing demand. Regional electricity grids are not interconnected, an inadequacy that hampers China's ability to move power around the country (6).

Oil and natural gas deposits have not been fully explored. The oil fields that produced most of China's oil in the past are being depleted rapidly. For example, although China's crude oil output did grow annually, the rate of growth declined in the first four years of the seventh five-year plan (32). Output was maintained by an increase in productive effort. Many of the major new discoveries being developed are in difficult locations such as offshore and in Xinjiang, far removed from major consumption centers. Natural gas has not been found in substantial quantities near consumption centers; major pipeline projects will be required to get it to market.

Although China is a relatively arid country, large hydropower resources remain unexploited. China's hydropower potential is considered the world's largest with some 380 GW of capacity (2). Again, most of these are distant from population centers; others, such as the now-deferred high dam project at the Three Gorges of the Yangtze, have serious social and ecological problems and doubtful economics. The nuclear power program is in its infancy with one reactor expected to be completed in 1991. China has areas with good solar, wind, and geothermal resources, but these are difficult to exploit in a capital-constrained and price-distorted economy.

Energy Policy for Modernization and for Rural Areas

China's stated energy policy for the next decade is to place equal emphases on the development and conservation of energy (5). Much of China's expected demand for energy, 1400–1700 million tce by the turn of the century, can be met by efficiency improvements throughout the energy sector. Price reform, a significant part of China's restructuring and modernization drive, must also play a strategic and catalytic role in energy policy.

China is embarking on a strategy to increase the share of electricity use in end-use sectors. This strategy would not only improve overall conversion efficiency, but would help alleviate two of the country's worst ecological problems—urban air quality and regional acid rain (33). As part of this strategy, China is committed to the construction of large power stations: mine-mouth coal-fired power plants and coal-fired plants in coastal areas and along railroad corridors; hydroelectric facilities along the Yellow, Yangtze, and Hongshui rivers; and nuclear power stations in developed areas that are short of exploitable sources of energy (34).

China has announced plans to add 70 GW of new coal-fired capacity during the 1990s (6). These additions will increase installed thermal capacity to 160 GW with annual generation of 900 TWh. Thermal power will continue to be at least 70% of total electricity generation in the year 2000. Use of hydropower, currently exploited at only 10% of its potential by China, is to be expanded. During the 1990s an additional 40 GW of hydroelectric capacity is planned for construction, bringing total installed hydropower to about 72 GW or 30% of total national capacity. In addition, China is beginning to develop its nuclear power industry. A 300 MW plant will soon operate, and there are reports of 3 GW of capacity in various stages of construction (6). In total China envisions 6 to 7 GW of installed nuclear power capacity by the year 2000. Investments are also planned for high-voltage transmission lines to move electricity from the west to the east and for rail networks to help move coal from the north to the south (6).

At present, about a quarter of China's coal output is used in power generation (5). With the expansion of thermal capacity, 50–60% of new coal supplies are expected to be used to produce electricity, bringing the proportion of coal used in power generation up from about 20 to 35% of total national coal consumption. Attempts are under way to reduce coal consumption in existing thermal power stations by better energy management and conservation and by technical efficiency improvements. New thermal units currently being installed in China are expected to operate at around 9600 to 10,250 kJ/kWh (5).

In addition to developing its power sector and promoting end-use electricity consumption, China is reforming its urban energy policies regarding direct combustion of coal in small- and medium-size industrial boilers and in single-building heating boilers (35). The low thermal efficiency of industrial boilers often results from the inferior quality of coal used as well as poor operation and maintenance. Improving the fuel used, conducting better maintenance, and replacing outmoded boilers with higher-quality equipment matched to specific applications are measures that are being implemented to reduce coal consumption and improve environmental conditions. The potentials for refurbishing some of these boilers to accommodate cogeneration and district heating are also being explored. Cogeneration (of heat and power) can be very cost-effective; however, district heating to replace single-building boilers requires the construction of expensive distribution systems that may limit its viability for many urban areas. In addition, replacing outmoded boilers with more advanced combustion processes (e.g. fluidized-bed combustion) is receiving attention (36). Many simple, low-cost options to conserve energy and improve efficiency can also be used: building energy audits, insulation, and financial incentives to encourage conservation.

To raise the quality of coal and reduce ash and sulfur, China is making greater use of coal washing. Currently, about 170 million tons of raw coal are washed each year. Numerous coal-gasification plants are also planned or being built to supply clean and convenient fuel for urban use. The energy losses in the conversion of coal to gas are partly made up in the greater efficiency of gas at the point of consumption. The environmental benefits are obvious; pollution is more easily controlled in the large facility and, in any event, it is moved away from concentrations of people. Coal is also being converted to briquettes for home and commercial stove use. The use of briquetted coal can improve combustion efficiency by 20–30%, and reduce emissions of CO by 70%, particulates by 60%, and polycyclic aromatic hydrocarbons by 50% (5). If a sulfur absorbent were added to the briquette, SO_2 emissions could be reduced by half. Moreover, convenience in use and transport costs are improved.

China's fuel-diversification policy also requires the continued exploitation of its oil and gas resources. Existing oil fields will continue to be developed and production expanded in eastern China. Exploration of oil and gas will continue in offshore areas (e.g. Bohai Bay) and in the west (e.g. Xinjiang area and Qinghai Province). More emphasis will be placed on developing natural gas and distribution facilities for the urban energy market. Because of the huge offshore potential and the large investments required, China is offering incentives (risk-reduction and financial) to attract foreign investment (37).

In rural areas, China appears to be committed to easing of energy shortages. Traditional fuels, whose use greatly degrades ecosystems, are receiving the most attention. China has in-place an ambitious program of disseminating fuel-efficient cooking stoves, which are claimed to be more than twice as efficient as traditional cooking devices. It is reported that more than 110 million rural families are using fuelwood-saving stoves with total savings of 300 million tons of fuelwood annually (35). The stoves also reduce the amount of crop residues used as fuel, making them available for use as mulch and compost. New designs of biogas digesters for households are also being actively promoted. However, the potential fuel savings from use of digesters are estimated to be only 10–20% of total household energy consumption (23). In combination with fuel-saving stoves and household-sized biogas digesters, China is actively reforesting large areas to check soil erosion and protect the natural environment as well as to provide new sources of fuelwood supply. To date, China has planted more than 2.5 million hectares specifically as a renewable source of fuelwood (22).

The development of rural township industries in China is a key component of rural economic growth. The creation of rural industry can raise standards of living and provide employment to the huge surplus rural labor force. The development and continued growth of the rural economy requires reliable and affordable modern sources of energy. It is China's policy to encourage rural collectives and individuals to develop small coal mines and to exploit small hydropower resources.[19] Small decentralized generation facilities including wind, geothermal, and solar have also been found to be viable in

---

[19]There are about 63,000 small hydropower facilities in-place with a total installed capacity of 25 MW generating about 29 TWh of electricity (27).

many rural areas because they can utilize simpler technology (much of which can be fabricated locally) and take advantage of abundant rural labor. Many of these decentralized projects have been built close to township industries thus avoiding costly distribution.

ENVIRONMENTAL POLICY AND ITS IMPLEMENTATION[20]

Prior to the 1980s, the first five "five-year economic development plans" had no provisions for environmental protection (38).[21] China followed a single economic goal of increasing industrialization and GNP; population control and environmental protection were ignored (39). Moreover, central planning of the economy provided no incentives for producers to make efficient use of primary energy and to adopt more efficient technologies (40), and under-priced energy resources encouraged wasteful consumption. In 1978, a provision for environmental protection was added to the Chinese Constitution. The following year the Environmental Protection Law was passed on a trial basis, and environmental protection and ecosystem degradation were incorporated as integral parts of economic and social planning at all levels of government.[22] During the 1980s, major legislation concerning pollution control and environmental management was also passed.[23]

China's environmental policy is grounded on the reality that poverty precludes massive capital investment for point source controls. Consequently, the focus is on prevention and management—seeking to exploit diverse interventions to limit health damage and environmental pollution and degradation at the least possible cost and disruption to the economy and society.

[20]Operation of environmental protection in China is unfamiliar to most non-Chinese audiences. What follows is a brief, impressionistic interpretation of the basic outlines of the principles and policies that represent the direction the Chinese central government is pursuing. In interpreting this, it is important to recognize the large extent to which China is decentralized and the high degree of autonomy that local and provincial authorities possess. Practices will vary.

[21]To be sure, the first National Conference on Environmental Protection was held in 1973. Subsequently, the Environmental Protection Group of the State Council was formed in 1974. The prevention of industrial pollution was placed on the agenda of the State and a series of prevention measures were adopted (39). However, this policy did not link pollution with ecosystem degradation nor did it link pollution with social and economic development.

[22]The Environmental Protection Law was revised and formally promulgated in December 1989.

[23]In addition to the Environmental Protection Law other major pollution control legislation in China includes: the Marine Environment Protection Act (August 23, 1982); the Water Pollution Control Act (May 11, 1984); the Air Pollution Prevention and Control Act (September 5, 1987); and the Noise Pollution Prevention and Control Regulation Act (September 1, 1989). There are also a Solid Waste Control Act and a Hazardous Waste Management Regulation Act in draft.

The first stage of implementing this strategy was to create an administrative and institutional structure that would place advocates for the environment throughout China.

As shown in Figure 4, the institutions of environmental protection are highly decentralized with confusing lines of responsibility and authority. In 1984, the Second National Conference on Environmental Protection established environmental protection as a fundamental national policy. In the same year, China established its National Environmental Protection Agency under the State Environmental Protection Committee of the State Council. The National Environmental Protection Agency has few executive or enforcement powers. It operates research facilities, drafts national standards and goals, promotes laws to be considered by the People's Congress, conducts China's international environmental activities, and serves as an advocate for environmental protection in the country at large and within the government. Environmental Protection Bureaus (EPB) were established at the provincial, city, county, and township levels of government. The provincial and local EPBs exert enforcement power through local administrative bodies. Now that the final version of the Environmental Protection Law has been passed, more enforcement powers exist (18). On a parallel track, each ministry (energy, heavy industry, chemicals, etc) has an environmental protection structure in place as well. The environmental administrative structure is virtually complete and a presence has been established, although the quality and training of local and county officials are sometimes poor. The first task of these officials has been to monitor and collect data on local conditions and, where appropriate, enforce environmental regulations.[24]

Centrally planned economies face the issue of whether actions to protect the environment are to be performed by a government agency (e.g. an agency that operates waste water treatment plants or puts controls on air emissions), or whether such actions are to be taken by those who emit the pollutants. China has explicitly chosen the latter, i.e. to follow the "Polluter Pays Principle," which puts the environmental authorities in the position of enforcer, not doer. The Chinese have three reasons for this choice: it affects relative prices (at least costs) and offers desirable incentives by internalizing costs; pragmatically, it places responsibility on income-generating enterprises with the cash flow to achieve needed change; and finally, it meets common standards of fairness.

In implementing the Polluter Pays Principle, China has also implicitly recognized that it is substantially easier politically, and as a matter of practice,

---

[24]It is a common observation that China collects more data about its pollution than any other country in the world, but these data are not managed or used very effectively. China has established a network of 5000 stations across the country to monitor the atmosphere, surface water, acid rain, radiation, and noise pollution. Another ecosystem network monitors grasslands, deserts, lakes, and oceans (41).

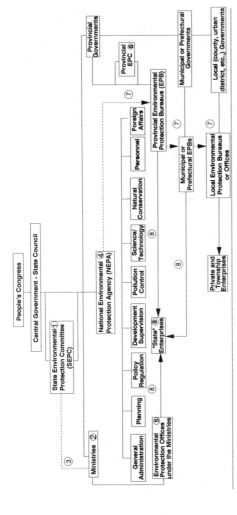

NOTES

(1) The State Environmental Protection Committee (SEPC) is not a permanent administrative branch under the State Council, rather, it is a coordinating committee. The SEPC meets every three months to discuss key issues and to coordinate interministerial actions.

(2) Ministries refer to the permanent administrative branches under the State Council, such as Planning Commission, Ministries of Energy, Agriculture, Water Resources, and Chemical Industry, etc.

(3) The members of SEPC represent over 20 ministries under the State Council and the military.

(4) National Environmental Protection Agency (NEPA) is a semi-ministry level government branch under the State Council. NEPA is responsible for implementing laws and regulations issued by the People's Congress or its standing committee. The Administrator of NEPA is also the deputy-chair of SPEC. There are nine main offices under NEPA.

(5) Most of the Ministries have their own office to deal with environmental issues. The scale of this office depends on the mission of the ministry.

(6) The Environmental Protection Committee at the provincial level functions in the similar way as the SEPC.

(7) The head of the provincial EPB is appointed by the corresponding provincial government, and NEPA gives policy and professional directions to provincial EPBs. The same relationship applies to the other two levels.

(8) Environmental protection responsibility for state enterprises can rest with Ministries, Provincial Governments, and Municipal or Prefectural Governments.

Figure 4 Environmental protection structure in China

usually cheaper, to achieve higher standards of control on new as compared to old plants. Environmental Impact Statements are required for major new facilities, and in general, new plants of a size or of an organizational type that come to the attention of environmental authorities are examined for consistency with environmental-quality goals. In practice, environmental authorities often influence the design of facilities, or sometimes their location, in order to lessen population or ecological risk. Laws and authorities exist to close or modify existing facilities, but these require a much stronger justification and burden of proof. Beyond the regulatory activities there exists the tort system. In China, those who can demonstrate direct harm (e.g. fish kills; crop damage) are entitled to damages.

The Chinese environmental authorities are seeking to inculcate the view that environmental concern should be integral to economic activity, not an afterthought. This view is characterized by the slogans used, such as "The Three Simultaneous Principle," which emphasize considering environmental along with economic and social aspects of planning, constructing, and operating new facilities, and of doing so at all points from conception of the project through production.

In terms of the level of attention given to different pollutants, the Chinese have exercised substantial discrimination. For example, in general, they have campaigned against persistent chemicals and heavy metals in contrast to organics and pathogens, which will either biodegrade, ultimately, or else will be rendered harmless when water is treated and boiled for cooking and drinking. In the same vein, they have categorized water bodies and air sheds into different quality goals, stressing protection (or nondegradation) of particularly vulnerable or high-impact locations. For example, China established three classes of ambient air quality standards similar in principle to those in the United States (42). Class I standards apply to nature reserves or areas of relatively pristine air quality. Class II standards apply to most areas of the country, while Class III standards apply to the most heavily polluted urban and industrial areas. The Chinese Class I standards are stronger than those of the United States, while their Class III standards for urban areas are weaker. The Chinese Class III standards for SO_2 and suspended particulates (250 and 500 $\mu g/m^3$, respectively) are also much less stringent than those of the World Health Organization (WHO) (100–150 $\mu g/m^3$ for SO_2 and 150–230 for particulates). The Chinese have stressed location standards, for example to avoid placing facilities upwind of population centers, as a reasonable compromise between ignoring environmental impacts and spending the capital and other expenses actually to prevent emissions.[25] This second-best plan

---

[25]In Tianjin, for example, the city moved more than 100 shops that discharged severe pollutants to the outskirts—an example of moving the hazard away from the people but not closing it down (43).

seeks to put the most serious pollutants where people are few, lessening the risk to public health. Chinese policy also calls for installation of very tall stacks to dilute the emissions before they reach the ground when no realistic prospect exists for controlling sulfur emissions. If all else fails, environmental authorities seek to have plants designed so that control retrofits can be added later when the cash flow allows.

Implementation of these broad thrusts of policy is largely decentralized, with formal authority often resting with committees of officials representing different bodies of government, including the Ministries whose plants emit pollutants (Figure 4). In other cases, the enforcement officials report to the same mayor or other country executive as does the manager who operates the plant, with the mayor, say, responsible both for the income level of the citizens and the amount of effluent released to streams. The inherent conflicts are obvious. These are made worse by the lack of a tradition of rule of law—much is left to the interpretation and judgment of officials, often responding to strongly expressed views of citizens.

What actually happens with regard to environmental protection is determined by the effectiveness of three instruments. The first of these is regulation, which mandates that certain equipment be installed and used, that emission standards be met, and so forth. Restrictions imposed pursuant to an Environmental Impact Statement are of a similar "command and control" type. Regulation and standards as they apply to a particular enterprise are always subject to negotiation, of course. In general, they are scarcely applied at all to small industries, and with various degrees of vigor to large state firms. (The Ministries themselves have the responsibility—and organizations—to meet environmental goals.)

The second instrument used in China is economic incentives. These are found in the form of tort liabilities, though such actions are rare and rarely successful except in egregious cases. There are also effluent and emission fees. These are mainly structured to be imposed on discharges above the standard. The plant has the incentive to reduce discharges to the extent the cost of doing so is less than the fee. The operation of this incentive depends, of course, on the decision makers in the facility actually bearing some of the burden. That does not occur so long as wages, employment, and amenities are guaranteed; the Chinese, by creating the "Responsibility System" whereby the incomes and other benefits from the factory are keyed to its economic performance, have strengthened the effectiveness of economic incentives.[26]

---

[26]The "Responsibility System" attempts to link national environmental policy with local implementation (44). Here, governors, mayors, and county heads are required to sign a responsibility contract that stipulates the target and tasks of environmental protection to be undertaken during their tenure in office. They are rewarded or disciplined based on their success or failure.

The final instrument is public opinion and changes in private behavior through persuasion. The Chinese government has been seeking to raise environmental consciousness among the citizenry for a number of years through organizations, schools, and the media. While designed in part to change individual behavior, these campaigns have also sensitized people to environmental problems. An activity announced in 1989 is instructive. City and provincial leaders enter into "contracts" specifying expected (promised) performance on an array of environmental indicators. These contracts are given wide publicity in the locality, and then the officials are held accountable (in terms of public opinion) for meeting these terms. This changes the trade-offs officials see when making choices and places the environment higher on priority lists. In a similar way, cities have been ranked according to environmental quality, with the officials of low-ranking communities clearly held accountable by a newly informed citizenry.

There is little doubt that progress in pollution control and environmental protection has been obtained not in spite of, but rather in conjunction with economic growth (40). Economic reforms as well as a growing awareness of the environment have clearly provided incentives to be more efficient. Investment in environmental protection has increased substantially in the 1980s. In 1982 pollution prevention and control investment was about 0.2% of national income, but by 1988 it had increased to 0.7% (35).[27] However, implementation and enforcement of this system is still woefully incomplete, a reflection of the pervasive lack of a "rule of law" that characterizes much of Chinese society.

INTEGRATING ENVIRONMENTAL AND ENERGY POLICIES

The gap between the existing and appropriate levels of environmental protection in China is thought to be large by most observers. However, it is important to set priorities for action on energy pollution based on Chinese conditions and not to presume that standards appropriate to other countries should be applied. The Chinese energy system has many demands on it: to increase industrial and agricultural output to meet basic needs of the population; to accumulate capital to promote growth in the future; and to improve amenities and provide educational, cultural, and recreational opportunities. Improved environmental quality competes with all of these for added resources. Therefore, application of environmental policy to the energy system requires careful study of the actual conditions in the energy sector and in the economy as a whole.

[27]In the United States investment in environmental protection is about 2%.

Capital is especially scarce in the Chinese economy, and actions to prevent pollution that require large capital investment are very difficult to accomplish. This reality suggests that emphasis on pollution prevention and improved management will be important. Improvements that lead to greater efficiency in energy use would enhance the environment and save capital that would otherwise go to energy production. These improvements would increase the prosperity of the country and make it healthier and cleaner at the same time.

Some improvements in management would occur automatically if energy prices were increased and the environmental responsibility system strengthened, as discussed above. Others, such as rationalizing the provision of space heating in urban areas, would require planning and policy changes. Still others, such as developing alternatives to direct residential raw coal burning in urban areas, would require changes in the delivery of energy services (gas, briquettes, distribution systems). Lastly, improvements such as better control of power plant and industrial emissions may require both better management and added capital. Experience in other countries after the world energy price increases shows, however, that major reductions in energy use are possible with little or no loss in industrial output or in consumer well-being over the long term. If this is true in China as well, a high priority should be given to management and planning for energy efficiency. This will bring economic, social, and environmental improvements simultaneously.

How can China protect the environment while continuing to increase its coal consumption to meet its economic goals? The answer requires a combination of strategies that begin with price reform; coordinating investment strategies for implementing energy and environmental policies; increasing financial and human resources for environmental control; strengthening environmental standards and enforcement; and investing in technical improvement in conversion technologies, end-use equipment, and alternative fuels.

Price Reform and Planning

The goal of price reform is to ensure that energy is produced and used so as to maximize its contribution to economic development and minimize its negative effects on health and environment. Coal price reform must include a gradual increase in prices adjusted toward the marginal cost of production and reflecting differences in heating values and qualities (ash content, sulfur content, washed, etc). Moving toward a freer price as a major basis for making decisions about energy is very difficult in China because the entire economy has adjusted to the previous fixed, low price. Experience in Western economies with the energy price shocks of the 1970s suggests that major disruption of agricultural and industrial output can be avoided if firms have a few years to adjust to a certain future price increase. Then they can make new investments based on the prices that will exist in the future. Energy-efficiency improvements in other processes are induced to take place as opportunities occur. At the consumer level, the reductions in energy subsidies can be gradual but certain so that decisions here, too, can be made for the future without large disruption in the present.

Reform may also be achieved through planning changes that would lead to more efficient energy use. For example, much more efficient electrical appliances have been developed over the past decade. If a decision were made that the Chinese economy would switch over to these advanced designs, energy use in the future could be reduced substantially, especially because the number of appliances is increasing so rapidly. Once the appliances are built, they last a long time and future adjustment is not practical. Even if energy-efficient appliances are somewhat more expensive to manufacture, the long-term energy savings may make them worthwhile. Another example is space heating. New building construction could be designed to use less energy. Further, it is much more efficient (and less polluting) for space heating to be provided from large boilers (block or district heating) that serve multiple buildings or sections of cities; prior planning could ensure that space heating was designed with long-term efficiency as a major factor. Block and district heating are now being encouraged in Beijing.

Finally, energy planning and price reform can serve as catalysts for substituting away from high-carbon and polluting coal, to low-carbon and less polluting fuels, such as gas. Energy price reform in noncoal energy sectors can help stimulate investment in resource development and the introduction of these fuels into China's urban areas. Where substitution for coal may not be feasible, price reform can improve the economics of coal washing and briquette manufacture.

A long-term solution to China's energy and environment problem will clearly require massive investment, but it is unlikely that China can devote enough of its scarce capital to this sector in the near term to meet all of the needs. This suggests the grave necessity to use existing facilities with greater efficiency and to ensure that new capital is invested where its contribution will be greatest. Adjustment of energy prices is a critical first step and by itself can do much to improve efficiency. When energy prices are equal to the true costs, they provide appropriate incentives to producers and consumers to conserve and encourage energy to be used where it produces the greatest value.

Priorities

Price reform will not be sufficient if China is serious about reducing energy-related pollution and environmental degradation. China must begin to set national and local priorities for investment in pollution abatement and control. At the national level China can refine policies on environmental regulations, strengthen emission standards, and promote more effective enforcement by increasing penalties and pollution-discharge fees. In addition, China can implement a variety of strategic steps and technical improvements to reduce the energy intensity of industry and increase energy efficiency. However, given the lack of capital funds, investments need to be made where the marginal benefits are greatest. This fact obviously suggests energy and environmental policies targeted at the industrial sector and urban household sector. These sectors are the least efficient, use the most coal, and are responsible for most of China's environmental pollution. Increased energy efficiency in these sectors would contribute to better air quality and improved public health, reduce global carbon emissions, and provide much of the energy for the expansion and modernization of China's economy.

As a first step, China should take active measures to improve the quality of coal it uses and match the uses to the appropriate final demands. This strategy would require the use of anthracite or carbonized coal in the urban residential market, better matching of coal to particular boiler designs, and greater use of coal washing to remove ash and inorganic sulfur. In addition, China should take active steps to begin to limit the burning of raw coal in cities, especially in the urban residential sector, and encourage the substitution of coal briquettes and gas. Moreover, the increase in small- and medium-scale industrial boilers should be restricted and outmoded boilers retired. There is also considerable potential to replace the backward and scattered small-scale heating systems and to develop cogeneration and block heating schemes. All of these steps would increase coal combustion efficiency and reduce emissions.

China's plans to diversify its primary energy structure—incorporating nuclear power, natural gas, and long-distance electricity transmission (minemouth generation and hydropower)—will also contribute to reduced pollution and greater energy efficiency. Of course, the environmental impacts at the mine site and the ecological damage from large hydropower projects must be considered. There is also considerable potential to develop and deploy advanced thermochemical combustion technologies as well as postcombustion control equipment (scrubbers and filter technology). Whether China will decide to devote a sufficient amount of its scarce capital to pollution control rather than to building power plants to raise the standard of living and to keep its industry operating is uncertain.

CONCLUSIONS

China's energy system has faced severe challenges arising from changing conditions. China has made substantial progress in adapting to the changed conditions that have caused shortages and led to pollution and degradation, and the country has sufficient energy resources to support the industrialization and growth in income that is planned. At present, however, the facilities to produce, transport, convert, and use energy are inadequate. Major expansion is required to provide enough energy to meet current needs and to support future growth. At the same time, many opportunities appear to exist to improve the efficiency of energy use. However, to achieve these desired outcomes, some policies may require strengthening, and adjustment of other policies may be desirable.

Economic development and environmental protection are both affected by the energy system. An economy must have adequate energy to support development, but energy production and use, if poorly managed, can harm the environment. Both development and environmental quality can be achieved through careful consideration of the needs of each, and proper coordination of policies. The successes of the Chinese in meeting these dual goals are beginning to be seen, but much remains to be done.

Recently China has chosen to slow its economy to restrain inflation and to put on hold or even reverse liberalizing reforms. The first of these changes could provide time to accommodate environmental concerns, but at the same time it will result in lower growth and thus fewer resources for investment in pollution control and ecosystem improvement. Environmental policies have changed little so far. Nevertheless, the new environmental law was passed in late 1989 to further solidify the institutional basis of protection; and the new policy placing more accountability for environmental quality on local and provincial officials has been adopted. The Chinese leadership continues to assert its dedication to environmental progress, and frequent press reports emphasize both successes and work yet to be done.

On the whole, the Chinese have demonstrated that they are serious about holding environmental degradation in check. The infrastructure to implement pro-environment policies is largely in place. New installations are less hazardous than those built a decade ago; some cities are noticeably cleaner and pollution growth has lagged behind increases in output. Public opinion is aroused and occasionally has led to plant closures and to community rejection of polluting industries. The problems, however, are enormous, and the resources to confront them are few.

Chinese leaders have emphasized pollution prevention and managerial interventions that would limit degradation and cause minimal reduction in economic growth. The key mechanism has been to create economic incentives for environmentally responsible behavior. These initiatives have already achieved modest success and have built the foundation for real progress in the future.

Energy and environmental policy for the 1990s must reinforce environmental administration and promote technological progress. China must increase investments that seek to use resources more efficiently. Moreover, China must continue to make investments in new energy technologies and antipollution projects. Industry must increase energy efficiency and reduce pollution. There must be regular supervision and inspections to see that progress is being made. New energy-saving and nonpolluting technologies as well as use of better-quality coals must be promoted and adopted.

In the global arena, China should be willing to cooperate on reduction in coal use since this is the major source of air pollution and acid rain. Any agreements on limiting the use of coal directly benefit China in terms of reduced acid rain and improved public health. However, the reality is that China will continue to develop and consume coal as its primary source of energy well into the next century. There are simply no other choices given its desire to expand its economy, the need to get power projects on-line, and its comparative advantage in coal production. The key is, of course, how much of China's existing production can be saved to fuel its economic growth. Whether China can increase the efficiency of coal use in the industrial sector and be successful in encouraging substitution in the residential sector remains to be seen. Improvements to date have by some accounts been largely in administrative in nature and relatively easy to implement. The technical improvements require capital of which China has very little. What effects price reform will have in motivating efficiency improvements and generating funds for investment in the energy and environmental control sectors is uncertain.

Literature Cited

1. *People's Daily*. Jan. 7, 1991
2. Ministry of Energy. 1989. *Energy in China*. Beijing
3. Bley, L. 1990. The impact of global warming. *Pet. Econ*. March
4. *China Today*. Feb. 1990
5. Huang, Y. C. 1990. Developing the energy industry and protecting the environment. *Int. Conf. Integration Econ. Dev. Environ. China*. Hainan, P. R. China
6. Sun, J. P. 1990. Present situation and outlook of China's electric power industry. *Advis. Group. Meet. Energy Electr. Demand Supply*. Vienna: Int. Atomic Energy Agency
7. Econ. Intell. Unit (EIU). 1990. *Country Report—China, North Korea*. 2:24
8. Int. Energy Agency. 1988. *Energy in Non-OECD Countries*. Paris
9. Owen, A. D., Neal, P. N. 1988. China's potential as an energy exporter. *Energy Policy* 17:485–500
10. British Petroleum. 1989. *Statistical Review of World Energy*. London
11. Int. Energy Agency (IEA). 1988. *World Energy Statistics and Balances, 1971–1987*. Paris
12. Global Energy Efficiency Initiative (GEEI). 1990. *Assessment of a Global Energy Efficiency Initiative*. Washington, DC: US Agency Int. Dev., Office of Energy
13. World Resourc. Inst. (WRI). 1988. *World Resources 1988–89*. New York: Basic
14. Smil, V. 1988. *Energy in China's Modernization*. Armonk, New York: East Gate, Sharpe
15. World Resourc. Inst. (WRI). 1990. *World Resources 1990–91*. New York: Basic
16. Huang Yuanjun, Zhao Zhongxing. 1987. Environmental pollution and control measures in China. *Ambio*. 16:257–61
17. Zhao, D., Sun, B. 1986. Atmospheric pollution from coal combustion in China. *J. Air Pollut. Control Assoc*. 36: 371–74
18. *China Environment News*. July 15,

1990. China's first communique on environmental state

19. Global Environ. Monitoring System (GEMS). 1989. An assessment of urban air quality. *Environment*. 31:6–13, 26–37

20. Chandler, W. U., Makarov, A. A., Zhou, D. Sept. 1990. Energy for the Soviet Union, Eastern Europe, and China. *Sci. Am.*

21. Barron, W. F., Hills, P. 1990. *Greenhouse Gas Emission: Issues and Possible Responses in Six Large Economy, Lower Income Asian Nations. Working Paper No. 5*. Cent. Urban Stud. Urban Planning, Univ. Hong Kong, Hong Kong

22. Minist. Agric. 1990. *China Rural Energy Construction System Its Development and Utilization*. Beijing

23. World Bank/United Nations Dev. Program. 1989. *Country-Level Rural Energy Assessments a Joint Study of ESMAP and Chinese Experts*. Washington, DC: World Bank

24. Han, G. J. *Beijing Rev*. June 25–July 1, 1990

25. *China Daily*. Aug. 21, 1990

26. Sonmez, A., Stott, P. 1990. Natural resource pricing and accounting in China's economic and environmental management. See Ref. 5

27. Granzer, R. April–May 1989. *OECD Observer*. Paris

28. Chen, X. April 21, 1990. *China Daily*

29. Huang, Y. C. 1989. China's energy development and problems. *New China Quarterly*

30. Gorst, I. 1988. Energy shortages becoming acute. *Pet. Econ.* Nov:363–64

31. *China Daily*. June 13, 1990

32. *Beijing Rev.* Nov. 5, 1990

33. Greer, C. 1990. *China Facts and Figures Annual, Volume* 13. Gulf Breeze, Florida: Academic Int.

34. Wen, T. S. 1990. China's energy resources: present and future. *China Today*. Feb.

35. Qu, G. P. 1990. Thoughts about policies for the coordinated development of China's economy and environment. See Ref. 5

36. JPRS Report. 1990. *Science and Technology, China: Energy*. Feb.

37. Zheng, Y. Z. 1987. China's offshore oil industry. *China's Foreign Trade*. April

38. Cheng Zheng-Kang. 1986. *A Brief Introduction to Environmental Law in China*. Boulder, Colo: Natural Resourc. Law Cent.

39. Jin, R. L., Liu, W. 1987. *Environmental Policy and Legislation in China. Proc. Sino-Am. Conf. Environ. Law*. Boulder, Colo: Natural Resourc. Law Cent.

40. Ross, L. 1987. Environmental policy in post-Mao China. *Environment*. 29:12–17, 34–39

41. *China Today*. Aug. 1990

42. Ross, L. 1988. *Environmental Policy in China*. Bloomington, Ind.: Ind. Univ. Press

43. *China Daily*. Dec. 7, 1990

44. Li, P. *Beijing Rev*. July 16–20, 1990

Annu. Rev. Energy Environ. 1991. 16:235–58

SAFETY AND ENVIRONMENTAL ASPECTS OF FUSION ENERGY

J. P. Holdren

Energy and Resources Group, University of California, Berkeley, California 94720

KEY WORDS: energy risk assessment.

INTRODUCTION

This article reviews the state of knowledge about the safety and environmental (S&E) characteristics of fusion energy technologies. Both magnetic-fusion energy (MFE) and inertial-confinement fusion (ICF) are considered, although necessarily with more attention to the former because of the larger body of literature on MFE safety.

One of the main incentives for investing in the development of fusion energy is the prospect that the S&E characteristics of fusion will be less troublesome in fact and in public perception than those of fission and fossil fuels have been (1–5). This expectation of S&E advantages for fusion is not a guarantee, however: tritium, neutrons, and neutron-activation products in fusion reactors represent radiological hazards similar in kind if not neccessarily in magnitude to those of fission reactors. Exploiting the potential for minimizing these hazards in fusion technology is a challenge that deserves high priority in fusion development efforts. Nor should the developers of fusion neglect the need to minimize non-nuclear risks—chemical hazards, exposures to magnetic fields and other non-ionizing radiation, thermal impacts, and so on.

The article begins with a brief historical overview of work on these issues, followed by summaries of the current knowledge with respect to both nuclear and non-nuclear hazards. Some concluding observations emphasize needs and priorities for future work.

1056-3466/91/1022-0235$02.00

HISTORICAL OVERVIEW

The first substantial analyses of the S&E characteristics of fusion reactors appeared in the period 1969–1970 and focused mainly on tritium hazards in routine operations and in accidents (6, 7), with secondary attention to problems posed by neutron-activation products (8). By the mid-1970s, activation-product hazards (occupational radiation doses, radioactive-waste problems, and possibilities for release in reactor accidents) were receiving attention comparable to that given to tritium (9–13), reactor-design studies featuring "low-activation" materials were being conducted (14, 15), and fusion S&E issues had become the focus of international working groups and reviews sponsored by the International Atomic Energy Agency (16, 17).

The evolution of fusion S&E studies in the second half of the 1970s reflected the influence of three trends that had been taking place in S&E assessments of other energy options.

1. A trend toward increasing comprehensiveness in the kinds of energy-related activities and environmental phenomena included in S&E assessments led to the inclusion, in fusion S&E assessments, of not only the most obvious radiological hazards but also the occupational and public hazards of fuel acquisition and transportation, accident risks to workers in component fabrication and power-plant construction, chemical hazards in fusion technology, and magnetic-field hazards (see, e.g., 18).
2. A trend toward comparative assessment of technologies with similar applications (owing to recognition that estimates of S&E risks and impacts are meaningful only in relation to the risks and impacts of alternative ways to obtain the same societal benefits) led to comparative S&E assessments of fusion and fission—most notably the study of the S&E characteristics of fusion and fission fast-breeder reactors conducted under the auspices of the International Institute for Applied Systems Analysis in 1975–1977 (19).
3. A trend, in the study of fission-reactor accident hazards, toward analysis of physically plausible pathways by which radioactive material could be released and estimation of radiation doses that could actually result (as opposed to indices based on inventory alone) (20–22) led to a similar focus on accident phenomenology in fusion-reactor safety assessments (23–26).

The findings of the fusion S&E assessments of the late 1970s, while obviously preliminary and incomplete, were not wholly reassuring. They suggested that:

1. tritium inventories as large as envisioned in early conceptual fusion-reactor designs could deliver significant off-site radiation doses in severe

accidents, and might be difficult to contain in normal operation to the extent needed to meet the likely regulatory limits on routine exposures at the plant boundary (27);

2. temperatures conceivably attainable in severe accidents involving lithium fires might be able to mobilize enough activation products from some of the candidate structural materials to create off-site dose potentials, in the absence of effective containment, greater than those of the tritium and not much smaller than those associated with severe fission-reactor accidents (28);

3. depending on the choice of materials, long-lived activation products in fusion-reactor structure could represent a radioactive-waste hazard sufficient to require deep geologic disposal of volumes of material not notably different than those associated with fission fuel cycles (see, e.g., 29).

These concerns were seen by many as calling into question the prospect that fusion would be able to escape the S&E difficulties that, by the end of the 1970s, were clouding the future of fission. The result was a considerably intensified interest in these matters in the fusion community—and at least a modest increase in the amount of related work.

The expanded efforts in fusion S&E assessment that took place in the 1980s embodied a number of elements proceeding in parallel:

1. more accurate characterization of the radiological hazards of fusion technology through use of more detailed reactor designs,[1] new experimental data on accident phenomenology (e.g. 32, 33), and more sophisticated analytical models (including computer simulation) of accident conditions, mobilization of radioactivity, and pathways to human exposure from reactors as well as from waste repositories (34–39);

2. use of this work to construct indices of relative hazard that strike a reasonable balance between clarity and comprehensiveness, that lend themselves to inter-design and inter-source comparisons, and that convey some information about probability of harm (albeit far short of what would be contained in a full probabilistic risk assessment) (35–40);

3. more extensive studies of the feasibility of low-activation structural materials (41, 42) and advanced (low-neutron) fuel cycles (43, 44);

4. systematic exploration of the trade-offs, in fusion-reactor design, among traditional performance criteria, minimization of tritium inventories, and maximization of safety margins against mobilization of activation products in accidents (45);

[1]Notable examples were the "Starfire" design (30) at the beginning of the decade and the ARIES design at the end (31).

5. integration of the results from the preceding elements, together with engineering-economic modeling of overall system performance and costs (46), to investigate the potential of diverse fusion-reactor concepts to achieve combinations of environmental, safety, and economic characteristics that are attractive compared to those of contemporary and advanced fission energy systems (47–52); and

6. analysis of the S&E characteristics of various tritium-burning experimental fusion machines that are in operation, under construction, or being designed (53–56).

At some risk of oversimplification, the thrust of the results of the more recent studies can be summarized as follows:

1. Fusion reactors of contemporary design would have advantages over fission reactors with respect to consequences of severe accidents and magnitude of radioactive-waste burdens, even in the relatively unfavorable case (for fusion S&E characteristics) of stainless-steel structure and liquid-lithium coolant/breeder.

2. The magnitude of these advantages becomes more impressive for fusion-reactor designs that employ lower-activation structural materials and/or contain less stored energy than the stainless-steel/liquid-lithium combination.[2]

3. For some fusion-reactor designs, it appears that avoidance of off-site deaths from acute radiation syndrome in severe accidents could be assured without reliance on active safety systems or containment buildings.

4. Fusion energy systems would present smaller problems than fission systems with respect to unwanted links to nuclear weaponry: fusion systems (other than fusion-fission hybrids) would contain no fissile material, and the introduction of means to produce it would be relatively easy to detect.

5. Increased experience in the past several years with the handling and control of tritium supports the belief that fusion will be able to meet the same tight standards on routine emissions with which fission reactors now comply.

6. A number of experimental fusion machines operating in the 1990s and beyond will be burning tritium and generating neutron-activation products at environmentally significant rates. While the neutron fluences in these machines (measured in megawatt-years of neutron energy per square meter of plasma-chamber wall) will be considerably lower than in commercial

---

[2]Examples of lower-activation materials, in order of increasing S&E attractiveness and, unfortunately, decreasing assurance of technological feasibility are: elementally tailored ferritic steel, vanadium alloy, silicon carbide. Coolant/breeders with less stored energy than liquid lithium include FLiBe molten salt, lead-lithium alloy, and helium/Li_2O.

fusion reactors, inventories of tritium and some activation products may not be.[3] Thus these experimental machines will require attention to the minimization of environmental and safety hazards no less than will the reactors to come.

7. Fusion will create certain non-nuclear risks and impacts similar to those of other energy sources (e.g. waste heat, hazards of materials transportation and facility construction, chemical hazards to workers) and at least one non-nuclear hazard not shared by other sources (intense magnetic fields in MFE reactors, powerful pulses of laser light or particle beams in ICF), but none of these issues currently appears to be serious enough to weigh importantly in societal decisions about fusion's attractiveness compared to other energy options.

The sections that follow provide more information about the reasoning and results on which these conclusions rest.[4]

NUCLEAR HAZARDS

The discussion that follows is divided into three subsections: tritium, activation products, and unwanted links to nuclear weaponry.

Tritium

Tritium (^{3}H, in the fusion literature often denoted T) is radioactive with a half-life of 12.3 years, emitting a beta particle of rather low energy (maximum 18 keV, average 6 keV) and yielding stable helium-3. Tritium will occur in necessarily sizable quantities as an ingredient of the fuel in fusion reactors operating on the deuterium-tritium (D-T) reaction, and in smaller quantities as the product of D-D reactions in "advanced-fuel" systems that contain deuterium. It is also formed as the result of neutron bombardment of light elements that may be present in fusion-reactor components.

The tritium in a D-T fusion reactor will reside in two quite different inventories. The "active" inventory consists of T in the plasma, plasma-facing components, breeding blanket, tritium extraction system, impurity-control and vacuum-pumping systems, fuel injectors, and any pipes and intermediate reservoirs linking these components. The magnitude of the active inventory

---

[3]The reactors that come later will be able to exploit low-activation materials and tritium-minimizing design innovations not available in time for the experimental machines.

[4]In addition to other references cited in what follows, a wealth of information on many of the relevant issues can be found in two particularly thorough reviews conducted for the US Department of Energy (57) and the International Atomic Energy Agency (58). Data presented here and not otherwise attributed are from these two studies.

depends on the characteristics of all these subsystems, including especially the mean residence time of T in the breeding blanket and the fraction of injected T that burns on each pass through the plasma. The "inactive" tritium inventory is that part kept in storage for use in the event of breakdown of the tritium extraction system or for eventual shipment to start up other reactors; its magnitude is a matter of choice, but since it can be kept separate from the fusion power core and very well confined and protected, it usually is not given much attention as a source of routine emissions or accidental releases.

Estimates of the active tritium inventory in various designs for D-T reactors range from around 200 grams to several kilograms (12, 13, 27, 45, 47, 48, 59, 60). The active tritium inventory in the International Thermonuclear Experimental Reactor (ITER)—a 1000-megawatt(thermal) fusion device scheduled to operate in 2005—is expected to be 4–5 kg (56). A D-T-fueled fusion reactor operating continuously at 4000 thermal megawatts—1200–1500 electrical megawatts—burns about 500 grams of tritium per day. At this burn rate, a fractional burnup of 0.1 corresponds to an injection rate of 5 kg/day of T and a fractional burnup of 0.01 corresponds to 50 kg/day.

Estimates of the amount of tritium in the breeding blanket have tended to fall over the years as designs of blankets and tritium removal systems became more sophisticated, but estimates of the inventory in MFE systems that recycle tritium from the plasma have been growing recently because of physics results that imply a low fractional burnup in tokamak reactors.[5] Use of graphite in plasma-facing components, moreover, as contemplated for the "physics phase" of ITER operation and for some commercial reactor designs, can increase the active tritium inventory by 1–2 kg or more (55, 56). Fusion reactors based on advanced fuels (D-D or D-He3), on the other hand, may be able to achieve T inventories below 100 grams (43, 47, 48, 52). Increasingly sophisticated dynamic models of tritium flows and stocks in fusion reactors are now coming into use and should permit improved estimates of tritium inventories as well as development of approaches that can reduce them (61–63).

An active tritium inventory in the range of 100 to 5000 grams corresponds, at tritium's specific activity of 9700 curies (Ci) per gram, to 1–50 megacuries (MCi). This tritium is of concern as a source of radiation exposures both to workers inside the plant and to members of the public off-site, in normal operation as well as in accidents.

International guidelines, as well as national regulations in most countries, limit occupational radiation exposures to an individual worker to 50 mSv (5 rem) per year (64–66). The corresponding Recommended Concentration

---

[5]In ICF reactors, estimated fractional burnups around 0.3 lead to lower tritium throughput and, therefore, lower inventories.

Guideline (RCG) for tritiated water in air[6] for continuous occupational exposure is 5 microcuries per cubic meter (64). It follows that 3 Ci of HTO—representing on the order of 10^{-6} to 10^{-7} of the active T inventory—would be enough to contaminate all the air in a 500,000 m^3 reactor building to the RCG. This figure suggests the degree of control over the tritium inventory that will be required to protect workers without protective suits. Even very modest leaks and spills clearly would necessitate the use of such suits. There is already considerable capability for and experience with tritium cleanup indicating that these problems are manageable, however (67–69).

National regulations typically limit radiation exposure, by airborne effluents, to a hypothetical member of the public who spends full time at the boundary of a nuclear-energy facility to 50 microsieverts (μSv) per year (66). For a boundary at a distance of 1 km from the reactor, and considering annual average meteorological conditions at typical sites, complying with such regulations implies restricting routine releases of airborne tritium to 100–200 Ci of HTO per day, which is on the order of 10^{-5} to 10^{-4} of the plant's active inventory (70, 71). (Releases of other radioactive isotopes, of course, would reduce the releases of tritium that could be permitted.) Recent experience with handling large quantities of tritium in fusion-technology test facilities—e.g. the Tritium Systems Test Assembly (69)—indicate that the needed degree of tritium control is feasible, but holding down the costs of achieving it in commercial reactors may require considerable ingenuity. (See also 72, 73.)

Tritium, either as HT gas or as HTO liquid or vapor, is highly mobile even under normal operating conditions, and under accident conditions it would be one of the more likely radioisotopes to escape in quantity. How large a release of tritium would be tolerable in an accident depends on the "threshold" dose, to the most-exposed member of the public, that one would not like to exceed. One such dose threshold used by regulatory authorities in many countries is a 50-year dose commitment of 100 mSv (10 rem) (59), which corresponds approximately to an increase of one percent in the exposed individual's pre-existing probability of dying of cancer. This dose could be received by an individual remaining for several days at the site boundary 1 km from an accidental release of 100–200 grams of tritium as HTO under highly unfavorable weather conditions (accounting for inhalation and skin absorption, resuspension of ground-deposited tritium, and a quality factor of 2 Sv/Gy in computing the dose per curie of tritium intake) (66, 70). Such a release represents from 5 to 100% of the active tritium inventory in recent fusion-reactor designs.

---

[6]HTO, about 20,000 times more hazardous per curie than HT gas, is usually assumed to be the form present. HT is converted to HTO in air with a variable characteristic time ranging from hours to days.

It is also of interest to ask how much tritium could be released in an accident without exceeding the 2 Sv critical dose[7] below which no early fatalities would be expected (47, 48). Given that the critical dose from an accidental exposure to tritium is about half of the 50-year dose commitment, the tritium release (as HTO) corresponding to a critical dose of 2 Sv to an individual at the site boundary 1 km away is 4 to 8 kg, neglecting the possibility of evacuation. If the individual leaves within the first few hours after the accident, the release needed to give a critical dose of 2 Sv is 2–3 times as large, say 10–20 kg. Thus, even a 100% release of the active tritium inventory in a fusion reactor could not produce an off-site dose large enough to lead to early fatalities unless that inventory exceeds 3–4 kg (no evacuation from the site boundary) and more probably 10–20 kg (most exposed individual leaves the site boundary after a few hours).

Activation Products

Neutron-activation products are formed when fusion neutrons (at 14 MeV from every D-T reaction and at 2.5 MeV from about half the D-D reactions) strike the main constituents and impurities in the reactor structure, coolant, tritium breeder (if any), neutron-multiplier materials (if any), and reactor-building atmosphere. The variety of neutron-activation reactions and resulting isotopes is very large when one takes into consideration the diversity of materials that may be found in fusion reactors, as well as the multistep reactions in which activation products or their decay products are themselves bombarded by further neutrons. On the other hand, the array of activation products that actually arises in a given reactor design can be controlled by the reactor designers to a considerable extent through their choices of structural and other materials to be used in the reactor (subject to other criteria these materials must satisfy). This flexibility to choose materials to minimize activation hazards is a key to fusion's potential to deliver major S&E advantages. It also represents a crucial challenge to the materials-science and fusion-reactor-design communities, for without a major collaborative effort to develop and test low-activation materials for use in fusion-reactor environments, the potential will not be realized.

The types and quantities of neutron-activation products that will be present in a fusion reactor can be predicted with considerable confidence using computer programs that have been cumulatively improved and checked against experiments over a period of many years.[8] These computer programs

---

[7]The critical dose is the whole-body dose received in the first 7 days of exposure plus half of that received in the 8th through the 30th days; it correlates well with acute radiation effects (20).

[8]One such set of computer programs, used at the Lawrence Livermore and Oak Ridge National Laboratories, is described in Refs. 74–77. Many variants exist at other institutions and in other countries.

take into account the detailed compositions and spatial relationships of different reactor components, they trace neutron-activation chains through several steps, and, similarly, they trace the decay of all the produced radioactive species over time in order to provide not merely a "snapshot" but a moving picture of the radioactive inventories that result from neutron activation. Such inventory calculations provide the basis for determinations of several characteristics important in S&E assessment, including: (a) the intensity of the radiation fields in the vicinity of activated reactor components (relevant to limiting worker exposures); (b) energy releases from radioactive afterheat (relevant to the possibility of structural damage and mobilization of radioactivity in accidents); (c) the hazard potentials associated with activated material that might be able to escape the reactor either in routine operation or in accidents; and (d) the characteristics of the radioactive wastes that will remain at the ends of the useful lives of fusion reactors and components.

The neutron-activation calculations reveal that typical D-T MFE reactors of contemporary design, built of metals such as austenitic or ferritic steels or refractory-metal alloys and sized to deliver 1200 electrical megawatts (MWe), would contain 0.6–4 gigacuries (GCi) of activation products when shut down after a long period of operation, and 3–10 times less a day after shutdown (30, 45, 47, 48). A low-activation D-T MFE reactor with SiC structure would have comparable activity to the metal-structure systems at shutdown but 10^2–10^3 times less than these more conventional systems a day after shutdown (48). The Cascade ICF reactor design, making extensive use of SiC components, has a calculated inventory of 0.24 GCi at shutdown and 0.008 GCi a day later (60). A fusion-reactor design utilizing the harder-to-ignite D-He3 reaction has been predicted to have an activation-product inventory of about 0.04 GCi at shutdown (48). (The neutrons in this case come from D-D side reactions.)

Between 15 and 60% of the activation-product inventory in typical fusion-reactor designs will be embedded in the solid "first wall," which separates the plasma chamber from the rest of the tritium-breeding and energy-absorbing blanket, and in other plasma-facing components such as the divertor plates found in typical tokamaks (45, 47, 48, 55–59). Most of the remainder of the inventory is found in the other solid blanket components and the radiation shield. Much smaller quantities, typically between 0.1 and 1% of the total, are found in the coolant—partly from activation of the coolant itself and partly from activation of corrosion products and the like—and in "erosion" dust mobilized from plasma-facing components by the constant bombardment of ions and electrons. Still smaller quantities are found in the reactor-building atmosphere. Some ICF D-T reactor designs achieve a substantial reduction in total activation-product inventory by interposing a flowing layer of low-activation liquid or granular coolant/breeder material between the reaction zone and the innermost solid structure (60, 78, 79).

The total activation-product radioactivity inventories for typical 1200 MWe D-T reactor designs at shutdown are 1.2–10 times smaller than the 5 GCi of fission products in a fission reactor with the same electrical output. Use of the D-D fuel cycle brings little or no advantage in the quantity of activation (19, 52, 60), but the inventory in a D-He3-fueled fusion reactor could be 3–100 times smaller than that of its fission counterpart (47, 48, 60). A comparison based on curies alone is not very informative about relative hazards, however. One needs to look as well at the characteristics of the radioactivity that govern the possibility of doses to people, including its mobility under normal and accident conditions and the doses it can deliver from outside and inside the human body.

HAZARD MEASURES FOR ACCIDENTAL RELEASES   Table 1 presents a number of measures of the accident hazard associated with the inventories of radioactivity in fusion and fission reactors. The measures are arranged in ascending order of informativeness, starting with numbers of curies. (The more informative the indicator, unfortunately, the more difficult it is to calculate.) The estimates shown are drawn largely from Refs. 37–39, 47, 48, 80, and 81. A few observations about the meaning of the measures follow.

1. The biological hazard potential in air (BHP_a) has units of volume and represents the amount of air needed to dilute all the radioactive isotopes in the inventory to the RCG for unrestricted (public) exposure (80).[9] This measure makes use of information about the relative radiological toxicities of different isotopes, but it is not readily translatable into an estimate of the doses that people could receive in an accident.

2. The complete-release critical-dose potential is the critical dose of radiation that would be received by an individual at the reactor-site boundary downwind of a (hypothetical) 100% release of the inventory under weather conditions tending to maximize this dose. It is calculated using sophisticated models of atmospheric transport and of mechanisms of radiation exposure (see, e.g. 37–39, 82–85), but it takes no account of the factors that would limit the release of most radioactive isotopes to much less than 100% of their inventories.

3. The complete-release chronic-dose potential is the whole-body dose (or, nearly equivalently, the dose to the bone marrow) that would be received from ground contamination 10 km downwind of a (hypothetical) 100% release by an individual who remained at this location for 50 years after the accident. It

[9]RCGs for all important fission products have long been available from official bodies (see, e.g., 64), but performing useful fission-fusion comparisons using BHPs became possible only after Fetter developed a consistently calculated set of RCGs for fusion-relevant isotopes not encountered in fission applications (37–39).

Table 1 Some quantitative measures of accident hazard[a]

Measure	Units	Fission/fusion ratios[b]
Radioactivity inventory	Ci	1–2, steel D-T
		1–9, V-alloy D-T
		2–3, SiC D-T
		100, V-alloy D-He3
Biological hazard potential	m³	5–20, steel D-T
		30–100, V-alloy D-T
Complete-release critical-dose potential	Sv	2–3, stainless steel D-T
		3–5, ferritic steel D-T
		4–20, V-alloy D-T
		10–50, SiC D-T
		400, V-alloy D-He3
Complete-release chronic-dose potential	Sv	0.3–0.4, stainless steel D-T
		1, ferritic steel D-T
		2–50, V-alloy D-T
		10–400, SiC D-T
		600, V-alloy D-He3
Partial-release critical-dose potentials, e.g. for:		
100% of T (as HTO) and noble gases	Sv	10–100, various D-T[c]
		300, V-alloy D-He3
maximum plausible release fractions[d] for all isotopes (giving maximum plausible dose, MPD)	Sv	2–3, stainless steel D-T
		2–5, ferritic steel D-T
		3–100, V-alloy D-T
		100–150 SiC D-T
		1000, V-alloy D-He3
Partial-release chronic-dose potentials, e.g. for:		
maximum plausible release fractions[d] for all isotopes (giving maximum plausible dose, MPD)	Sv	0.6–0.8, stainless steel D-T
		1–2, ferritic steel D-T
		2–100, V-alloy D-T
		100–900, SiC D-T
		1500, V-alloy D-He3
Integrated risk (probability × population dose summed over all possible accidents)	man-Sv per reactor-year	not yet available

[a] Sources: (47, 48, 81)

[b] Based on characteristics of liquid metal fast breeder reactors (LMFBRs) in the numerator and characteristics of indicated fusion-reactor types in the denominator. In many cases a range of values is given.

[c] Assumes active T inventories from 200 to 2000 g and 0.2 Sv critical dose at 1 km per kg of T released as HTO.

[d] Based on division of elements into five mobility categories with maximum plausible release fractions of 1.0, 0.3, 0.1, 0.03, and 0.01, for both fusion and fission; some fusion designs (and some future fission designs) may warrant smaller MPRFs for given elements (47, 48, 55, 56).

is calculated using models like those described in Refs. 37–39 and 82–84, taking into account inhalation of resuspended material and, in some instances, ingestion in contaminated food and water (86, 87).

4. Partial-release dose potentials are derived from the complete-release potentials, critical and chronic, using specified estimates of release fractions.

It is informative, for example, to consider the consequences of accidents in which only the most mobile isotopes escape (mainly tritium in a fusion reactor, noble gases in a fission reactor), since such accidents presumably will be more frequent than those severe enough to mobilize less volatile isotopes. The maximum plausible doses (MPD), by contrast, are partial-release dose potentials based on specifying for all isotopes present the maximum plausible release fractions—that is, the fractions that could escape under the most severe conditions of temperature, chemical reactivity, and structural disruption that cannot be convincingly ruled out.

5. The integrated risk is the product of probability times consequences summed over all of the types of accidents that a reactor could suffer. The consequences may be expressed in terms of population dose of radiation by all dose pathways or in more disaggregated units distinguishing, for example, between critical doses and long-term doses.

The integrated risk is by far the most informative hazard measure but also by far the most difficult to obtain: very detailed design information and extensive operating experience are needed in order to catalog all the possible accident modes and to predict their probabilities of occurrence, and separate consequence calculations must be carried out for all the accident modes (20). Although it will not be possible to determine convincing estimates of the absolute values of the integrated risks from fusion reactors until such reactors have been built and operated, the associated technique of probabilistic risk assessment (PRA, sometimes also called probabilistic safety assessment, PSA) will nonetheless be useful in the meantime for identifying accident mechanisms and revealing relative advantages and disadvantages of different approaches to fusion-reactor design (23, 24, 88–91).

The maximum plausible doses (MPD) from "worst-case" accidents are of considerable interest (including to the public and to licensing authorities) and are easier to calculate than the integrated risk, but doing so convincingly still requires detailed analysis of specific designs combining sophisticated models of accident phenomenology and experimental data on mobilization of radioactivity under extreme conditions (21, 22, 32–35, 92–96). Rough estimates of MPDs can be obtained by dividing the radioactive inventories into categories according to the relative mobilities of the associated elements under accident conditions and associating each category with a maximum plausible release fraction (MPRF) based on whatever mobilization data and analysis are available for the relevant materials and configuration (47, 48). For the most preliminary comparisons of different reactor types, one may choose to use a common, conservatively defined "envelope" of MPRFs for all of the designs considered (as in Table 1).

In addition to such approaches to estimating MPDs, it is useful in preliminary safety assessments to be able to characterize in at least a general way the relative ease or difficulty of accident prevention in different reactor designs. Relevant in this connection are the amounts of energy stored in various forms (e.g. nuclear, thermal, chemical, magnetic) in the reactor, the pathways by which these energy forms could be released rapidly enough to do damage, and the reactor-materials properties that affect susceptibility to damage. Much useful work has been done on characterizing these aspects in ways that facilitate comparisons among different fusion-reactor designs and materials, and between fusion and fission (35, 36, 47, 48, 55, 56, 66, 92–96). Some important conclusions are as follows.

1. The amount of fusion fuel in the reaction chamber at one time is small, so the magnitude of energy releasable through excess reactivity is quite limited compared to the situation in fission. Nonetheless, the plasma conditions at the chosen operating point for a fusion reactor may not be those corresponding to maximum reactivity, in which case unwanted increases of power level—temporary and fuel-limited, but still perhaps sufficient to damage the reactor—would be possible. Preventing these will require active or passive means for burn control and emergency plasma shutdown (97–99).

2. The large amounts of energy stored in the plasma current in tokamak reactors and in the plasma's thermal energy in all reactors represent a potential for significant structural damage (through thermal energy deposition, runaway electrons, and electromagnetic forces)—with potential safety consequences—in the event of large-scale plasma disruptions (56). Fusion reactors will have to be designed to minimize the frequency and magnitude of such disruptions and to resist their consequences.

3. The chemical energy of liquid lithium is a troublesome source of stored energy in D-T fusion reactors that use this material as the coolant and tritium breeder (28, 35). This contribution to accident risk can be greatly reduced by using lithium-lead alloy in place of liquid Li (100), and can be avoided altogether by using solid or molten-salt Li compounds as the tritium breeder (or by use of advanced fuel cycles that do not need to breed tritium). Liquid Li should only be used in combination with structural materials highly resistant to the release of activation products at temperatures attainable in lithium fires. Lithium-lead unfortunately creates some activation problems (47, 48, 100).

4. Use of graphite "armor" on plasma-facing components presents a combustion hazard unless contact between the heated graphite and air or water can be ruled out under accident conditions (55, 56). Combustion of the graphite would readily mobilize the considerable quantities of tritium such components would contain.

5. Another potentially troublesome source of stored energy sources in fusion reactors is the decay heat from structural activation products. Decay heat from the radioactivity in fission reactors is the major source of concern about core melt in loss-of-coolant accidents (LOCAs) and is the reason for

requiring engineered emergency-core-cooling systems in fission reactors of current types. For most fusion-reactor designs and materials, however, the afterheat power levels and power densities are enough lower than those in fission to ease the emergency cooling problem substantially. With suitable attention to this point in materials selection and design, in fact, D-T fusion reactors can be designed so that passive mechanisms of heat removal alone suffice to prevent structural damage and significant radioactivity releases in LOCAs (40, 93, 101, 102); and D-He3 fusion reactors should be able to achieve this desirable characteristic with ease.

6. The energy stored in the confinement system (magnets for MFE reactors, lasers or particle-beams in ICF reactors) and its power supplies will be sufficient to damage the reactor if released suddenly; such releases could generate projectiles capable of breaking coolant pipes or damaging other safety-related barriers (103, 104). It may be possible, however, to design these systems with passive mechanisms to dissipate their energy gradually in accidents (48, 60).

A useful way to summarize some of the most important safety-related characteristics of a given fusion-reactor or fission-reactor design is the level-of-safety-assurance (LSA) concept developed by Piet (40) and applied in the US ESECOM study (47, 48). LSA evaluations are based on the extent and nature of dependence on passive versus active design features for assurance of public safety—perhaps most importantly for assurance that early off-site fatalities from release of radioactivity are precluded.

An LSA value of 1 means that safety is assured by the sizes of the radioactivity inventories and stored energy sources alone, without reference to any specifics of reactor configuration or accident scenario; an LSA of 4 means that assurance of safety requires demonstrating that active safety systems will perform satisfactorily; and the intermediate LSA values mean assurance of safety depends on demonstrating that passive design features can maintain the large-scale configuration of the reactor (LSA = 2) or that they can maintain both large-scale and small-scale configuration (LSA = 3). Although LSA = 4 systems may be adequately safe, lower LSA values mean that safety is easier to prove and, therefore, that siting and licensing should be easier.

The ESECOM study concluded that many D-T fusion reactor designs could achieve LSA = 2 or 3 and that use of very-low-activation materials (such as silicon carbide) or advanced fuels (such as D-He3) could bring LSA = 1 within reach (47, 48). Similar conclusions were reached in the TITAN and ARIES fusion reactor studies, which also employed the LSA concept in their safety assessments (31, 105).

Arriving at an LSA rating for a particular fusion reactor design requires a substantial amount of design-specific analysis by safety specialists. To help guide choices of structural and tritium-breeding materials in the early stages of reactor design, before detailed safety analysis is practical, Piet and colleagues (106) recently developed a framework that characterizes the intrinsic accident-hazard potential of each of the first 83 elements in the periodic table (i.e. through bismuth) when these are used in fusion-reactor materials. The method is based on calculation of the complete- and partial-release dose potentials and afterheat generation per gram of each element after bombardment by 14-MeV neutrons at a fluence of 20 MW-yr/m^2.

The resulting figures then can be integrated in appropriate combinations to characterize the hazard potential of actual engineering materials, including impurities. Based on the accident-safety figures of merit computed in this work, the leading candidates for fusion structural materials, ranked from best to worst, are silicon carbide, vanadium alloys, ferritic steels, nickel-based austenitic steels, and manganese-based austenitic steels.

HAZARD MEASURES FOR RADIOACTIVE-WASTE MANAGEMENT   Radioactive wastes from fusion reactors will consist mainly of activated structural material, part of it leaving the reactor when fusion-core components are replaced at intervals during the reactor's operating lifetime and part of it resulting from the dismantling of the reactor at the end of its service. This material will total between 400 and 3000 cubic meters over a 1200-MWe reactor's nominal life-cycle of 30 years, compared to perhaps 400 m^3 of "high-level" ($>$10,000 Ci/m^3) fission-product and structural waste from an LMFBR of the same output (19, 37, 47, 56, 60). Associated with the reprocessing plant and fuel-fabrication plant for the LMFBR, however, will be another 100–400 m^3 per year of "medium-level" (0.1 to 10,000 Ci/m^3) and transuranic-contaminated wastes (19, 107), which will require long-term management. Both the fusion and LMFBR fuel cycles will generate solid and liquid low-level wastes ($<$0.1 Ci/m^3) amounting to perhaps 1000 m^3 per year (19, 49), which are more easily managed.

In any case, other measures convey much more information about the relative difficulty of the waste-management task than volumes do. Two such measures are (37, 48):

1. the integrated biological hazard potentials (IBHP$_w$), obtained by dividing curies of each isotope by its RCG for public water supplies, multiplying by its mean life, and summing over all isotopes in the wastes (units are m^3 per reactor-year);
2. the annualized intruder hazard potentials (IHP), obtained from standardized scenarios of the consequences of inadvertent intrusion into shallow burial sites after periods of 100, 500, and 1000 years (rem-m^3 per reactor-year).[10]

---

[10]The standardized scenarios are part of US federal regulations governing the conditions under which radioactive wastes are eligible for disposal by shallow land burial (108).

Table 2 Some measures of radioactive-waste hazard[a]

Measure	Units	Fission/fusion ratios[b]
Integrated BHP_w	m^3-yr	5,000 stainless steel D-T
		10,000 ferritic steel D-T
		100,000 V-alloy D-T
Annualized intruder	rem-m^3	200–300 stainless steel D-T
hazard potential	per yr	200–50,000 ferritic steel D-T[c]
		2,000–70,000 V-alloy D-T
		20,000–200,000 SiC D-T
		3,000,000 V-alloy D-He3

[a] Sources: (37, 47, 48, 81)
[b] Based on LMFBR characteristics in the numerator and characteristics of indicated fusion-reactor types in the denominator. In many cases a range of values is given.
[c] The lower figure relates to a design with a lithium-lead blanket; the upper (better) figure relates to HT-9 ferritic steel modified to replace molybdenum with tungsten (48).

Some ratios of these indices for fission and fusion reactors are presented in Table 2.

Use of the shallow-burial scenarios to characterize relative hazard does not necessarily imply that shallow burial will be deemed appropriate for such wastes; it is simply a systematic and standardized way to account for the relative mobilities and toxicities of the constituents of wastes of different types. Regulations governing shallow burial of radioactive wastes differ from country to country (55, 56). The ESECOM study (47, 48) found that two of eight fusion-reactor designs considered could meet current US shallow-burial standards—which are based on the maximum "intruder dose" never exceeding 5 mSv/yr (500 mrem/yr)—for all of their wastes without dilution, while three more of the designs could meet these standards after twofold dilution of $0.3–1.2 \ m^3$ of first-wall waste per reactor-year.[11] European studies based on modeling waste-escape pathways and dose potentials for specific shallow- and deep-disposal sites, on the other hand, indicate that meeting the 0.1 mSv/yr (10 mrem/yr) maximum-exposed-individual guideline widely used in Europe to determine acceptable disposal practice would be very difficult with shallow burial of the most activated components of fusion wastes, irrespective of reactor design (109); deep burial, however, would easily meet the guideline at a variety of site types, and with much larger safety margins than those characterizing fission wastes at the same sites (51, 110).

[11] For comparison, according to Refs. 47 and 48, meeting the US shallow-burial standards with fission wastes would require dilution by a factor of 70,000 for 4 m^3 of high-level fission-product wastes per reactor-year.

HAZARD MEASURES FOR ROUTINE EMISSION AND EXPOSURES   Most of the activation products in a fusion reactor remain embedded in solid structure during routine operation. Only the far smaller quantities of activated material in the coolant and the building atmosphere have any chance of escaping to the environment under these normal conditions, and various studies of the matter have indicated that the doses expected from these sources can easily be held within the range of 10 to 50% of the total-exposure guidelines (50 μSv/yr each from airborne and aqueous effluents) for the most-exposed individual at a 1-km site boundary (49, 57, 58, 111). This would leave room for the anticipated levels of tritium emissions without exceeding the guidelines.

Activation products will pose larger problems for the control of worker exposures inside the power plant. Approximate calculations of the contact dose rates to be expected at the surfaces of various fusion-reactor components (48) give values exceeding the usual guidelines for unlimited hands-on maintenance (0.025 mSv/hr contact dose rate) by seven to ten orders of magnitude at the first wall at 1 hour after shutdown and by four to nine orders of magnitude at 1 year after shutdown. Even at the shield, the factors are four to eight orders of magnitude at 1 hour and three to seven orders of magnitude at 1 year. Clearly, all maintenance in the immediate vicinity of the fusion-power core will have to be done remotely. Proper choice of coolant, however, may allow hands-on maintenance of some parts of the coolant system; and use of low-activation structural material can also expand the scope for limited contact maintenance.

Unwanted Links to Nuclear Weaponry

An electricity-supply system based on fusion would be less likely than a fission-energy system to contribute to the acquisition of nuclear-weapons capabilities by subnational groups, and would also be easier to safeguard against clandestine use for fissile-material production by governments (47, 48, 112). Except for the special case of hybrid breeders, fusion reactors need not produce or contain any fissile material, and a fusion-based electricity-supply system would not circulate any. Because fusion reactors could be modified to produce fissile material, they will need to be subjected to international safeguards. On the other hand, using fusion reactors for fissile-material production would require prolonged control over the reactor (unlikely for subnational groups) and would be easier to detect by international safeguards (hence less likely to be judged an attractive option by governments) than is diversion from fissile fuel cycles.

In principle, tritium in D-T fusion reactors could be stolen by subnational groups or diverted to weapons programs by governments, but the consequences of this vulnerability are less profound than those of fissile-materials vulnerability in fission energy systems: the sophistication in nuclear weapons design and construction needed to make use of tritium is likely to remain beyond the reach of subnational groups, and for governmental nuclear-weapons programs tritium is useful but it is not indispensable in the sense that fissile materials are (112, 113).

The development of an ICF capability by countries not possessing thermo-nuclear weaponry could convey insights and calculational capabilities rele-vant to the design of such weapons (112, 114). This, too, is a possible concern with respect to governments but not with respect to subnational groups, and its importance even in the case of governments has been the subject of considerable disagreement (115).

NON-NUCLEAR HAZARDS

Previous studies of the fuel cycles of coal, fission, fusion, and renewable energy sources have indicated that, aside from emissions and accident risks from power-plant operations, the biggest energy-associated risks to health and safety tend to be those of more-or-less routine accidents in fuel and materials acquisition, processing, manufacturing, and transport. These hazards for fusion energy systems appear to be in the same range as those of fission and the most widely used renewable electricity sources of the present time, hydropower and wind, and well below those of coal (116, 117). Occupational hazards of particularly toxic materials, such as beryllium, deserve attention, but they seem unlikely to be as difficult to handle as the radiological occupa-tional hazards considered above; this seems to be true also for exposures to electric and magnetic fields (57). Fusion reactors and fuel-cycle operations would not release significant quantities of greenhouse gases or acid-rain precursors, and fusion's land-use requirements would be smaller than those of most fossil and renewable energy sources (116, 117). Thermal pollution from D-T fusion reactors would not be significantly different in proportion to electrical output than that from fission- or coal-based power plants. Adv-anced-fuel fusion reactors with a high proportion of the reaction energy carried by charged particles have the potential for significantly increased conversion efficiencies and, thus, markedly reduced thermal pollution (43).

NEEDS AND PRIORITIES FOR FUTURE WORK

Fusion energy has the potential to deliver S&E benefits large enough to be regarded as a major part of the rationale for the fusion-development effort (1–5, 48, 51, 56, 101, 118, 119): fusion would have no counterpart to the problems of mining, air pollution, acid rain, and climate-change associated with coal use, and it offers the prospect of increased safety from major accidents, diminished weapons linkages, and a smaller waste-management task compared to fission.

Achieving the full S&E potential of fusion will not happen automatically, however (48, 101, 118). The safety benefits, especially, appear to range from modest to enormous, depending on the materials used to construct the reactors, other aspects of the reactor design, and, ultimately, the particular fusion reaction that is harnessed. Unfortunately, the materials, designs, and reaction that are closest to practical realization today are not the ones that offer the greatest S&E potential, and the effort being devoted worldwide to the task of pursuing the more advantageous alternatives is inadequate (48, 56, 118).

What is needed above all is a greatly increased and redirected program in materials science for fusion technology, in order to develop and certify structural and breeder materials that can provide a combination of low neutron activation, low tritium inventory, and high integrity under accident conditions (5, 41, 48, 51). This effort must include the earliest possible construction of a high-fluence, 14-MeV neutron test facility, without which the capacity of materials to function in a fusion-reactor environment cannot be confirmed. The materials effort should also include a greatly expanded program of testing the behavior of candidate materials under the thermal and chemical conditions that could be encountered in fusion-reactor accidents.

A second priority is for a broader program of plasma physics research in pursuit of confinement schemes that could improve the prospects for advanced-fuel fusion (43, 44, 48). With suitable materials and reactor designs, D-T fusion can offer major advances over existing energy technologies; but even greater potential for drastically reduced tritium inventories, further reductions in neutron activation, and increased conversion efficiencies lies with D-He3 and perhaps other advanced fuels.

Third, the community of fusion S&E researchers and the resources available to them are simply too small to carry out the range of studies required even to properly support and benefit from the current and next-generation fusion machines, not to mention the generation of full-scale reactors to follow. As a result, opportunities to help steer fusion development in a timely fashion toward the technology's highest potential are likely to be missed. The resources being devoted to fusion S&E research of all kinds—from the most fundamental laboratory experiments to the integration of safety, environmental, and economic considerations in reactor design—need urgently to be increased (48, 56, 101).

Finally, the size of the challenges and the inadequacy of the resources underline the importance of increased international coordination and collaboration in the fusion S&E field. While the degree of international communication and cooperation on fusion S&E topics already is, and for a long time has been, greater than in any other field of S&E work, there is still the potential for better coordination to avoid duplication of effort and for more collaborative efforts to optimize the utilization of diverse national resources (48, 56, 118, 119). We must seize this potential if we are to achieve that of fusion.

ACKNOWLEDGMENTS

Various drafts of this article have benefitted from comments by Sandra Brereton, William Hogan, Mujid Kazimi, Bob Krakowski, Steve Piet, and Ken Schultz. Research assistance was provided by Paul Hibbard, Ann Kinzig, and Denis Sarigiannis. This work was supported by the Office of Fusion Energy, US Department of Energy, under Contract DE-FG03-89ER52154.

Literature Cited

1. Holdren, J. P. 1978. Fusion energy in context: its fitness for the long term. *Science* 200:168–80
2. Logan, B. G. 1985. A rationale for fusion economics based on inherent safety. *J. Fusion Energy* 4:245
3. Holdren, J. P. 1985. How safe is safe enough? The relation of environmental characteristics and economic competitiveness in fusion-reactor design. *Fusion Technol.* 8:1625–30
4. US Congress, Office Technol. Assess. 1987. *Starpower: The US and the International Quest for Fusion Energy.* Washington, DC: Govt. Printing Off. 236 pp.
5. Dean, S. O. 1988. Commercial objectives, technology transfer, and systems analysis for fusion power development. *J. Fusion Energy* 7:25–47
6. Morley, F., Kennedy, J. W. 1969. Fusion reactors and environmental safety. *Nuclear Fusion Reactor Conference Proceedings*, pp. 54–65. London: British Nuclear Energy Society
7. Fraas, A. P., Postma, H. 1970. *Preliminary Appraisal of the Hazards Problems of a D-T Fusion Reactor Power Plant. ORNL-TM-2822.* Oak Ridge, Tenn: Oak Ridge Natl. Lab.
8. Lee, J. D. 1970. Some observations on the radiological hazards of fusion. *Fusion Technology: Presentations from the 5th Intersociety Energy Conversion Eng. Conf., Las Vegas, 1970*, pp. 58–60
9. Steiner, D., Fraas, A. P. 1972. Preliminary observations on the radiological implications of fusion power. *Nuclear Safety* 13:353–62
10. Kulcinski, G. L. 1974. Fusion power: an assessment of its potential impact in the USA. *Energy Policy*. June: pp. 104–25
11. Dudziak, D. J., Krakowski, R. A. 1975. Radioactivity induced in a theta-pinch fusion reactor. *Nuclear Technol.* 25:32–40
12. Conn, R. W., Sung, T., Abdou, M. A. 1975. Comparative study of radioactivity and afterheat in several fusion reactor designs. *Nuclear Technol.* 26:391–400
13. Watson, J. S., Wiffen, F. W. 1976. *Proc. Internatl. Conf. Radiation Effects & Tritium Technology for Fusion Reactors. CONF-750989.* Washington, DC: US Energy Res. Dev. Admin.
14. Powell, J. R., Miles, F. T., Aronson, A., Winsche, W. E. 1973. *Studies of Fusion-Reactor Blankets with Minimum Radioactivity Inventory and With Tritium Breeding in Solid Lithium Compounds. BNL-18236.* Upton, NY: Brookhaven Natl. Lab.
15. Rovner, L. H., Hopkins, G. H. 1976. Ceramic materials for fusion. *Nuclear Technol.* 29:274
16. Mitchell, J. T. D. 1974. Fusion and the environment. *Proc. IAEA Workshop of Fusion Reactor Design Problems, Culham, 1974. IAEA Publ. STI/PUB/23.* Vienna: Int. At. Energy Agency
17. Flakus, F. N. 1975. Fusion power and the environment. *At. Energy Rev.* 13:587–614
18. Young, J. R., ed. 1976. *An Environmental Analysis of Fusion Power to Determine Related R&D Needs.* BNWL-2010/2028 (19 vols.). Hanford, Wash: Battelle Pacific Northwest Lab.
19. Haefele, W., Holdren, J. P., Kessler, G., Kulcinski, G. L. 1977. *Fusion and Fast Breeder Reactors. RR-77-8.* Laxenburg, Austria: Int. Inst. Appl. Systems Anal. 506 pp.
20. US Nuclear Regul. Comm. 1975. *Reac-

tor Safety Study. WASH-1400 (NUREG-75/014). Springfield, Va: Natl. Tech. Inf. Serv.

21. Bayer, A., Ehrhardt, J. 1984. Risk-oriented analysis of the German prototype fast breeder reactor SNR-300: off-site accident consequence model and results of the study. Nuclear Technol. 65:232

22. US Nuclear Regul. Comm. 1985. Reassessment of the Technical Basis for Estimating Source Terms. NUREG-0956. Springfield, Va: Natl. Tech. Inf. Serv.

23. Kastenberg, W. E., Okrent, D., eds. 1978. Some Safety Considerations for Conceptual Fusion Power Reactors. EPRI ER-546. Palo Alto, Calif: Electr. Power Res. Inst. 332 pp.

24. Okrent, D., Kastenberg, W., Botts, T. E., Chan, C. K., Ferrell, W. L., Frederking, T. H. K., Sehnert, M. J., Ullman, A. Z. 1976. On the safety of tokamak-type, central station fusion power reactors. Nuclear Eng. Design 39:215–38

25. Kazimi, M. S., ed. 1977. Aspects of Environmental and Safety Analysis of Fusion Reactors. MITNE-212. Cambridge, Mass: MIT Dept. Nuclear Eng.

26. Dube, D. A., Kazimi, M. S. 1978. Analysis of Design Strategies for Mitigating the Consequences of Lithium Fire Within Containment of Controlled Thermonuclear Reactors. MITNE-219. Cambridge, Mass: MIT Dept. Nuclear Eng.

27. Kabele, T. J., Johnson, A. B., Mudge, L. K. 1976. Tritium Source Terms for Fusion Power Plants. BNWL-2018. Hanford, Wash: Battelle Pacific Northwest Lab.

28. Holdren, J. P. 1981. Contribution of activation products to fusion accident risk: Part I. A preliminary investigation. Nucl. Technol./Fusion 1:79–89

29. Maninger, R. C., Dorn, D. W. 1982. Issues in Radioactive Waste Management for Fusion Power. UCRL-87786. Livermore, Calif: Lawrence Livermore Natl. Lab.

30. Argonne Natl. Lab., McDonnell-Douglas Astronautics Co., General Atomic Co., The Ralph M. Parsons Co. 1980. STARFIRE: A Commercial Tokamak Fusion Power Study. ANL/FPP-80-1. Argonne, Ill: Argonne Natl. Lab.

31. Najmabadi, F., Conn, R. W., eds. 1989. The ARIES Tokamak Reactor Study. UCLA-PPG-1274. Univ. Calif., Los Angeles

32. Jeppson, D. W. 1982. Scoping Studies—Behavior and Control of Lithium and Lithium Aerosols. HEDL-TME-80-79. Hanford, Wash: Hanford Eng. Dev. Lab.

33. Piet, S. J., Kraus, H. G., Neilson, R. M., Jones, J. L. 1986. Oxidation/volatilization rates in air for candidate fusion reactor blanket materials. J. Nucl. Materials 141–43:24–28

34. Tillack, M. S., Kazimi, M. S. 1982. Modeling of lithium fires. Nucl. Technol./Fusion 2:233

35. Piet, S. J. 1982. Potential consequences of tokamak fusion reactor accidents: the materials impact. PhD thesis. Mass. Inst. Technol. 645 pp.

36. Piet, S. J., Kazimi, M. S., Lidsky, L. M. 1984. The materials impact on fusion reactor safety. Nucl. Technol./Fusion 5:382

37. Fetter, S. A. 1985. Radiological hazards of fusion reactors: models and comparisons. PhD thesis. Univ. Calif., Berkeley. 791 pp.

38. Fetter, S. 1985. A calculation methodology for comparing the accident, occupational, and waste-disposal hazards of fusion reactors. Fusion Technol. 8:1359

39. Fetter, S. 1987. The radiological hazards of fusion reactors. Fusion Technol. 11:400–15

40. Piet, S. J. 1986. Approaches to achieving inherently safe fusion power plants. Fusion Technol. 10:7–30

41. Conn, R. W., Bloom, E. E., Davis, J. W., Gold, R. E., Little, R., Schultz, K. R., Smith, D. L., Wiffen, F. W. 1983. Report of the DOE Panel on Low Activation Materials for Fusion Applications. UCLA-PPG-728. Univ. Calif., Los Angeles

42. Hopkins, G. R., Cheng, E. T., Creedon, R. L., Maya, I., Schultz, K. R., Trester, P., Wong, C. P. C. 1985. Low activation fusion design studies. Nuclear Technol./Fusion 4:1251

43. Miley, G. H. 1980. Potential and status of alternate fuel fusion. Proc. 4th Topical Meet. Technol. Controlled Nuclear Fusion, King of Prussia, Penn., pp. 905–10. Hinsdale, Ill: Am. Nuclear Soc.

44. Miley, G. H. 1989. Advanced fuels for heavy-ion beam fusion. Nucl. Instr. Methods Phys. Res. A278:281–87

45. Smith, D. L., Baker, C. C., Sze, D. K., Morgan, G. D., Abdou, M. A., Piet, S. J., Schultz, K. R., Moir, R. W., Gordon, J. D. 1985. Overview of the blanket comparison and selection study. Fusion Technol. Suppl. 8(Part 1):10–30

46. Sheffield, J., Dory, R. A., Cohn, S. M., Delene, J. G., Parsly, L. F., Ashby, D. E. T. F., Reiersen, W. T. 1986. Cost

assessment of a generic magnetic fusion reactor. *Fusion Technol.* 9:199

47. Holdren, J. P., Berwald, D. H., Budnitz, R. J., Crocker, J. G., Delene, J. G., Endicott, R. D., Kazimi, M. S., Krakowski, R. A., Logan, B. G., Schultz, K. R. 1988. Exploring the competitive potential of magnetic fusion energy: the interaction of economics with safety and environmental characteristics. *Fusion Technol.* 13:7–56

48. Holdren, J. P., Berwald, D. H., Budnitz, R. J., Crocker, J. G., Delene, J. G., Endicott, R. D., Kazimi, M. S., Krakowski, R. A., Logan, B. G., Schultz, K. R. 1989. *Report of the Senior Committee on Environmental, Safety, and Economic Aspects of Magnetic Fusion Energy. UCRL-53766.* Livermore, Calif: Lawrence Livermore Natl. Lab. 345 pp.

49. Comm. Eur. Communities. 1986. *Environmental Impact and Economic Prospects of Nuclear Fusion. EURFU BRU/ XII-828/86.* Brussels

50. Yamamato, K., ed. 1986. *Investigation of Feasibility of Nuclear Fusion.* Report by the Japanese Atomic Energy Society to the Atomic Energy Commission of Japan. (In Japanese)

51. Pease, R. S., Darvas, J., Flowers, R. H., Gouni, L., Grieger, G., Koeberlein, K., Roncaglia, A. 1989. *Environmental, Safety-Related, and Economic Potential of Fusion Power.* Brussels: Comm. Eur. Communities

52. Brereton, S. J., Kazimi, M. S. 1988. A comparative study of the safety and economics of fusion fuel cycles. *Fusion Eng. Design* 6:207

53. Gordon, C. W. 1987. The tritium safety programme at JET. *Fusion Reactor Safety. IAEA-Tecdoc-440,* pp. 63–72. Vienna: Int. At. Energy Agency

54. Holland, D. F., Lyon, R. E. 1989. *Potential Off-Normal Events and Associated Radiological Source Terms for the Compact Ignition Tokamak. EGG-FSP-7872.* Revision 1. Idaho Falls: Idaho Natl. Eng. Lab.

55. Raeder, J., Piet, S. 1989. *ITER Safety and Environmental Impact: Concepts and Technical Information. ITER TN-SA-9-1.* Vienna: Int. At. Energy Agency

56. Raeder, J., Piet, S., eds. 1990. *ITER Safety Analyses.* ITER Documentation Series No. 36. Vienna: Int. At. Energy Agency

57. Cannon, J. B., ed. 1983. *Background Information and Technical Basis for Assessment of Environmental Implications of Magnetic Fusion Energy. DOE/ER-0170.* Washington, DC: Off. Energy Res., US Dept. Energy

58. Int. At. Energy Agency. 1986. *Fusion Safety Status Report. IAEA-TECDOC-388.* Vienna

59. Raeder, J., Gulden, W. 1988. NET safety analyses and the European safety and environmental programme. *Proc. 15th Symp. Fusion Technology, Utrecht, The Netherlands.* Vienna: Int. At. Energy Agency

60. Hogan, W. J., Kulcinski, G. L. 1985. *Advances in ICF Power Reactor Design. UCRL 91647.* Livermore, Calif: Lawrence Livermore Natl. Lab.

61. Abdou, M. A., Vold, E. L., Gung, C. Y., Youssef, M. Z., Shin, K. 1986. Deuterium-tritium fuel self-sufficiency in fusion reactors. *Fusion Technol.* 9:250–85

62. Gabowitsch, E., Spannagel, G. 1989. Computer simulation of tritium systems for fusion technology. *Fusion Technol.* 16:143–48

63. Sarigiannis, D. 1990. *Dynamic simulation of the tritium inventory in fusion power plants.* MS thesis. Energy and Resources Group, Univ. Calif., Berkeley

64. Off. Fed. Register. 1984. *Code of Federal Regulations: Title 10. Energy.* Part 20. Washington, DC: US Govt. Printing Off.

65. Off. Fed. Register. 1985. *Code of Federal Regulations: Title 40. National Emission Standards for Hazardous Air Pollutants.* Part 61. Washington, DC: US Govt. Printing Off.

66. Raeder, J., Piet, S., Seki, Y., Topilski, L. N. 1989. Safety and environment work performed for ITER. *Proc. IAEA Workshop on Fusion Reactor Safety, Jackson, Wyoming.* Vienna: Int. At. Energy Agency

67. Anderson, J. L., Bartlit, J. R., Carlson, R. V., Coffin, D. O., Damiano, F. A., Sherman, R. H., Willms, R. S. 1988. Experience of TSTA milestone runs with 100 grams-level tritium. *Fusion Technol.* 14:438–43

68. Hircq, B. 1988. Tritium activities for fusion technology in Bruyeres-le-Chatel Center, CEA, France. *Fusion Technol.* 14:424–28

69. Carlson, R. V. 1989. Five years of tritium handling experience at the Tritium Systems Test Assembly. *Proc. IAEA Workshop on Fusion Reactor Safety, Jackson, Wyoming.* Vienna: Int. At. Energy Agency

70. Casini, G., Ponti, C., Rocco, P. 1985. *Environmental Aspects of Fusion Reactors. Technical Note 1.04B1.5.156.* Italy: Ispra Joint Res. Cent.

71. Devell, L., Edlund, O. 1986. Environmental radiation doses from tritium re-

leases. *Fusion Reactor Safety. IAEA-Tecdoc-440*, pp. 317–26. Vienna: Int. At. Energy Agency

72. Watson, J. S., Easterly, C. E., Cannon, J. B., Talbot, J. B. 1987. Environmental effects of fusion power plants. Part II: Tritium effluents. *Fusion Technol.* 12: 354–63

73. Easterly, C. E., Hill, G. S., Cannon, J. B. 1989. Environmental effects of fusion power plants. Part III: Potential radiological impact of environmental releases. *Fusion Technol.* 16:125–36

74. Plechaty, E. F., Kimlinger, J. R. 1976. *TARTNP: a Coupled Neutron-Photon Monte Carlo Transport Code. UCRL-50400*, Vol. 14. Livermore, Calif: Lawrence Livermore Natl. Lab.

75. Blink, J. A., Dye, R. E., Kimlinger, J. R. 1981. *ORLIB: a Computer Code that Produces One-Energy-Group, Time- and Spatially Averaged Neutron Cross Sections. UCRL-53262*. Livermore, Calif: Lawrence Livermore Natl. Lab.

76. Blink, J. A. 1985. *FORIG: a Computer Code for Calculating Radionuclide Generation and Depletion in Fusion and Fission Reactors. UCRL-53633*. Livermore, Calif: Lawrence Livermore Natl. Lab.

77. Croff, A. G. 1980. *ORIGEN2: Isotope Generation and Depletion code—Matrix Exponential Method. ORNL/TM-7175*. Oak Ridge, Tenn: Oak Ridge Natl. Lab.

78. Pitts, J. H. 1986. A high-efficiency ICF power reactor. In *Laser Interaction and Related Plasma Phenomena.* Vol. 7. New York: Plenum

79. Pitts, J. H., Maya, I. 1985. *The Cascade Inertial-Confinement-Fusion Power Plant. UCRL-92558*. Livermore, Calif: Lawrence Livermore Natl. Lab.

80. Holdren, J. P., Fetter, S. 1983. Contribution of activation products to fusion accident risk: Part II. Effects of alternative materials and designs. *Nuclear Technol./Fusion* 4:599–619

81. Ho, S. K., ed. 1990. *Status Report: Code Development Incorporating Environmental, Safety, and Economic Aspects of Fusion Reactors. UC-BFE-009*. Berkeley Fusion Eng., Univ. Calif., Berkeley

82. Moore, R. E., ed. 1979. *AIRDOSE-EPA: A Computerized Methodology for Estimating Environmental Concentrations and Dose to Man from Airborne Releases of Radionuclides. ORNL-5532*. Oak Ridge, Tenn: Oak Ridge Natl. Lab.

83. Porter, L. J. 1989. *Upgrade of a Fusion Accident Analysis Code and its Application to a Comparative Study of Seven Fusion Reactor Designs. PFC/RR-89-*

10. Cambridge, Mass: Plasma Fusion Cent., Mass. Inst. Technol. 112 pp.

84. Fetter, S. 1988. *Internal Dose Conversion Factors for 19 Target Organs and 9 Irradiation Times and External Dose-Rate Conversion Factors for 21 Target Organs for 259 Radionuclides Produced in Potential Fusion Reactor Materials. EGG-FSP-8036*. Idaho Falls: Idaho Natl. Eng. Lab.

85. Brereton, S. J., Piet, S. J., Porter, L. J. 1989. Offsite dose calculations for hypothetical fusion facility accidents. *Proc. IAEA Workshop on Fusion Reactor Safety, Jackson, Wyoming.* Vienna: Int. At. Energy Agency

86. Rood, A. S. 1988. *Environmental Transport Concentration Factors for the FUSECRAC Fusion Reactor Safety Code. EGG-ESE-8033*. Idaho Falls: Idaho Natl. Eng. Lab.

87. Abbott, M. L., Rood, A. S. 1990. *Concentration Factors for Fusion-Related Radionuclides Calculated using the Food-Chain Model FUSEMOD. EGG-EST-9223*. Idaho Falls: Idaho Natl. Eng. Lab.

88. Piet, S. J. 1986. Implications of probabilistic risk assessment for fusion decision making. *Fusion Technol.* 10:31–48

89. Djerassi, H., Rouillard, J. 1986. Fusion reactor blanket and first wall risk analysis. *Fusion Reactor Safety. IAEA-Tecdoc-440*, pp. 123–32. Vienna: Int. At. Energy Agency

90. Fujii-e, Y., Kozawa, Y., Nishikawa, M., Yano, T., Yanagisawa, I., Kotake, S., Sawada, T. 1990. Development of general methodology of safety analysis and evaluation for fusion energy systems. *Fusion Eng. Design* 12:421–47

91. Sarto, S., Cambi, G., Zappellini, G. 1986. A methodological safety analysis for NET accident scenarios. *Fusion Reactor Safety. IAEA-Tecdoc-440*, pp. 185–202. Vienna: Int. At. Energy Agency

92. Holland, D. F. 1986. Computer codes for safety analysis. *Fusion Reactor Safety. IAEA-Tecdoc-440*, pp. 177–84. Vienna: Int. At. Energy Agency

93. Massida, J. 1987. *Thermal design considerations for fusion reactor passive safety.* PhD thesis. Mass. Inst. Technol.

94. Holland, D. F., Cadwallader, L. C., Brereton, S. J., Herring, S. J., Longhurst, G. R., et al. 1989. *Fusion Safety Program Annual Report. EGG-2561*. Idaho Falls: Idaho Natl. Eng. Lab.

95. Piet, S. J. 1989. Fusion activation product behavior and oxidation-driven volatility. *Proc. IAEA Workshop on Fu-*

sion Reactor Safety, Jackson, Wyoming. Vienna: Int. At. Energy Agency

96. Piet, S. J., Neilson, R. M., Smolik, G. R., Reimann, G. A. 1989. *Initial Experimental Investigation of the Elemental Volatility from Steel Alloys for Fusion Safety Application. EGG FSP-8459.* Idaho Falls: Idaho Natl. Eng. Lab. 148 pp.

97. Piet, S. J., Watson, R. 1990. *Need for Emergency Shutdown and Burn Control.* Presented at ITER Plasma Operational Control Workshop, Garching, Federal Republic of Germany, July

98. Perkins, L. J., Campbell, R. B., Haney, S. H., Ho, S. K. 1990. *ITER Emergency Shutdown.* Presented at ITER Plasma Operational Control Workshop, Garching, Federal Republic of Germany, July

99. Haney, S. W., Perkins, L. J., Mandrekas, J., Stacey, W. M. 1991. Active control of thermonuclear burn conditions for the International Thermonuclear Experimental Reactor. *Fusion Technol.* In press

100. Jeppson, D. W., Muhlestein, L. D. 1985. Safety considerations of lithium-lead alloy as a fusion reactor breeding material. *Fusion Technol.* 8:1385–90

101. Piet, S. J., Holland, D. F. 1989. Fusion safety and environmental performance goals. *Fusion Technol.* 15:793–802

102. Kazimi, M. S., Massidda, J. E., Brereton, S. J. 1989. Thermal limits for passive safety of fusion reactors. *Fusion Technol.* 15:827–32

103. Thome, R. J., Czirr, J. B., Schultz, J. H. 1986. Survey of selected magnet failures and accidents. *Fusion Technol.* 10:1216–22

104. Zimmerman, M., Kazimi, M. S., Siu, N. O., Thome, R. J. 1988. *Failure Modes and Effects Analysis of Fusion Magnet Systems. PFC/RR-88-17.* Cambridge, Mass: Plasma Fusion Cent., Mass. Inst. Technol. 149 pp.

105. Najmabadi, F., Conn, R. W., Krakowski, R. A., Schultz, K. R., Steiner, D. 1985. *The TITAN Reversed-Field-Pinch Reactor Study—The Final Report. UCLA-PPG-1200.* Univ. Calif., Los Angeles

106. Piet, S. J., Cheng, E. T., Porter, L. J. 1990. Accident safety comparison of elements to define low-activation materials. *Fusion Technol.* 17:636–57

107. Haug, H. O. 1975. *Anfall, Beseitigung, und relative Toxizitaet langlebiger Spaltprodukte und Actiniden in den radioktiven Abfaellen der Kernbrennstoffzyklen. KfK 2022.* Kernforschungszentrum Karlsruhe, Federal Republic of Germany. (In German)

108. US Nucl. Regulatory Comm. 1982. *Final Environmental Impact Statement on 10 CFR Part 61, Licensing Requirements for Land Disposal of Radioactive Waste. NUREG-0945.* Springfield, Va: Natl. Tech. Inf. Serv.

109. Broden, K., Bergstrom, U., Hultgren, A., Olsson, G. 1987. Geological disposal of fusion wastes. *Fusion Reactor Safety. IAEA-Tecdoc-440,* pp. 271–79. Vienna: Int. At. Energy Agency

110. Smith, K. R., Butterworth, G. J. 1989. Radiological impact of fusion waste disposal. *Proc. IAEA Tech. Meet. Fusion Reactor Safety, Jackson, Wyoming.* Vienna: Int. At. Energy Agency

111. Cannon, J. B., Easterly, C. E., Davis, W., Watson, J. S. 1987. Environmental effects of fusion power plants. Part I: Effluents other than tritium. *Fusion Technol.* 12:341–53

112. Holdren, J. P. 1989. Civilian nuclear technologies and nuclear weapons proliferation. In *New Technologies and the Arms Race,* ed. C. Schaerf, B. Holden, D. Carlton, pp. 161–98. London/New York: Macmillan

113. Cochran, T. B., Arkin, W. M., Hoenig, M. M. 1984. *Nuclear Weapons Databook. Vol. I: U.S. Nuclear Forces and Capabilities.* Cambridge, Mass: Ballinger. 340 pp.

114. Holdren, J. P. 1978. Fusion power and nuclear weapons: A significant link? *Bull. At. Sci.* 34:4–5

115. Keisch, B., Auerbach, C., Fainberg, A., Fiarman, S., Fishbone, L. G., Higinbotham, W. A., Lemley, J. R., O'Brien, J. 1986. *Long-Term Proliferation and Safeguards Issues in Future Technologies. BNL-52010.* Upton, NY: Brookhaven Natl. Lab.

116. Holdren, J. P., Anderson, K. B., Deibler, P. M., Gleick, P. H., Mintzer, I. M., Morris, G. P. 1983. Health and safety impacts of renewable, geothermal, and fusion energy. In *Health Risks of Energy Technologies,* ed. C. Travis, E. Etnier. Boulder, Colo: Westview

117. Kazimi, M. S. 1983. Risk considerations for fusion energy. *Nuclear Technol./Fusion* 4:527

118. Conn, R. W., Holdren, J. P., Sharafat, S., Steiner, D., Ehst, D. A., Hogan, W. J., Krakowski, R. A., Miller, R. L., Najmabadi, F., Schultz, K. 1990. Economic, safety, and environmental prospects of fusion reactors. *Nuclear Fusion* 30:1919–34

119. Colombo, U., Jaumotte, A., Kennedy, E., Lopez Martinez C., Popp, M., Reece, C., Schopper, H., Spitz, E., Troyon, F. 1990. *Fusion Programme Evaluation Board Report.* Brussels: Comm. Eur. Communities. 94 pp.

Annu. Rev. Energy Environ. 1991. 16:259–73

A PRELIMINARY ASSESSMENT OF CARBON DIOXIDE MITIGATION OPTIONS

D. F. Spencer

Electric Power Research Institute, Palo Alto, California

KEY WORDS: global strategy, minimum cost options, developed and developing countries, carbon storage or sequestering, policy analysis.

ABSTRACT

The purpose of this paper is to place the potential needs to control global carbon dioxide emissions in perspective. In order to limit carbon dioxide levels in the earth's atmosphere to no more than twice pre-anthropogenic levels, it will be necessary to limit carbon emissions to approximately 10 gigatons per year by 2050. The implications of such a constraint to the developed countries, developing countries, and international community are assessed. It is clear that international priorities must be established and specific approaches developed in the first quarter of the 21st century to define the necessary, minimum-cost mitigation strategies. Because of the complexity of establishing a meaningful policy approach, imposition of an arbitrary carbon tax is unlikely to provide the constraints necessary to achieve a satisfactory earth atmosphere–carbon dioxide equilibrium state.

INTRODUCTION

Over the past 100 years, anthropogenic emissions of carbon dioxide from fossil fuel use have added approximately 150 gigatons of carbon to the earth's atmosphere. This represents approximately a 25% increase over the pre-anthropogenic level. The present global fossil fuel flux is approximately 6

259

1056-3466/91/1022-0259$02.00

gigatons of carbon annually and the airborne fraction is approximately 0.6. Thus, we are adding 3.6 gigatons of carbon to the atmosphere annually from fossil fuel use. Recent data indicate that deforestation may be a source term of 0.5–5.0 gigatons of carbon annually, also in the form of carbon dioxide. It is unclear what the sink for this possible additional CO_2 flux may be, as the fossil fuel airborne fraction can account for the full net increase in CO_2 within the atmosphere.

In order to estimate the size of the potential fossil fuel CO_2 source term, we must take into account that stored fossil resources in the earth contain approximately 5–10 times the present atmospheric carbon loading (1). In addition, the ocean contains 50–60 times the atmospheric carbon loading, primarily in the form of dissolved inorganic carbon. Thus, potential carbon releases to the atmosphere are many times the present level, and the earth's overall carbon cycle must be carefully assessed and controlled.

Although the exact consequences of increasing the concentration of carbon dioxide in the atmosphere are unknown, most experts predict significant global climate change when the CO_2 partial pressure reaches 560 parts per million; i.e. double the pre-anthropogenic concentration. With increased worldwide use of fossil fuels growing at approximately 3% per annum, this doubling could occur by 2040 if carbon emissions are unabated and the world economy grows as projected. Should the recent CO_2 increases from deforestation be as high as reported, and the unknown sink for this CO_2 become saturated in its absorption capability for CO_2, doubling could occur earlier than 2040. The combined addition of carbon from fossil fuel use and from deforestation is growing at an increasingly alarming rate, and requires our close monitoring and appropriate mitigation planning.

The consequences of CO_2 doubling are generally the focus for climatic-impact modeling studies; however, climatic changes may already be occurring as a result of increases in the concentration of CO_2, as well as of other greenhouse gases. Whether temperature increases in the earth's atmospheric have been induced by higher CO_2 levels, or whether these temperature increases result from natural processes, is presently an issue of significant controversy. A decade or more of detailed measurements coupled to more specific models will likely be required to determine clear effects of CO_2 increases. Until we obtain better coupled atmosphere-ocean models, better spatial resolution, a much better understanding of CO_2 sources and sinks, and a better understanding of biogeochemical, cloud, and other feedbacks, the uncertainties in the effects of increasing CO_2 in the atmosphere will remain.

Further, the costs associated with controlling CO_2 emissions or sequestering atmospheric CO_2 are much more easily estimated than the benefits. Modest increases in atmospheric temperature (1–3°C), sea level rises of less than 1 meter, increased precipitation, etc, may be acceptable. Careful economic analyses of acceptable, consequential, and potentially cataclysmic effects of carbon dioxide increases must parallel the global climate change predictions. A near-term focus on evaluating alternative climatological impacts (e.g. depth and frequency of pressure systems, wind speeds, precipitation patterns, etc) would be highly beneficial in assessing "acceptable" CO$_2$ atmospheric concentrations.

With these considerations in mind and based on the projected growth in world population and energy demands, it is expected that CO$_2$ levels will reach approximately 500 ppm by the mid-21st century, even with substantial pressure to reduce fossil fuel use and to conserve. Although we do not know that this concentration of CO$_2$ in the atmosphere is acceptable; i.e. we are not certain of the concomitant global change implications, I have chosen a target control level of 500 ppm of CO$_2$ by 2050.

Assuming only a net increase in CO$_2$ from fossil fuel use, this level is consistent with emissions of approximately 10 gigatons of carbon in 2050. This increase over today's world levels has been projected to account for the planned increase of coal use within the Republic of China alone; i.e. all other countries would have to maintain CO$_2$ emissions at today's levels. Obviously, equitable budgets for carbon emissions must be developed.

The author is very aware of the arbitrary establishment of a 10-gigatons-of-carbon emission target by 2050. This was done to force consideration of the real implications of such a constraint. In addition, the author is aware of the uncertainties surrounding each estimate of potential carbon displacement or sequestering and associated costs. However, again the intent of this paper is to force specific consideration of each of these mitigation strategies, which requires specifying the expected effectiveness, limitations, and potential costs of each. All too often, each strategy is discussed out of context, not on a comparative basis, where it can better be assessed relative to competing approaches.

Beyond 2050, we must seek to eliminate net CO$_2$ emissions by developing a global strategy to produce and use equal amounts of carbon annually. Such a strategy would require us to discontinue fossil fuel use, and would likely take another 50 years to implement fully. The CO$_2$ concentration of the atmosphere would then reach a maximum of approximately 600–650 ppm by the end of the 21st century.

Because equalizing the amounts of carbon produced and used is such a prodigious scientific and technical challenge, I believe we must be developing the framework for such a strategy now. The first steps towards creating such a framework are to assess the various CO$_2$ mitigation options, their potentials for reducing or eliminating CO$_2$ emissions, and their approximate costs per ton of carbon displaced or sequestered. The remainder of this paper focuses on these considerations.

In addition, with the significant uncertainties that presently exist relative to the effects of increasing concentrations of CO$_2$ as well as other greenhouse gases such as nitrous oxides, methane, and chlorofluorocarbons, we must continually reassess whether our target CO_2 control levels are acceptable. Strategies that curtail CO_2 concentrations to below our target levels and their concomitant costs should be assessed. However, as this paper shows, achieving a 500 ppm CO_2 concentration in the atmosphere by 2050 will be extremely difficult.

CO_2 MITIGATION POTENTIAL

0. Perspective

Recent international attention has been focused on a potential carbon tax to limit CO_2 emissions. For example, one analysis, Ref. 2, indicates that long-run equilibrium values of such a tax would be approximately $250 per ton of carbon removed from the fossil-fuel source term, with greater costs to the United States and Soviet Union/Eastern European countries in the early 21st century. The analysis in Ref. 2 projects that such a tax would lead to an estimated carbon-emission level in 2100 of approximately 8 gigatons per year. Since the airborne fraction is only 0.6, the effective cost of reducing airborne carbon is $400 per ton of C.

To place these costs in perspective, a control cost of $100 per ton of carbon removed from the fossil fuel source term—60% less than the proposed $250 tax—would increase the costs of coal-based electricity by approximately 50%. I believe this is a maximum acceptable increase in energy costs due to CO_2 mitigation requirements. The value of carbon actually removed from the atmosphere, i.e., sequestered, is greater—a value of $160 per ton of carbon sequestered would correspond to $100 per ton removed from the fossil-fuel source term.

Therefore, I adopt a target control-cost range of $100 per ton of reduced fossil fuel emissions of carbon and $160 per ton of carbon sequestered. Obviously the lowest-cost options are preferred, and are limited only by their estimated CO_2 mitigation potentials.

Numerous approaches have been proposed for reducing carbon dioxide emissions. In this paper we classify the various control strategies as applicable to (a) developed countries, (b) developing countries, and (c) the entire international community. It is clear that certain mitigation approaches are more appropriate than others to the state of economic development of a particular country/region.

1. Mitigation Options for Developed Countries

The United States, Organization for Economic Cooperation and Development (OECD), Soviet Union, and Eastern European countries essentially represent the economically developed world and presently produce approximately two thirds of worldwide CO$_2$ emissions. Present trends suggest that energy use within developed countries will continue to grow by approximately 1.5% per annum into the foreseeable future. Thus, energy use in the developed countries would double by 2050, resulting in fossil fuel emissions of approximately 8 gigatons of carbon per year. It is expected, lacking CO$_2$ emission contraints, that these countries will represent approximately 30% of CO$_2$ emissions by 2050. If we adopt a target global constraint of 10 gigatons of carbon emissions by 2050, the developed world would be "allocated" 3 gigatons of carbon in 2050—a reduction of approximately 63% from uncontrolled levels or 33% below today's annual carbon emissions.

The primary approaches for reducing carbon emissions available to developed countries include:

1. Major enhancement in end-use efficiencies of electricity, oil, and gas.
2. Enhancement in electricity-generation efficiency.
3. Power plant controls of CO$_2$ emissions including (a) substitution for carbonaceous fuels; e.g. by nuclear or solar technologies, (b) scrubbing, compression, and sequestering of CO$_2$ from stack gases, and (c) partial carbon recycling.

ENHANCEMENTS IN END-USE EFFICIENCIES   Since the oil embargo of 1973, energy intensity—the amount of energy required to produce a dollar of US gross national product—has fallen by 28% (3). Estimates of the potential for further reductions in consumption resulting from improvements in end-use efficiency range from 20 to 75% within the United States (3, 4). In addition, electricity is becoming an increasingly important component of our energy system, now representing 40% of our annual energy resource use and projected to grow to well over 50% in the future. Increased electrification is particularly desirable where substitution enables use of more efficient end-use technologies: e.g. microwave ovens and freeze concentration. Although not all developed countries have the same potentials to improve efficiency, increases in world energy prices, such as those caused by the recent Iraqi invasion of Kuwait, provide major incentives to conserve.

In the United States, the growing gap between oil production and demand presents a major energy security threat and an incentive for significant energy savings. Between 1974 and 1988, transportation's share of US petroleum consumption rose from 51% to more than 63%, and transportation continues to be almost totally dependent on petroleum-based fuels. Many believe that the technical and economic potentials to double the fuel economies of cars and light trucks exist. Cars twice as efficient as current-generation cars are projected to cost only $200–800 more (5). Although these first costs would be more than recovered through fuel savings over the lives of the cars, manufacturers probably will not install these features in the short term. Over the next 20–25 years, however, these efficiency improvements should be realized, producing a potential savings of 30–35% in energy use with current growth rates of annual vehicle-miles.

Although other OECD countries do not have this same efficiency improvement potential, they do have opportunities to increase fuel efficiency somewhat. In addition, some economies may result from improved processing of crude oil into petrochemicals and further improvements in use of oil and gas for industrial process heat and combined electricity generation. However, it is not expected that the total improvements in oil and gas use in other OECD countries will be more than 5–10%.

The biggest savings in electricity can be attained in a few areas: lights, motor systems, and the refrigeration of food and rooms. In the United States lighting consumes about 25% of electricity, about 20% directly, plus another 5% in cooling equipment. In a typical existing commercial building, lighting uses about 40% of all electricity directly, or more than 50% including cooling load.

The Electric Power Research Institute (EPRI) believes that electricity use in the United States could be reduced by as much as 55% through cost-effective means. Reaching this potential electricity savings at the lowest possible cost requires exploiting the full menu of efficiency opportunities. This entails not only installing new technologies, but also re-engineering whole systems, paying meticulous attention to detail.

The US natural gas industry and gas consumers are at the threshold of a great period of challenge and opportunity. Further research and development can lead to technologies that enhance gas usage. Two areas of particular interest in this regard are gas-fueled cooling technologies (e.g. gas heat pumps) and gas-fueled cogeneration systems (e.g. small gas turbines or fuel cells).

Significant cost savings and energy-efficiency improvements are also possible for industrial gas customers by (a) using advanced gas-fueled processes for manufacturing, (b) using improved burner systems, controls, and components, (c) incorporating advanced high-temperature materials, and (d) applying advanced technologies and materials to recover and reuse waste energy.

During the 21st century, natural gas usage thoughout the world will be greatly expanded; therefore these end-use technology improvements are crucial to overall energy-efficiency improvement.

Without carrying out a specific nation-by-nation analysis, we adopt an expected realizable target for energy-efficiency improvement of 25–50% by 2050 in developed countries. Such an improvement could provide a potential carbon-emission reduction of 2–4 gigatons annually (Table 1).

The costs for end-use efficiency improvements vary with specific situations; however, most inexpensive approaches have already been adopted.

Table 1 Potential carbon-emission reduction for developed countries

Mitigation approach	Potential carbon displacement/ sequestering by 2050	Approximate cost (equivalent) $/ ton of C	Approximate net cost (equivalent) $/ ton C
End-use efficiency improvement	2.0–4.0	40–60	–140 to +20
Generation efficiency improvement	0.2–1.2	20–60	0–40
Substitution for carbonaceous electricity production			
Nuclear	0.2–2.0	—	35–70
Solar and other	0.05–0.1	—	25–200
CO$_2$ scrubbing/compression (sequestering costs not included)	1–3	100–200	100–200

Studies conducted for EPRI indicate an approximate initial cost of $400–600 per kWe of installed electric end-use technology. Although this may not pertain to other energy systems, we use it to obtain an approximate estimate of the cost of reducing carbon emissions. These capital costs are equivalent to approximately $40–60 per ton of carbon displaced (Table 1).

In this analysis, I differentiate between the estimated cost of avoiding the production of carbon, through increased efficiency, etc, and the "net cost to society." The net cost is the difference between the cost of CO$_2$ reduction and the costs that would have been incurred to meet the energy or electricity demand. For purposes of analysis, I have utilized the costs associated with a coal-fired power plant as the costs that would have been incurred if the efficiency improvement, or fuel-substitution approach, had not been adopted.

If we credit the entire equivalent cost of the generation and transmission system and of the coal, we have a credit of $180 per ton. If we only credit the conservation system with the coal cost, the credit is approximately $40 per ton of carbon. Thus, the net cost for enhanced end-use efficiency is minus 140 to plus 20 dollars per ton of carbon (Table 1). Thus, there is a tremendous economic incentive to find additional opportunities to increase end-use efficiency in the developed countries.

ENHANCEMENT OF GENERATION EFFICIENCY Here we focus on electricity-generation efficiency improvements. If we assume that approximately 50% of the developed countries' energy use will be in the form of electricity over the next 60 years, the electricity system will be responsible for approximately 50% of the carbon emissions. If the conservation measures discussed above are achieved, the total carbon emissions will be 4–6 gigatons. Over this period, we can anticipate power-generation efficiency improvements from such advanced systems as coal gasification–combined cycle or fuel cells, magneto hydrodynamic power generation, or other advanced coal systems in the range of 20–40%. If one half of the coal plants in 2050 use advanced coal technology, we have the potential to reduce carbon emissions by 0.2–1.2 gigatons (Table 1).

In order to achieve the higher efficiencies, these advanced coal technologies are expected to have 10–30% higher capital costs than conventional coal. This translates into a $20–60 per ton additional cost for carbon control (Table 1). The improved efficiencies result in net costs of $0–50 per ton of carbon.

SUBSTITUTION FOR CARBONACEOUS MATERIALS   The electricity sector also provides great fuel flexibility from the generation standpoint. Nuclear, solar, biomass, and hydro are all potential substitutes for carbonaceous materials. The substitution potential for nuclear power is difficult to estimate; however, it could provide the preponderance of electricity generation within the developed world. The technology is available; however, costs, public concerns, and waste disposal are limiting factors. We could certainly envisage anywhere from 100,000 to nearly 1,000,000 MWe being deployed over the next 60 years. This range results in a carbon-substitution potential of 0.2–2.0 gigatons per year. The costs of nuclear power plants are expected to become nearly competitive with advanced coal; however, presently we estimate their costs could be substantially greater—perhaps $500–1000 per kWe; however, fuel costs are approximately one-half that of coal. These translate into an equivalent net carbon cost of $35–70 per ton.

Direct solar and other non-nuclear noncarbonaceous energy systems are expected to have only a limited role within the next 50–60 years in the developed countries. (Potential biomass substitution is discussed in Section 3.) The fundamental limitation is the low average solar availability in these countries and the resulting poor economics of solar systems compared with other energy technologies. Further, most of the solar systems, including wind, are limited to low average annual capacity factors. I can project no more than 50,000–100,000 MWe of capacity with an average capacity factor of 0.3. This translates into a potential carbon substitution of 0.05 to 0.1 gigatons. The capital costs of these systems are likely to be $2000–3000 per kWe greater than comparable fossil-fired units; however, these fossil units consume natural gas or distillate at prices of $2–5 per million Btu. The approximate net cost range is $35–200 per ton of C.

CO_2 SCRUBBING/COMPRESSION/SEQUESTERING   EPRI has conducted a number of studies to estimate the net power loss and costs associated with scrubbing carbon dioxide from coal-fired power plants. These studies indicate a net power loss of approximately 35% and plant incremental capital costs of $1000–2000 per kWe (6). This translates into a net cost of carbon control of $100–200 per ton. Coal consumption for power production in the developed countries is approximately 1.5 billion tons per year and will likely double over the next 60 years. The degree to which the carbon-dioxide scrubbing, compression, and sequestering process is applicable to the various types of coal-fired power plants is very uncertain; however, I estimate that 1–3 gigatons of carbon could be controlled in this manner. Further, it is unclear just how the CO$_2$ would be sequestered; although recent research indicates the potential for sequestering CO$_2$ in a clathrate form at ocean depths greater than 3000 feet.

The scrubbing/compression/sequestering process is clearly uneconomical, with limited applicability, and certainly would be a "last resort." Other technological approaches being considered are partial carbon recycle through photocatalytic processes, forest waste substitution for coal, etc. It is extremely difficult to estimate any significant contribution from these sources.

ASSESSMENT   It is clear from Table 1 that the least-cost approach to CO$_2$ mitigation is enhanced end-use efficiency and improved generation efficiency to the extent achievable. If a major commitment were made by the developed countries to conservation and these potential carbon displacements could be realized, there would be great global carbon benefit.

Returning to our allocation of 3 gigatons of carbon to the developed world by 2050, versus an unconstrained level of 8 gigatons, this could conceptually be achieved within the least-cost options in Table 1, at net costs of $70 per ton or perhaps less. This certainly provides us a basis for focusing on these options; however, a major commitment to a policy emphasizing these options is required. The maximum annual cost to the developed world would be $350 billion, well within its economic capability.

2. Mitigation Options for Developing Countries

The global carbon control situation is much different from the perspective of the developing countries (DC). Unconstrained growth in the DCs is projected to produce a total of nearly 20 gigatons of carbon by 2050. If we again are attempting to limit carbon emissions to 10 gigatons by 2050, the "allocation" to the developing world would be 7 gigatons or a limitation in growth by 65%. In addition, far fewer options are available to the developing world than to the developed world.

ENHANCEMENTS IN END-USE EFFICIENCIES   Although end-use efficiency improvements have long-range potential, in many DCs, not even today's state-of-the-art commercial products are available nor is the average individual per-capita income available to purchase high-efficiency, high-cost appliances or systems. Our ability to have these markets served by state-of-the-art appliances and systems is essential, in order to limit energy growth with minimal impact on the economic development of DCs.

Although a careful analysis is necessary to verify these estimates, it is projected that a 12.5% reduction in energy growth can be made with little effect on the economies of these countries, particularly if the primary energy use is in the form of electricity. Beyond this level of conservation, the gross developed products (GDPs) of individual DCs are likely to be affected.

In order to curtail CO_2 emissions, I establish an overall target of 25% reduction in energy growth in the developing countries, from 4% per annum now to 3% per annum in 2050. This would translate into a savings of five gigatons of carbon annually by 2050. For purposes of estimating the cost of this "control strategy," I assume that to achieve conservation beyond the 12.5% level, GDP falls in tandem with conservation (energy/GDP = 1.0), so that by 2050 GDP growth would have been curtailed by 12.5% in meeting the 25% conservation target.

If this approach could be achieved, it would cost developing countries about \$2.5–3.0 trillion of GDP by 2050, or a cost of \$800–1000 per ton of carbon. This clearly is an unacceptable cost. Thus, an energy-conservation strategy that meets global CO_2 control needs, but does not tax the lesser developed countries inordinately, must be carefully designed. At this point, energy conservation in developing countries cannot be expected to be much greater than 12.5%, or a reduction from 4% per annum growth to 3.5%. Thus, energy conservation will only decrease carbon emissions from the uncontrolled projection by 2.5 gigatons annually.

ENHANCEMENT OF GENERATION/TRANSMISSION/EFFICIENCY   Here we focus on efficiency improvements that could be made in electricity generation and transmission systems. If we assume that 50% of the developing countries' energy use will be in the form of electricity by 2050 (4), the electricity system will be responsible for approximately 50% of carbon emissions. If the conservation measures that have little impact on the economy are achieved, the total carbon emissions in the DCs would be 17.5 gigatons. In addition to the advanced coal-generation efficiency improvements discussed previously, which provide potential efficiency improvements of 20–40%, there are opportunities to improve transmission efficiencies by another 10–20%. If one half of the coal plants in 2050 use advanced coal technology, we have the potential to reduce carbon emissions by 1.75–3.5 gigatons.

Assuming the costs of these plants are comparable to those within the developed countries, the costs of carbon control are \$20–60 per ton, with net costs of \$0–50 per ton.

SUBSTITUTION FOR CARBONACEOUS MATERIALS   Nuclear, solar, biomass, hydro, etc are potential substitutes for carbonaceous fuels for producing electricity. It is difficult to estimate the nuclear potential in developed countries, but even more so in developing countries. Most of the developing countries that will contribute to major increased energy use—China, India,

Thailand, Indonesia, etc—have large coal or other carbonaceous fuel resources. The use of these resources is clearly their "low-cost option."

As in the developed countries, one can envisage nuclear power substitution of 100,000–1,000,000 MWe in the developing countries during the next 60 years. It may behoove the developed world to subsidize such an approach, since major nuclear substitution plays a significant role in reducing carbon emissions. On the other hand, the potential for nuclear proliferation is greatly increased, and may be a key limiting factor. Further, in the DCs, nuclear plant cost premiums may be greater than in the developed world, so we estimate additional capital costs of $1000–1500 per kWe. Fuel costs may be comparable to coal costs in these countries; therefore, net carbon costs will be $80–120 per ton.

In general, solar and other non-nuclear noncarbonaceous energy systems have a greater potential use in the developing countries owing to their geographical location. A major exception to this statement is China, with its large coal resources. However, unless significant cost reductions are made in solar systems, their high specific capital costs per unit energy output will remain a major limitation. Therefore, I believe that solar and other noncarbonaceous energy systems will have only a limited role within the next 50–60 years. (Biomass substitution is discussed in Section 3.) The solar contribution is expected to be comparable to that in the developed world and should have comparable cost premiums, or perhaps higher. Thus, carbon substitution of 0.05 to 0.1 gigatons is estimated, at net costs of $50–200 per ton of C.

CO₂ SCRUBBING/COMPRESSION/SEQUESTERING   This option has a great carbon-control potential within the DCs; however, its very high costs (Table 2) make this approach even less likely to be accepted by developing than developed countries.

ASSESSMENT   It is clear from Table 2 that achieving 13 gigatons of carbon reduction from unabated energy development in the DCs is impossible at costs less than $100 per ton (maximum potential approximately 8 gigatons). Even the 8-gigaton estimate assumes strong nuclear substitution and adoption of advanced coal technologies. A more likely estimate is 6.0 gigatons. If our carbon control budget is to be achieved, it appears that a global mitigation strategy to sequester 5.0–7.0 gigatons of carbon annually must be considered, if costs of less than $100 per ton of carbon can be achieved.

3. International/Global Mitigation Options

The primary mitigation approach available on a global basis is ecological macro fixation of carbon. Terrestrial or oceanic phytomass species may be used. The enhancement of net primary productivity and ultimate sequestering

Table 2 Potential carbon-emission reductions for developing countries

Mitigation approach	Potential carbon displacement/ sequestering by 2050 (gigatons of carbon)	Approximate cost (equivalent) $/ ton of C	Approximate net cost (equivalent) $/ ton C
Conservation/end use efficiency improvement	2.5	Essentially no cost	Essentially no cost
	2.5	800–1000	800–1000
Generation/transmission system efficiency improvement	1.75–3.5	20–60	0–50
Substitution for carbonaceous electricity production			
Nuclear	0.2–2.0	—	80–120
Solar and other	0.05–0.1	—	50–200
CO_2 scrubbing/compression (sequestering costs not included)	5.0–7.5	100–200	100–200

or substitution of biomass for fossil fuels provide a significant carbon-mitigation potential. However, even enhanced biomass production results in required areas of approximately one million square miles per gigaton of carbon fixed. Of course, if biomass is utilized for carbon fixation, it must substitute for fossil fuel use in order to reduce net CO_2 fluxes to the atmosphere. Therefore, harvesting, transportation, and utilization must be considered in estimating the realizable CO_2 reduction potential.

TERRESTRIAL PHYTOMASS   EPRI has recently supported two assessments of the potential to store carbon in terrestrial systems, from increases in forest growth and area and from production of halophytes in marginal lands such as salt marshes and deserts. Kulp (7) has estimated that increased forest growth could store 1.1 gigatons of carbon by 2050 and plantations to increase forest area could store an additional 2.5 gigatons (Table 3). The analysis indicates the world could meet increased demand for wood products and energy by that time, with half the wood going into long-term storage and half for energy. Delivered costs of such products are highly dependent on harvesting and transportation costs, but are likely to be in the range of $30 to more than $100 per ton of carbon.

Halophytes can be grown as managed systems in marshes, coastal deserts, and inland salt deserts (8). It is estimated that perhaps 0.5 million square miles could be managed to store and sequester 0.5 gigatons of carbon. These halophytes could be sequestered initially by turning them into the soil; however, as carbon levels in the soil increase, they would have to be

Table 3 Ecological macro fixation

	Potential carbon storage or sequestering by 2050	Approximate cost to combustion/ conversion site ($/ ton C)
Terrestrial phytomass[a]		
Increased forest growth	1.1	30 to >150
Plantations—increased forest area	2.5	
Halophytes	0.5	150–250
Oceanic phytomass[a]		
Macro-algal species		
Continental Shelf		
Harvested	0.5	300[b]
Open ocean		(30–70)[c]
Natural system stimulation	1–2	50–100
Artificial systems	1–5	30–210
Phytoplankton[d]	1	50–100

[a] Area requirements approximately 10⁶ square miles per gigaton of carbon stored
[b] Present state-of-the-art being practiced in the Far East
[c] Projected long-term based on open ocean cost estimates
[d] Area requirements approximately 18×10^6 square miles

harvested and used for fuel. Halophyte biomass for power generation would cost $100–160 per ton of carbon, not including transportation costs. Total costs are likely to be $150–250 per dry ton delivered.

OCEANIC PHYTOMASS   The open ocean area of the earth covers more than 100 million square miles. Although marine primary productivity in the ocean is estimated to be 25–50 gigatons of carbon annually (1), the net carbon flux to the bottom of the ocean is only 2–3 gigatons annually. By artificially stimulating production and harvesting or sequestering the oceanic phytomass, very significant amounts of airborne carbon could be stored. (Although it is expected that most of the carbon fixed in the oceanic phytomass will be atmospheric carbon, there is some uncertainty as to the fraction that may be dissolved inorganic or organic carbon contained within the sea water.)

A wide variety of macroalgal species, such as *Macrocystis, Laminaria, Gracilaria,* and *Euchema,* are produced naturally in the ocean. Most of the large kelp beds are found on the continental shelves. It is estimated that 5% of the continental shelves could be planted with various macroalgal species (8), storing approximately 0.5 gigatons of carbon annually. These most likely would be harvested and converted to specialty chemicals, synthetic natural gas, etc. Based on today's yield and harvesting approaches, production costs are estimated at $300 per ton of carbon. These estimates appear very high and should have great potential for reduction as our understanding of artificial macroalgal farms develops.

With the production limitations on the continental shelves, open ocean–stimulated macroalgal growth is clearly required for substantial carbon fixation to occur. A recent workshop sponsored by EPRI (9) reviewed the open ocean production and sequestering potential of macroalgal systems. Initially, the focus for carbon fixation would be to stimulate growth of natural macroalgae in major currents and gyres such as the Sargasso Sea. Plant disposal may be by sinking or conversion to fossil-fuel substitutes, e.g. methane, and possibly sequestering carbon as CO_2 hydrates (clathrates). The sequestering approaches would result in real carbon extraction of 1–2 gigatons, if major productivity increases can be achieved. The costs associated with this stimulation would mainly be for "seed" farms and nutrient addition if the carbon is sequestered by sinking. Minimum nutrient supply from upwelling would cost approximately $30 per ton of carbon, and sequestered costs would be $50–100 per ton.

H. Wilcox has estimated a range of costs for an open-ocean orbiting tensioned grid system (9) of $40–220 per ton, not including upwelling or nutrient costs. In this case, valuable chemicals, synthetic natural gas, etc, are produced. If we include the costs of upwelling and estimated value of the SNG produced, costs are estimated to be $30–210 per ton of carbon substituted. These estimates are derived from an early conceptual design and, of course, require much more rigid analysis to establish them with any confidence.

Finally, studies to stimulate phytoplankton production have been conducted (10, 11). If iron addition can be made uniformly and function as projected, up to one gigaton of carbon could be sequestered annually with this approach. However, the entire southern Atlantic and Pacific Oceans (some 18 million square miles) would have to be "fertilized" with iron, and once this fertilization experiment is initiated, it must be continued for hundreds of years to keep the carbon sequestered. In addition, the frequency and effectiveness of the iron addition has yet to be established; thus, no accurate overall costs have been estimated. However, costs of $50–100 per ton of C are consistent with costs estimated for a 400 km^2 experiment. If this experiment establishes the feasibility of the approach, these estimates should be approximately valid.

CONCLUSIONS

Controlling net carbon emissions in 2050 to 10 gigatons per year will be a prodigious task, if the world population and energy demands grow as projected. The developed world must focus on major end-use and generation-system energy improvements. A major nuclear power plant substitution policy may be necessary. If clear priorities and directions are established, the developed countries could reduce CO$_2$ emissions to their fair prorata share at reasonable costs ($350 billion per year or less).

The developing world cannot both meet its economic growth objectives and comply with global CO$_2$ emission limitations at reasonable cost. The cost of energy conservation to these countries becomes totally unacceptable by the mid-21st century. Even with major generation efficiency improvements and nuclear power substitution, CO$_2$ emissions will be nearly twice the desired level.

It appears that a major international focus on increasing both terrestrial and marine phytomass is necessary to achieve global stability in CO$_2$ emissions. This focus is necessary not only to meet the CO$_2$ emission objectives targeted here, but also to provide a firm basis for substitution of annual produced carbon for fossil carbon by the end of the 21st century. This is the only approach for achieving long-term stability in the atmospheric carbon burden.

Further, the imposition of a carbon tax, absent a specific plan for CO$_2$ mitigation, is not likely to achieve a satisfactory carbon limitation in the earth's atmosphere. The problem is far too complex to be solved by a simple tax. Careful attention should be paid to the proper incentives, world CO$_2$ cost sharing, and other factors in order to limit atmospheric carbon levels significantly.

Finally, this assessment is based on current information relating to potentially acceptable atmospheric CO$_2$ burdens. As additional information becomes available, this analysis and attendant mitigation strategies should be reviewed and modified. However, the framework and analysis results provide both a policy analysis approach and a methodology to establish meaningful research, development, and commercial mitigation strategies.

Literature Cited

1. Sundquist, E. T., Broecker, W. S., eds. 1985. *The Carbon Cycle and Atmospheric CO$_2$: Natural Variations Archean to Present. Geophysical Monograph 32.* Washington, DC: Am. Geophysical Union

2. Manne, A. S., Richels, R. G. 1990. *Global CO$_2$ Emission Reductions—The Impact of Rising Energy Costs.* June

3. Fickett, A. P. Gellings, C. N. 1990. Efficient use of electricity. *Sci. Am.* Sept.

4. Starr, C., Searl, M. F. 1990. Global energy and electricity futures, *Energy Policy Book,* Chapter VI. The Atlantic Council. 29 June (revised Oct. 2)

5. Chandler, W. U., Geller, H. S., Ledbetter, M. R. 1988. *Energy Efficiency: A New Agenda,* Am. Counc. Energy-Efficient Econ. July

6. EPRI. 1991. *Engineering and Economic Evaluation of CO$_2$ Removal From Fossil Fuel Powerplants,* Vol. I. *Pulverized Coal Fired Powerplants. EPRI Report IE 7365,* Vols. I and II

7. Kulp, J. L. 1990. *The Phytosystem as a Sink for Carbon Dioxide. EPRI EN-6786 Special Report.* May

8. EPRI. 1991. *Seaweeds and Halophytes to Remove Carbon From the Atmosphere. EPRI N-7177.* Feb.

9. EPRI. 1990. *A Summary Description of the Second Workshop on the Role of Macroalgal Oceanic Farming in Global Change.* Newport Beach, Calif., 23–24 July (internal report)

10. Martin, J. H. 1990. Glacial-interglacial CO$_2$ change: the iron hypothesis. *Paleoceanography* 5(1):1–13.

11. Sarmiento, J. L. 1990. *Modeling Biological and Chemical Controls of Carbon Dioxide, Symp. Global Change Systems, Am. Meteorol. Soc.* Feb.

Annu. Rev. Energy Environ. 1991. 16:275–94

ELECTRICITY GENERATION IN ASIA AND THE PACIFIC: HISTORICAL AND PROJECTED PATTERNS OF DEMAND AND SUPPLY

Hossein Razavi

World Bank, Washington, DC

Fereidun Fesharaki

East-West Research Center, Honolulu, Hawaii

KEY WORDS: Asia energy, Asia fuel oil, Asia natural gas, Asia nuclear power, Asia investments.

INTRODUCTION

The Asia-Pacific region contains about 17% of world commercial[1] energy resources and produces about 16% of energy supply but consumes about 20% of world energy demand. The gap between production and consumption varies for different types of fuel. In nonoil energy, the region maintains a balance between demand and supply. The gap is therefore almost totally reflected in the difference between production and consumption of petroleum, of which the region produces 10% but consumes 18% of the world's total production. The net import of oil into the region is about 4.5 million barrels per day (mmbd), which still indicates a significant improvement compared with 5.6 mmbd in 1976 and 5.7 mmbd in 1980.

---

[1]The term commercial energy is used here to refer to oil, gas, coal, hydro, nuclear, and geothermal energy.

1056-3466/91/1022-0275$02.00

The reduction in oil imports has been achieved through major efforts by almost all countries in the region to reduce their dependence on petroleum. The drive towards such an objective started in the mid-1970s, and the results started to show up in the early 1980s. Neither the drive itself nor its results have yet come to an end. The trend to replace oil is expected to continue well through the 1990s.

A major avenue for shifting away from oil use is to change the fuel mix in the power sector. This possibility has been utilized to its full extent in many Asian countries, resulting in drastic changes in sources of electricity generation. The share of oil in total electricity production declined from 43% in 1975 to 18% in 1987, while the share of coal increased from 29% to 39% and nuclear power and natural gas were introduced and grew to account for 15% and 11%, respectively, of total generation by 1987. The fuel shares are expected to change further as the ongoing and planned investments in power generating capacity become operational through the next 11 years.

The purpose of this study is to assess the future mix of fuels required for power generation in the Asia-Pacific region through the year 2000. Such an assessment is based on the following steps:

1. Project future electricity demand.
2. Review the present generating capacity to estimate the shortfall in meeting future demand.
3. Appraise the capital and operating costs of electricity generated by oil, coal, gas, hydropower and nuclear power.
4. Use the results of steps 1–3 to arrive at a "least-cost" mix and level of generating capacity that would meet future demand.
5. Calculate the level of electricity generation from each type of capacity.
6. Estimate the quantity of each type of fuel that will be needed to generate the above level of electricity generation.

The study presents the results for each country separately. A detailed, country-by-country analysis is reported elsewhere (1). This paper presents the regional profile by reviewing the power demand, supply, and fuel mix for the entire Asia-Pacific region, including 16 countries: Japan, Korea, Taiwan, Singapore, the Philippines, Thailand, Malaysia, Indonesia, China, India, Pakistan, Bangladesh, Sri Lanka, Nepal, Australia, and New Zealand.

HISTORICAL AND PROJECTED DEMAND FOR ELECTRICITY

Although most electricity demand forecasts have not survived the test of time, they are the only means available to draw up plans for investments in expanding the generation, transmission, and distribution capacities. It is therefore important to acquire a feel of the magnitude of the future power load before assessing the fuel requirements of the power sector.

The growth in electricity consumption in a country is influenced by several factors, including:

1. the rate at which national income grows;
2. the increase in access to electricity—electrification policy;
3. the rate of electricity penetration—the rate at which electricity replaces other forms of energy;
4. the extent to which rising prices of electricity induce a slowdown in electricity use; and
5. the extent to which electricity consumption may approach a saturation point resulting in slower growth in electricity use.

The countries in the Asia-Pacific region are in various phases of economic development, and thus their electricity use represents an entire spectrum of a classical pattern, from an early stage in Nepal and Bangladesh to a saturation stage in New Zealand, Australia, Japan, Taiwan, and Singapore.

Total electricity consumption in the Asia-Pacific is presently about 1,500,000 GWh per annum (p.a.), equivalent to about 7 mmbd of crude oil. Electricity consumption grew at a high rate of about 15% p.a. during the 1960s and 10% p.a. during the first half of the 1970s. The growth slowed substantially to about 6% p.a. in the second half of the 1970s and to only 5% p.a. in the first half of the 1980s. Since 1985 there has been a noticeable recovery to about 7.5% p.a.; the rate is expected to stabilize over the long term at about 5.8% from 1990 to 2000.

Sectoral Distribution of Electricity Demand

The industrial sector was the dominant consumer of electricity in the late 1960s and early 1970s, accounting for about 58% of the total electricity demand (see Table 1). By 1980 this share was reduced to about 50%, and it has subsequently remained at that level. The share of the residential and commercial sectors grew in the 1970s. The residential sector's share (17% in 1975) reached about 20% in 1980. The sectoral distribution of electricity consumption has remained stable since 1980 and is expected to remain at the current mix during the 1990s.

Regional Patterns of Electricity Consumption

The regional distribution of electricity consumption presents some interesting patterns in both the total and the per-capita use of electricity. The total consumption of electricity is concentrated among the four largest users:

Table 1   Historical and projected electricity demand by consumer group, 1965–2000

Consumer group	1965	1975	1980	1987	1995	2000
Industrial						
GWh	165,939	393,029	554,778	728,363	1,137,701	1,491,649
% of total[a]	58	49	50	51	51	50
Residential						
GWh	54,743	134,527	198,759	300,116	453,726	605,737
% of total[a]	19	17	19	21	21	20
Commercial and other						
GWh	63,872	276,280	347,045	400,670	603,528	854,658
% of total[a]	23	34	31	28	28	29
Total consumption (GWh)	284,554	803,836	1,100,582	1,429,149	2,194,955	2,952,044

[a] Percentages may not total 100 because of rounding error.

Japan, China, India, and Australia. Japan consumed more than 50% of the region's electricity demand until the late 1970s (see Table 2). Its share has now declined to less than 40%. China's share of electricity consumption has grown from about 17% in 1970 to 27% in 1987, while India's and Australia's shares have remained around 9% and 7%, respectively. The four countries together consumed about 90% of the region's electricity in the 1970s and still use about 80% of the total power demand. The decline in the combined share of the largest four consumers has been taken over by several other countries, notably Korea and Taiwan. Korea's electricity consumption increased from 2,463 GWh in 1965 to 63,410 GWh in 1987, while Taiwan's consumption increased from 5,673 GWh to 59,175 GWh. As a result, these countries have become the fifth and sixth largest users of electricity, each representing more than 4% of the total demand.

The ranking of Asian countries in terms of per-capita consumption of electricity is substantially different from their position in total electricity use. Based on 1987 consumption figures, New Zealand, with a per-capita consumption of 7,800 kWh per annum ranks first, followed by Australia (6,500 kWh), Japan (5,500 kWh), Singapore (4,000 kWh), Taiwan (2,600 kWh), and Korea (1,600 kWh) (see Table 2). China and India, with per-capita consumptions of 420 kWh and 260 kWh, rank ninth and eleventh, respectively.

Projected Demand to the Year 2000

The eventual objective of reviewing past trends is to arrive at a reasonable forecast for future growth in electricity demand. Most Asia-Pacific utilities

Table 2 Historical and projected distribution of electricity demand in Asian countries, 1965–2000

Country	Share of total consumption (%) 1965	1975	1987	2000	Average annual growth rate 1987–2000 (%)	Per-capita electricity consumption (kWh)
Japan	50.6	53.9	38.3	26.1	2.6	5,500
Korea	0.9	1.8	4.4	5.0	6.8	1,600
Taiwan	2.0	2.9	4.1	4.8	7.0	2,600
Singapore	0.3	0.4	0.7	0.7	5.2	4,000
Philippines	1.2	1.2	1.3	1.7	7.5	400
Thailand	1.4	1.1	1.6	1.8	6.6	510
Malaysia	0.4	0.5	1.0	1.1	7.0	1,012
Indonesia	0.3	0.4	1.2	2.8	13.0	176
China	19.9	19.5	26.7	31.1	7.0	420
India	8.5	7.7	9.5	15.9	10.0	260
Pakistan	—	0.7	1.2	1.5	7.0	250
Bangladesh	0.1	0.1	0.2	0.6	14.0	40
Sri Lanka	—	0.1	0.2	0.2	8.1	150
Nepal	—	—	—	—	9.6	20
Australia	10.5	7.2	7.3	5.2	2.9	6,500
New Zealand	2.8	2.1	1.9	1.2	2.5	7,800

	1965	1975	1987	1995	2000
Total electricity demand (GWh)	284,554	803,836	1,429,149	2,194,955	2,952,044

undertake a continuous effort to update their past data and revise their projections of future demand. The forecasts utilized in this study are based on the latest available information for each country. Each utility uses a different type of demand forecasting methodology. However, the final assumption regarding the demand forecast is made based on planners' judgement while taking account of the results from various models. The models comprise (a) statistical relationships between electricity demand, per-capita income, electricity prices, and the prices of alternative forms of energy; and (b) end-use relationships which assess the power demand by each consumer category (2). Statistical relationships are regularly revised to take account of new data. The basic equations are included in demand forecasting documents and results reflected in occasional publications (3–5). End-use models are often utilized to provide a "second-opinion" about the future growth in demand. These models comprise detailed functions which aim at a bottom-up aggregation of electricity demand. The results normally provide a "feel" of growth in power use rather than actual forecasts. The power demand figures used in this study reflect the latest forecast made by utility planners in the region. In most countries, forecasts have been reviewed and officially accepted by government agencies responsible for allocating investment funds to power utilities.

Table 1 contains the projections of electricity demand for the region's utilities. The overall growth is forecast at an annual rate of 5.8% p.a. This is substantially lower than the rates experienced in the 1960s and 1970s but somewhat higher than the rate of 5% p.a. during 1980–1985. The sectoral distribution of demand is expected to remain rather stable through the 1990s.

Regional distribution of the power demand is expected to change substantially. Electricity demand in Japan is projected to grow at 2.6% p.a. during 1987–2000 (see Table 2). As a result, Japan's share of electricity consumption in the regional context is expected to decline from 38% in 1987 to 26% in 2000. Similarly, Australia's electricity demand is projected to grow at 2.9%, resulting in a decline in its share from 7% to 5%. Demand of both of the other two large consumers—China and India—is expected to grow faster than the average rate for the region. China's growth rate is projected at 7% p.a. while India's growth rate is forecast at 10% p.a. Korea and Taiwan are expected to maintain their ranks, respectively, as the fifth and sixth largest consumers. Korea's growth of electricity demand is projected at 6.7% p.a., while Taiwan's growth rate is forecast at 7% p.a.

In summary, electricity demand is expected to grow at substantially different rates (from 2.7% p.a. in Japan to 10% p.a. in India and 13% in Indonesia) in different countries of the region, but aggregate regional demand is forecast to grow at a rate of 5.8% p.a.—very close to the average rate experienced during the past 10 years.

EXISTING GENERATING CAPACITY

A review of the quantum and types of present power generating capacity is needed before one can assess the possible shortfalls and the need for future expansion (see Table 3). The total installed capacity of power utilities in the Asia-Pacific region increased from 77,000 MW in 1965 to 388,000 MW in 1987, representing a growth rate of about 8% p.a. The growth during 1965–1975 was substantially higher—11% p.a.—than the growth rate of 5.5% p.a. experienced during the 1975–1987.

The mix of the region's generating capacity has changed substantially in the past two decades. During the 1965–1975, the overall tendency was to increase the use of oil-based and gas-based capacity and to cut back on investments on hydroelectric and coal-based power plants. The main reason was, of course, the lower petroleum prices of the 1960s and early 1970s, which made oil- and gas-based electricity generation less costly than other options. As a result, between 1965 and 1975 the shares of hydropower capacity declined from 32%

Table 3   Regional distribution of installed electricity-generating capacity at end-1987 (MW)

Country	Hydro	Oil	Gas	Coal	Nuclear	Total
Japan	26,123	43,900	30,000	12,000	25,098	137,122
Korea	2,232	4,822	2,550	3,700	5,716	19,020
Taiwan	2,558	4,932	0	3,955	5,144	16,589
Singapore	0	3,371	0	0	0	3,371
Philippines	2,124	2,365	894[a]	405	0	5,788
Thailand	2,250	1,042	2,720	885	0	6,877
Malaysia	1,316	2,557	900	0	0	4,773
Indonesia	1,262	3,957	378	930	0	6,527
China	28,000	9,000	0	52,000	0	89,000
India	16,196	168	1,178	30,394	1,330	49,266
Pakistan	2,800	2,000	2,000	0	125	6,800
Bangladesh	130	,408	1,069	0	0	1,607
Sri Lanka	868	270	0	0	0	1,138
Nepal	140	36	0	0	0	176
Australia	7,143	1,200	2,400	22,100	0	32,843
New Zealand	4,324	0	2,000	1,019	0	7,433

[a] Geothermal plants.

to 27%, coal-based capacity declined from 38% to 30%, whereas oil- and gas-based capacity increased from 29% to 40% (see Table 4).

The rapid rise of petroleum prices in the 1970s reversed the trend by stimulating an intensive effort to reduce the dependency on petroleum for power generation. The result was a sharp drop in the share of oil-based capacity from 31% in 1980 to 20% in 1987. The oil share was initially replaced by nuclear power, representing 10% of generating capacity in 1987 compared with 6% in 1980. At a later stage, coal-based power is becoming an even more important source of replacing oil. Its share in total capacity increased from 30% in 1980 to 33% in 1987. The major effect of the move towards coal is yet to come, as the share of coal-based capacity is projected to increase to about 40% by the late 1990s.

PROJECTED GENERATING CAPACITY

Projections of electricity demand and the review of existing facilities indicate that the generating capacity of the power sector needs to be expanded by about 360,000 MW during 1989–2000. The most important investment decision that the utility companies will make is how much of this additional capacity should be based on coal, gas, hydropower, nuclear, and oil. The decision will be based on two important considerations: first, the availability of domestic sources of energy; second, the economics of building and operating each type of plant.

Table 4   Total installed power-generating capacity in the Asia-Pacific, 1965–1987

Type of capacity	1965	1970	1975	1980	1985	1987
Total capac-ity (MW)	69,561	117,756	200,856	281,857	365,217	288,329
Hydroelectric						
Capacity (MW)	22,545	34,919	54,020	71,760	90,974	97,466
% of total	32	29	27	25	25	25
Oil-based						
Capacity (MW)	20,350	35,822	63,139	83,860	86,650	80,028
% of total	29	30	31	30	24	20
Gas-based						
Capacity (MW)	258	10,412	17,761	24,705	34,646	46,069
% of total	—	9	9	9	9	12
Coal-based						
Capacity (MW)	26,408	35,383	60,298	84,690	120,709	127,478
% of total	38	30	30	30	33	33
Nuclear						
Capacity (MW)	0	1,220	5,640	16,842	32,238	37,288
% of total	0	1	3	6	9	10

Indigenous Sources of Energy

The Asia-Pacific region has a wide base, though not evenly distributed, of sources of energy. Table 5 contains the latest (1987) estimates of energy resources by country, compared with total resources worldwide (3, 6). The region has 17% of the world's energy resources. From an approximate total recoverable energy reserve of 132 billion tons of oil equivalent (toe), about 95% is located in three countries alone: China (48%), Australia (31%), and India (16%). In addition to this uneven regional distribution, the resources are unevenly distributed in form. Oil and gas reserves of the region amount to only 5.4% and 6.3%, respectively, of the world's resources, whereas recoverable coal reserves represent more than 20% of the world's resources. The region's coal reserves are likely to be far greater than the 228 billion tons included in the table, but they have not yet been brought into the estimates of recoverable reserves.

The above distribution of indigenous resources reveals that: (a) despite the availability of energy resources in the region, most countries need to import their energy requirements either from a few countries within the region or from outside the region; and (b) a large part of oil requirements has to be imported from outside the region. Most energy-importing countries of the

Table 5   Most recent estimates of economically recoverable energy resources in the Asia-Pacific

	Crude oil (million barrels)	Natural gas (trillion ft³)	Coal (billion short tons)	Total (million toe)
Japan	100	1.1	1.1	610
Korea	0	0.0	0.7	379
Taiwan	0	0.6	0.2	124
Singapore	0	0.0	0.0	0
Philippines	15	—	0.3	164
Thailand	100	7.4	0.4	426
Malaysia	2,800	49.4	0.03	1,715
Indonesia	8,300	49.4	1.4	3,243
China	18,400	30.0	108.9	62,441
India	4,200	17.6	37.7	19,870
Pakistan	100	18.7	0.1	560
Bangladesh	18	12.7	—	179
Sri Lanka	—	—	—	—
Nepal	—	—	—	—
Australia	1,700	18.7	72.4	39,975
New Zealand	200	5.7	0.3	360
Others	1,500	17.4	2.1	2,000
Total Asia-Pacific	37,500	228.7	225.6	132,046
World total	699,800	3,632.3	1,049.3	764,051
% of world	5.4	6.3	21.5	17.2

region appear to view the supply of resources within the region as more reliable than the supply from outside. And, of course, they all view the supply of coal to be more reliable than oil. As a result, the energy policy of most countries has aimed at shifting away from oil to coal, with the understanding that coal would be available at relatively stable prices from Australia, China, and possibly Indonesia.

Economics of Various Fuel Options

The utility planning and decision-making environment in the Asia-Pacific region has changed substantially during the past 15 years. In the early 1970s a move was initiated to modernize the planning practice. Thereafter, most utilities in the region equipped themselves with least-cost planning models (7–9), such as the WIEN Automatic System Planning Package (10), and convinced their policy-makers that adhering to the solutions of these models was a symbol of objectivity. This mentality has changed quite rapidly since the early 1980s, as decision-makers have realized that a "least-cost" solution may not really be least-cost when things do not happen as assumed. Policy-makers have become increasingly convinced that there is a considerable risk in any decision they take, and they are willing to pay a bit more, in terms of cost, to have the flexibility to reduce the risk. This has, to a large extent, served as a kind of backlash to the use of formal modelling—or at least to the "black-box" approach for minimizing costs. Instead, what is becoming increasingly popular is to choose an investment strategy intended to "minimize future regrets" (11). To minimize future regrets, the utility develops a variety of alternative investment programs, each of which is associated with a particular scenario of future events. The utility then attempts to identify the similarities among these programs and to arrive at a set of decisions that should perform well under all plausible scenarios.

Many utilities in the region combine the results of least-cost analysis with subjective judgments about various types of risks to devise their generation expansion programs. Indeed, the policy adopted by Japan, Korea, and Taiwan (12) of proceeding with nuclear power construction and the policy pursued by China and India of keeping a balanced (hydroelectric-coal) generation mix are aimed more at preserving flexibility rather than minimizing costs.

Future Generating Capacity

The power-generating capacity of the region is expected to almost double by the end of the century, reaching some 850,000 MW. The shares in total capacity of hydropower, nuclear, and gas-based capacity are expected to remain relatively stable, and only the oil and coal shares will change significantly. Between 1987 and 2000, the oil share is expected to decline from 18% to 7%, whereas the coal share is projected to increase from 30% to 40% (see Table 6). These are continuations of trends that began in 1980 when both oil and coal had equal shares (30%). During 1980–1987 the declining oil share was overtaken by three alternatives—coal, natural gas and nuclear power—although in the future the oil share is expected to be replaced almost entirely by coal-based capacity.

The total installed capacity of hydropower plants is expected to increase by about 118,000 MW, from 97,466 MW in 1987 to 216,165 MW in 2000. The additional capacity will keep the share of hydropower generation in the total power generation stable at about 17%. From the added capacity of 118,000 MW, about 48,000 MW will be built in China, 39,000 MW in India, and 10,000 MW in Japan. The remaining 21,000 MW will be built in the rest of the region, notably in Australia, Korea, and Taiwan.

The total installed capacity of oil-fired power plants is expected to decline from 80,000 MW in 1987 to 64,000 MW in 2000. No significant new investments in oil-based capacity are planned. A small share of the existing

Table 6   Projected levels of total power-generating capacity in the Asia-Pacific, 1987–2000

	1987	1990	1995	2000
Total capacity (MW)	388,329	450,826	574,854	775,176
Hydroelectric				
Capacity (MW)	97,466	116,593	150,143	216,165
% of total	25	26	26	28
Oil				
Capacity (MW)	80,028	76,710	71,398	64,199
% of total	21	17	12	7
Gas				
Capacity (MW)	46,069	58,995	75,068	83,803
% of total	12	13	13	11
Coal				
Capacity (MW)	127,478	154,938	221,955	327,630
% of total	33	34	38	42
Nuclear				
Capacity (MW)	37,288	43,590	56,290	83,380
% of total	10	10	10	11

capacity is being converted to use natural gas. Most of the existing capacity is being run for intermediate and peaking purposes. No replacement is planned for oil-fired plants as they gradually retire. Most of the decline of 16,000 MW during 1987–2000 in total capacity is due to this retirement. Japan will retire 10,000 MW of oil-fired plants during this period, Korea 2,000 MW, Taiwan 2,000 MW, and Malaysia 1,000 MW.

The total installed capacity of gas-fired power plants is expected to increase from 46,000 MW in 1987 to about 84,000 MW in 2000. This increase will leave the share of gas-based power capacity at about 11% in 2000, almost the same as its 1987 level. In the interim, however, the share of gas-based power capacity is expected to increase to about 13% by the mid-1990s and then gradually decline back to 11%. The notable difference between the past and future growth of gas-based generation is that a major part of past capacity was built for the use of imported liquefied natural gas (LNG). Most of the future gas-based plants will be built by countries that own gas. The 38,000 MW of capacity additions comprise about 11,000 MW being built in Japan, 3,500 MW in Malaysia, 2,700 MW in Thailand, 2,500 MW in Indonesia, and the remainder in Pakistan and Bangladesh.

The total installed capacity of coal-fired power plants is projected to increase from 127,000 MW in 1987 to 328,000 MW in 2000. This huge increase in capacity will raise the share of coal-based power in total electricity generation from 39% to 50%. Almost all countries in the region are planning to undertake substantial investments in coal power plants. The roughly 200,000 MW of additions to coal-based capacity will come from China (84,000 MW), India (60,000 MW), Japan (11,000 MW), Korea (11,000 MW), Indonesia (10,000 MW), Taiwan (8,000 MW), Thailand (4,000 MW), Australia (4,000 MW), and the Philippines (3,000 MW). China, India, Indonesia, and Australia plan to supply the fuel requirements of these plants from domestic coal supplies, whereas Japan, Korea, Taiwan, the Philippines, and Thailand will be dependent on imports for the major portion of their additional coal needs.

The total installed capacity of nuclear power plants is expected to increase from 37,000 MW in 1987 to 83,000 MW in 2000. This increase will keep the share of nuclear power stable at about 10% of total installed capacity and 15% of total generation. More than half of the addition of 46,000 MW during this period will be built by Japan (28,000 MW), and the remaining 23,000 MW will be built by India (8,000 MW), Korea (5,000 MW), and Taiwan (4,000 MW).

Regional Distribution of Future Generating Capacity

The regional distribution of generating capacity is expected to shift somewhat, as the share of Japan in total power-generating capacity will decline from about 35% in 1987 to about 24% in 2000 and the share of Australia will decrease from 8% to 6% (see Table 7). The declining shares of Japan and Australia will be offset by increases during this period in the shares of China (from 23% to 29%) and India (from 13% to 20%).

FUEL REQUIREMENTS OF THE POWER SECTOR

The total generation of electricity in the Asia-Pacific region is expected to double by the year 2000—reaching 3,500,000 GWh, equivalent to 15 mmbd of crude oil (see Table 8). Power generation in 1987 was 18% dependent on fuel oil, 11% on natural gas, 39% on coal, 17% on hydropower, and 15% on nuclear power. The share of oil is expected to pass to 8% and coal to increase to 50% by 2000. The shares of hydropower, gas, and nuclear are expected to remain stable during 1990s.

Utilities use their own estimates of thermal efficiency for each type of plant and the heat content for each type of fuel to determine the fuel requirements of electricity generation. We use the same procedure, but we calculate the physical quantities of coal on the basis of heat content of 12,000 Btu/lb, gas

Table 7   Regional distribution of future electricity-generating capacity

	1987	1990	1995	2000
Total capacity (MW)	388,329	450,826	574,584	775,176
Share of individual countries (%)				
Japan	35.3	32.6	28.7	23.9
Korea	4.9	4.7	4.4	4.4
Taiwan	4.3	3.8	4.1	4.0
Singapore	0.9	0.8	0.8	0.7
Philippines	1.5	1.5	1.5	1.5
Thailand	1.8	1.6	1.9	1.8
Malaysia	1.2	1.2	1.1	1.0
Indonesia	1.7	1.9	2.4	2.8
China	22.9	25.5	27.4	28.6
India	12.7	13.1	15.4	20.0
Pakistan	1.7	2.2	2.0	2.1
Bangladesh	0.4	0.5	0.7	0.8
Sri Lanka	0.3	0.3	0.2	0.2
Nepal	—	—	—	—
Australia	8.5	8.1	7.2	6.1
New Zealand	1.9	1.9	1.8	1.7

1,000 Btu per cubic foot, and fuel oil 18,500 Btu/lb, to make the fuel requirements of different countries comparable.

Fuel Oil Demand

The fuel oil requirement of the region's power sector is expected to increase slightly during the next two years and then to revert to its trend of gradual decline. The total requirement of fuel oil in 1987 was about 87 billion liters, equivalent to 83 million tons or 1.5 mmbd. This is expected to decline to about 82 billion liters in 1995 and 75 billion liters in 2000 (Table 9). The impact of this decline on total demand—along with cutbacks in other uses of heavy fuel oil—is of critical importance in determining the price of fuel oil in the region.

The share of fuel oil in total consumption of petroleum has consistently declined during the past eight years. In 1987 the region's consumption of major oil products totalled 7.6 mmbd—32% of which was heavy fuel oil. Between 1988 and 1995 demand for total petroleum products is expected to grow at 2.3% p.a. The strongest growth is projected for middle distillates (3.8% p.a.) followed by light products (2.7% p.a.). Demand for heavy fuel oil is expected to decline at 0.1% p.a., which would further reduce the demand share of heavy fuel oil to 26% by 1995.

Table 8 Projection of electricity generation by type of fuel for the Asia-Pacific

	1987	1990	1995	2000
Total generation (GWh)	1,755,565	1,961,536	2,553,366	3,521,203
Hydro				
Generation (GWh)	297,965	334,308	435,808	584,204
% of total	17	17	17	17
Oil				
Generation (GWh)	313,332	334,385	301,387	281,749
% of total	18	17	12	8
Gas				
Generation (GWh)	191,962	239,044	304,750	357,589
% of total	11	12	12	10
Coal				
Generation (GWh)	685,506	774,062	1,124,591	1,746,943
% of total	39	39	44	50
Nuclear				
Generation (GWh)	265,026	277,504	383,246	547,197
% of total	15	14	15	15

The decline in demand for heavy fuel oil may result in an increase in the differential between its price and the prices of light and middle distillate products. The differential has in recent years fluctuated substantially as the surplus of heavy fuel oil has kept its price depressed. With a fluctuating crude oil price and fluctuating product price differentials, there may be periods in which the heavy fuel oil price will drop considerably below the price of imported coal. Utilities that have the flexibility to switch to heavy oil will then benefit.

Natural Gas Demand

The power sector's demand for natural gas in 1987 was about 4,686 mmcfd, equivalent to 0.8 mmbd of crude oil. This demand is expected to increase to 7,440 mmcfd in 1995 and 8,730 mmcfd in 2000. From the 1987 consumption of 4,686 mmcfd, about 3,400 mmcfd (73%) was used in Japan's power sector. Other large users were Korea, Thailand, Australia, and Pakistan.

Japan is the largest consumer of natural gas with an annual consumption of about 1,600 bcf, of which about 75% is used in the power sector and the remaining 25% is used in the industrial and residential sectors. Japan imports about 94% of its gas needs—from Indonesia (51%), Malaysia (19%), Brunei (18%), United Arab Emirates (7%), and the United States (3%). Gas use in

Table 9 Fuel requirements of the power sector

Country	Fuel oil 1987 (million tons)	Fuel oil 2000 (million tons)	Natural gas 1987 (bcf)	Natural gas 2000 (bcf)	Coal 1987 (million tons)	Coal 2000 (million tons)
Japan	51.8	43.2	1,220	1,534	35.4	59.4
Korea	1.1	1.3	83	128	7.2	27.3
Taiwan	1.6	0.8	0	88	8.2	25.2
Singapore	2.9	4.4	0	0	0.0	0.0
Philippines	2.3	2.0	0	0	0.9	9.6
Thailand	1.5	0.5	83	236	2.0	12.2
Malaysia	2.4	0.2	26	265	0.0	1.2
Indonesia	1.9	1.0	5	127	1.8	21.5
China	15.5	12.2	0	0	106.5	271.3
India	0.0	0.0	30	0	42.6	133.7
Pakistan	2.3	5.0	50	268	0.0	0.0
Bangladesh	0.2	0.4	37	175	0.0	0.0
Australia	1.0	3.0	105	135	31.2	41.2
New Zealand	0.0	0.0	39	59	0.5	2.1
Total	84.5	74.0	1,678	3,015	236.3	604.7

the Japanese power sector is expected to increase to about 1,500 bcf by the year 2000. The additional requirement will be imported in the form of LNG from the same sources.

Korea and Taiwan are the only other countries that will increase their gas use by importing it in the form of LNG. Korea started to import LNG from Indonesia in 1986. It currently uses the LNG in 2,500 MW of power plants that were converted from oil to gas in 1985. If Korea proceeds with its current plans to build new gas-based power plants, its gas use in the power sector will increase from 83 bcf in 1987 to 128 bcf in 2000. This requirement, along with the anticipated use of gas in nonpower sectors, would necessitate the negotiation of a second LNG contract by the mid-1990s. Taiwan is moving along a similar path, preparing to use gas in the power sector and expecting to have about 2,600 MW of gas-based power plants by 2000.

Thailand increased its use of domestic natural gas very rapidly by converting some of its oil-fired plants to burn gas. It also promoted gas use in the industrial sector. It will further increase the use of gas in the power sector by installing combined-cycle power plants. There is considerable uncertainty with regard to the timely availability of domestic gas, as production from current fields is expected to fall sharply in the late 1990s and plans for new exploration and development are not being implemented.

Malaysia has embarked on a rather aggressive plan to utilize its large resources of natural gas. Its increased use of gas in the power sector will be supplied easily by domestic resources, and it may become more actively involved in exporting gas to other countries. Its present contract to sell 6 million tons p.a. of LNG to Japan is likely to be extended in the mid-1990s with an increase in quantity. Its proposed sale of piped gas to Singapore is expected to start by 1992–1993.

Indonesia is the largest producer of gas in the region. It exports about 80% of its gas to Japan and Korea and has a contract to supply gas to Taiwan. It is also the most likely candidate for meeting the increased demand for LNG in the 1990s. Indonesia currently uses only a small amount of gas in its power sector but will build about 2,900 MW of gas-based power plants by the year 2000. The nonpower use of gas is substantial and is expected to make up a large part of demand through the 1990s.

China and India produce relatively large quantities of gas but use all of it in the nonpower sector.

Pakistan and Bangladesh produce considerable quantities of gas, most of which are used in the nonpower sector. In recent years, however, both countries have embarked on aggressive plans to increase gas use in the power sector. Pakistan is planning to install 7,000 MW of gas-based power plants by 2000. Bangladesh is aiming to build about 3,000 MW of gas-fired power stations during the same period.

Australia and New Zealand produce about 550 bcf and 113 bcf of gas, respectively, of which about 20% is used in the power sector. For these countries, gas is not the dominant fuel in the power sector.

Coal Demand

The power sector's demand for coal in 1987 was about 233 million tons, equivalent to about 3 mmbd of crude oil. This demand is expected to increase to about 600 million tons by 2000. The largest producer and consumer of coal in the region is China, which produces about 900 million tons of coal, of which about 15% is used in the power sector. China's utilization of coal to generate electricity is expected to increase by 150% during the next 12 years. All the coal requirements will be met by domestic production.

India, the second-largest producer and consumer of coal in the region, produces about 180 million tons, of which 25% is used in the power sector. India's use of coal to generate electricity will triple during the next 12 years. The country is planning to meet the additional demand by domestic supply.

Japan, the largest importer of coal in the region and in the world, imports coal from Australia (47%), Canada (19%), the United States (13%), South Africa (10%), the Soviet Union (6%), and China (4%). Japan's consumption of coal was about 105 million tons in 1987, of which 35% was used in the power sector. The coal needs of the power sector are expected to increase by about 70% by the end of the century. The additional requirement is expected to be imported from the same sources that currently supply coal to Japan.

Korea, the second-largest importer of coal in the world, imports coal from Australia (40%), South Africa (25%), Canada (16%), and the United States (14%). Korea's use of coal in the power sector will quadruple during the next 12 years. All the coal requirements of the power sector will be supplied from increased imports.

The Philippines and Thailand produce some low-quality coal, a large proportion of which they use in the power sector. However, the coal used to generate electricity will substantially exceed domestic coal production and will have to be supplemented by imports. These countries are not yet large importers of coal but consider Australia and China as the main potential sources of coal supply.

Indonesia is not currently a large producer or consumer of coal. Its coal reserves, however, are believed to be much larger than the current estimate of 1.3 billion tons, and its production capacity is expected to expand rapidly. The government expects not only to supply the future coal demand from domestic production but also to export to other countries. This expectation may not be realistic, however, in view of the projected substantial increase in demand for coal by the power sector.

Australia is the world's largest exporter of coal. It produces about 133 million tons of coal, of which it exports 90 million tons. Its domestic use of coal is about 40 million tons, of which 75% is used in the power sector. The production of coal is expected to increase substantially to enable the country to meet anticipated exports.

CONCLUSION

The usual perception is that power utilities choose their capacity expansion plans so that they will be able to minimize the discounted value of their future investment and operating costs. While this remains their eventual objective, most utilities in the Asia-Pacific deviate from a strict "deterministic" least-cost development plan to take account of the risk associated with the projection of future prices of fuels. The deviation is either in terms of the choice between primary (hydro and nuclear) versus fossil fuel (oil, gas, and coal) power or the choice within the fossil-fuel group.

Nuclear power is by most accounts more costly than coal-, gas-, and even oil-based electricity. Our cost analyses indicate a per-unit cost of more than US¢6/kWh for nuclear power compared with US¢4.5/kWh for coal-based and US¢5.1/kWh for oil-based electricity. Nevertheless, several countries—particularly Japan, Korea and Taiwan—are proceeding with considerable investments in nuclear power generation. The justifications for pursuing such a policy is to keep up with nuclear power technology, to utilize the expertise already developed in nuclear industry, and more importantly, to cope with the supply and price uncertainties of international fuel prices. The latter reason is also offered to justify investments in some hydropower projects that are more costly than coal or oil power plants.

Within the fossil-fuel group, almost all utilities in the region are moving towards increasing use of coal. As a result, the coal requirements of the power sector will increase from about 233 million tons in 1987 to about 600 million tons in 2000. From this 600 million tons, about 450 million tons will be used in China, India, and Australia, which are expected to supply their requirements from their own domestic resources. Most of the remaining 150 million tons should be supplied by imports. The majority of utilities use the cost and quality of Australian coal in their planning practices, assuming that the coal will be available when they need it. The coal reserves of Australia and other coal-exporting countries seem to be sufficient to meet the long-term demand of coal-importing countries. However, a rapid increase in demand for coal may create a tight market in the 1990s, and production and shipping constraints may cause a sharp rise in the price. If, at the same time, heavy fuel oil remains depressed due to cutbacks in the power sector's demand, then the coal option may be much more costly than oil.

The policy implication of the above discussion is that, while some investments in hydroelectric and nuclear generation may be justified based on the decision-makers' desire to preserve a flexible balance and to avoid certain risks, the next logical extension of the same theory would be not to lock out the oil option entirely.

Dual-firing (and in certain cases triple-firing) power plants provide substantial flexibility in the economic operation of power systems. Japan and New Zealand already have some fuel-switching capabilities among oil, gas, and coal, and Thailand is building triple-firing plants. However, the current motive for maintaining or building dual-firing plants is to be able to cope with possible supply problems rather than economic operation of the generating system. In our view, fuel-switching capabilities offer an effective tool to deal with fuel price uncertainty, and the region's power utilities should investigate the viability of expanding these capabilities.

The fuel price uncertainty should be viewed from two different angles: (*a*) the differential between the prices of coal and petroleum, and (*b*) the differentials among petroleum product prices, in particular the price of heavy fuel oil relative to that of other products. Present projections indicate that (*a*) the price of coal, after adjusting for capital cost differential and heat content, remains lower than that of oil, and (*b*) heavy fuel oil remains some 10–15% cheaper than crude oil. Both patterns are subject to considerable uncertainty. The price of coal relative to that of oil may increase beyond current expectations due to constraints in coal production and shipping. The relative price of fuel oil may also behave differently from the projected path due to cyclical surpluses and deficits of heavy oil in the regional petroleum products balance. The region's refinery configuration is not presently capable of absorbing all the surplus fuel oil—in particular, high-sulfur residual fuel. This surplus will continue to depress the price of heavy oil but, at the same time, will trigger investments in refinery upgrading. As a result, the region's refinery configuration will become capable of absorbing more heavy fuel oil, leading to stronger fuel oil prices and smaller incentives for further refinery upgrading. This cyclical pattern in fuel oil prices is an inherent phenomenon in the economics of refinery upgrading.

A power utility with reasonable dual-firing capability can switch between fuels if the long-term trends in fuel prices develop differently than assumed at the planning stage. In addition, the utility can take advantage of cyclical patterns of fuel oil prices and random fluctuations of petroleum and coal prices. The switching capability will be utilized in the day-to-day dispatch decisions of the power system in order to minimize the cost subject to the operational constraints. Furthermore, for utilities with access to natural gas, fuel-switching capability provides a means of coping economically with environmental constraints. Presently, some US and European utilities use a minimum amount of gas during certain periods of the year as a means of keeping plant pollutant emissions within prescribed standards. This practice offers good potential for many utilities in the region that are under increasing pressure to reduce the environmental impacts of their power plants.

Fuel-switching capabilities can often be installed at a small cost if planned at an early stage. Most utilities in the region are planning substantial additions to gas-fired and coal-fired plants. The gas-fired plants are either conventional steam or higher efficiency combined-cycle. Both types can operate with oil, but a conventional steam plant would use heavy oil whereas a combined-cycle plant would have to use distillate fuel. Steam plants would be perfect candidates for dual-firing. The capital cost of a dual (oil/gas)-fired steam plant would be only 1% higher than a single-firing plant. But since the price of distillate fuel is expected to remain substantially higher than that of heavy oil, combined-cycle plants are not viewed as good candidates for dual-firing. The policy implication would then be to install combined-cycle gas-firing plants only when there is a firm supply of inexpensive gas, and to install dual (oil/gas)-fired steam plants when there are uncertainties in the stability of supply or when the gas price is within a limited margin of the fuel oil price. In the latter case, the utility can minimize its cost by adjusting the proportion of gas and oil use in its day-to-day operation.

Dual-firing between coal and oil is a somewhat different matter than that of gas and oil. If a plant is intended to use oil, it would be quite costly (about 30–40% increase in capital cost) to make it a dual (oil/coal)-firing plant.

However, if the plant is being designed to use coal, it would be quite inexpensive (1–2% additional capital cost) to make it dual (oil/coal)-firing. In the Asia-Pacific region, almost all the candidates for oil/coal-firing fall into the latter category. Most utilities are planning vast additions to their coal-firing capacity, and they can add the oil-firing option at a small cost.

Literature Cited

1. Fesharaki, F., Razavi, H. 1990. *Electricity in Asia Pacific*, London: Econ. Intell. Unit
2. World Bank. 1989. *World Tables*. Washington
3. Asian Development Bank. 1989. *Energy Indicators of Major Developing Member Countries of ADB*. Manila
4. Japan Electr. Power Inf. Cent. 1988. *Electric Power Industry in Japan*. Tokyo
5. Kadir, A., Kim, Y. H. 1985. *The Electric Future of Singapore*. Honolulu: Resour. Systems Inst., East-West Cent.
6. United States Dept. Energy. 1975–1988. *International Energy Annual*. Washington
7. Albouy, Y. 1985. *Assessment of Electric Power System Planning Models. Energy Department Paper No. 20*. Washington: World Bank
8. Anderson, D. 1972. Models for determining least cost investment in electricity supply. *Bell J. Econ.* 2(1)
9. Caraminas, M. C., Schweppe, F. C., Tobors, R. D. 1987. *Electric Generation Expansion Analysis System (EGEAS)*. Palo Alto, Calif: Electric Power Res. Inst.
10. Jenkins, R. T., Joy, D. S. 1974. *WIEN Automatic System Planning Package (WASP)*. Oak Ridge, Tenn.
11. Pindyck, R. S. 1988. *Options, Flexibility and Investment Decisions*. Cambridge, Mass: Cent. Energy Policy Res., Mass. Inst. Technol.
12. Kawamura, T. 1988. Latest evaluation of power generating cost. *Energy in Japan*.

Annu. Rev. Energy Environ. 1991. 16:295–335

TRANSITIONS IN HOUSEHOLD ENERGY USE IN URBAN CHINA, INDIA, THE PHILIPPINES, THAILAND, AND HONG KONG*

Jayant Sathaye and Stephen Tyler[1]

International Energy Studies Group, Lawrence Berkeley Laboratory, Berkeley, California

In collaboration with Qiu DaXiong, Ruben Garcia, Peter Hills, J. Krishnayya, and Amara Pongsapich

KEY WORDS: fuel ladder, electric appliances, surveys.

INTRODUCTION

Recent transitions in the demographic and industrial trends of developing countries have sparked a renewed interest in urban household energy consumption. Historically, work on the developing world's energy sector has focused on establishing new energy supply strategies and on improving the methodologies used for national energy forecasting and planning. Studies carried out on household energy consumption have focused primarily on the

---

*The US government has the right to retain a nonexclusive royalty-free license in and to any copyright covering this paper.

[1]Dr. Sathaye is Co-Leader of the International Energy Studies Group at Lawrence Berkeley Laboratory, California. Dr. Tyler was formerly a Research Associate at the Laboratory and is presently a Regional Program Officer with the Canadian International Development Research Centre, Singapore.

use of biomass fuels (fuelwood, agricultural residues, and dung) in rural households. This work has led to a greater appreciation of the use of traditional energy sources and has highlighted the importance of the transition from traditional to modern fuels at the national level. As a result, many developing-nation governments have introduced fuel subsidies enabling poorer households to gain access to higher-quality, modern fuels.

In recent years, however, the energy issues facing developing countries have shifted. Various factors have led to these transitions. First, the rapid urbanization of the developing world has fueled the expansion of residential energy demand and increased the reliance on modern energy sources, particularly eletricity. Second, international financial conditions have changed, placing constraints on efforts to satisfy this growing demand. Hence, many developing-world governments, saddled with colossal debt burdens, are unable to finance the construction of capital-intensive electricity-generating systems. In addition, economic restructuring under the guidance of international financial institutions, such as the International Monetary Fund (IMF) and the World Bank, has placed heavy pressure on governments to eliminate residential fuel subsidies. Third, an emerging new perception of energy in the international arena has compelled developing world governments to re-address energy issues under new terms. International concern has mounted over global environmental and resource issues, many of which (e.g. acid precipitation, climate change) relate directly to energy use in developing nations. The combination of these factors has emphasized the need for a greater understanding of energy consumption in the developing world's urban households.

This paper identifies the most important results of the Urban Energy Network Project, a recent comparative analysis of household energy use in Asian cities. Home to some of the world's most dynamic economies and rapidly growing urban populations, Asia has experienced a number of trends in recent years that have major implications on both a regional and a global scale. This paper describes the Urban Energy Network Project, examines energy use (primarily fuel use)[2] for cooking, water heating, and space heating as well as household electricity consumption disaggregated according to its primary end-uses in the surveyed cities. Based on the analysis of residential energy use patterns, this paper presents a range of policy options aimed at promoting more efficient energy use in urban Asian households.

---

[2]Electricity accounts for a small to negligible fraction of the energy used for cooking, water heating, and space heating in the surveyed cities. Therefore, the sections on these activities focus primarily on fuel use (non-electric power). Where relevant, the analysis discusses electricity use, such as in the case of the widespread use of rice cookers in Thailand. The tables in this section, however, include only non-electric energy sources.

THE URBAN ENERGY NETWORK PROJECT

Despite the household energy surveys that have been conducted in different parts of the world, good comparative data on urban household energy use in developing countries remains largely unavailable. This five-country project emanated from the recognized need to fill this gap.

Background

Financed by Canada's International Development Research Centre (IDRC), research groups in China, Hong Kong, India, the Philippines, and Thailand agreed on a consistent set of basic issues for analysis and on a similar structure for survey instruments and data coding. Using these methods, the researchers carried out surveys on households in 12 cities between mid-1989 and mid-1990. The results of the interview questionnaires provide a substantive and consistent data set for an international comparison of urban residential energy use.[3] In the cases of China and India, because similar surveys were carried out several years earlier, this work generated information on long-term household energy use trends. Separate reports provide more detailed analyses of the results of urban surveys for India, China, Thailand, and Hong Kong (1–4).

The surveys specifically examine issues related to fuel choice, household income, urban scale, and the energy decision-making process. One principal question guided the research: What is the nature and extent of the fuel transition in urban households? The research examines the adoption of modern fuels for cooking, water heating, and space heating and the penetration of electricity and electric appliances in relation to the socio-economic characteristics and locations of households. The researchers focused particularly on energy use in low-income households,[4] and examined a variety of factors that may serve to limit poor households' access to modern fuels. In order to evaluate the influence of the size and weather conditions of an urban area on household energy use, the study looked at households in urban centers of varying sizes and climates within the same country.

Study Objectives

The primary objective of the Urban Energy Network Project is to obtain better information on household fuel choices and energy consumption levels. This information is to be used in determining the need for socially targeted energy

---

[3]The results from the Philippine surveys are not included in the totals because an earthquake and political upheaval served to delay the surveys. Wherever appropriate, this report does incorporate the results of an earlier Philippines' survey.

[4]Throughout this paper, low-income households refers to the lowest of five income groups and high-income households refers to the highest of five income groups.

policies in developing Asian countries. Past studies have examined energy use for cooking in urban households across a wide range of countries and cities (5–13). The results of these studies consistently indicate a strong correlation between household income levels and the types and amounts of fuel used for cooking. Based on this observation, researchers have developed the notion of a "fuel/income ladder" to explain the shift to more convenient and higher-quality fuels as household incomes pass certain thresholds (14). The surveys reported in this paper break down energy consumption according to income levels in order to observe this relationship more carefully.

Relative fuel prices do not consistently appear to constrain household fuel choices. Various studies have concluded that many poor households actually pay more for their fuels, on an energy-content basis, than do higher-income households (5). The first costs of appliances, liquefied petroleum gas (LPG) bottles, and/or electrical meter connection fees, however, often restrict the fuel choices of low-income households. Poor consumers often lack the credit or adequate cash required to purchase large quantities of fuel (such as a 13-kilogram bottle of LPG) or lack access to fuel distributors (15).

In addition to cost factors, the size of cities in which households are located influences fuel choices. As urban areas expand, various changes occur in access to fuels, infrastructure, market diversity, housing choices, and household behaviors, all of which appear to determine the use of household fuels independent of the effects of changes in household incomes (16, 17).

Past studies suggest not only that each fuel type has its own pattern of socio-economic incidence, but also that the types of fuels used have different implications for the consumer. For example, poor households have limited access to modern fuels and therefore tend to rely primarily upon traditional fuels. As a result, these households face far greater health risks from the pollutants generated from biomass fuels (18) and suffer from far greater inconveniences because of the lower energy density and combustion efficiency of biomass, the attendant dirt and ash-disposal problems, and the difficulty of fuel storage (particularly in cramped urban quarters). In most developing countries, women shoulder the brunt of the added burdens. For example, traditional fuel use typically entails more time and effort for fuel purchase and collection (even in urban areas), fire maintenance, and food preparation.

The difficulties poor households have in obtaining modern fuels exacerbate the wide socio-economic disparities already existing among urban households in the developing world. Only by fully understanding the existing conditions and trends in urban households and by identifying the nature of the relationship between fuel choices and energy use in those households can researchers and governments determine what types of policy interventions will be effective.

Sampling and Survey Methodology

Researchers in each study country collected data by conducting survey interviews in urban households. Each research team developed its survey according to a common question format and then prepared the survey in the local language. Where necessary, survey teams presented the questionnaire in any of several languages spoken regionally.

In most nations, samples were geographically stratified, i.e. researchers first divided cities geographically by housing type or neighborhood district, both of which were verified in the field, then chose a regular sampling interval for the city and an appropriate number of households to be interviewed from given blocks or neighborhoods. China's research team took a slightly different route; it made use of a recent official household census sampling frame to generate an income-stratified random sample. India's research objectives included resurveying a number of the same households that participated in a similar 1982 study. These households, which constituted a 20% share of the total India sample, were randomly selected in the original 1982 survey.

An examination of the limited available housing and income data on urban centers suggests that the chosen households in Thailand, India, and China broadly represent the nations' urban populations as a whole. Hong Kong's sampling focused mainly on public housing estates, where access is carefully income controlled. These estates constitute about 43% of the nation's households.

Repeated return visits ensured low nonresponse rates. In some countries (where culturally appropriate), the research teams used incentives such as food coupons or wax candles to promote participation. The Philippines' survey, which had the highest nonresponse rate, ended up with a sample skewed towards low-income participants because high-income households often refused to participate in or complete the survey and could not be swayed through the use of incentives. In the case of the Philippines, the unavailability of accurate data for comparing survey results with the characteristics of the actual urban population made it impossible to quantify the effect of the distortion; however, the Philippine results still are useful, inasmuch as the slant towards poor consumers corresponds with the emphasis of the research. Table 1 provides the sample sizes in each city surveyed.

This study devoted particular attention to rectifying some of the methodological problems encountered by other organizations during past household energy surveys (19, 20). Researchers took special precautions, for example, to reduce potential biases due to cultural patterns of deference to the gender, socio-economic status, or authority of the interviewer (21).

The developers designed the survey instruments to enable internal consistency checks and the confirmation of key data. For example, researchers obtained and scaled household income data differently depending on the

Table 1 Urban population estimates and sample sizes

	Total population (millions)	Sample size (number of households)
China		
Beijing	8	500
Nanning	1	200
Hong Kong	5	486
India		
Ahmednagar	0.20	166
Pune	1.70	265
Telegaon	0.03	100
Philippines		
Baguio	0.5	300
Cebu	2	300
Manila	8	300
Thailand		
Ayutthaya	0.05	200
Bangkok	6	420
Chieng Mai	0.14	200

cultural sensitivity of this information. In all cases, however, they verified the data by comparing it with reported expenditures and observed household characteristics, such as the conditions of the home, furnishings, or clothing of respondents.

Respondents typically reported fuel use data according to the amount of the expenditure, physical quantity measures, and unit prices as compared with prevailing retail levels. Electricity consumption data were often difficult to obtain, either because electricity distribution authorities were unwilling to provide them or because of difficulties determining in the field which meter belonged to which housing unit (in the cases where an electrified housing unit actually had an official meter). When this information was unavailable, researchers estimated electricity consumption according to reported monthly expenditures on electricity using the published tariffs in effect at the time of the survey.

The Survey Cities

The cities surveyed encompassed a wide range of living conditions, income levels, economic activities, and national policies.

CHINA   Researchers selected the two survey cities in order to display the contrasts in the nature of Chinese urban centers. Beijing, the metropolitan national capital, has cold winters, diverse energy and consumer goods markets, and high levels of income, education, and industrialization. Nanning, on the other hand, located in the interior of the Guangxi-Zhuangzu province, has a tropical climate, far lower incomes, and a less diverse regional economy.

HONG KONG   The city was selected to illustrate the nature of energy use in a modern high-income city with no constraints on the availability of fuels and electricity and with access to all types of light bulbs and appliances. Household energy use in Hong Kong was of particular interest to Chinese researchers because it was viewed as a potential role model for other Chinese cities.

INDIA   The three Indian cities, all located in the state of Maharashtra, differ primarily in size. The two smaller cities, Ahmednagar and Telegaon, historically have depended highly on local agriculture; in both cities, however, and particularly in Telegaon, a growing share of the population now commutes long distances to the regional metropolitan center (Pune) for employment. Pune, the third Indian city surveyed, has witnessed far more rapid population growth over the past two decades than Ahmednagar or Telegaon. Pune's diverse manufacturing- and service-based economy continues to offer employment opportunities to its growing population.

THE PHILIPPINES   Manila's somewhat precarious socio-economic conditions stem from the threat of deterioration of the city's physical infrastructure and the city's continued political instability. High levels of foreign debt and ever-declining confidence in the Philippines' political durability continue to jeopardize the national economy despite the considerable expansion of manufacturing production. This unpredictability has rendered Manila's urban households far more vulnerable to economic misfortune than households in the other study cities; urban residents across all income levels in Manila have great concerns about survival and security. Of all the cities surveyed, Manila encompassed the widest socio-economic disparities, had one of the highest birth rates, and had one of the largest average household sizes. The second largest city in the Philippines, Cebu, was chosen as the second city to survey. It is located on a different island and has different traditional and cultural patterns. The third city chosen to survey, Baguio, was selected because of its cooler climate and much smaller size.

THAILAND   The researchers in Thailand chose to survey Bangkok, the nation's only large metropolitan center; Chieng Mai, Thailand's second largest city; and Ayutthaya, the center of an historically agriculture-based region located less than 100 kilometers from the center of Bangkok. Chieng Mai's location in the hills provides its residents with easy access to large regional forests and, therefore, abundant and inexpensive fuelwood supplies. The climate in this northern city differs from that of Bangkok and the rest of the country; it is cooler there during the evenings and winter months. Ayutthaya is located just outside of Bangkok's peripheral manufacturing belt; as a result, the city has witnessed the spread of "bedroom-type" housing developments for commuters along with changing employment patterns.

The economies of the largest urban centers in all five study countries are strong and dynamic. The survey data suggest that urban residents increasingly participate in these cities' rapidly expanding economic activities. The growth of manufacturing activities, driven by international trade and external investments, has raised household incomes in most of these cities, and thereby has fueled higher demand for consumer goods. In Thailand, for example, the total national demand for refrigerators has grown by nearly 30% per annum over the past several years according to estimates by appliance manufacturers (22). In all of the cities studied, the combination of migration to the cities and the youthful demographic profiles of urban households ensure continuing high urban growth rates in the future even as national birth rates fall.

The Survey Households: Socio-Economic Characteristics

A comparison of energy use patterns across cities requires a thorough understanding of the nature of the households being examined. The surveys for each city included collection of data on income levels, household sizes, and education levels. These characteristics help to explain why households choose certain fuels, why they use certain amounts of electricity, and why they purchase certain types of appliances. In particular, changes in household incomes and sizes explain the character of fuel and electricity use in the surveyed households.

Household incomes reflect the economies where the cities are located (Table 1a). In US dollars, average income in Beijing and Nanning is an order of magnitude lower than in Hong Kong. However, exchange rates do not adequately reflect the purchasing power of the surveyed households. Expressed in terms of purchasing power parity (PPP) (22a), average income of the Chinese households becomes about one third that of Hong Kong. Similarly, in US dollar terms, Pune households on the average appear to have incomes twice as high as Chinese ones, but they have about the same purchasing power. Moreover, lower-income households in Pune are decidedly worse off than those in the other cities. Income distribution is more even in Beijing, where average income in the highest quintile is about three times that in the lowest quintile. The ratio is six or eight to one in the other cities.

Household sizes range from an average of 3.5 members in the Chinese cities, where government policies restrict the number of children per house-

Table 1a Monthly household incomes

	Household income level				Exchange rate	1988 PPP/ US$ ratio[a]
	Low	Medium	High	Average		
Monthly income (US$)						
Beijing	76	118	189	118	3.7	
Hong Kong	256	970	1914	1120	7.8	
Nanning	63	92	126	92	3.7	
Pune	85	214	536	254	16	
Thailand[b]	160	384	1296	512	25	
Monthly income (PPP)						
Beijing	430	664	1066	666	3.7	5.65
Hong Kong	380	1437	2834	1658	7.8	1.48
Nanning	356	517	714	519	3.7	5.65
Pune	241	610	1530	723	16	2.85
Thailand[b]	373	895	3019	1193	25	2.33
Percentages of households in different income groups						
Beijing	38	40	22			
Nanning	35	35	30			
Pune	35	50	15			
Thailand	21	60	20			
Hong Kong	11	65	24			

[a] PPP values based on Ref. 22a.
[b] Weighted average values for Bangkok, Chieng Mai, and Ayutthaya.

hold, to 5.5 members in the Indian cities. Data for China, Thailand, Hong Kong, and the Philippines (23) indicate that household income levels increase alongside household sizes (Table 1b). In the urban areas of these nations, lower-income households often consist of young couples at the beginning of their careers with no or few children. As families progress and acquire higher incomes, family sizes also increase. This pattern characterizes a dynamic society in which adult household members can find employment easily and in

Table 1b Household sizes (persons/household)

	Income level			
	Low	Medium	High	Average
Beijing	2.84	3.45	3.84	3.31
Hong Kong	4.04	4.72	5.56	4.85
Nanning	3.16	3.56	3.81	3.49
Pune	5.80	5.30	4.50	5.26
Thailand[a]	3.44	4.29	5.12	4.28

[a] Weighted average value for Bangkok, Chieng Mai, and Ayutthaya.

which basic social features (e.g. family size) are changing rapidly. In contrast, in the Indian cities, incomes decrease with the growth of household sizes. As a result of limited employment opportunities and poor income distribution and mobility, most Indian households face far more stagnant economic circumstances.[5]

Higher-income households have either more working members per household or higher average incomes per worker than low-income households. In Hong Kong, the number of working members per household rises from 1.4 in the low-income group to 3.7 in the high-income category. In the Indian households, the middle-income group has the fewest household members employed. In Nanning, the number of household members employed remains stable across income groups. Higher incomes in Nanning households result from higher earnings of individual working members, rather than from additional working household members. In Thailand, household sizes increase for each income group, but not as rapidly as average household incomes.

The growing share of women in the labor force also bears significant implications for household behavior and fuel use patterns. Across the survey cities, female household members were active in the labor force. The number of working women per household declines in Pune and Ahmednagar from the low- to mid-income groups and then increases in the high-income bracket. In general, women in surveyed households had relatively high education levels. In the Philippines, for example, female household members had higher average educational levels than their male counterparts. In India, however, working women in the low-income group tend to be less educated and their income forms a necessary part of the family income. Hong Kong has twice the number of working women per household as India; and the number increases with income. In Thailand, even in the lowest-income group, each household averages more than one woman working and the number of working women per household increases alongside household incomes.

ENERGY USE FOR COOKING, WATER HEATING, AND SPACE HEATING

Developing-country households channel the majority of their energy use towards two activities: cooking and water heating. An earlier survey of semi-urban homes in Pondicherry, India, for example, reported that cooking consumed more then 90% of the total energy used in both low- and middle-income households (24). Because these two activities tend to be intertwined, it is difficult to disaggregate the amounts of fuel used for each activity. Space

[5]Longitudinal data from a 1982 survey of the same 75 households in Pune that were surveyed again in 1989 provide a clue to the changes in income and household size over time. Household incomes increased by 12% in the lowest-income group, by 20% in the middle-income group, and by 45% in the highest-income group.

heating accounts for a significant portion of the household fuel use in Beijing, China. The following section analyzes the types of fuels used to satisfy each end-use, the costs of these fuels, and the amount of energy consumed by the major end-use activities.

Fuel Availability and Costs

LPG, kerosene, piped gas (coal gas or natural gas in Beijing and naphtha-based gas or LPG in Hong Kong), coal, and biomass in various forms constitute the primary fuels used by the surveyed households. The availability and prices of each fuel vary across the surveyed cities (Table 2). Respondents reported adequate availability of all fuels during the period of the survey.

The Indian and Chinese governments administer the supply of certain fuels. The Indian government limits the amount of kerosene available at subsidized prices. Consumers with legal residences may purchase subsidized kerosene supplies using ration cards provided to them by the government. Because the government does not give out ration cards to households located in illegal squatter settlements, these low-income consumers must purchase kerosene at higher prices on the open market or use alternate fuels such as biomass.

Table 2 Fuel prices[a] per gigajoule (GJ)

	LPG	Kerosene	Coal	Charcoal	Wood	Town gas
Beijing						
Yuan/GJ	5.3		2.2			
US$/GJ	1.4		0.6			
Hong Kong						
HK$/GJ	114	63				128
US$/GJ	14.6	8.1				16.4
Pune						
Rupees/GJ	92	64	87	100	126	
US$/GJ	5.7	4.0	5.4	6.2	7.1[b]	
Thailand[c]						
Bahts/GJ	225	254		189	90	
US$/GJ	9.0	10.2		7.6	3.9	

[a] 1 US$ = 3.7 China yuans; 1 US$ = 7.8 Hong Kong dollars (HK$); 1 US$ = 16 India rupees (Rs.); 1 US$ = 25 Thai baht (B)
[b] While charcoal typically costs more than wood, wood prices in Pune surpass charcoal prices because the former are set by nonresidential users.
[c] Weighted average values for Bangkok, Chieng Mai, and Ayutthaya

Government restrictions limit each household to one subsidized LPG cylinder, which when emptied can be exchanged for another.

Chinese households have limited quantities of LPG and coal available. The government quota allows each household to purchase 10–12 cylinders of LPG per year at a subsidized price of 3.6 yuan/15 kilogram (kg) cylinder (this price is one-third the production cost). Chinese households can purchase an additional four cylinders at a price of 9.9 yuan per 15 kg cylinder. Based on the official exchange rate, at $1.4 per GJ in China, LPG costs several times as much on the international market as in China. Coal briquettes cost 32 yuan/ton. Local governments subsidize the producers. In the Philippines and Thailand, the national governments subsidize LPG and kerosene for household use.

Cooking and Water Heating

COOKING FUEL CHOICE AND COSTS   Two fuel transitions characterize the choice of cooking fuels in urban households: urban energy users are increasingly replacing biomass and solid fuels with kerosene and kerosene with LPG and/or piped gas. In many metropolitan areas, these transitions occur simultaneously. Various factors affect the speed at which these transitions occur: income levels, the availability of fuels, government policies, cuisine, and household activity patterns. A statistical analysis of the data drawn from Thai cities indicates that factors such as increases in female employment often dictate cooking fuel transitions more than changes in household incomes.

According to the survey results, biomass use has declined from previously high levels to low or insignificant levels in all of the survey cities. Coal use remains minimal except in China; coal gas, natural gas, and LPG are the fuels of choice of higher-income households in Beijing. LPG and kerosene dominate the cooking fuel mix in India, LPG and charcoal in Thailand, and piped gas in Hong Kong (Table 3).

Fuel shares vary according to income. The share of solid fuels (coal and biomass) in the cooking fuel mix tends to drop as income levels rise. Higher-income households tend to employ larger portions of liquid and gaseous fuels. For example, coal accounts for 31–32% of cooking fuel use in Beijing's low- and middle-income households. At higher income levels, the use of coal drops to 10%. Thai households, a notable exception, tend to use charcoal for grilling meat and steaming glutinous rice; while charcoal use remains highest in low-income households in Thailand, its share is substantial across all income levels.

At present, LPG and kerosene remain far more prevalent across the study cities than piped gas. Where it is available, however, households prefer piped gas. Middle-income households in Beijing, for example, have increasingly

Table 3   Delivered fuel shares (excluding electricity)[a] by income (%)

City	Fuelwood	LPG	Piped gas[b]	Kerosene	Coal	Charcoal
Beijing		48	24		27	
Low income		52	17		31	
Middle income		43	25		32	
High income		55	35		10	
Nanning		6			94	
Low income	1	2			97	
Middle income		5			95	
High income		13			87	
Pune	5	50		44		
Low income	9	26		65	1	
Middle income	4	58		38		
High income		78		22		
Thailand[c]	4[d]	67				28
Low income	10	43				46
Middle income	5	67				28
High income	2	73				25

[a] These figures represent the useful energy delivered to households from each type of fuel. Efficiencies vary widely for different fuel types. Hence, these numbers do not reflect the number of households using each type of fuel (see Table 5).
[b] Refers to the use of coal gas and natural gas in Beijing
[c] Weighted average values for Bangkok, Chieng Mai, and Ayutthaya
[d] Households in Chieng Mai consume far more biomass than those in Bangkok and Ayutthaya.

replaced LPG with piped gas in recent years to complement their coal supplies. At higher income levels, Beijing residents rely on both LPG and piped gas and reduce their coal consumption. In China, single-family dwellings typically cannot afford the first costs of grid connections for piped gas and, therefore, must rely on other fuel types. In multi-family dwellings, however, the institution pays for the connection costs as opposed to the individual household. For this reason, piped gas has a wider presence in multi-family dwellings.

Earlier surveys on household energy use report high levels of biomass consumption, particularly in low-income households (11). For example, past surveys of two Indian cities, Hyderabad (1982) and Bombay (1973), showed that 41 and 10–30% of all low-income households respectively use biomass for cooking. Our survey suggests surprisingly lower levels of biomass use in urban areas today (aside from the use of charcoal in Thailand). In Indian cities, less than 10% of low-income households use biomass for cooking and in China, a negligible share of low-income households use biomass.

Data from a 1982 survey of households in Pune allow for an examination of fuel share shifts during the 1980s. Table 4 shows the changes in LPG, kerosene, and biomass use for the same 75 households between 1982 and 1989 adjusted for the efficiency of cook stoves. While biomass—defined here as fuelwood, dung cakes, charcoal, agricultural residues, and coal—accounted for just under one quarter of the energy used for cooking in 1982, the share of biomass dropped to 8% by 1989. Simultaneously, the share of LPG increased.

Households in all income categories commonly use multiple fuels for cooking. In Hong Kong, for example, a typical household uses electricity for rice cookers, kerosene for slow cooking of soups, and LPG and/or town gas made from naphtha for all other cooking activities. Similarly, many Thai households employ electricity, LPG, and charcoal depending on the type of food being cooked.

Table 5 shows the percentage of households in five of the surveyed cities using each type of fuel for cooking. While the data for Beijing indicates that a relatively small proportion of all households use coal for cooking, coal's share in Beijing's cooking fuel mix is considerably higher (Table 3). Similarly, while 32% of all low-income households in Pune use LPG for cooking, LPG accounts for only 26% of Pune's household cooking energy demand.

Although electricity saturation is ubiquitous in most urban areas, electricity continues to account for a negligible share of the cooking energy mix in the survey cities. In Thailand, for example, only one of the 820 households surveyed lacked electricity service (and that one household was waiting for the local utility to install its electric meter), and 90% of the surveyed households owned electric rice cookers. The amount of electricity used by rice cookers, however, accounts for a minimal share of the total energy used for cooking.

Few households use other types of electric cooking appliances and, in many cities, even the use of rice cookers is rare. Surprisingly, rice cookers are not owned by households in Beijing and Nanning. Three factors account for this low saturation of rice cookers in China: (a) the Chinese diet favors wheat products over rice; (b) many Chinese consumers who do eat rice purchase it already prepared from street vendors; and (c) appliances that draw a heavy

Table 4   Shares of useful energy for cooking in Pune (%)

	1982	1989
LPG	24.7	41.4
Kerosene	51.8	50.7
Biomass	23.4	7.9

Table 5 Percentage of households using each fuel for cooking by income

City	Biomass[a]	LPG	Piped gas[b]	Kerosene	Coal	Charcoal	No fuel
Beijing[b]		62	33		8		
Low income		64	29		11		
Middle income		61	34		8		
High income		60	40		4		
Hong Kong			100				
Low income			100				
Middle income			100				
High income			100				
Nanning[c]	1	16		3	91		
Low income	1	4			96		
Middle income		13		3	96		
High income		32		5	83		
Pune[c]	3	68		28	1		
Low income	9	32		58	2		
Middle income	1	78		19	1		
High income		92		8			
Thailand[d]	7	83		2		38	4[e]
Low income	13	59		5		55	
Middle income	6	86		1		36	
High income	3	94		1		32	

[a] Excluding charcoal
[b] Refers to the use of town gas or LPG in Hong Kong and coal gas or natural gas in Beijing
[c] Data shows shares of the dominant fuel used for cooking
[d] Weighted average values for Bangkok, Chieng Mai, and Ayutthaya
[e] These households have rice cookers (and therefore electricity) but use no fuels.

electric current cannot be operated easily under standard Chinese 5-ampere wiring. One third of the surveyed households in Beijing purchase wheat staple products from restaurants or bakeries, a far more convenient practice than cooking these products at home.

Commercial foods, convenience foods, street vendors, and restaurants serve as important sources of prepared meals in Chinese and Thai cities and in Hong Kong. Across the cities the types of food purchased outside the home vary considerably. Thai households have opposite food preparation patterns of Chinese households: in Thailand, while most households prepare their rice at home, more than one quarter of the survey respondents purchase prepared "convenience" foods from street vendors or stalls to satisfy all or most of their meals. In Hong Kong, the majority of households surveyed indicated that most members ate at least two meals a day outside the home. As a result, some of the energy used for food preparation shifts from the residential to the commercial sector.

These results reflect two major trends: a drop in household sizes and an increase in the share of women entering the labor force. These two patterns, discussed in the above section on socio-economic indicators, have led many urban households increasingly to value convenience and to rely on commercial food preparation for a larger share of household meals.

Two factors determine the costs of cooking: stove type and fuel type. On a first-cost basis, LPG stoves cost far more than simple biomass stoves (traditional three-stone biomass stoves can be assembled at virtually no cost). Table 6 shows the prices and efficiencies of various stove types in India and Thailand according to an index where LPG stoves are given the value of "1" and all other stoves are shown relative to LPG. In India, kerosene stoves cost about 10% of the price of LPG stoves and the costs of all other stoves equal a small fraction of LPG stove costs. The high prices of LPG stoves combined with the deposit required for LPG cylinders serve as a disincentive to low-income consumers.

Once acquired, stoves have varying fuel efficiencies which largely influence cooking costs from that point forward (Table 6). Stove efficiencies increase as households move from solid to gaseous fuels. While the efficiencies of coal and wood stoves tend to vary considerably because of the nonstandardized fuel quality and stove designs, estimates indicate that coal stoves have an average efficiency of 25%. Estimates for biomass stoves are lower, ranging from 7 to 15% (19).

Fuel prices generally move in the opposite direction of stove prices; gaseous fuels usually cost far less than solids. While governments subsidize coal, kerosene, and LPG prices, market pressures usually determine the costs

Table 6   Cost of cooking by fuel type in India and Thailand (indexed with LPG stove and fuel costs given the value of "1")

	LPG	Kerosene	Coal	Charcoal	Wood
Stove efficiency (%)[a]	1.00 (45)	0.89	0.56	0.56	0.33
India					
Stove price (Rs.)	1.00 (1500)	0.10	0.03	0.03	0.01
Fuel price (Rs./GJ)	1.00 (0.34)	0.78	3.41	3.91	16.48
Life-cycle cost (Rs./GJ)	1.00 (0.23)	0.73	1.52	1.75	3.65
Thailand					
Fuel price (baht/GJ)	1.00 (0.23)	1.13		0.84	0.44
Life-cycle cost (baht/GJ)	1.00 (450)	1.90		1.69	2.13

[a] Efficiency figures based on values in Ref. 19.

of biomass fuels in urban areas. In addition, the increasing scarcity of biomass supplies in many cities, due primarily to the loss of forests surrounding urban areas, has served to push up the cost of these fuels. Hence, the efficiency-adjusted price of charcoal in the surveyed cities in India surpasses that of LPG by a factor of four and the price of wood exceeds that of LPG by a factor of 16. A recent Indian study reveals that the price of biomass has increased more rapidly than the prices of both LPG and kerosene in recent years, thereby broadening the already wide gap between the cost of using biomass and the cost of using LPG (25). As a result, lower-income households, which use the majority of biomass, shoulder the double burden of relying on fuels that cost more per unit of delivered heat and having to spend a higher proportion of their incomes for fuel.

Table 6 also shows the life-cycle cost for each stove type. Fuel costs, rather than stove costs, make up the bulk of the life-cycle cost for every type of stove. In India, for example, fuel accounts for 90% of the total life-cycle cost of an LPG stove and an even higher proportion of the total costs for every other stove type. As the numbers indicate, the fuel costs of providing the same amount of useful energy from an LPG stove are far below those for a wood stove. Poor households, which tend to rely on biomass and kerosene as opposed to LPG, generally pay more for each unit of useful energy. In Pune, low-income households pay an average of 27% more than high-income households for each unit of useful energy. Similarly, in Thailand, fuelwood and charcoal cost more than LPG, except in Chieng Mai, where large forests provide plentiful and inexpensive fuelwood.

In addition to stove and fuel costs, several other factors influence which fuel a consumer chooses for cooking. As mentioned above, the inability of Indian consumers without a legal residence to purchase subsidized kerosene leads them to opt for other fuels. In Hong Kong, suppliers sometimes use the marketing technique of lowering the first cost of obtaining town gas connections (town gas costs more than LPG); this practice sways many consumers to make the transition to this fuel type.

WATER HEATING FUEL CHOICES AND COSTS   The practice of heating water for bathing and laundry appears to be a function of cultural traditions. While Indian households prefer to use hot water for these practices, Philippino households rarely do. While individual households in China rarely heat water, Chinese citizens commonly use heated public baths which are often provided at their place of work. Bathing styles also differ among and within cities. Therefore, the amounts of fuel used for these purposes vary.

Fuel choice patterns for water heating mirror those for cooking with two important exceptions. First, a far larger number of high-income households rely on electricity for water heating than for cooking. About three quarters of the high-income Pune households surveyed satisfy their water-heating requirements with electricity. In contrast, for cooking activities, 92% of all high-income households primarily use LPG. Second, many more low-income households use biomass for water heating than for cooking. In Pune, 30% of all low-income households use biomass for water-heating activities (Table 7) versus 9% for cooking. While the share of kerosene remains the same for cooking and water heating, biomass use is three times higher for water heating than for cooking. The limited availability of subsidized kerosene and the ease of using biomass in bulk amounts leads to the increased use of biomass for water-heating activities.

During the past 10 years, about 65% of the surveyed Pune households switched to electricity for heating water. While data for Hong Kong reflects the same trend of increasing electricity use for water heating, unlike households in India and Thailand, Hong Kong residents rely on gas as their main water-heating source. In contrast, households in China rely most heavily on coal and gas; constrained by the 5-ampere wiring, a negligible share of Chinese households use electric water heaters.

TRANSITIONS IN COOKING AND WATER-HEATING FUEL USE   Recent years have witnessed major changes in the cooking and water-heating energy mixes. In order to understand these fuel transitions, we have used two modes of analysis. The first technique relies on household responses to questions explicitly targeted at determining why the interviewees switched fuels, and the second on a statistical analysis of the sample to reveal the most significant factors attributed to fuel switching.

According to the India survey, 10% of all respondents who shifted fuels for cooking over the past 5 to 10 years shifted from charcoal and/or wood to kerosene and 90% shifted from various fuels to LPG. About 80% of the latter group made the transition from kerosene to LPG. In these same households, the fuel transition for water-heating activities differed substantially. While 14% of the households relied on electricity to power their water-heating

Table 7   Percentage of households using each fuel for water heating

City	Wood	Other biomass	LPG	Kerosene	Piped gas[a]	Electricity
Hong Kong[b]					79	21
Pune						
Low income	27	3	2	53		10
Middle income	13	2	14	23		44
High income	2	0	10	12		76

[a] Refers to the use of town gas or LPG in Hong Kong and coal gas or natural gas in Beijing
[b] Average

activities a decade ago, today 21% of all households use electricity for water heating in Pune. The fuel switch was aided by moves to different living quarters—primarily to modern apartments that have separate wiring for high-power appliances.

The survey respondents overwhelming identified convenience as the driving force behind their shifts. About two thirds of the LPG users cited convenience as the reason they shifted from earlier fuels to LPG. About 5–6% of the respondents cited cost and availability as the reason for changing fuels. While LPG-fueled cooking appliances tend to cost more than traditional charcoal stoves in Thailand, very few charcoal users (2% of the surveyed households) in Thailand reported this initial cost as a barrier to switching to LPG. Of the 15% of all Thai households that did not use LPG, almost half (7%) stated they simply preferred other fuels.

A statistical analysis of the Thai data reveals that the important explanatory factors with statistical correlations to LPG or charcoal share are: the value of women's time, the price of charcoal, whether the residence is used for commercial food preparation, and the city in which a household is located. Household income does not correlate strongly with fuel choice for cooking relative to these factors (22). The single most statistically significant factor influencing fuel choice in Thai urban households proved to be the average per capita income of women employed outside the home. The share of LPG in the cooking fuel mix rises alongside the per capita income levels of female household members. Assumably, women working outside of the home have less time to devote to household duties such as food preparation. For these women, LPG proves to be more convenient than biomass. While overall household income similarly has a positive effect on the share of LPG in cooking activities, the effect is statistically insignificant when the value of women's time is included in the analysis.

COOKING AND WATER-HEATING ENERGY CONSUMPTION   Researchers based estimates of total fuel (excluding electricity) delivered for cooking and water-heating on a combination of survey responses and either metered data (for piped gas) or purchase receipts (for LPG and kerosene). In China, respondents felt more confident providing information on the number of coal briquettes purchased.

The following calculations of useful energy were computed using the efficiency figures provided in Table 6. Because the research teams made no independent assessment of stove efficiencies, the useful energy figures for the low- and middle-income households, which rely more heavily on solid fuels and kerosene, may differ more widely than Table 8 suggests. The useful energy figures for high-income households, which use primarily LPG or piped gas, are more robust.

In China, household delivered energy use declines as income levels rise and

Table 8   Cooking fuel consumption (without electricity) (GJ/hh/year)

City	Low	Medium	High	Average
		Household income level		
Beijing				
Delivered (Index)	4.1 (1)	3.9 (0.93)	2.9 (0.69)	3.7
Useful (Index)	1.5 (1)	1.4 (0.92)	1.2 (0.79)	1.4
Hong Kong[a]				
Delivered (Index)	3.6 (1)	1.7 (0.46)	1.7 (0.46)	1.9
Useful (Index)	1.3 (1)	0.7 (0.51)	0.7 (0.57)	0.8
Nanning				
Delivered (Index)	7.2 (1)	7.0 (0.96)	5.9 (0.81)	7.0
Useful (Index)	1.4 (1)	1.5 (1.00)	1.4 (0.92)	1.4
Pune[a]				
Delivered (Index)	2.1 (1)	2.4 (1.13)	2.5 (1.17)	2.3
Useful (Index)	0.7 (1)	0.9 (1.34)	1.0 (1.52)	0.9
Thailand[b]				
Delivered (Index)	2.7 (1)	2.4 (0.89)	3.0 (1.11)	2.6
Useful (Index)	0.7 (1)	0.8 (1.21)	1.1 (1.62)	0.8

[a] Includes fuel used to heat water for bathing
[b] Weighted average values for Bangkok, Chieng Mai, and Ayutthaya

households use higher-efficiency stoves and fuels. Delivered energy demand remains relatively constant between China's low- and mid-income households, varying at most by 7%. The major drop occurs between these two income categories and the high-income group. In Beijing, where coal supplies only 10% of the energy used by high-income households, household delivered energy demand drops by 31% and useful energy declines by 21%. In Nanning, where coal use remains somewhat higher in the top income categories but LPG and gas use play a larger role than in the lower-income categories, delivered energy decreases by 19%. Part of the reason for these declines in Chinese households may be the increased purchases of precooked staples from restaurants as income levels rise, particularly in Beijing. In both Pune and the Thai cities, useful and delivered energy consumption increase alongside income as households increasingly can afford to cook more meals and to heat water.

Of the surveyed cities, Beijing and Nanning consume the largest amounts of delivered and useful energy[6] (Table 8). The prevalent use of coal and the inefficient coal stoves both contribute to this high fuel use. Chinese households commonly use coal briquettes placed in sleeves in stoves. Because

[6]*Delivered energy* is energy delivered at the point of use. *Useful energy* is energy used for satisfying a service or need, the efficiency of which is factored in.

lighting the briquettes often proves difficult, households often keep their stoves burning all day. A number of research efforts currently under way are examining the potential for improving the efficiencies of coal stoves. The development of improved stove-lighting practices could save large quantities of coal in China and put an end to these wasteful practices.

The high figures for cooking fuel consumption per capita partially reflect the small size of Chinese households relative to those in the other study countries. To a certain degree, cooking fuel consumption per capita serves as an appropriate measure of comparison because cooking requirements increase with household sizes. However, incremental increases in fuel use are not directly proportional to the number of additional persons per household; hence, smaller households often appear to use more fuel on a per capita basis than larger households to meet equivalent cooking needs.

The amount of energy used for water heating depends largely on the extent of hot water use for bathing. The Beijing survey indicates that while low-income respondents use hot water for private baths 36 times each year, hot water use for bathing among high-income respondents averaged 70 times per year. Public bath use averaged about 40 times per year among all Chinese respondents. On average, bathing in China consumes three times more coal than gas. The amount of useful energy per capita derived from each fuel hardly varies for the various private bath fuels. However, public baths appear to consume considerably more energy per capita than private ones. The widespread use of showers or buckets in the latter case largely accounts for this disparity. Better managing hot water use in public baths could reduce the amount of fuel used for water heating significantly.

Space Heating

SPACE-HEATING FUEL CHOICE   Most of the survey cities, located in tropical regions, require little or no space heating. However, in China's northern regions, home to 41% of the Chinese population, space-heating activities constitute a significant energy end-use. The government controls the use of fuels for space heating by allocating limited amounts of coal for each boiler. Authorities in Beijing permit the use of central heating for about four months of the year (from mid-November to mid-March) and allow coal heating for five months of the year. Despite these restrictions, fuel use for space heating often accounts for as much as half of total household energy consumption.

Beijing's households employ two different types of space-heating systems. Independent dwellings generally use coal stoves, and apartment buildings typically use individual or district-heating boilers. Residents of individual homes have lower incomes than those living in apartments. The survey revealed that 84% of the households in the lowest-income category lived in individual homes and relied on coal stoves for heating versus only 33% of those in the highest-income category (Table 9).

Table 9   Space heating in Beijing

Indicator	Household income level			
	Low	Medium	High	Average
Percentage of households (hh)				
Single-family dwellings	84	44	33	54
Apartment dwellings	16	56	67	46
Heated area (m²)				
Single-family dwellings	18.8	18.5	21.1	19.7
Apartment dwellings	27.5	30.3	33.0	29.0
Fuel consumption (GJ/hh/yr)				
Single-family dwellings	16.1	18.4	19.6	17.8
Apartment dwellings	27.4	30.2	32.9	29.2
Delivered energy (MJ/hh/m²/heated day)				
Single-family dwellings	5.7	6.6	6.2	6.0
Apartment dwellings	7.4	7.4	7.4	7.4
Useful energy (MJ/hh/m²/heated day)				
Single-family dwellings	4.6	5.3	5.0	4.8
Apartment dwellings	4.4	4.4	4.4	4.4

In both types of dwellings, the heated areas increase with rising incomes. Apartment dwellings have on average 50% more heated area than independent homes. While the size of apartment dwellings in Beijing corresponds closely with those in Hong Kong's public estates, they remain far smaller than average dwellings in industrialized nations.

SPACE-HEATING ENERGY CONSUMPTION   The estimates of fuel consumption shown in Table 10 assume that the amount of coal used annually for central heating equals the officially allocated amount of 34 kg of coal equivalent per square meter of heated area. Because the government distributes this amount regardless of the consumer's income, the increases that occur in fuel consumption for centrally heated dwellings as income levels rise actually stem from increases in household floor areas.

Beijing's centrally heated households consume an average of 29.2 GJ per household each year, or about 64% more fuel than coal-stove households. In contrast, cooking fuel consumption in Beijing consumes an annual average of only 7 GJ.

Assuming an 80% efficiency for households (hh) with coal stoves and a 60% efficiency for central boilers, coal stoves provide 4.8 MJ/hh/m²/heated day of useful energy whereas central-heating systems produce 4.4 MJ/hh/m²/heated day. Given the efficiency assumptions, the use of central boilers increases fuel consumption while providing less useful energy over a shorter period of time. Central boilers, however, provide a much safer and less polluting way of supplying heat to urban dwellers.

ELECTRICITY USE FOR LIGHTING AND APPLIANCES.

Electricity demand has increased rapidly in all the survey cities in recent years. Higher appliance saturations and more intensive lighting have spurred this growth. The total demand for household electricity depends on the unit electricity consumption of major electric appliances, appliance ownership, and usage patterns. This section examines the availability and price of electricity in the survey cities, reports on the saturation of electricity end-uses—lighting, television sets, refrigerators, water heaters, air conditioners, and washing machines—and analyzes the total amount of electricity consumed by the surveyed households.

Electricity Availability and Costs

Virtually every household in the surveyed cities has access to electricity. In India, only 8% of the surveyed households lacked electricity in each of the three cities. Elsewhere, nearly all households had an electricity connection. Despite the prevalence of electricity throughout these urban regions, the quality of the electricity supply varies considerably among the surveyed cities. India and China suffer from frequent power outages, and occasional power surges strong enough to burn out television sets have been reported in some cities.

While most of the households surveyed receive their electricity supplies directly from the utility companies, about 9% of the Indian sample had illegal connections, more than half of which were for low-income households. In Thailand, about 4.5% of all surveyed households received electricity without a meter; the proportion ranged from 8% in Chieng Mai to 2.6% in Bangkok.

Most of those households with illegal connections in India receive their electricity through legally connected households with metered electricity supplies. Households with illegal electricity connections typically pay between two and four times more for their electricity than average. However, considering the high initial investment and tedious paperwork required to get a connection, receiving electricity through another household's meter often proves cheaper in the short term. In Thailand, most nonmetered households obtain their electricity through a single meter serving several households owned by their landowner. The Pune surveys revealed that the incidence of illegal connections has declined in recent years; 60% of those households with illegal connections have maintained their connections for more than 10 years.

All of the surveyed cities, aside from Hong Kong, have inverted electricity tariffs. In Thailand, the residential electricity tariff increases from 0.70 Baht/kWh for consumption levels below 10 kWh per month to 2.43 Baht/kWh for consumption levels above 800 kWh per month. Beijing's government institutes a far steeper tariff, which serves as a major disincentive to electricity use beyond basic levels. While electricity costs 0.16 yuan/kWh for consumption levels of 80 kWh or less per month, at higher levels the tariffs are not uniform and increase five-to ten-fold. In the three Indian cities, the tariff rises little from Rs. 0.85/kWh under 30 kWh to Rs. 1.00 between 30 and 150 kWh and Rs. 1.10 beyond that level of electricity consumption.

Lighting

Because of the high degree of electrification in all the cities surveyed, a negligible share of households use kerosene for lighting. Electricity's quality and cost advantages ensure its preferred use for lighting whenever it is available. Households commonly use both fluorescent and incandescent light-bulbs in living rooms, dining rooms, and often in kitchens.

The penetration of both types of bulbs increases alongside income (Table 10). The number of bulbs per household ranges from 1.1 to 6.7 for incandescent bulbs and from 0.8 to 4.1 for fluorescent bulbs. Beijing has an unusually high penetration of both types of bulbs; this pattern indicates that these households target particular types of bulbs for specific applications. Thai households use fluorescent lighting far more than incandescent lighting. Even low-income households in Thailand have an extremely high penetration

Table 10 Percent of households with incandescent and fluorescent bulbs

	Household income level			
City	Low	Medium	High	Average
Beijing				
Incandescent	83	91	91	88
Fluorescent	90	94	98	93
Manila				
Incandescent	55	82	97	85
Fluorescent	73	92	97	88
Pune				
Incandescent	65	94	96	86
Fluorescent	58	92	100	84
Thailand[a]				
Incandescent	52	56	64	57
Fluorescent	98	98	100	98

[a] Weighted average values for Bangkok, Chieng Mai, and Ayutthaya

of fluorescent bulbs. This result suggests that the first costs of tubes do not prevent their purchase, despite the fact that tubes cost about 10 times more than incandescent bulbs to purchase and install. Apparently, Thai consumers perceive high light output and economy of operation as benefits that outweigh the high costs. The survey results also indicate widespread use of incandescent bulbs, particularly at higher income levels.

Electric Appliances

After households obtain lighting and fans, irons and television sets tend to be among the first appliances purchased. Refrigerators are generally next in line, followed by washing machines, electric water heaters, and air conditioners depending on climatic conditions, income levels, and cultural traditions.

TELEVISION SETS   Historically, the type of set purchased has depended upon the availability of color broadcasts and the ability of households to afford more expensive color televisions. The penetration of color televisions typically increases alongside income. Thus, the penetration of black-and-white sets remains lowest in high-income households (Table 11). In Beijing, however, color television ownership does not vary much with income. Beijing's households have twice as many color televisions as black-and-white sets. This disparity may result from the rapid increase in TV ownership after the Chinese government began to implement its program of economic liberalization in the 1980s (Table 12). At the time of this liberalization, color broadcasts were widely available and, as a result, many households chose to purchase color TVs. In contrast, television sets began to penetrate Indian

Table 11   Percent of households with television sets

City	Low	Medium	High	Average
Household income level				
Beijing				
Black & white	43	40	42	42
Color	70	82	80	77
Hong Kong				
Black & white	6	1	3	2
Color	96	99	100	98
Manila				
Black & white	42	52	23	41
Color	11	59	93	55
Pune				
Black & white	42	49	23	40
Color	6	40	74	38

Table 12   Percent of households with television sets, 1980–1989

	1980	1982	1984	1986	1988	1989
Beijing						
Black & white			97	69	58	42
Color	66[a]	86[a]	8	51	70	81
Pune[a]	—	25	—	—	—	68

[a] Total for both black & white and color sets

households widely in the early 1960s, but color broadcasts were not introduced until the 1970s.

REFRIGERATORS   Consumers in the developing world tend to purchase relatively small refrigerators because of generally low income levels, frequent-shopping habits, and small house sizes. The majority of Indian, Chinese, and Thai households surveyed owned single-door, manual defrost type refrigerators, which average about 170 liters (about 6 cubic feet) in size.

While the penetration of refrigerators remains high throughout the high-income bracket households in all the survey cities, penetration levels vary in the low- and medium-income households. In Pune and Manila, 7 and 21% respectively of all low-income households have refrigerators compared to 79% in Beijing and 90% in Hong Kong (Table 13). An important reason for the higher penetration of refrigerators is the higher purchasing power of consumers in the latter two cities. In each city, households pay less than 10%

Table 13   Percent of households with refrigerators and electric water heaters

	Household income level			
City	Low	Medium	High	Average
Refrigerators				
Beijing	79	89	95	87
Hong Kong	90	98	99	98
Manila	21	65	93	93
Nanning	44	56	63	54
Pune	7	40	81	40
Thailand[a]	59	85	98	82
Electric water heaters				
Beijing	2	0	0	1
Hong Kong	10	16	23	17
Pune	7	40	84	41
Thailand[a]	1	3	16	5

[a] Weighted average values for Chieng Mai, Bangkok, and Ayutthaya.

Table 14   Percent of households with refrigerators, 1982–1989

City	1982	1984	1986	1988	1989
Beijing	3	15	62	81	88
Pune	13	—	—	—	23

of their income for housing subsidized by the government. The ownership of vehicles is restricted because of high taxes and unavailability of parking in Hong Kong public estates and because of unavailability of vehicles in Beijing. Appliances are the only items available for consumers to purchase as evidenced by their high penetration in our surveys.

One in every eight Thai households owns more than one refrigerator. In the high-income bracket, one in three Thai households has two refrigerators. About half of the households possessing multiple refrigerators double as shops (most often grocery and food stores), thus creating the greater need for refrigeration.[7]

Historical data indicates that refrigerator penetrations have increased considerably over the past decade (Table 14). In Pune, 13% of the households owned refrigerators in 1982 versus 23% in 1989. Beijing witnessed a more remarkable ownership surge. Refrigerator penetration rose from 3% in 1982 to 88% in 1989. Assuming Beijing has about 2.5 million households, then over the seven-year time period consumers bought about 2.1 million refrigerators or 730 refrigerators every day. The increased availability of these appliances coupled with rapid increases in income during the 1980s led to this growth.

Manufacturers provide scant information on the performance of refrigerators. In most instances, refrigerator labels indicate voltage, frequency, and occasionally the power rating of the compressors. None of the manufacturers report the electricity consumption of refrigerators, and most developing countries lack the calorimeter rooms to test under standard conditions.

In Bangkok, Beijing, and Pune the surveys included the metering of several refrigerators over one-week periods to determine their in situ electricity consumption. Surveyors also noted household sizes and refrigerator sizes and types and, in China, indoor temperatures. These tests shed light on the potential for reducing refrigerator electricity consumption through better maintenance and/or the introduction of more efficient refrigerators. A statisti-

---

[7]The surveys in Thailand revealed frequent small-scale commercial activities within the homes. On average, 41% of the surveyed households carried out commercial activities in the home. Grocery stores and food shops constitute about 40% of these in-home commercial activities, and tailoring, hair dressing, and furniture making make up another 30%. These practices lead to increases in both appliance ownership and fuel use in Thai households (Ref. 22).

Table 15 Manual refrigerator electricity consumption indicators[a]
(kWh/month)

	Constant	Household size (persons)	Refrigerator size (liters)
Coefficient	−21.5	8.5	0.18
Standard error		2.4	0.10
T-statistic		3.5	1.84
R-squared	0.33		

[a] Combined data for Bangkok, Beijing, and Pune

cal analysis of the pooled data on electricity use by manual-defrost refrigerators in Bangkok and Pune indicates a positive correlation between electricity consumption and household and refrigerator sizes (Table 15). Thai researchers noted increased consumption in refrigerators that showed condensation around the door seals, which indicates the importance of the condition of a refrigerator.

ELECTRIC WATER HEATERS   Among the survey countries, only households in India, China, and Hong Kong heat water for bathing purposes. Despite their higher first and life-cycle costs, electric water heaters are common in Indian households primarily because of their convenience. Households in Hong Kong and China prefer piped gas water heaters (Table 13).

According to historical data for Pune, storage water heaters have increasingly replaced the flow-through type alongside rising incomes. While only 1% of Pune's households owned storage heaters in 1982, by 1989, this share had increased to 9% while that of flow-through types had declined slightly.

Surveyors in Pune metered electric water heaters (both storage and flow-through heaters) over a one-week period in order to determine their electricity use. The results revealed considerable variance, with a range from 18 to 195 kWh of electricity consumption per month. Table 16 provides the results of a

Table 16 Electricity consumption in water heaters, Pune, India

	Constant	Household size (persons)	Capacity (kW)
Coefficient	−34.1	15.6	26.4
Standard error	44.6	5.9	16.1
T-statistic		2.64	1.64
R-squared	0.37		

Table 17 Percent of households with air conditioners and washing machines

| | Household income level | | | |
City	Low	Medium	High	Average
Air conditioners				
Beijing	0	0	0	2
Hong Kong	26	52	60	51
Nanning	0	0	0	0
Pune	0	1	9	2
Thailand	0.7	11	40	15
Washing machines				
Beijing	81	86	86	84
Hong Kong	55	92	95	89
Nanning	65	82	82	76
Pune	5	8	13	8
Thailand	3	18	43	20

regression analysis to examine the influence of household size and heater wattage on electricity use. Household size appears to influence electricity consumption more than wattage. Electricity use increases 15.6 kWh per month for every additional household member. The increased consumption of higher-wattage water heaters may be because of higher standby losses and the tendency to use more hot water.

ROOM AIR CONDITIONERS AND WASHING MACHINES   The penetration of room air conditioners and automatic clothes washers, two energy-intensive appliances,[8] corresponds closely with income levels (Table 17). In both Pune and Thailand, high-income households show the greatest penetrations of air conditioners. Although penetrations of washing machines in Thai urban areas currently far surpass those in Pune, over the past two years washing machines have gained popularity in India and replaced TV sets as the most commonly purchased household appliance (26). As incomes rise in urban households and as residents of smaller cities attempt to emulate the life-styles of their large-city neighbors, both air conditioners and washing machines are likely to gain popularity.

The surveys revealed a striking difference between the penetration of air conditioning in Bangkok and the two smaller Thai cities. Climatic differences alone do not account for this variation. Although Chieng Mai has a sub-

---

[8]In general, clothes washers are energy intensive because they usually include in-line electric water heaters to provide hot water for washing. Piped hot water in homes is rare. In Beijing, however, washers rarely have heating systems, thereby making them far less energy intensive. Regression coefficients suggest that Thai households often use the water-heating features.

stantially cooler climate than Bangkok, Ayutthaya, which has essentially the same climate as Bangkok, has far fewer air conditioners than its big-city neighbor. In Thailand, two factors do appear to influence significantly the penetration of air conditioners: housing design and construction practices. Air-conditioner ownership correlates highly with concrete homes (22), largely because these homes are much more difficult to ventilate than traditional Thai homes, which are built on platforms raised off the ground with walls made of light wood or woven bamboo. Concrete construction is far more common in Bangkok, where rapid economic growth has led to a boom in single-family, town house, and apartment-style homes, than in the smaller Thai cities. Other factors particular to Bangkok, such as lower wind speeds in more developed areas, poor air quality, and smaller average room sizes, also may predispose households to higher levels of artificial ventilation and cooling.

In China, households commonly own electric fans but rarely use air conditioners. On the other hand, the majority of Chinese households, including those in the smaller and less modernized city of Nanning, own small washing machines.

Household Electricity Use

APPLIANCE CONSUMPTION ESTIMATES FROM REGRESSION ANALYSIS The relative contribution of each appliance to aggregate residential electricity consumption may be determined either by metering appliances or by estimating the average electricity consumption of each type of appliance under typical patterns of household use. Direct metering of specific appliances proves time consuming, costly, and intrusive. More commonly, researchers estimate these values by using a form of conditional electricity demand analysis (27). The following estimates of appliance electricity use in Thai households compare the results using both these methods.

Statistical estimates of major appliance electricity use were derived by regressing the total reported monthly hours of utilization of each appliance type on the monthly electricity consumption in watt-hours. The resulting coefficients were directly interpretable as the mean power demand of the appliance (or, more accurately, as the mean consumption in watt-hours per hour). This technique enables the estimates to incorporate directly the effects of multiple appliances and of varying operating conditions (e.g. on/off cycling).

The Thai research team submetered individual refrigerators, clothes washers, televisions, and rice cookers in 38 Thai households. In each given household, only three appliances were metered. Researchers metered refrigerators in all of the households and two of the other three appliances. Table 18 compares estimates of monthly appliance electricity use generated by the different techniques (for analytical details see Ref. 22). In the case of

Table 18 Appliance electricity use in Thailand

Appliance type	Estimated average consumption/hour (watt-hour)	Regression estimate consumption/month (kWh/appliance)	Survey metered consumption/month (kWh/appliance)	Regression estimated share (%)
Air conditioner	700	96.0		10
Refrigerator	118	81.9	86.5	30
Rice cooker	882	28.4	14.3	12
Washer	574	19.5	6.4	2
Color TV	174	21.8	13.2	9
Electric fan	40	7.1		7
Fluorescent bulb	44	7.7		23
Incandescent bulb	40	3.8		3
Other		36.0		4

rice cookers, television sets, and clothes washers, the statistical estimates were much higher than the metered electricity use.[9]

The analysis suggests that refrigerators consume by far the largest share of household electricity in Thailand. Fluorescent lighting accounts for the second largest share of electricity demand. Given the high penetration of other types of appliances in Thai households, the predominance of lighting in electricity demand comes as a surprise. The frequency of light use and the large number of light fixtures in Thai homes accounts for lighting's large portion of electricity demand. Previous estimates far exceed this study's calculations of air-conditioning electricity use (28). In coming years, however, as household incomes rise, air-conditioner penetrations are likely to increase and account for a larger share of household electricity demand.

In contrast to the Thai refrigerator consumption of electricity, which averaged 88 kWh per month, refrigerators metered in Beijing averaged only 22 kWh per month. The age of refrigerators in the surveyed Thai households averaged 10 years compared to less than 5 in Beijing, and the average size was 190 liters compared to 165 in Beijing. The past decade has witnessed notable improvements in refrigerator technologies in Thailand. It appears that the earlier refrigerators consume far more electricity than newer models.

[9]A number of plausible explanations may account for this variance. Multicollinearity contributed to high standard errors for the estimates, which may explain some of the variation. Variation in the metered appliances was also large, and the sample size was at most 38 (for refrigerators). Respondent exaggeration of utilization time could lead to higher statistical estimates for some appliances, or user awareness of the large kilowatt-hour meters installed in homes may have inhibited some discretionary use of the appliances being monitored. In this regard, it is striking that estimates for refrigerators, appliances whose use is largely nondiscretionary, were very close to the statistical estimates.

HOUSEHOLD ELECTRICITY CONSUMPTION   The amount of electricity consumed by households varies considerably from city to city (Table 19). Despite the high penetration of certain appliances, households in China consume far less electricity than those in the other study cities. For example, 79% of the low-income households in Beijing and 44% in Nanning have refrigerators compared to only 7% in Pune, yet low-income households in Pune consume more electricity than those in either of the former two cities. The explanation for China's lower electriciy consumption may lie in the fact that the electrical wiring in Chinese households has a capacity to carry only a 5-ampere current. (US households often have 100-ampere wiring and Indian ones have 15-ampere wiring for small appliances and 25-ampere wiring for high-wattage ones). This limits the simultaneous use of appliances and lighting to 1000 watts. The limitation forces households to purchase appliances that are small and efficient, and to use them in a frugal manner. Chinese households are urged to turn lights off while watching television for example. An earlier survey of Indonesia reported a similar limitation on household wattage in that country to less than 1000 watts.

In each of the survey cities, electricity consumption increases at higher income brackets. Variations in electricity consumption correspond with levels

Table 19 Household electricity consumption (kWh/year/household)

City	Household income level			
	Low	Medium	High	Average
Beijing	430	500	470	489
Index[a]	1.00	1.16	1.09	
Hong Kong	2740	3060	3920	3232
Index[a]	1.00	1.12	1.43	
Nanning	280	370	440	359
Index[a]	1.00	1.33	1.56	
Pune	470	1120	1970	1130
Index[a]	1.00	2.41	4.21	
Thailand[b]	1390	2560	5080	2817
Index[a]	1.00	1.84	3.66	

[a] Indexed so that low-income households are given the value of "1" and medium- and high-income households are shown in relative terms.

[b] Weighted average values for Bangkok, Chieng Mai, and Ayutthaya.

of appliance ownership, unit electricity consumption, and frequency of appliance use. Table 19 shows the increases in electricity consumption with income levels. The sharpest variations in electricity use among income categories occur in Pune and the Thai cities; variations remain far less extreme in the two Chinese cities and Hong Kong. The more even penetration of appliances in the latter two cities partially explains the lower percentage increase across income levels.

AGGREGATE ENERGY CONSUMPTION AND EXPENDITURES

Table 20 shows the aggregate consumption of fuels and electricity per capita in some of the surveyed cities. As income levels rise, useful fuel consumption declines in Beijing, remains flat in Nanning, and increases in Pune and the Thai cities. The substitution of LPG and piped gas for coal accounts for the declining trend in Beijing. In the other cities, higher-income households

Table 20 Fuel and electricity consumption per capita (GJ/year/capita)

City	Household income level			
	Low	Medium	High	Average
Beijing[a]				
Useful fuel	1.5	1.4	1.2	1.4
Delivered fuel	4.1	3.9	2.9	3.7
Electricity	0.5	0.5	0.5	0.5
Hong Kong				
Useful fuel		0.7	0.7	0.7
Delivered fuel		1.6	1.6	1.6
Electricity	2.4	2.3	2.5	2.4
Nanning				
Useful fuel	1.5	1.5	1.4	1.4
Delivered fuel	7.2	7.0	5.9	7.0
Electricity	0.3	0.4	0.4	0.4
Pune				
Useful fuel	0.7	0.9	1.0	0.9
Delivered fuel	2.1	2.4	2.5	2.3
Electricity	0.3	0.8	1.6	0.8
Thailand				
Useful fuel	0.7	0.8	1.1	0.8
Delivered fuel	2.7	2.4	2.9	2.6
Electricity	1.5	2.2	3.6	2.4

[a] Excludes space heating

consume more useful energy than lower-income homes because they are able to afford more cooking and water-heating activities.

In terms of delivered fuel use, consumption levels in Chinese households also surpass those elsewhere, primarily because of the heavy reliance on coal, which households tend to burn continuously because it is difficult to relight. The prevalence of coal raises household energy consumption levels in China far beyond what would be necessary to satisfy cooking and water-heating needs if other fuels were used.

In contrast, electricity consumption per capita remains much lower in Chinese cities than in the other survey cities. For example, high-income households in Thailand consume nine times as much electricity as those in China. The high penetration of air conditioners in Thailand partially accounts for this difference.

As income levels rise, households tend to use more fuel and electricity; therefore, energy expenditures increase as well. However, energy expenditures as a proportion of either income or total household expenditures decline as incomes rise. While the surveys confirm this pattern, they also reveal a great variance in energy expenditures across the cities. Energy expenditure as a fraction of income ranges from only 1% in Beijing to almost 10% in Pune (Table 21). The heavier reliance on purchased biomass, which is more expensive than other fuels, for cooking and water heating in Pune partially accounts for this difference. Beijing has relatively lower fuel prices, explaining the lower fraction of household incomes channeled towards energy purchases.

Historical data for Beijing indicates that fuel expenditures have gradually declined as a share of household incomes; while in 1984, Beijing households devoted 1.6% of their incomes to fuel purchases, by 1989, this share had

Table 21 Energy expenditures as a share of income (%)

City	Low	Medium	High	Average
	\multicolumn{4}{c}{Household income level}			
Beijing				
Fuels	0.9	0.6	0.3	0.7
Electricity	1.8	1.6	1.2	1.6
Hong Kong				
Fuels	3.7	1.2	0.5	1.3
Electricity	4.0	1.5	1.4	1.7
Pune				
Fuels	9.8	4.4	2.3	5.5
Electricity				

dropped to 0.74% (Table 22). Simultaneously, electricity expenditures as a fraction of total income have increased from 0.7 to 1.6%. Total energy expenditures have fluctuated. In the early part of the decade, because fuel expenditures declined faster than electricity expenditures increased, total energy expenditures dropped. In more recent years, however, electricity expenditures have increased sharply, thus offsetting earlier declines in total energy expenditures.

CONCLUSIONS AND POLICY IMPLICATIONS

Asia's urban households have witnessed dramatic increases in modern fuel use in recent years. The results of the household surveys suggest that changing patterns of activity and livelihood in these households underlie this growth. The five countries examined in this study, while at different stages of development, have experienced a number of similar energy-related transitions. Most striking, electricity has come to penetrate most households. In all but the poorest households surveyed, appliance penetrations are correspondingly high. Biomass fuels, traditionally the dominant household energy source in these five countries, currently play only limited roles in most urban household activities. Even the use of kerosene, an inferior modern fuel, appears to be diminishing.

The survey responses indicate that this increased reliance on modern fuels stems from a preference among consumers for more convenient and readily available fuels. The low price of many modern fuels relative to biomass prices also leads consumers to favor the former. In many cities, as opposed to rural areas, households cannot simply self-produce or gather biomass fuels, but rather must purchase them. Because of rapid deforestation and the increasing scarcity of fuelwood, biomass fuels have become more costly than modern fuels on the urban markets surveyed.

In certain regions and at certain income levels, however, urban biomass use persists. The poorest Asian households cannot afford the first costs of stoves, tanks, and/or regulators needed for LPG use. As mentioned earlier, the poorest Indian households often lack the formal legal address required to

Table 22 Energy expenditure as a % of total expenditures

	1984	1985	1986	1987	1988	1989
Beijing						
Fuels	1.58	1.33	1.15	1.07	0.85	0.74
Electricity	0.69	0.82	0.99	1.20	1.18	1.59
Total	2.27	2.15	2.14	2.27	2.03	2.33

obtain kerosene rations at subsidized prices. In these cases, low-income consumers have two main options: to gather waste wood or biomass or to purchase more costly fuelwood, charcoal, or kerosene on the open market. Thus, in absolute terms, the poor pay more than the rich to purchase the same quantity of energy. As a fraction of household income, the disparity is even greater. The results for China and Hong Kong indicated that because of the lower efficiencies of the fuels used, the delivered and useful average fuel consumption per capita of low-income households actually surpasses that of higher-income groups.

While the surveys document a dramatic decline in the use of biomass for cooking, biomass is commonly used for water heating in low-income households in Pune and, in the form of charcoal, for specialized cooking even in higher-income Thai households.

Space heating is important in northern China. Coal stoves are used for space heating in single-family dwellings and boilers in apartment dwellings. The survey estimates show that boilers provide less useful heat than coal stoves in Beijing households. The government has embarked on a program for the construction of apartment dwellings. While the use of boilers has health advantages, from strictly an energy perspective, the coal use per household will increase while providing less useful energy, and perhaps less comfort, unless steps are taken to ensure that the building's heating is done efficiently.

High appliance penetrations have resulted from a combination of more affordable equipment and higher household incomes. In many of the cities surveyed, certain appliances, such as televisions and refrigerators, have nearly reached saturation. While the growth of these appliances in urban areas may be limited, urban households will continue replacing smaller and older models with larger and improved technologies. Enormous financial and environmental burdens accompany the rapid growth of electricity demand, as electricity sector planners have come to recognize. The extent of these burdens will increase alongside the growing demand. Thus, policy efforts aimed at reducing the growth rate of electricity demand could offer substantial benefits and savings.

Local manufacturing subsidiaries or joint ventures use different types of components than international manufacturers. Only recently, for example, have local manufacturers begun to use efficient compressors and high-quality insulation materials. For this reason, the quality of the existing refrigerator stock in developing countries appears far lower than similar-vintage equipment in industrialized nations (31).

Efforts to slow the expansion of electricity should focus on the technology of household appliances. All the large international appliance manufacturers have made dramatic improvements in energy efficiency over the past decade (29). The surveyed Asian countries must focus on one major policy challenge in this realm: ensuring that domestic manufacturers, which account for the vast majority of all appliance sales, adopt the best available international technologies for their products.

Various barriers currently hinder the incorporation of the most energy-efficient technologies in the manufacture of household products. In some cases, international joint venture partners of domestic manufacturers have vested interests in licensing their older technologies, which no longer meet the quality or energy standards of industrialized countries, in order to continue earning revenues. Developing-nation governments often try to protect domestic suppliers through the implementation of high tariffs on imported equipment or components, which in many cases are more energy efficient than those produced at home. In some countries, restrictions on sourcing components (e.g. domestic content regulations) limit manufacturers' choices in specifying components for their large, energy-intensive appliances.

In recent years, appliance markets have become more international. Most developing-country producers prefer to export to large high-value markets in the United States, Western Europe, and Japan. All of these markets have de facto quality and energy-efficiency standards. This tightening of appliance standards likely will reverberate to producers in Asian developing nations, who seldom manufacture completely different domestic and export models. Hence, the improvements required for the Western markets will result in a significant upgrading of local equipment as well.

The survey results indicate the efficiency gains may be most easily achieved in lighting and refrigeration systems. These two end-uses account for most of the electricity consumed in developing world households. Both have witnessed major technological improvements in recent years. In many developing nations, these better technologies have yet to be adopted.

Both fluorescent and incandescent lighting offer opportunities for greater savings. Preliminary estimates of mean power consumption by fluorescent lights in Thailand suggest that the common core-coil ballasts used may be highly inefficient, leading to unnecessary heating and high power consumption. Almost all Thai households could use improved ballasts and high-efficiency fluorescent tubes (36- or 18-watt standard tubes) with no loss of service quality and at very low incremental cost.

The high penetration of incandescent bulbs presents an excellent opportunity for introducing more efficient compact fluorescents (CFLs). CFLs require less than one quarter of the electricity used by incandescent bulbs, last about eight times longer, and have lower life-cycle costs. They have one major drawback: higher first costs. A recent analysis of the economics of CFLs versus incandescents in the Philippines indicates that using CFLs costs far less than generating peak electricity using gas turbines (23). Bombay recently undertook a program to increase the penetration of CFLs as a lesser-cost alternative to generating electricity (30).

Great potential exists for reducing electricity consumption in refrigerators (although metered consumption of electricity was much lower in Chinese than in Indian or Thai refrigerators). However, researchers cannot determine the most cost-effective means for improving refrigerator efficiencies as long as data on electricity consumption in existing refrigerators remain unavailable. Historical data that are available indicate significant decreases in unit electricity consumption over time; for example, Korean manufacturers report that the electricity consumption of a 200-liter refrigerator declined by about 75% between 1980 and 1988, from 56 to 20 kWh per month. With the implementation of efficiency improvements, the survey cities could witness similar declines and limit the contribution of refrigerators to the future growth of household energy demand.

The testing and labelling of major domestic appliances, the aggressive promotion of test results, and the comparison of performance levels in different models could provide a major incentive for manufacturers in developing nations to improve the efficiency of their equipment. In a voluntary appliance-testing program in Brazil, the availability of test results alone prompted some manufacturers to make rapid improvements in their refrigerator models (32).

Consumers in these Asian cities are sensitive to the quality and sophistication of appliance technologies; given credible information on appliance performance, these consumers are likely to make wise choices. In order to provide consumers with this opportunity, these nations will need funding to develop suitable appliance-testing facilities. Manufacturers, standards institutes, and electricity authorities then will need to agree on an acceptable technical testing procedure. Ideally, this procedure will conform with either the procedures used in the United States or Japan thereby enabling international comparisons of the results. Finally, a public agency or electric utility must take responsibility for promoting the results of the tests aggressively by publishing comparative lists of appliance performance levels. In 1990, a nongovernment organization published a list of appliance performance levels according to manufacturers' specifications (in the absence of test data) (33).

None of the countries surveyed in this study had explicit policies or programs to improve the efficiency of appliances and lighting. Procedures and calorimeter rooms are in place in China to test refrigerator electricity consumption, and a test room is in place to test air-conditioners and another will be ready in June 1990 to test refrigerators in the Philippines. Programs to set standards and provide rebates for appliances are being considered in the Philippines.

The current household energy use patterns hold both positive and negative implications for the environment. The decline of biomass consumption may alleviate some of the heavy pressure on Asian forests. Given previous highly extractive practices of fuel wood consumption and the vast improvements in the efficiencies of LPG- and kerosene-burning technologies, these changing patterns may actually reduce, for a time, the net contribution of household energy use to greenhouse gas concentrations (even if carbon emissions increase).

On the other hand, all of the countries surveyed rely heavily on coal for electricity generation. The surge in consumer appliance penetration and the energy-intensity of emerging popular appliances (particularly air conditioners and water heaters) bodes ill for the carbon emissions, acid precipitation, and localized environmental impacts of coal mining and coal-based thermal electric generation.

These comparative survey data demonstrate the extent to which urban households in the world's most populous region have adopted modern forms of energy to support modern household activities. Major household appliances have penetrated households in all of these cities rapidly. Consumers across all income levels have shown a willingness to pay for equipment that serves to improve the convenience and comfort of domestic life. The environmental effects of this fuel shift appear ambiguous. Although the overall environmental ramifications may be positive, the influx of inefficient electrical appliances in millions of Asian households gives cause for concern. Recent years have witnessed great progress in increasing the efficiency of major household electricity-consuming appliances. By establishing appliance-testing and energy consumption measurement standards, these nations can stimulate market-based competition and the adoption of new, more efficient technologies by the rapidly expanding consumer appliance market. These policy measures could offer a simple but effective means of responding to the rapid growth in residential electricity use.

ACKNOWLEDGMENTS

The authors would like to thank Zhigang Qin for the analysis of Chinese data and for assisting in collation of the data and information for other countries and Willy Makundi for the analysis and collation of the India data. The original surveys and research data upon which the information in this article is based were supported by the International Development Research Centre (IRDC), Canada and undertaken by the following research teams: in China, the Institute for Techno-Economic and Energy Systems Analysis at Tsinghua University, Beijing, led by Professor Qiu Da Xiong; in India, the Systems Research Institute, Pune, led by Dr. J. G. Krishnayya; in Thailand, the

Chulalongkorn University Social Research Institute, Bangkok, led by Dr. Amara Pongsapich; in Hong Kong, the University of Hong Kong's Centre of Urban Planning and Environmental Management, led by Professor Peter Hills; and in the Philippines, the University of the Philippines, College of Engineering, Quezon City, led by Professor Ruben Garcia. Their generosity in providing access to original data and research reports made this article possible.

This work was supported by the Office of Environmental Policy of the US Department of Energy under Contract No. DE-AC03-76SF000098. Additional support was provided by IRDC, Canada, and the University-wide Energy Research Group of the University of California.

Literature Cited

1. Hills, P. 1991. *Household Energy Consumption in Hong Kong. Draft Final Report No. 3-P-88-0123-03.* Cent. Urban Planning and Environ. Manage., Univ. Hong Kong

2. Qiu, D., Ma, Y., Lu, Y., Zhu, Z. 1990. *Urban Household Energy Consumption and Air Pollution Survey in China. Report No. IDRC-3-P-88-0123-02.* Inst. Techno-Economic and Energy Systems Analysis, Beijing, China

3. Pongsapich, A., Wongsekiarttirat, W. 1991. *Urban Household Energy Consumption and Indoor Air Pollution.* Soc. Res. Inst., Chulalongkorn Univ., Bangkok, Thailand

4. Kulkarni, A., Deshpande, J., Sant, G., Krishnayya, J. 1991. *Urbanization in Search of Energy.* Energy Res. Group, Systems Res. Inst., Pune, India

5. Bhatia, R. 1988. Energy pricing and household energy consumption in India. *Energy J.* 9:71–105

6. Cecelski, E., Dunkerley, J., Ramsay, W. 1979. *Household Energy and the Poor in the Third World.* Washington, DC: Resour. Future

7. Leach, G. 1988. Residential energy in the Third World. *Annu. Rev. Energy* 13:47–65

8. Behrens, A. 1986. *Household Energy Consumption in Rio de Janeiro Shanty Towns. Manuscript Report 134e.* Int. Dev. Res. Cent., Ottawa, Canada

9. Meyers, S., Sathaye, J. 1989. Electricity use in the developing countries: changes since 1970. *Energy* 14(8):435–41

10. Newcombe, K. 1979. Energy use in Hong Kong: part IV. Socio-economic distribution, patterns of personal energy use and the energy slave syndrome. *Urban Ecol.* 4:179–205

11. Sathaye, J., Meyers, S. 1985. Energy use in cities of the developing countries. *Annu. Rev. Energy* 10:109–33

12. Ulluwishewa, R. 1989. A case study of energy use for domestic cooking by urban dwellers in Colombo City. *Energy* 14(6):341–43

13. Reddy, S. 1990. *The energy sector of the metropolis of Bangalore.* PhD. thesis. Dept. Manage. Stud., Indian Inst. Sci., Bangalore, India

14. Hosier, R., Dowd, J. 1987. Household fuel choice in Zimbabwe: an empirical test of the energy ladder hypothesis. *Resources and Energy* 9:347–61

15. Jannuzzi, G. 1989. Residential energy demand in Brazil by income classes: issues for the energy sector. *Energy Policy* 17(3):254–63

16. Jones, D. 1989. *Urbanization and Energy Use in Economic Development. Report No. ORNL-6432.* Oak Ridge Natl. Lab., Oak Ridge, Tenn.

17. Meyers, S., Leach, G. 1989. Biomass fuels in the developing countries: an overview. *Economia delle Fonti di Energia* 37–38:35–90

18. Smith, K. 1987. *Biofuels, Air Pollution, and Health: A Global Review.* New York: Plenum

19. Leach, G., Gowen, M. 1987. *Household Energy Handbook: An Interim Guide and Reference Manual. Technical Paper Number 67.* Washington, DC: World Bank

20. Leitmann, J. 1988. *Some Considerations in Collecting Data on Household Energy Consumption. Energy Series Paper No. 3.* Washington, DC: Industry and Energy Dept., World Bank

21. Tyler, S. 1990. *Household Energy Surveys in Third World Cities: Some*

Methodological Issues. Int. Dev. Res. Cent., Ottawa, Canada

22. Tyler, S. 1991. *Urban household energy use in Thailand.* PhD thesis. Energy Resour. Group, Univ. Calif., Berkeley

22a. Summers, B., Heston, A. 1988. A new set of international comparisons of real product and price levels estimates for 130 countries, 1950–1985. *Rev. Income Wealth.* May: pp. 1–25

23. Sathaye, J. 1990. Personal communications with appliance manufacturers and electric utility in the Philippines.

24. Gupta, C. L., Rao, K., Vasudevraju, V. 1980. Domestic energy consumption in India (Pondicherry Region). *Energy* 5:1213–22

25. Bhatia, R. 1988. Energy pricing of household energy consumption in India. *Energy J.* 9:71–105

26. Increasing popularity of washing machines. *India Today.* Nov. 15, 1990

27. Sebold, F., Parris, K. 1989. *Residential End-Use Energy Consumption: A Survey of Conditional Demand Estimates.* Research Project 2547-1. Final Contractor Report prepared by Regional Econ. Res. Inc. Palo Alto, Calif: Electr. Power Res. Inst.

28. Thailand Load Forecast Working Group (LFWG). 1989. *Load Forecast for the Thailand Electric System.* Natl. Energy Policy Office, Bangkok, Thailand

29. Schipper, L., Hawk, D. 1989. *More Efficient Household Electricity Use: An International Perspective. Report No. LBL-27277.* Lawrence Berkeley Lab. Berkeley, Calif.

30. Gadgil, A., Jannuzzi, G. 1989. *Conservation Potential of Compact Fluorescent Lamps in India and Brazil. Report No. LBL-27210.* Lawrence Berkeley Lab., Berkeley, Calif.

31. Meyers, S., Tyler, S., Geller, H., Sathaye, J., Schipper, L. 1990. *Energy Efficiency and Household Electric Appliances in Developing and Newly Industrialized Countries. Report No. LBL-29678.* Lawrence Berkeley Lab., Berkeley, Calif. Dec.

32. Geller, H. 1990. *Electricity Conservation in Brazil: Status Report and Analysis.* Report prepared for US Environmental Protection Agency and Office of Technology Assessment, US Congress, Washington, DC. Nov.

33. Int. Inst. Energy Conservation. 1990. *The Consumer Guidebook for Energy-Efficient Appliances in Thailand.* Bangkok, Thailand. Oct.

Annu. Rev. Energy Environ. 1991. 16:337–64

UNCONVENTIONAL POWER: ENERGY EFFICIENCY AND ENVIRONMENTAL REHABILITATION UNDER THE NORTHWEST POWER ACT

Kai N. Lee‡

Williams College, Williamstown, Massachusetts

KEY WORDS: energy conservation, Pacific Northwest, sustainable development, demand management, environmental quality.

INTRODUCTION

By 1982, when electric power celebrated its centennial, the industry that grew from Thomas Edison's legendary Pearl Street station faced a deep uncertainty: whether the demand for electricity would continue to grow[1]—and if it did not,

‡Member representing the state of Washington, Northwest Power Planning Council, 1983–1987.

[1]In 1982 total sales of electricity declined over the previous year, and the long-term trend looked ominous as well:

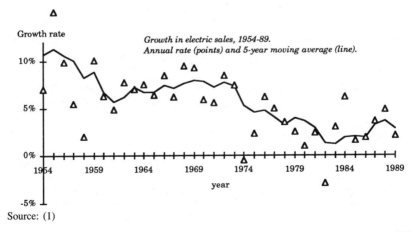

Growth rate

Growth in electric sales, 1954-89.
Annual rate (points) and 5-year moving average (line).

Source: (1)

what that would mean for the goals and profitability of utilities. For nearly a decade, the demand for energy in general had faltered, as two oil price shocks worked their way through the international economy. Then a deep recession hit. Nuclear and coal, the backbone of plans for the future, faced high costs, public criticism, and technological uncertainty. As sales stagnated, rates increased and ratepayers and regulators threatened to rebel. The electric centenary was a time for worry.

The Pacific Northwest is a region highly sensitive to the fortunes of its electric power industry; in 1983, the largest municipal-bond default in US history was triggered by the cancellation of two uncompleted nuclear power plants in Washington state. Yet in the crucible of institutional crisis and recession, the Northwest also innovated, pioneering electric-energy conservation programs and environmental rehabilitation on an unprecedented scale. The vehicle through which these inventions occurred is a statute enacted in 1980 by the US Congress (2), called here the Northwest Power Act. This article reviews these developments and comments on their implications for energy more broadly.

BACKGROUND

Historical

Like much of the American West, the states of Idaho, Montana, Oregon, and Washington lie largely within the administrative domain of the national government. Since the Great Depression federal policies have been a shaping force in the region's affairs (3–6). With Grand Coulee and Bonneville dams, two major projects of the 1930s, the federal government transformed the economy of the Pacific Northwest by harnessing the Columbia River. When the New Deal development of the Columbia basin ended in the 1970s,[2] three-quarters of the power in the Northwest came from falling water and electricity rates were the lowest in the nation. Irrigated agriculture, control of once-devastating floods, riverborne commerce, basic manufacturing, and the structure of the regional economy had all been shaped by the federal program of public works (8).

The central presence was the Bonneville Power Administration (BPA),[3] an agency within the US Department of the Interior[4]—whose soul and substance

---

[2]Supporters of irrigated agriculture cling to hope for the eventual completion of the Columbia Basin Project, a million-acre irrigation development, of which half has been built and put into operation (see 7). The hope is quixotic until federal deficits and the long-festering problems of national farm policy abate, however. It is interesting, nonetheless, that with warming climate a secure source of irrigation water may become an important asset in the next century.

[3]Figures on Bonneville are for the 1989 operating year and are taken from Ref. 16.

[4]With the formation of the Department of Energy in 1977, Bonneville was transferred to that department.

nonetheless lay in the Northwest (9). Originally created to market the enormous output of the federal dams on the Columbia, Bonneville became something more, a regional economic force whose influence extended beyond its 15,000-mile transmission system or the Federal Columbia River Power System it ran, generating nearly half the electricity produced in the four-state region. Although the dams themselves are operated by the US Bureau of Reclamation and the Army Corps of Engineers, BPA power sales have usually governed river operations. If it was no Tennessee Valley Authority, Bonneville's influence did rival that of state governments, and appointment to head BPA was seen by many as the region's most important political plum.

Bonneville is largely invisible to the retail consumer (see Figure 1), marketing more than half of its power to utilities (the nongenerating and generating publics in the figure), which then supply businesses and households (see Figure 2). Two fifths goes directly to Northwest industrial customers. Most of these are large aluminum refiners, which make up 40% of the nation's aluminum production capacity. Aluminum refining is a baseload demand

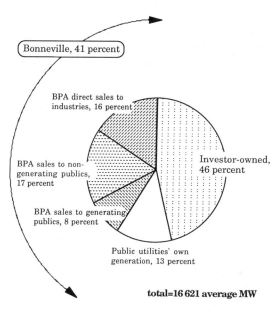

Figure 1 Firm electric power market shares in the Pacific Northwest, 1988. BPA sales are to industrial customers and to two categories of publicly owned utilities, those that rely on Bonneville for all their power, the "non-generating publics," and those that provide some of their own supply, the "generating publics." The figures above are for firm power; when additional power is available because of abundant water for hydro generation, the additional supplies are marketed to investor-owned utilities within the Pacific Northwest and to customers outside the region, primarily in California. Source: Ref. 10, p. 19

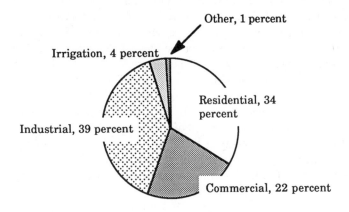

total=16 621 avg MW

Figure 2   Firm electricity use in the Pacific Northwest in 1988, by sector. Source: Ref. 10, p. 20

amenable to interruptions, so that it is highly compatible with the operations of a large power system. Since the beginning of aluminum production in the region during the Second World War, the industry and BPA have forged a quiet, close-knit symbiosis.

When Bonneville was not invisible it was benevolent. Power supplied by the federal dams was cheap, because of federal financing, the low wages of the Depression, and the large inherent capabilities of the river and the sites at which dams were built. BPA's wholesale rates did not change from 1937, when power was first delivered, until 1965; it remained for some time afterward the least expensive power anywhere. Low cost meant that New Deal activism could draw upon cheap power as a tool for social change, notably in rural electrification (9). More recently, BPA's cost advantages have indirectly funded conservation and fish and wildlife.

The New Deal left another monument, public power. Slightly less than half the retail electricity in the region is sold by municipalities, public utility districts, and cooperatives—121 agencies (Ref. 16, pp. 39–40) that are accorded priority access to federally generated hydroelectric power. Seven of these public agencies generate a portion of their own power. Throughout most of the United States, investor-owned utilities are dominant in electricity supply and generation. In the Northwest, as in the Tennessee Valley, another legatee of the New Deal, privately owned power companies are but one of several interest groups.

Not all change was progress. Logging and mining practices, water withdrawals for irrigation in the Columbia's tributaries, overfishing, and—perhaps most of all—the closing of the dams had seriously diminished the once-bountiful runs of salmon and steelhead trout in the river (Ref. 11, Appendix E). With their decline, and the fitful stewardship of federal policies, the Indian tribes of the Columbia basin suffered too (12).

Federal policies do not necessarily finish what they start. The economy of the Willamette Valley and Puget lowlands, running from south of Portland, Oregon, to north of Seattle, Washington, attained self-sustaining growth. The rest of the region remained an economic hinterland, controlled by external forces and markets. By the recession of the early 1980s the result was a dual region: a western, well-watered corridor that was prosperous, populous, and rapidly suburbanizing—a part of the country suddenly touted for its "livability"; and two areas—the semiarid, rural interior and the coastal areas of Oregon and Washington—both losing population and jobs as forestry, mining, and agriculture continued to stagnate.

Motivations for Congressional Action

From the turn of the century, when steam-driven turbines came into service, until 1970, electric power was a story of economies of scale and declining marginal cost (Ref. 13, pp. 363–64). Growth in demand meant that the total cost went down per unit sold: growth benefitted producer and consumer alike. By the 1970s, however, this conventional wisdom was under attack, as costs of construction, borrowed money, and environmental damage all rose ominously. In the Pacific Northwest, the puzzling and contentious manifestations of this nationwide trend led industry leaders and their critics to turn to the federal government for solutions. The product of their political labors was the Northwest Power Act.

FEDERAL HYDROPOWER ALLOCATION   As the Northwest economy grew, so did its demand for power, until it seemed obvious that the limits of the hydropower-based generating system were being reached. In 1968, led by Bonneville, an ambitious Hydro-Thermal Power Program was announced, to build dozens of new power stations operated by both public utility agencies and private, investor-owned utility firms. Immediately, and then persistently, controversy erupted, as the various interests with stakes in the regional power system jockeyed for advantage (Ref. 8, Chap. 3–5).

The most difficult issue was the equitable allocation of federal hydropower. If BPA's trove of low-cost power was no longer going to be enough to supply the region's growing needs, who would get how much of it? It was much cheaper than any alternative, and in the suddenly aroused energy consciousness of the 1970s low-cost, secure supply was worth fighting over. The utilities were prepared to fight hard—an apportionment determined administratively by Bonneville was likely to be litigated. That meant prolonged uncertainty for everyone, a worrying prospect to an industry used to timelines for future growth stretching decades into the future. For these interests, Congressional action was needed to referee a factional dispute.

NEW RESOURCES AND THE WPPSS PROBLEM   A second major problem of the 1970s was building new resources to augment hydropower and the existing supply of thermal power. Publicly owned utilities could issue municipal bonds, but lacked the economic motivation to do so because of their legal claims to federal hydropower. Yet the federal government had financed so much of the generating plant and transmission system that privately owned utilities lacked the equity to finance all the power plants thought to be needed.

Pressed by Bonneville and other utilities, public power did launch five nuclear power plants sponsored by the Washington Public Power Supply System (WPPSS), a public utility consortium based in Washington state but drawing upon the creditworthiness of well over 100 utilities throughout the Northwest. Plagued by cost overruns, high interest rates, and swiftly falling demand for power in the 1980s, WPPSS completed one plant, mothballed two, and canceled the other two (14). That cancellation put at risk $2.25 billion in municipal bonds, money borrowed to construct and finance the two terminated plants. In July 1983, WPPSS and the utilities behind it defaulted on the bonds, and one of the nation's largest securities lawsuits was launched (15). The suit was finally settled out of court in 1989, with repayment to bondholders and their attorneys of approximately $700 million of the original $2.25 billion loaned.[5]

As utility leaders foresaw, WPPSS was only one aspect of the institutional problems that needed to be solved. The key appeared to lie in joint public-private action, backed by federal guarantees to facilitate financing. Although this was done informally during the early 1970s, the rising cost of thermal power plants made an explicit, federally organized financing capability imperative (Ref. 8, Chap. 3). That would take an act of Congress. The Northwest Power Act passed, but too late to prevent the WPPSS default.

Even as Congress attempted to preserve the low-cost power of the Northwest, the ground was slipping out from under the utilities. The costs of new

---

[5]The complex WPPSS litigation spawned a newsletter aimed at the legal and utility communities enmeshed in the lawsuit. *Clearing Up,* published by NewsData of Seattle, contains a chronicle of the lawsuit and associated developments. The newsletter has outlived its subject, becoming a forum for those involved in regional energy matters.

power plants began to come due,[6] and rates increased rapidly—a rise made more dramatic by the low historical base. Between 1979 and 1984 the Bonneville wholesale rate increased more than 700%, and the retail price of power followed, more than doubling on average (16). At the same time, high interest rates, together with a worldwide economic slowdown triggered by the oil crisis of 1979, depressed the Northwest economy, hitting its energy-intensive industries with gale force. In the rural hinterlands, layoffs and skyrocketing utility bills stirred rebellion (17).

By 1982, as the Northwest Power Act was in the early stages of implementation, the expected shortage that had driven its design and passage had evaporated. Demand was far below expectations because the economy was in recession, and sales were expected to stay well below earlier projections, because rates were rising rapidly and conservation gained plausibility as energy costs mounted. In place of a deficit, power supply was in surplus and stayed that way through the 1980s.[7] Only one small hydroelectric plant, in Idaho Falls, Idaho, was built under the terms of the act; all the rest of BPA's acquisition efforts have been in energy efficiency. Instead of a financing mechanism to build new power plants, the Northwest Power Act became, faute de mieux, the blueprint for a laboratory of energy and environmental conservation.

INDIAN TREATY CLAIMS   A third threat to the region's power supply and economy loomed in the 1970s: the claims of Indian tribes to a productive natural environment. With the revitalization in the 1960s of Indian law (18), tribal attorneys began to seek judicial intervention to benefit their often-dispossessed clients. In the Northwest, the argument centered on the meaning of treaties concluded in the 1850s between the US government and tribes in what is now Oregon and Washington. The result of lengthy litigation was that treaty tribes won the right to half the harvest of salmon and steelhead trout (19, 20).

More threatened. Litigation initiated in the 1970s suggested that the courts might also determine that the tribes had a right to enjoy a productive natural

---

[6]In traditional utility practice, the cost of a new power plant is not counted as part of the "rate base," or invested capital, until it goes into service. Ratepayers are consequently shielded from the cost impact of utility managers' investment decisions until power is delivered; at that time, it is generally too late to affect the project substantially. In the Northwest, the costs of the first three WPPSS nuclear plants were financed by Bonneville, but the cost impacts were delayed by the financing arrangements until 1979. Since then, even though only one of the plants is in service, the costs of all three are being recovered through Bonneville's rates.

[7]The principal exception was the suburbanizing service territory surrounding Seattle, served by Puget Sound Power and Light, where the prolonged boom of Boeing's commercial airplane business and the rise of computer software, led by Microsoft, powered sustained expansion of jobs, population, and economic activity.

system, so that there would be enough fish to ensure a reasonable standard of living by historical benchmarks. No one knew what this would mean in practice, but its implications were clearly large: salmon runs had suffered major declines as the Columbia and other Northwest rivers were harnessed for hydropower and other industrial ends (12). Reversing that history would be costly even if it were feasible.[8]

A Revised Framework

After three years of lobbying and hearings, Congress enacted the Northwest Power Act in the waning days of the Carter administration. The act reflected the patient shepherding of Senator Henry M. Jackson, Democrat of Washington, whose chairmanship of the Senate committees overseeing the Bonneville Power Administration had long given him the dominant voice in government on matters affecting the region's power system. The act was Jackson's last hurrah in natural resource policy: the Republicans took over the Senate's leadership posts in 1981 and Jackson died in office in 1983.

The Northwest Power Act is festooned with complex compromises. When it passed, a lobbyist for the state of Oregon met with his governor and announced, "The good news is, we got what we wanted. The bad news is, so did everybody else." Yet in essence the act embodied a familiar strategy of governance, to define a new process, so that a stream of choices can be made by participants in that process without further appeals to legislative or judicial fora. The strategy has been successful thus far: no additional legislation has been enacted or even proposed in Congress; after an initial flurry of litigation over the meaning of the act for power sales contracts and other arrangements, the judicial front has been calm as well; and the massive litigation spawned by the WPPSS municipal-bond default was eventually settled without disturbing the regime established by the Northwest Power Act.[9] In 1990 petitions were filed under the Endangered Species Act, starting a process that could affect the implementation of the power act's fish and wildlife provisions (21), though the prospect of new legislation remains distant.

[8]The intersection of Indian, natural resources, and energy law in the Pacific Northwest has been followed by Professor Michael C. Blumm of the law faculty at Lewis and Clark. His articles, appearing mostly in the Lewis and Clark law review, *Environmental Law*, are complemented by more detailed commentary in the *Anadromous Fish Law Memo*, published by Oregon Sea Grant, Corvallis, which he edited until 1990. The final Memo (21) provides an overview and bibliographic listing. These writings and the sources cited in them, complemented by the indispensable article by Wilkinson & Conner (12), provide a good starting point for mapping this complex area of policy and law.

[9]One measure of the minimal impact of the bond default is the payment made by Bonneville to the settlement fund of $20 million in 1988. The settlement of the lawsuit permitted WPPSS to return to the bond market to refinance bonds on its first three plants, which are financed by BPA's ratepayers. The refinancing is estimated to save $70–80 million per year, several times the entire cost of the settlement to Bonneville (16).

The centerpiece of the new process is the Northwest Power Planning Council and the two plans it develops with wide public involvement.

THE COUNCIL   Chartered by the four Pacific Northwest states, the council comprises two members from each state, appointed by the governor under procedures set in state law. Yet the council's plans and powers can, under some circumstances, restrict or redirect the actions of federal agencies—a rare authority in a system of national supremacy. The council has been described as an interstate compact (22), a form of governmental organization that partakes of both state and federal authority.

In the complex world of energy policy, influence can be more important than authority. The council's influence has three bases: the credibility of its members, the analyses of its staff, and its standing among the public. Organized during the crises of the early 1980s, the council began with impressive strength in all three areas (23). Yet as energy has faded from the public agenda, and as governors have tended increasingly to treat appointments as political rewards, it has been the strength of its staff that has kept the council at the forefront.[10]

It is a frontier largely shaped by the council's planning. The two regional plans described below are the product of extensive public discussion. The public that participates has narrowed substantially, as electric power rates stabilized in the mid-1980s. Many divergent interests are at stake in the energy-dependent sectors of the economy and in the environmental management of the Columbia River basin, however, and the spectrum of those who pay attention is still wide. As a result, the pluralist approximation to democracy—characterized by one political scientist as majority rule by the minority that cares (24)—arguably remains valid (see also 25, 26). The council's plans, by this standard, make an important contribution: they articulate a public interest that goes beyond the claims of utilities or other interest groups. There is, perhaps surprisingly, little disagreement about this among observers.

The Columbia River basin fish and wildlife program was assigned to the council as a secondary task, but it has been of rising importance. Although the fish and wildlife language was added at the end of the legislative bargaining process, the statutory mandate to protect and enhance the fish and wildlife of the Columbia basin elevated the importance of fisheries qualitatively in the continuing competition for the river's resources. The council has formally adopted three versions of its program since 1982, to be discussed below.

[10]The council staff, an energetic and bright group of professionals with little prior experience in utilities or fisheries agencies, has altered the organized ecology in which the council operates. Resentment of the staff's disregard of tradition is mixed with admiration for their inventive competence. The mix of these reactions, together with the degree to which other organizations hire professionals of similar style, will substantially affect the durability of the council's policies.

POWER PLAN   The council's primary task is to formulate a plan to guide electric power development, including energy conservation. The act contains two principal guidelines: to choose resources that are "cost-effective," and, in ranking resources, to choose conservation first, then renewable and high-efficiency resources, and thermal power plants last (Ref. 2, §839b(e)(1)). These two rules can come into conflict with one another, but conservation has turned out to be the least costly resource so far, and others are not yet relevant for acquisition. Three plans have been issued by the council (27, 28, 28a).

The capital intensity of electricity production makes it essential to know the future. Power supply cannot be expanded easily or cheaply, and when capital is costly building excess supply can be financially disastrous—as illustrated by WPPSS. Conversely, undersupply can inflict economically unacceptable damage on customers, in an industry subject to extensive government regulation.

The rapidly changing energy economy of the period 1973–1986 made forecasting difficult, however. Consumer demands changed rapidly as energy prices rose to levels not observed in modern times. Government policy sought to affect consumer behavior directly by exhortation, regulation, and subsidy. The cost of new supply was difficult to estimate as regulatory requirements, cost of capital, and changes in technology all experienced high turbulence. Techniques were undergoing rapid development too, as econometric methods and engineering-based end-use analyses were increasingly adopted by forecasters.

In its first plan, completed in 1983, the council used four forecasts, each representing a different scenario of regional economic development over the next 20 years. The council resisted using a single or best-case forecast, insisting that the planning environment was too dynamic to stake multi-billion-dollar decisions on a fragile projection.[11]

Multiple forecasts frame an approach to planning that curbs the momentum of capital investment (Ref. 27, Chap. 3). Nearly half the time needed to build a new thermal power plant is taken up with planning, design, and regulatory compliance, but as little as 5% of the cost is incurred at this stage. Then, with the start of construction, costs mount swiftly. The council proposed that Bonneville's financing authority be directed to develop more projects than

[11]The decision to use four scenarios was taken consciously, to avoid having three forecasts, of which the medium would be taken to be the "real" expectation of the council.

Since 1988, Bonneville and the council have done their forecasting jointly, a reflection of the declining controversy over the future need for power. The insistence on avoiding a median forecast is still maintained by the council (Ref. 10, pp. 10–11). In practical effect, however, the decision to use a probabilistic system planning model in the 1986 regional plan (Ref. 28, Vol. 2, Chap. 8) has superseded such doctrines.

were likely to be needed, delaying a final decision on which options to exercise until the need for them was closer and clearer (29).

The regional approach emphasized a portfolio perspective: the essential point was not any single power plant or conservation program, but the whole set of alternatives available. The portfolio was to be managed on a regionwide basis, using cost-effectiveness as a guideline. Cost-effectiveness meant that the least costly options should be exercised, and that costs should be reckoned on a regional, lifecycle basis, discounted to their present values. It was a businesslike and slightly idealistic gospel to preach to a utility industry riven by factional rivalries and facing political confrontation from once-docile ratepayers.

At the time, the recession-hobbled economy was not demanding much energy. Wary of issuing a plan that would languish unused, the council identified action steps that needed to be taken: creating conservation programs, developing the options process, preserving the suspended WPPSS plants so that they could be called upon if needed. The lowest-cost items in the regional portfolio turned out to be conservation.

ENERGY EFFICIENCY

Although the perspective of the regional power plan is centralized, its key innovation—the development of energy efficiency as a source of power—is radically decentralized in concept. Conservation entails changes in structures, appliances, motors, and lighting equipment, and a host of other shifts in technology and procedure among a good share of the nine million ratepayers of the Pacific Northwest. Decentralization, in turn, makes implementation and effective intermediary institutions crucial to the development of conservation as a resource.

Consider buildings: owners need to be motivated to participate, structures need to be audited to locate conservation opportunities, contractors and financing need to be assembled to install conservation measures, and monitoring of savings and collecting of costs need to be conducted for some time after installation. What is done for buildings needed to be extended likewise to commercial equipment like lighting and refrigeration, industrial processes that use electricity, and other uses such as street lighting by city governments. None of this fell within the traditional scope of electric utility functions; all of it needed to be organized, as illustrated by the Hood River project described below.

Background

Conservation makes economic sense: electric energy had been cheap for nearly half a century in the Pacific Northwest, and power was used more

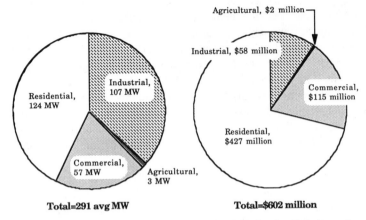

Figure 3 BPA acquisitions of energy efficiency, 1981–1989. Source: Ref. 40, p. 6

freely than is warranted at higher rates. That meant that existing buildings and technologies could be retrofitted economically and new ways of doing things could be instituted, while saving money for both customers—who would see their bills decline—and utilities—who would be able to defer building costly generating facilities.[12]

The question was, particularly in the absence of a power shortage, would utilities and ratepayers do what was economically rational (see 31)? The council called for rapid action, arguing that conservation was a largely untried method of obtaining power resources. It was imperative, accordingly, to build the "capability" to develop and acquire conservation (Ref. 27, Action Plan). Capability-building replaced resource acquisition in the first regional plan as the central priority assigned to BPA and the utilities. As shown in Figure 3, both results and expenditures were impressive for a period of surplus: long-term demand reductions equivalent to a medium-sized coal plant, at about half its cost.

The benchmark for the capability-building effort was the Model Conservation Standards (MCS; Ref. 27, Appendix J). MCS is a building code for residential and commercial structures, recommended by the council to local governments, state legislatures, and utilities in the region. MCS translates the idea of cost-effective energy efficiency into the practical details of building construction—the size of lumber needed to frame a house with extra thicknesses of insulation, the ventilating fan capacity required to minimize lifecycle costs of operating the building, and so forth.

---

[12]This logic also means that customers who do not participate benefit as well, since their rates do not rise as rapidly as they would otherwise. Moreover, nonparticipants are free riders (30): they run no risk and need not undergo the stress of change. Inertia is rational in this case—a reality facing policy implementation.

MCS is useful for purposes beyond building codes. Utilities have used MCS as a guideline for subsidies to builders participating in a voluntary program called Super Good Cents, which pays the difference between conventional and efficient construction. The commercial MCS is used by municipalities to develop energy-efficient design improvements in large office buildings not covered by an explicit code. In addition, MCS led inadvertently to determining the legal status of the council, when a lawsuit challenging the council's authority to promulgate MCS was decided by the US court of appeals (32). The court ruled that the unusual circumstances of state powers constraining federal actions had been agreed to by Congress and the states and was therefore constitutionally permissible.

Hood River

Because of its dispersed nature, conservation poses the problem of lost opportunity: the need to affect energy users' investment behavior at the time they decide. Conservation is mostly investment—in better-insulated windows, more efficient motors, or the like. Thus conservation is often more capital-intensive than is generation, because there are few or no operating costs. The investments are made, however, by firms and households, rather than utilities. Conservation cannot be sized or scheduled readily to meet utility needs. Worse, the power system can easily misjudge the targeting, intensity, or form of the incentives needed to recruit investors. For example, fewer than 10% of the electrically heated houses built in the Pacific Northwest in the past decade have all the efficiency measures that are cost-effective (Ref. 33, p. 3). The rest are a lost opportunity: their remaining potential for efficiency can only be partly recaptured by later retrofitting; even where subsequent action is technologically feasible, it may no longer be cost-effective.

Although the problem of lost opportunities was recognized in the original power plan, it remains a difficult challenge to affect the behavior of large numbers of people who are ignorant of and indifferent to electric utilities. Responding to a suggestion of the environmentalist Natural Resources Defense Council, Bonneville and two utilities attempted to market energy efficiency under laboratory conditions (34). The locale chosen was Hood River, Oregon, a small city located on the banks of the Columbia, as the river cuts a scenic gorge through the Cascade mountain range east of Portland. Hood River County is served by Pacific Power, the region's largest investor-owned firm, and by an electric cooperative with access to Bonneville's low-cost federal hydropower. The town also combines urban and rural consumers and experiences weather that mixes the mild marine climate west of the Cascades and the colder winters of the interior.

The three-year, $20-million project proved to be highly successful in penetrating the market, but was less successful in saving energy. With advance research followed by a full-scale marketing effort, the project persuaded the owners of 85% of the residential structures to participate, retrofitting nearly 3000 buildings. In participating houses, 83% of the efficiency measures recommended were adopted. These high figures reflect both the fact that the program required no payments from homeowners, and the intensity of the marketing effort.[13]

An analysis of post-retrofit energy savings (35) found that savings were lower than anticipated. Net savings per house averaged 2300 kWh in 1985–1986 (after retrofit). The project cost-effectiveness limit was set at $1.15 per kWh estimated to be saved in the first year (Ref. 34, Vol. 1, p. 11). This implies $2645 per house would have been cost-effective, but actual expenditures averaged $4820 per house (Ref. 36, p. 48)—an 82% cost overrun in hindsight.[14] Hirst, Goeltz, & Trumble (35) suggest that the discrepancy can be largely explained by the choice of base year. In 1982–1983 rates were undergoing a steep increase, and this price effect depressed electric power consumption. This is the basis of the planning council's defense of its cost-effectiveness estimates (Ref. 37, Section 4). Nonetheless, statistical analysis of the Hood River data indicates that only 40% of the discrepancy can be explained this way.

The difference between expected and actual demand reductions seems worrying. Yet the Hood River project had the characteristics of both a research effort and a social program. Conservation was a new and unfamiliar activity, not a well-tried engineering option. Monitoring of energy uses in about one tenth of the houses weatherized in Hood River, together with surveys, statistical analyses, and field observation of the social dynamics, cost $5.6 million, about half the cost to install the energy efficiency measures. The evaluation lasted three times as long as the project itself. These ratios are typical of those found in social experimentation, and the expenditures supported well-designed research that produced important insights into and documentation of the social process that lies behind using energy more wisely. It was not cheap, however.

Perhaps the chief lesson learned is the critical importance of people and organization. Managerial flexibility was the cardinal virtue: the project needed to be revised repeatedly, as the complications of acting in a real

---

[13]While the program was being implemented, concern rose sharply about deteriorating indoor air quality in houses that had been tightened for energy conservation. The Hood River project included energy-efficient heat-exchanger ventilators in its package of recommended measures. These increase costs as they produce a more healthful indoor environment, but reduce energy savings.

[14]Even excluding administrative costs, which may not scale with program size, and indoor-air quality measures—an issue that emerged during the program—the overrun is 43% (figures from Ref. 36, p. 48). It is unrealistic to assume no administrative cost, of course.

community with real contractors became apparent. Project staff discovered that they were reinventing some wheels. Retrofitting windows and insulation is a kind of remodeling, and hiring a general contractor would have cut the administrative workload substantially. It was also important to enlist cooperation by people of divergent interests. An oversight committee, with members from environmental groups and utilities, wrestled with problems of implementation, hammered out compromises—and put aside their differences to act as advocates for the project when budgets or external difficulties needed attention (Ref. 34, Vol. 1).

These observations round out the picture conveyed by the lower than expected energy savings. The fact remains that a misestimate of costs by 50% is enough to reverse the priorities that define a "soft" rather than "hard" path. The art of verification deserves more emphasis than it has been given (38).

Decentralized Action and Resurgent Load Growth

Energy efficiency can be acquired, on a basis similar enough to generation to warrant its inclusion in regional plans; that is the point reaffirmed by the Hood River Conservation Project. But complications remain (see 39).

PROBLEMS IN CONSERVATION   Bonneville's conservation staff tried repeatedly through the 1980s to design and carry out an effective conservation effort in commercial structures. By its own admission, "None of these efforts proved able to capture significant amounts of savings" (Ref. 40, p. 10). In the city of Seattle, which counted on commercial-sector conservation to achieve overall levels of savings equivalent to the council's MCS, performance levels were only half what had been expected.

Conservation expenditures must generally be made during renovation or construction, times when operating costs may not be a high priority. Energy-efficient technology generally costs more to produce and install, so its cost-effectiveness depends on how long it is used, but commercial spaces are renovated in response to short-term market forces. Moreover, commercial firms operate under leases, franchises, and administrative arrangements that frequently separate financial accountability from operational control of energy use (Ref. 40; Ref. 41, Chap. 4). These difficulties of capturing lost opportunities all highlight the advantages of a direct but politically infeasible method, raising rates, which would presumably induce compliance by consumers worried about rising bills. Using utility and government programs to anticipate and forestall higher rates can be much less effective technically, though much more acceptable to ratepayers.

A different problem is presented by lost opportunities in new buildings. Efficient building technology faces two hurdles: (a) if energy prices are increasing, one should build the structure to be cost-effective at the expected marginal cost of energy, not its current average cost; (*b*) builders are often different from owners, and owners, different from occupants—and only the last benefit directly. These characteristics suggest the use of building codes like the MCS.

A building code must be enacted, however by governments sensitive to political pressures, especially those exerted by well-organized constituencies. Builders of residential housing are formidably well-organized. They are also responsive to the leadership of mass market builders, who aim at first-time home buyers. These are families just barely qualifying for a mortgage, for whom any increase in the price of the home may be unaffordable. Builders claiming to represent these aspirants to the American dream have thwarted adoption of statewide codes in Idaho and Montana, and they delayed upgrading of codes in Washington and Oregon for eight years. At this writing, residential codes equivalent to the MCS are scheduled to go into effect in the latter two states in 1991.

The power surplus has also left utilities ambivalent about conservation, which has the immediate effect of lowering revenues when demand is already slack.[15] In a comprehensive evaluation, the council's staff concluded that "outside Bonneville and Puget Power, few programs are in place" (Ref. 33, p. 20). Figure 4, which is based on that study, bears comparison with the market shares in Figure 1. Puget Power was the only utility facing a supply deficit, so that it had an incentive to acquire conservation as a way to slow politically unpopular rate increases. With its surplus, BPA's devotion to conservation is also political rather than economic in origin—a noteworthy instance of government efficiency.

A CONSERVATION-CENTERED APPROACH . . .   The difficulties of conservation in practice do not vitiate its economic and social virtues, but they do put them into perspective:

1. Energy efficiency is cost-effective. Follow-up evaluations have shown good performance that endures over time in residential structures. Commercial efficiency improvements have been more troublesome, as described above.
2. Conservation programs tend to follow economic activity, particularly in new construction and the commercial sector. This automatic regulation feature tends to merge conservation investments with other capital spending.

---

[15]In 1989 state regulatory commissions in New England began to implement rules for conservation programs designed to overcome the problems of revenue losses. See, e.g. (42). This innovation relies upon the growing credibility of conservation as a resource that can be acquired on terms similar to generation.

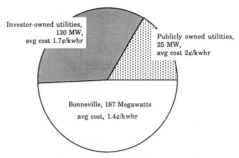

Totals: 350 avg MW, $1.1 billion

Figure 4   Regional effort in conservation, 1978–1987. Source: Ref. 33, Appendix

3. Energy efficiency avoids the environmental costs of power generation, though it carries its own risks. Conservation does not require transmission, nor are effluents created as kilowatts are saved. But tighter sealing of buildings—a principal tenet of efficient construction—can impair indoor air quality and expose occupants to buildups of radon, a carcinogen and significant public-health hazard. These hazards are manageable with readily available technology, and regulatory requirements for these remedies are included in the MCS (Ref. 43, pp. 39–40).

4. Conservation is politically popular, both with consumers, who support more efficient use of resources and enjoy direct subsidies for home improvements, and with utility managers eager to be seen as benevolent members of their communities.[16]

5. Although popular in general, conservation practices require significant social change and face implementation barriers such as the opposition of homebuilders. This is one area in which the pluralist approximation to democracy fails: many builders regard additions to building codes as unwarranted intrusion onto their turf. Instead of compromise, there was prolonged political deadlock.

The conservation efforts of Pacific Northwest utilities and governments amount to a proof of principle. It is possible to obtain conservation savings at costs that compete favorably with new generation, and to do so within a highly fragmented utility industry and governmental structure. Achieving all

---

[16]Political popularity has affected which programs are implemented; in 1987 the council pressed Bonneville to lower its expenditures on its residential weatherization program, a perennial favorite, because the energy acquired from existing homes was not needed and would be available later anyway.

cost-effective savings is not yet possible, however, because of the difficulties of matching program action to the decentralized incentives facing energy users.

. . . FOR A DWINDLING SURPLUS   The Pacific Northwest economy has outperformed the national economy since 1986 and the outlook is for that trend to continue (Ref. 10, pp. 4–5). Bonneville's most recent annual report announced brusquely, "Surplus Gone" (Ref. 16, p. 8). What to do next has become the central question of the council's new regional plan. While the potential for conservation is still substantial, it is not enough by itself to cover the range of possibilities that prudent planning needs to take into account.

Through the surplus of the 1980s utilities focused on near-term problems, many spawned by the high fixed costs of the surplus. With the momentum inherent in a large network, there were few immediate consequences in this foreshortened perspective. Yet the near future presents the Northwest with important, largely unrecognized challenges.

The most serious are financing of additional resources and the management of the Columbia River hydropower system. Although the Northwest Power Act envisioned a business environment in which Bonneville would continue to be a dominant supplier of power, that has not happened. Utilities that generate part or most of their own requirements, including all the Northwest's investor-owned firms, have gone their own ways, seeking to unload their surpluses and to develop high-voltage transmission networks to reach out-of-region markets for both sales and purchases of electricity. Since the mid-1980s, BPA has ceased to count the anticipated needs of investor-owned utilities in its forecasts, although there are power sales contracts in place that could rapidly increase Bonneville loads. The generating utilities, for their part, focus on the near term and expect their power-marketinag skills to provide enough breathing room to cover deficits.

These policies represent sharp departures from tradition in a capital-intensive industry with long leadtimes. In a pinch, the ability of the Columbia to produce additional hydroelectric energy could become the practical alternative. But that will mean that claimants to the river's nonpower benefits—particularly fish and wildlife—will have to give way.[17]

Articulating and attempting to resolve these economic and environmental conflicts remain critical responsibilities of policy makers and political leaders.

---

[17]See Ref. 21 for a discussion of tightening pressures on river flows associated with the long-term disposition of storage capacity in dams that impound water in the upper reaches of the Columbia drainage.

ECOSYSTEM REHABILITATION

Low-cost hydropower produced in the Columbia drainage has long been of instrumental importance to the Pacific Northwest, compensating for the distances its goods must travel to reach buyers. With the rise of environmental awareness, backed by the legal force of Indian treaty rights, the energy economy of the region has turned to funding environmental enhancement. In the Northwest Power Act, Congress declared a responsibility to "protect, mitigate and enhance" the fish and wildlife of the Columbia River basin to the extent they have been damaged by the development of hydropower. The costs were to be borne by the ratepayers of the Bonneville Power Administration (Ref. 2, §839b(h)(10)(A)). This language authorizes the largest program of ecosystem rehabilitation on the planet.

Background

The Northwest Power Planning Council (Ref. 11, Appendix E) has estimated that the Columbia basin once produced 10–16 million adult salmon and steelhead each year, returning from their two- to four-year periods of growth in the North Pacific Ocean.[18] The prolific fish runs were the economic mainstay of tribal populations of approximately 50,000 humans (44), who used the fish for food and as trade goods. Today's human population is roughly 100 times larger, whereas the fish runs have declined to about 2.5 million per year, one-fourth to one-seventh their primeval abundance.

These declines have many causes, but the foremost is the construction of dams. High dams—Chief Joseph on the mainstem Columbia; Hells Canyon on its principal tributary, the Snake—block the migration of fish entirely, destroying some populations outright. Even low dams alter the river habitat and expose fish to risks to which they are not adapted. The degree of responsibility to allocate to hydropower is necessarily imprecise. The Northwest Power Act limits the contribution of ratepayers to damages attributable to hydroelectric power generation—a limit the council set at between 8 and 11 million adult fish per year. The practical meaning of this constraint is unclear, however, since the biological capability of the remaining habitat and technically feasible hatchery sites may fall substantially below 8 million fish. As with energy efficiency, the conceptually clear boundaries do not provide detailed guidance.

---

[18]There are six species of Pacific salmon native to the Columbia drainage, of which the chinook (*Oncorhynchus tsawytscha*), steelhead (*O. mykiss*), and coho (*O. kisutch)* are of economic significance today.

Fish and Wildlife Program

That guidance is formulated in the council's Columbia River basin fish and wildlife program (11), initially adopted in 1982. The program identifies measures to be taken by the Indian tribes and government agencies that exercise management responsibility for fish and wildlife, by hydropower project operators and regulators, and by those charged with land and water management. Program measures are mostly funded by the Bonneville Power Administration, with a total economic cost running at about $130 million per year (Ref. 16, p. 14). This is about 7% of the agency's $2 billion annual revenue stream, comparable in size to Bonneville's most recent rate increase. The sum is unprecedented for fish and wildlife, however, as is the program's spatial scope, overlapping four states and covering a land area the size of France. About $50 million is spent directly by BPA through contracts let mostly to the Indian tribes and fish and wildlife agencies. The remainder of the cost is incurred as lost revenues—earnings foregone because water was released to benefit fish rather than power users.

WATER BUDGET   The keystone of the program is a change in the flow of the Columbia that, for the first time, gives substantial meaning to the idea that fish deserve "equitable treatment" (Ref. 2, §839b(h)(11)(A)) among the claimants to the river's water. Before the dams were built, the Columbia's flow was heavily concentrated in the spring, when mountain snow rapidly melts. Spring floods carried juvenile salmon and steelhead to the ocean, a trip that usually took several days, instead of weeks now required to move through sluggish reservoirs. Traversing the reservoirs so slowly exposes the juvenile fish to predators for a longer time and it stresses their physiology: a biochemical changeover must be made in going from fresh to salt water; if arrival in salt water is delayed, the fish can revert to a freshwater constitution, stop growing, and never reach harvestable size.

For years the fish and wildlife managers and Indian tribes had requested higher flows in the springtime migration season. But the requests were just that, and the dams, controlled by utilities and the Corps of Engineers, were run to optimize power revenues most of the time (21).

Led by its first chairman, Daniel J. Evans, a former governor of Washington state and later a US senator, the planning council took seriously its mandate to provide "protection, mitigation, and enhancement of anadromous fish at, and between, the region's hydroelectric dams" (Ref. 2, §839b(h)(2)(C)). After extensive negotiations with utility experts, tribal and state agency leaders, and checking the pulse of the public, the council created a "water budget" that restored part of the spring flood (45). The water budget is a quantity of water, approximately equal in energy capacity to a medium-sized coal plant, which can be controlled by fisheries managers. During the late spring, when fish are observed to be migrating in the river, the fisheries managers call for water budget flows to commence. Most of the time, once the augmented flows have started, they continue until the budgeted water is used up. Water flowing in the river can be used by power managers to generate power, but they do not control its timing. For this reason, the water budget loses money for the power system, a sum that averages about $40 million per year.

The effectiveness of the water budget is difficult to estimate, even in principle. Its benefits lie in lower predation and healthier fish—which are difficult to pick out from the background noise of natural fluctuations. The size of a fish run and the quantity of water available in the river are observed at the rate of once per year, so accumulating statistically reliable observations requires patience. In practice, the water dedicated to the water budget in the Snake River drainage is often unavailable, a result of limited water-storage potential and the reluctance of utilities and irrigators to yield much short of a direct order. Moreover, the Corps of Engineers asserts its legal claim to be the final arbiter of river operations with stubborn tenacity. In combination with lingering fears among the fisheries advocates, this has produced recurrent, but inconclusive, disputes.[19]

What is important in the interim is that a substantial reallocation of resources—similar in size to all the energy conservation savings acquired under the Act—has been made in the name of tribal tradition and environmental values.[20]

SYSTEM PLANNING   The Northwest Power Act directs the council to treat the Columbia River and its tributaries "as a system" (Ref. 2, §839b(h)(1)(A)). Though this could have been ignored easily, it became instead the rationale for an extraordinary planning effort.

The Northwest Power Act gave the lead role in formulating the fish and wildlife program to the Indian tribes and fish and wildlife agencies with legal responsibility to manage the living resources. Yet the state fisheries agencies and tribes had been bitter adversaries in the treaty fishing rights disputes of the 1970s. By 1982, when the first Columbia basin program was adopted, these two groups were still living under a fragile truce (20). Not until the mid-1980s did a stable integrated management system begin to emerge (19). As Indian

[19]Recent improvements are discussed by Muckelston (46) and Blumm (21). Their overall evaluation of the water budget is considerably more skeptical than the account given here.

[20]The commercial value of the 2.5 million Columbia River salmon and steelhead is likely to be greater than $50 per individual at the margin, but it remains difficult to justify the cost of the inputs from the power system—which exceed $50 per individual on average—in the context of marketplace economic values. Expert observers differ sharply on the importance of economic comparisons in this issue (21, 47, 48).

and non-Indian resource managers cautiously felt their ways toward shared jurisdiction in the early 1980s, their decision rule was usually consensus. That meant that their proposals to the council simply conflated earlier requests and set few priorities. The result was undigestible, both to utilities wary that the program might rapidly turn into a multimillion dolar giveaway, and to a council that sought to fashion rational responses to contentious issues (49).

The council asserted a shaping role as planner, thrusting BPA into an unfamiliar and initially awkward role as funding agent. The council's boldness in defining the water budget gave it credibility with the fisheries interests—who saw a major change in the way the river's water would be used—and simultaneously with the utilities—who worked with council staff on the calculations of cost and operational constraints (45). That early victory provided the council and its staff an endowment of political capital, which was invested in a continuing effort called system planning (Ref. 11, Section 200). System planning meant trying to think about the interactions among the hundreds of activities affecting the abundance and health of the basin's fish and wildlife, including changes outside the scope of the program. System planning sought to govern the Columbia drainage as an ecosystem, with neither the authority nor resources to enforce its writ. Surprisingly, it has worked so far.

A three-year effort to gather data so as to form an explicit picture of the Columbia basin's 31 biological provinces is being completed at this writing. This phase of system planning is intended to specify projects and priorities for implementation over the next several years. The process of planning has included unprecedented attempts to inform local constituencies and to engage them in planning. While this fit naturally with the council's mandate for public participation, it had a more deliberate purpose: to make program implementation irreversible, because there would be widespread support for projects that would keep the program moving forward, even if tightened power economics or other unforeseen conditions appeared. Given the initial resistance of the power interests to the program, this has turned out to be a successful piece of institution-building. But, as discussed below, the biological effectiveness of this approach remains to be proved.

ADAPTIVE MANAGEMENT   As it did in the power plan, the council responded to scientific uncertainty by fashioning an approach that takes account of the gaps in knowledge. Adopted first in the 1984 revision of the Columbia basin program, the method, called adaptive management (50, 51), became the guiding philosophy of system planning.

Information on environmental variables such as river flow or the size of a migrating population accumulates slowly. Theories built on such sparse data are limited and often wrong, especially in cases where human manipulation has moved an ecosystem far from its natural configuration. Adaptive management proceeds from a simple premise: If human understanding of nature is imperfect, then human interactions with nature should be experimental (52; see also 53–56). Thus, policies should be designed and implemented as experiments probing the behavior of the natural system. Experiments often surprise, but if resource management is recognized to be inherently uncertain, surprises are opportunities to learn rather than failures to predict. Knowledge, too, is a lost opportunity—one that can be captured by adaptive management.

From this perspective, planning became a way of assembling a picture of the ecosystem to be managed. Information is collected and organized by a model, forming both a database—an index to facts gathered—and a set of hypotheses—a set of beliefs about how those facts relate to each other. Adaptive management tests those hypotheses against the experience of rebuilding the fish and wildlife of the Columbia. Although virtually all policy designs take into account feedback from action (57), the idea of using a deliberately experimental design, paying attention to the choice of controls and the statistical power needed to test hypotheses, is one rarely articulated and usually honored in the breach (58, 59). It is for this reason that the explicit adoption of an adaptive policy in the Columbia River basin is noteworthy.

Like the rationalism of cost-effective energy policy, the adaptive perspective has an idealistic ring in a political setting where blood feuds are recent memory (59a). Adaptive management relies on ample funds to pay for monitoring and analysis, and social flexibility to adjust to mistakes and to try new ideas. These have been in good supply during the 1980s. Yet as the inevitable mistakes become evident, and the inevitable natural fluctuations in the abundance of fish depress catch from time to time, preserving an approach to management that tolerates error and looks to the long term may prove difficult. For now, all that can be said is that the first project to be designed under the premises of adaptive management—a major hatchery serving the Yakima and Klickitat rivers—is proceeding well (60).

PROTECTED AREAS   A fourth noteworthy element of the fish and wildlife program is the designation of more than 40,000 miles of streams, mainly in tributaries, as off-limits to hydropower development.

With the energy crises of the 1970s came a reawakened interest in hydropower, and thousands of potential sites in the Northwest were filed with the Federal Energy Regulatory Commission (FERC), the agency that licenses the development of hydroelectric projects. Many of these projects came under spirited critique from environmentalists and natural resource agencies, who invoked potential damage to migratory salmon as a reason to forestall development. Like the Army Corps of Engineers on the mainstem, FERC showed little inclination to accommodate opponents of the power interests.

But the legal context had been changed by the Northwest Power Act, which directs hydropower project operators and regulators "to the fullest extent practicable" to take into account the Columbia basin program (Ref. 2, §839b(h)(11)(A)). Beginning in 1984, the council collected information and held public meetings to stimulate negotiations among project developers, Indian tribes, state agencies, and citizen groups. These efforts succeeded in 1988, when the council designated more than 40,000 miles of streams in the Northwest as "protected areas"—areas in which FERC is advised not to approve hydroelectric project applications. This protection is limited to hydroelectric projects; however, given the deference FERC appears ready to accord the council's program, the designation has already made a difference in project proponents' expectations.

CONCLUSIONS

"To direct attention today to technological affairs is to focus on a concern that is as central now as nation building and constitution making were a century ago," proclaims the historian Thomas Parke Hughes (Ref. 13, p. 1) at the beginning of his study of electric power networks. Whether or not such a claim is sustained by the historians of the future, it is apparent that the changes of direction in the Pacific Northwest echo with significance in our own time.

Major changes of agenda, symbol, and substance have been achieved. The notion that a major electric power system should have changed both its goals and its behavior to encompass demand reduction (energy conservation) and the reallocation of its principal capital asset (fisheries enhancement using Columbia River water) was ludicrous only 15 years ago, when the Bonneville administrator of the time attacked environmentalists as "prophets of shortage" hell-bent on denying the Northwest its rightful patrimony of economic growth (Ref. 8, Chap. 3).

Innovation has been shaped by institutional setting and circumstance. What has occurred is historical rather than rational: economic circumstance has conspired with a wide—if still shallow—turn in public opinion to shift the agendas of large organizations with surprising speed. Higher power rates, the product of miscalculation by utility managers, have become the mainspring of efficient use of energy and more magnanimous use of river water. These results were not planned; those who lived through the times did not often understand the directions in which the events were carrying them. Even today, the apportioning of causal responsibility among driving forces remains unclear.

Bounded crises provide important—but circumscribed—opportunities for innovation. It has proved easier to improve the substantive rationality of the regional power system than to make lasting procedural reforms. More efficient use of electric energy, together with a new balance in the uses of Columbia River water, have become official goals and daily practices of the operating agencies of the Pacific Northwest. This is an important pair of achievements, but there is a shortfall that should be noted.

Power generation was long characterized by declining average costs and its structural correlate, natural monopoly. The shift toward rising average costs signals a fundamental need for new economic organizations. The exploration and development of conservation is the first of what may turn out to be a series of frontiers to be crossed as the industry develops new forms. Yet the Northwest Power Act, by buttressing the role of the Bonneville Power Administration, preserved existing, perhaps outmoded forms of economic organization even as it pushed these institutions into new territory.

Making conservation a resource has created a new wing of the bureaucracy, rather than a new industry. Although the existing agencies claim to be more open to surprise and innovation (40), conservation has been engineered, treated as a form of power generation—at least from the standpoint of contractual arrangements—and forced into institutional molds formed by other, very different technologies. This is not all bad, of course: feeding the new in the womb of the old provides adequate resources, access to existing markets and information channels, and at least a measure of legitimacy. Most important, cost-effective planning seeks to forestall the higher average costs that would encourage energy efficiency on their own. So long as regulatory officials are unwilling to acknowledge the economic force of rising average costs with prices based on marginal cost, directed innovation is likely to remain the best available option.

Similarly, the fundamental political realignment in the governance of Columbia River water has empowered new claimants to the porkbarrel—when one might have hoped instead to change the nature of the game. There are now more interests to be accommodated, and they are more vocal than in the past. But the process is more similar than different, and there is little recognition of the far higher uncertainties that lie in the path to rehabilitation of a large ecosystem.

These judgments are not directed to those who have participated in the processes of change. By their lights, much has moved in the last decade—a fair conclusion. Yet one can puzzle over the intractability of procedures and organizational cultures. As other observers [e.g. Selznick (61), Mazmanian &

Nienaber (62), Worster (4), Morone & Woodhouse (63)] have noted, large technological systems created amid visions of democratic control often exhibit a stubborn resistance to pressures for redirection.

Thus, resumption of load growth reopens questions that might have been thought closed by the historic changes of the last decade. As the stock of low-cost, readily obtained conservation dwindles, as the cost of providing the water budget mounts, the economics of electric power clash increasingly with the environmentalism that reigns today. So far, there is no backsliding, however.

Perhaps the most durable lesson of the Pacific Northwest experience is the inherent inefficiency of large-scale systems. Electric power grids take decades to build, thousands of people to operate, and affect electorally significant numbers of citizens; for all these reasons, simple optimizing algorithms fail to describe either the course of development or its endpoint. That academic conclusion has an important practical corollary: there is almost always slack (64)—"inefficiencies" that can be tapped to respond to opportunities and crises. The emergence of energy efficiency as a segment of the utility business illustrates a wider pattern: opportunities that were lost in placid times can be identified and harnessed in times of stress. Slack fosters innovation because it often provides low-cost ways to change. Whether the innovation lasts is another matter.

Resurgent load growth suggests that the continued prominence of conservation as a supply option may yet be in doubt. The ambitious fish and wildlife program is shadowed by stiffening competition for river water. Environmentalist objectives have been adopted in the Pacific Northwest to a degree that is both unprecedented and substantial. But whether those objectives can be institutionalized over the long run remains uncertain.

ACKNOWLEDGMENTS

Work supported in part by Washington Sea Grant and by the Ministry of Education of Japan. I am indebted to Professors Gordon Orians and David Olson, as well as to V. Crane Wright, all of the University of Washington, for intellectual challenge and collegial support in my studies of the Pacific Northwest. I am grateful to Professors Tsuneo Tsukatani and Takamitsu Sawa for the hospitality of the Kyoto Institute of Economic Research, Kyoto University, where this paper was written. Research assistance from the Northwest Power Planning Council and the Bonneville Power Administration is gratefully acknowledged. I thank David Bodansky, Jean Edwards, Randall Hardy, James J. Jura, Robert O. Marritz, and John Palensky for informative and helpful discussions, and Rick Applegate, Jon Koomey, Keith W. Muckleston, Marion E. Marts, Edward Sheets, John M. Volkman, and John N. Winton for comments on an earlier version of this article.

Literature Cited

1. US Dept. Energy, Energy Inf. Admin. 1989. *Annual Energy Review 1989. DOE/EIA-0384*. Off. Energy Markets and End Use
2. US Congress. 1980. *Pacific Northwest Electric Power Planning and Conservation Act*. P.L. 96-501, 96th Congress, 94 Stat. 2697-736, codified at 16 US Code §§839 et seq., Dec. 5. Referred to here as the Northwest Power Act
3. McKinley, C. 1952. *Uncle Sam in the Pacific Northwest*. Berkeley: Univ. Calif. Press
4. Worster, D. 1985. *Rivers of Empire. Water, Aridity, and the Growth of the American West*. New York: Pantheon
5. Volkman, J. M., Lee, K. N. 1988. Within the hundredth meridian: Western states and their river basins in a time of transition. *Univ. Colo. Law Rev.* 59:551–77
6. Reisner, M. P. 1986. *Cadillac Desert*. New York: Viking
7. Janeway, K. A. 1985. Of time and the river: the debate over completion of the Columbia River Basin Irrigation Project. *Northwest Environ. J.* 1:97–113
8. Lee, K. N., Klemka, D. L., with Marts, M. E. 1980. *Electric Power and the Future of the Pacific Northwest*. Seattle: Univ. Wash. Press
9. Tollefson, G. 1987. *BPA and the Struggle for Power at Cost*. Washington, DC: US Gov. Printing Off.
10. US Dept. Energy, Bonneville Power Admin. 1989. *Forecast of Electricity Use in the Pacific Northwest. DOE/BP-1262*. Aug.
11. Northwest Power Planning Counc. 1987. *Columbia River Basin Fish and Wildlife Program*. Portland, Ore.
12. Wilkinson, C. F., Conner, D. K. 1983. The law of the Pacific salmon fishery. *Kansas Law Rev.* 32:17–109
13. Hughes, T. P. 1983. *Networks of Power. Electrification in Western Society, 1880–1930*. Baltimore: Johns Hopkins Univ. Press
14. Chasan, D. J. 1985. *The Fall of the House of WPPSS*. Seattle, Wash: Sasquatch
15. Leigland, J., Lamb, R. 1986. *WPPSS: Who is to Blame for the WPPSS Disaster*. Cambridge, Mass: Ballinger
16. US Dept. Energy, Bonneville Power Admin. 1989. *1989 Financial Summary. DOE/BP-1318*. Dec.
17. Sugai, W. H. 1987. *Nuclear Power and Ratepayer Protest: The Washington Public Power Supply Crisis*. Boulder, Colo: Westview
18. Wilkinson, C. F. 1987. *American Indians, Time and the Law*. New Haven: Yale Univ. Press
19. Harrison, P. H. 1986. The evolution of a new comprehensive plan for managing Columbia River anadromous fish. *Environ. Law* 16:705–29
20. Cohen, F. G. 1986. *Treaties on Trial*. Seattle: Univ. Wash. Press
21. Blumm, M. C. 1990. Anadromous fish law—1979–90. *Anadromous Fish Law Memo*, issue 50, Aug.
22. Hemmingway, R. 1983. The Northwest Power Planning Council: Its origins and future role. *Environ. Law* 13:673–97
23. Evans, D. J., Hemmingway, L. H. 1984. Northwest power planning: Origins and strategies. *Northwest Environ. J.* 1:1–21
24. Wildavsky, A. 1979. *Speaking Truth to Power*. Boston: Little, Brown
25. McFarland, A. W. 1987. Interest groups and theories of power in America. *Br. J. Polit. Sci.* 17:129–47
26. Halbert, C. L., Lee, K. N. 1990. The Timber, Fish, and Wildlife Agreement: Implementing alternative dispute resolution in Washington State. *Northwest Environ. J.* 6:139–75
27. Northwest Power Planning Counc. 1983. *Northwest Conservation and Electric Power Plan*. Portland, Ore.
28. Northwest Power Planning Counc. 1986. *Northwest Conservation and Electric Power Plan*. Portland, Ore.
28a. Northwest Power Planning Counc. 1991. *Northwest Conservation and Electric Power Plan*. Portland, Ore.
29. Lee, K. N. 1983. The path along the ridge: Regional planning in the face of uncertainty. *Wash. Law Rev.* 58:317–42
30. Olson, M. 1971. *The Logic of Collective Action*. Cambridge, Mass: Harvard Univ. Press
31. Ruderman, H., Levine, M. D., McMahon, J. E. 1987. The behavior of the market for energy efficiency in residential appliances including heating and cooling equipment. *Energy J.* 8:101–24
32. *Seattle Master Builders Association et al vs Pacific Northwest Electric Power and Conservation Planning Council 1986*. 786 F.2nd 1359, decided 10 April, US Court of Appeals, 9th Circuit
33. Northwest Power Planning Council. 1989. *Assessment of Regional Progress toward Conservation Capability Building*. Issue paper 89-8. March 13
34. US Dept. Energy Bonneville Power Admin. 1987. *The Hood River Story*.

Vol 1, *How a Conservation Project Was Implemented*, ed. K. Schoch. *DOE/BP-11287-12*. Sept. Vol. 2, *Marketing a Conservation Project*, ed. S. Kaplon. *DOE/BP-11287-13*. Sept.

35. Hirst, E., Goeltz, R., Trumble, D. 1987. *Electricity Use and Savings*. Final report, Hood River Conservation Project. *ORNL/CON-231. DOE/BP-11287-16*

36. Philips, M., Khawaja, M., Engels, D., Peach, H. G. 1987. *Cost Analysis*. Final report, Hood River Conservation Project. *DOE/BP-11287-8*. April

37. Northwest Power Planning Counc. 1987. *A Review of Conservation Costs and Benefits. Five Years of Experience under the Northwest Power Act*. Oct. 1

38. Brewer, G. D. 1983. Some costs and consequences of large-scale social systems modeling. *Behav. Sci.* 28:166–85

39. US Gen. Account. Off. 1987. *Federal Electric Power: A Five-Year Status Report on the Pacific Northwest Power Act. RCED-87-6*. Feb.

40. US Dept. Energy, Bonneville Power Admin. 1990. Big savings from small sources. How conservation measures up. *Backgrounder*. March

41. Koomey, J. G. 1990. *Energy efficiency choices in new office buildings: An investigation of market failures and corrective policies*. Unpublished PhD thesis. Energy and Resources Group. Univ. Calif., Berkeley

42. Stipp, D. 1990. States use profit motive to induce utilities to conserve. *Asian Wall Street Journal*. Thurs., Nov. 8, p. 22

43. Northwest Power Planning Counc. 1990. *New Resources: Supply Curves and Environmental Effects*. Staff issue paper 90-1. Feb. 28

44. Schalk, R. 1986. Estimating salmon and steelhead usage in the Columbia Basin before 1850: the anthropological perspective. *Northwest Environ. J.* 2:1–30

45. Lawrence, J. 1983. *The Water Budget: a step towards balancing power and fish in the Columbia River Basin*. Unpublished MS thesis. Univ. Wash. Dept. Civil Eng.

46. Muckleston, K. W. 1990. Salmon vs. hydropower: Striking a balance in the Pacific Northwest. *Environment* Jan/Feb: 10–15, 32–36

47. Lothrop, R. C. 1986. The misplaced role of cost-benefit analysis in Columbia Basin fishery mitigation. *Environ. Law* 16:517–54

48. Lee, D. C., Kneese, A. V. 1989. Fish and hydropower vie for Columbia River waters. *Resources* No. 94, Winter

49. Mahar, D., Riley, M. 1985. Interview: Chuck Collins. *Northwest Energy News*. Jan/Feb:6–9

50. Lee, K. N., Lawrence, J. 1986. Adaptive management: Learning from the Columbia River Basin Fish and Wildlife Program. *Environ. Law* 16:431–60

51. Lee, K. N. 1989. The mighty Columbia: Experimenting with sustainability. *Environment* July/Aug: 6–11, 30–33

52. Holling, C. S., ed. 1978. *Adaptive Environmental Assessment and Management*. New York: Wiley

53. Larkin, P. A. 1974. Play It Again Sam—An essay on salmon enhancement. *J. Fisheries Res. Board Can.* 31:1433–56

54. Environ. Soc. Syst. Anal., Ltd. 1982. *Review and Evaluation of Adaptive Environmental Assessment and Management*. Ottawa: Environment Canada

55. Walters, C. 1986. *Adaptive Management of Renewable Resources*. New York: Macmillan

56. Orians, G. H. 1986. The place of science in environmental problem-solving. *Environment* 28:9 (Nov.) 12–17, 38–41

57. Lindblom, C. E. 1959. The science of 'muddling through.' *Public Admin. Rev.* 19 (Spring):79–88

58. Rivlin, A. M. 1971. *Systematic Thinking for Social Action*. Washington, DC: Brookings Inst.

59. Natl. Res. Counc. 1986. *Ecological Knowledge and Environmental Problem-Solving. Report of the Committee on the Applications of Ecological Theory to Environmental Problems*. Washington, DC: Natl. Acad. Press

59a. Lee, K. N. 1991. Rebuilding confidence: Salmon, science and law in the Columbia Basin. *Environ. Law* 21:In press

60. Northwest Power Planning Counc. 1990. *Yakima-Klickitat Fish Production Project*. Staff issue paper. June 1

61. Selznick, P. 1949. *TVA and the Grass Roots*. Berkeley: Univ. Calif Press

62. Mazmanian, D. A., Nienaber, J. 1979. *Can Organizations Change? Environmental Protection, Citizen Participation, and the Corps of Engineers*. Washington, DC: Brookings Inst.

63. Morone, J. G., Woodhouse, E. J. 1989. *The Demise of Nuclear Energy?* New Haven: Yale Univ. Press

64. Cyert, R. M., March, J. G. 1963. *A Behavioral Theory of the Firm*. Englewood Cliffs, NJ: Prentice-Hall

Annu. Rev. Energy Environ. 1991. 16:365–77

ENERGY IN POLAND: PRESENT STATE AND FUTURE OUTLOOK

Z. A. Nowak

Central Mining Institute, Katowice, Poland

KEY WORDS: coal, energy intensity, Polish economy, pollution.

BACKGROUND

The history of Poland in this century has been very dramatic both politically and economically beginning with the regaining of independence in 1918 after 140 years of partition.

World War II also had a dramatic impact on Poland, both in large population losses and heavy damages. The National Product dropped and did not recover to its 1938 level until the late 1950s. The energy sector was also damaged badly. However, the energy sector was actually a help in rebuilding the country, because of the abundance of Polish coal, the only domestic energy source.

Poland rebuilt and strengthened its energy sector following the War, starting from having per capita energy consumption less than one-third of today's and no country-wide electricity system, and developing the sector into a source of strength that supplied the country both with power and with foreign currency from exported coal. However, over time, the bituminous coal and lignite that account for more than 80% of the total primary energy supply have also caused some problems, particularly to the environment.

It is important to understand that in the centrally planned socialist economies of Central Eastern Europe energy supply was built into the social system with the same priority as food supply. Energy has become a political rather than economic category, and low energy prices a common practice. Energy consumption is thus relatively high compared with Gross Domestic Product

365

1056-3466/91/1022-0365$02.00

(GDP) as can be seen from Figure 1. The figure shows energy consumption and GDP for several countries: a comparison of Poland and Czechoslovakia, adjusted to account for discrepancies connected with GDP estimation for centrally planned economies, yields the conclusion that Poland consumes 2–3 times more energy related to GDP.

Changing attitudes about the value and reasonable prices for energy is one of the most difficult tasks in restructuring the economy and in initiating an appropriate energy policy. Other main challenges for Poland's energy economy in the near future include: diversifying the fuel supply, increasing electricity use, and improving the environment, which is strongly related to a serious energy conservation program.

The author has gathered the results of discussions and planning meetings carried out in his country and abroad, and presents here his views concerning the present state and future outlook for energy in Poland.

STATE AND STRUCTURE OF THE ECONOMY

The Polish economy has been in a state of crisis for almost 15 years, with its deepest decrease in National Income, of about 20%, in 1990. The annual

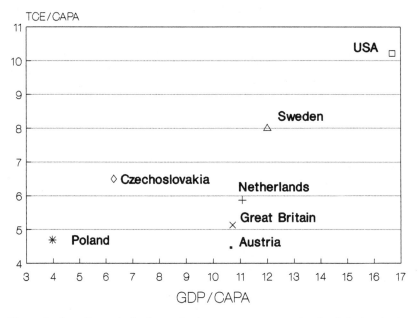

Figure 1 Gross Domestic Product and primary energy consumption for Poland and other countries—1985. Abbreviations used: GDP, Gross Domestic Product in US$ (based on purchasing power parity); TCE, tons of coal equivalent; /capa, per inhabitant and year.

changes of the Net Material Product for the past two decades are presented in Figure 2.

Among the main factors responsible for the poor state of the economy are:

1. improper structure and functioning of the economy,
2. high energy and raw materials consumption,
3. very low participation in international trade and cooperation,
4. poor state of the environment with critical environmental loads exceeded in some regions.

A more detailed illustration of the structure of the National Product is given in Figure 3, which shows the estimated shares of sectors in the generation of the GDP. A very important feature that is not shown in Figure 3 is the high share of energy-intensive production such as steel manufacturing and minerals mining. One of the future outlooks has been included in the figure to show that planners have not taken a very progressive approach in shaping the future shares of various activities in the economy.

One can also follow the various periods of the economy according to Figure 2. The early 1970s was a period of high investment, including of foreign capital, and was the first time the economy was really open to the Western

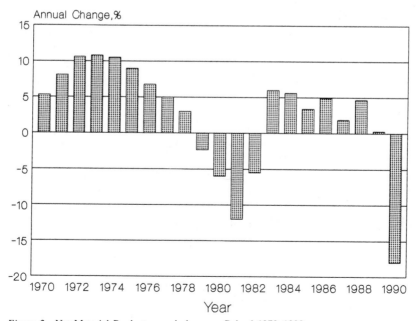

Figure 2 Net Material Product annual changes—Poland 1970–1990.

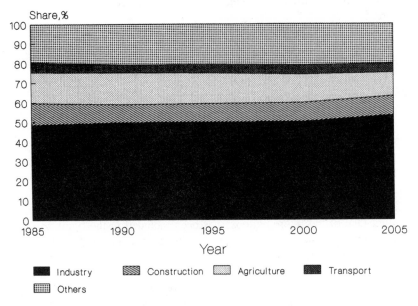

Figure 3 Structure of National Product—present state and possible future outlook. Y-axis:share of various sectors in the National Product.

market. The first symptoms of the approach of a deep crisis—both economic and social—can be seen already in the period from the second half of the 1970s to the strikes of 1980 and imposition of martial law in 1981. Efforts to revive the economy during the 1980s can also be seen, but with only minor results achieved because of the unfavorable political situation. The "Political Round Table" discussion of 1989 with all political groups led to the start of reforms but unfortunately also to a period of hyperinflation.

It is worthwhile to show the role of the energy sector in shaping the economic picture of Poland. Energy production—almost exclusively of coal—has always played a significant role in the economy, covering the majority of internal demand and also being a source of hard currency through export. Unfortunately, as can be seen from Figure 4, the negative balances occurred in the most difficult periods for the economy. During the years 1975–1978, when the gap between supply and demand increased dangerously, there was also not yet a proper understanding of or the means for an energy conservation policy.

ENERGY SUPPLY/DEMAND AND ENERGY INTENSITY

Energy supply and consumption for 1988—the recent normal year in Poland's economy (see Figure 2)—are shown in Figure 5. Primary energy—the sup-

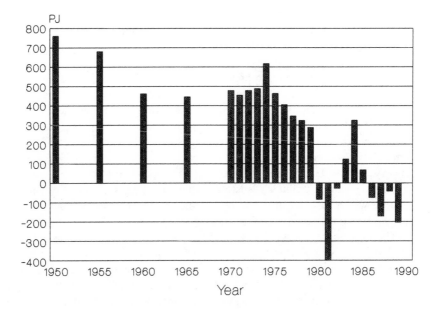

Figure 4   Energy import/export balance of Poland. Abbreviation used: PJ, petajoules.

ply—was 80% solid fuels and totaled about 5450 PJ. This was converted with an efficiency of about 65% to a total of 3530 PJ of final energy, which was characterized by

1. its very high solid fuel share—about ⅓—resulting in poor energy service,
2. a very low share of high-value energy sources such as electricity, gas, and oil products, and
3. a relatively good share of cogeneration in heat consumption.

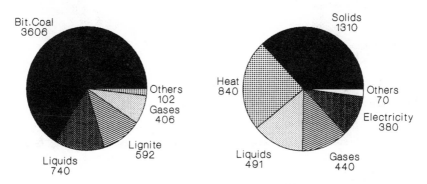

Figure 5   Primary and final energy consumption of Poland—1988. *Left:* Primary energy = 5450 PJ. *Right:* Final energy = 3530 PJ.

The energy structure shown in Figure 5 combined with the income structure presented in Figure 3 justify to a certain extent the poor position of Poland on the energy/economy graph in Figure 1. Figure 6 shows the actual fuel structures for various sectors of the economy in 1988.

The existing fuel structure presented in Figures 5 and 6 is a result of the historical development of the energy supply, which is shown in Figure 7. For the first two decades after World War II, the main goals of the energy sector were rebuilding and development of heavy industry and providing electricity based on local fuel, with less attention paid to the development of road transport. Coal, and later a small amount of gas, were the only domestic energy sources. During the 1970s, the international energy crisis and growing internal economic problems led to an autarchic energy/raw material supply policy, which resulted in a low rate of oil and gas imports—about 20% in 1989—for some years.

The only energy sector developing dynamically was electricity generation as shown in Figure 8. Power generation has been based almost exclusively on bituminous coal (60%) and lignite (37%). The average growth factor of more than 6% per year for 1960–1989 is higher than the GDP growth factor. Per capita consumption is low, however, compared with other countries, as can be seen from Figure 9. This figure is presented mainly to show the dramatically low oil and gas consumption. Comparison with the Netherlands shows how great the changes would have to be to restructure the economy towards a

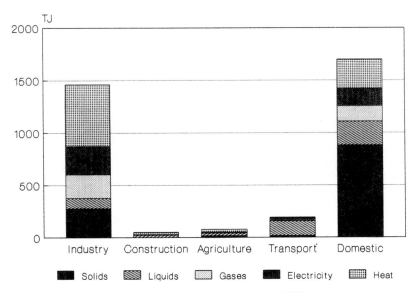

Figure 6   Consumption of final energy by economic sectors—1988.

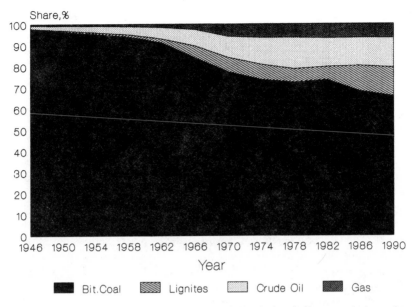

Figure 7 Long-term fuel supply structure of Poland. Y-axis:share in energy equivalent units.

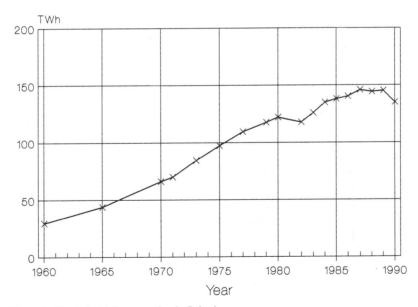

Figure 8 Total electricity generation in Poland.

sustainable development philosophy with the proper respect for environmental issues.

Economic growth and energy consumption for the past 30 years are graphed in the left portion of Figure 10. The economic growth over the first two decades does not look bad at first glance, but it is obvious now that the increasing gap between energy use and economic growth in the decade 1970–1980 was responsible for the crisis at the end of the 1970s. The energy/economy crisis was superimposed on the political crisis, and its influence is still being felt. One measure of the depth of the economic crisis is the fact that the GDP level of 1978 has still not been regained.

Looking at the primary energy picture it is interesting to note that the only growth is in the domestic sector with very little change in the other setors. The decrease of energy consumption at the end of the 1980s is mainly a result of the low economic activity of the country, but is also due to some energy-conservation measures taken after the hard lessons learned in the previous period. The GDP curve shows the various development periods described in the previous section.

The aggregate energy intensity resulting from the yearly data presented in Figure 10 decreased in the first two decades, because energy growth was lower than GDP growth in the 1960s and much lower in the 1970s. The changes since 1970 are shown in Figure 11 relative to the base year 1970. The figure shows the recent dramatic periods of the economy with no return to the lowest point of 1977.

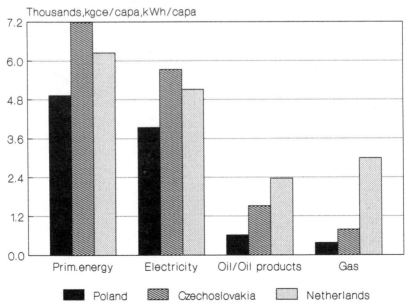

Figure 9 High-quality energy sources in final energy consumption.

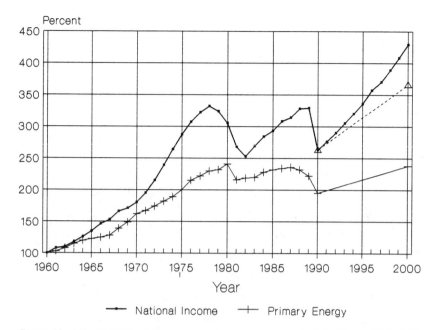

Figure 10 Poland's National Income and primary energy growth. *Left:* History 1960–1990. *Right:* Outlook 1990–2000.

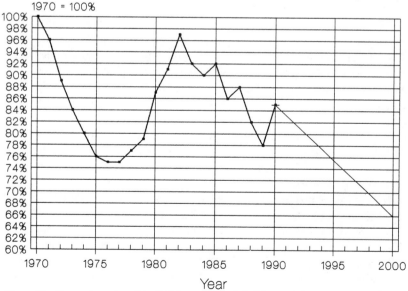

Figure 11 Energy intensity of the Polish economy. *Left:* History 1970–1990. *Right:* Outlook 1990–2000.

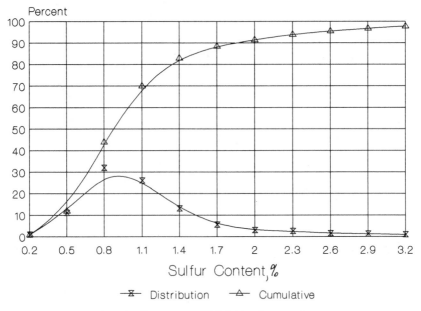

Figure 12   Distribution of sulfur content of Polish bituminous coal reserves.

ENVIRONMENTAL IMPACTS OF THE ENERGY SECTOR

The energy sector, mainly because of its fuel structure (Figure 7), has been the largest contributor to pollution in Poland. The main pollutants are solid waste from coal mines and fly ash disposed on land, salted mine water discharged into the river, dust, sulfur and nitrogen oxides emissions from power plants, and other pollutants connected with a coal-based energy system. Some of the pollutants and their quantities are presented in the following figures.

Figure 12 shows the distribution of sulfur content in Polish bituminous coal. Polish coal is rather low in sulfur compared with other coal, but the large quantities consumed lead to high pollution. About 70% of the total reserves of coal contains less than 1% sulfur.

The environmental impacts of the Polish energy sector should be analyzed on a background of the main polluters in central Europe. The very large Polish share is only part of the serious problem. Three quantities—the amount of emission from a particular country, the amount of pollutants deposited on a country's own territory, and the total deposition in a country resulting from both imported and its own emissions—have been presented in Figures 13 and 14. Both for sulfur oxides and for nitrogen oxides the main conclusion is that only collective international action can improve the very bad environmental

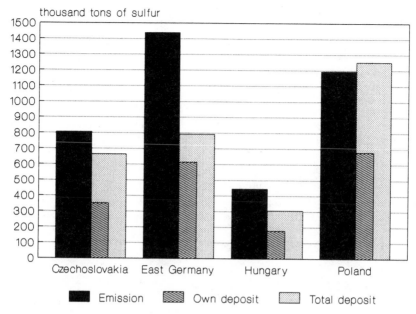

Figure 13   Sulfur emissions and deposits in Central Europe.

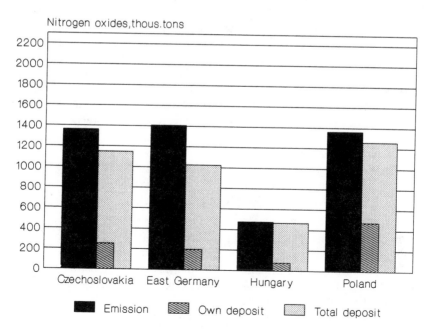

Figure 14   Nitrogen oxides emissions and deposits in Central Europe.

situation in this region, because of the high level of imported and exported emissions of the pollutants among the four neighboring countries.

Evaluation of emissions of carbon dioxide, a "greenhouse gas" responsible for global warming, is much more difficult. On the basis of the primary energy mix a value of about 120 million tons of carbon per year is estimated. The intended changes in fuel structure and the very large potential for energy saving can over the long term greatly reduce this number. However, in the near future, carbon emissions may remain high if production is built up and efficiency improvements in the energy sector are not rapid enough because of financial and technical constraints.

Great efforts are being made today to improve the environmental situation through preventive and clean-up measures. Among these efforts, the coal cleaning/desulfurization program, which is part of the National Environment Protection Program, should be mentioned.

ENERGY CONSERVATION AND FUTURE ENERGY OUTLOOK

The state of the economy today in Poland and the ongoing political, economic, and social changes make for great uncertainty about the future of the economy, including the energy supply and demand picture. Energy forecasting is very difficult if at all possible today in Poland.

One thing that is certain is the existence of great potential for energy savings. Preconditions to implementing such conservation are the following.

1. Stronger administrative and economic measures must be taken, using a pro-active incentive approach based on enterprise and personal self-interest.
2. A clear energy pricing policy related to the cost structure of the whole economy must be implemented very quickly, to give proper long-term price signals both for internal and for foreign investment in energy efficiency and industrial restructuring.
3. Technology transfer must be organized and supported by an active regulatory and financial policy enabling flow of capital and technology, mainly on the basis of joint ventures utilizing the great energy-saving potential in Polish industry.

Energy-efficiency programming must be given higher priority than energy supply planning in order to reach the goal of reviving the economy and achieving a reasonable economic growth rate very soon without falling again into the trap of rising energy use. An Energy Conservation Agency is being organized in 1991 to control energy policy implementation. This agency could provide hope for development without energy consumption exceeding the top level reached in the 1980s as has been shown in the right part of Figure 10 as one of the optimistic outlooks for the near future. Figure 10 shows two possible growth rates of the National Product; the optimistic one of 5% per year and a realistic value of 3.5%.

The energy intensity outlook for 1990–2000 shown in Figure 11 assumes the higher GDP growth rate. Even though the result is not spectacular—only 12% below the 1977 level—to reach this result is a great challenge not possible without considerable foreign cooperation. The benefits from energy saving are a good offer for this cooperation, needed for the following steps:

1. implementation of the coal cleaning program as the first and cheapest measure for efficiency and environmental improvement,
2. replacement of energy-intensive equipment in industry,
3. reducing electricity transmission losses, which are twice as high as in developed countries,
4. development of cogeneration in urban areas, and
5. application on a wider scale of insulation in existing and new buildings.

The total potential saving level is of the order of one-third of today's consumption. This could be achieved in a period of 10–15 years provided a very active energy policy is implemented, parallel to the restructure of the industrial sector to increase the share of less energy-intensive subsectors. Institutional, financial, and technical measures will be required.

On the supply side the constraints are more serious. For the time being, coal is Poland's only dependable, proven domestic energy resource. All other fuels must be imported. The main options for Poland are:

1. the clean coal option, based on the implementation of clean coal conversion technology.
2. the nuclear energy option, the social and political difficulties of which have already led to the closure (while under construction) of the only nuclear plant in Poland, and
3. the clean energy import option, based mainly on gas and oil imports.

The most attractive option from the standpoints of efficiency and the environment is natural gas, which if imported to Poland could lead to significant improvements in both efficiency and environment. A return to the nuclear option could be considered in light of the improved safety factor of the future generation of reactors. Both the gas and nuclear options would require heavy imports of fuel and technology.

The clean coal option is ongoing and based on Poland's own fuel. This option requires a positive cost structure and a well-developed infrastructure in industry and research.

This dilemma cannot be solved today. The solution will depend, among other factors, on the efficiency of the future Polish economy and its place in the world market economy.

Annu. Rev. Energy Environ. 1991. 16:379–400
Copyright © 1991 by Annual Reviews Inc. All rights reserved

POLICIES TO INCREASE US OIL PRODUCTION: LIKELY TO FAIL, DAMAGE THE ECONOMY, AND DAMAGE THE ENVIRONMENT

Robert K. Kaufmann and Cutler J. Cleveland

Department of Geography, Center for Energy and Environmental Studies, Boston University, 675 Commonwealth Avenue, Boston, Massachusetts 02215

KEY WORDS: yield per effort, economic efficiency, environmental impacts of oil production, oil policy, resource scarcity.

Abstract

This analysis evaluates the costs and benefits of a policy that seeks to reduce US dependence on imported oil by increasing domestic supply. Statistical analyses of oil discoveries, additions to proved reserves, and rates of production are used to evaluate the benefits: increased domestic supply. The results indicate that high rates of exploratory or developmental drilling will not lead the oil industry to discover or develop oil in significant quantities and that higher prices will not reverse the current decline in production. The costs incurred by the failure of such policies will be exacerbated by negative effects on the economy and the environment. Previous increases in drilling effort directed large amounts of capital to the oil industry but did not increase the industry's contribution to national economic production. Policy that increases effort could duplicate such inefficiencies. Finally, attempts to increase US production will speed global warming by increasing the amount of carbon dioxide released to produce a barrel of oil.

Introduction

The Iraqi invasion of Kuwait, uncertainty regarding supply, and the subsequent rise in crude oil prices returns US dependence on imported oil to the top

379

1056-3466/91/1022-0379$02.00

of policy debate. The return to prominence of energy security comes when US dependence on imports is at an all-time high and the US balance of trade is near an all-time low. During the first six months of 1990, the United States imported 50% of its oil. If this level continues, US import dependence will eclipse the previous mark of 47% in 1977. The high level of dependence is responsible for the largest component of the US trade deficit. In 1989, the energy-import bill accounted for $42 billion of red ink out of a total of $109.4 billion and each dollar increase in the price per barrel raises the US trade deficit another 3.3 billion dollars.

The US and world economies at risk due to political upheaval in the Middle East raises the specter of a threat recently thought to be vanquished. From 1977 to 1985, oil imports declined from 47 to 27% of US oil consumption. This decline was achieved by a series of demand and supply responses to the first two oil shocks. Oil demand declined 14.7% due to recession, fuel switching, and improved efficiency. US oil production increased 8.8% from 1977 to 1985. The collapse of oil prices in 1986 reversed both of these trends and redirected the United States down the road of greater dependence on imports. Lower oil prices increased domestic demand 9.7% from 1985 to 1989. On the supply side, oil production declined 9% from 1985 to 1989, due in large part to a decline in output from the lower 48 states. The rapid reversal in US dependence on imported oil indicates that some of the attempts to wean the economy from imported oil were temporary or illusory.

The two causes of increased dependence on imports since 1985, higher demand and lower supply, suggest two strategies for reducing the nation's dependence on imported oil: cutting oil demand or increasing domestic supply. Many analysts argue that oil demand can be reduced using current technology with little effect on life-styles (1). Automobiles that travel more than 50 miles per gallon can be purchased from several manufacturers. This level is more than double the fuel efficiency of the average automobile sold in 1989 and is nearly triple the fuel efficiency of the US auto fleet. Similarly, there are a host of fuel-efficient models of household appliances such as refrigerators and air conditioners that can reduce significantly the amount of energy used to keep food and people cool. Nevertheless, the availability of technical fixes or behavioral changes that reduce demand for oil does not guarantee that cutting demand will be the policy instrument by which the United States attempts to reduce oil imports. Nearly 20 years of experience indicates that there are significant roadblocks to the penetration of oil-efficient technologies. Consumers are sometimes reluctant to purchase fuel-efficient models, even when the expectation that energy prices would remain high offered large money savings over the lifetimes of use (2). Furthermore, it does not appear likely that politicians will exert the effort needed to correct such roadblocks. For example, President Carter's praise for conservation and his efforts to give it an equal chance in the market were derided by critics and won little support among voters. Conversely, the Reagan and Bush administrations successfully fought attempts to mandate improvements in the fuel efficiency of new motor vehicles.

Increasing domestic oil production is another strategy for reducing US dependence on imports. Despite the decline in production, many analysts argue that there are substantial undeveloped oil resources to be tapped within our own borders. According to this line of thought, increased exploration and developments of the domestic resource base will increase production and reduce imports. Based on this logic, proposals to ease the tax burden and environmental regulations on the oil industry now appear more frequently. In his address to Congress concerning the Iraqi invasion of Kuwait, President Bush called for "initiatives to encourage more domestic drilling." This rhetoric is supported by proposals to ease the oil industry's tax burden by several billion dollars. Such proposals probably will find support in both houses of Congress, where political action committees that represent the major oil companies and the independents donate large sums to politicians from both parties.

Despite the political power behind it, there is good reason to believe that policy to increase conventional exploration and development by the US oil industry will not increase production significantly, and therefore will not reduce significantly US dependence on imported oil. Historical analyses of the factors that determine rates of discovery, development, and production indicate that the US oil resource base is depleted to the point at which a large increase in effort would produce little extra oil. The potential for drilling to augment proved oil reserves through the discovery of new oil fields and the development of existing fields is limited. The costs of discovering, developing, and producing oil from domestic fields have been increasing steadily for decades. As a result, a large and sustained price increase is needed to reverse the ongoing decline. Furthermore, a policy aimed at reducing US dependence on imports by increasing domestic production will inflict high costs on the economy and the environment. Trying to squeeze more oil from a waning resource base siphons large amounts of investment capital from other sectors without a commensurate increase in the oil industry's contribution to the nation's economic output. Increasing production from a depleted resource base increases the amount of carbon dioxide and other pollutants that are released per unit of oil produced, which will speed global warming and exacerbate other environmental problems. Opening the Arctic National Wildlife Refuge for exploration and reversing President Bush's recent moratorium on offshore drilling puts some of the nation's most important and sensitive ecosystems at risk for at most a few years' additional supply of oil.

This paper describes the small chance for success of a policy that seeks to reduce US dependence on imports by increasing domestic production, and the large economic and environmental costs associated with such an effort. We do so by presenting empirical analyses that document the rising costs of finding and producing oil from a depleted resource base, and some of the important economic and environmental costs associated with attempts to boost production.

Increasing Effort Will Not Reverse the Decline in US Production

Oil pumped from the ground depends on the successful completion of several steps. Exploratory drilling identifies new oil fields and new reservoirs in existing fields. Developmental drilling delineates the exact size of fields and increases the capacity to produce oil. Finally, geological, economic, political, and technical conditions must permit operators to pump oil from the ground. The fraction of national economic resources directed to the domestic oil industry and how that effort is apportioned among exploration, development, and production depends on the return on investment (ROI). The ROI to the oil industry and the ROI associated with each of the steps does not depend only on the amount of oil discovered, developed, or produced. The economic conditions under which firms operate also influence the ROI. The government can alter the tax code and increase the ROI to the oil sector relative to non-oil sectors. Such a policy will redirect investment towards the oil industry. But the ultimate national goal is to increase production of oil, not economic activity by the industry. This goal defines a criterion by which the benefits of such a policy can be measured: the amount of oil discovered, added to proved reserves, or produced that is generated by the increased effort. This criterion suggests that the analysis of policy that aims to increase oil supplies must look beyond the effect on effort expended by the industry and must evaluate the degree to which increased effort by the oil industry increases the rate at which the industry discovers, develops, and produces oil.

The ability of the United States to reduce oil imports by increasing domestic discovery and production depends on the level of resource depletion and the degree to which increased effort causes diminishing returns. In the long run, the amount of oil discovered, developed, or produced per unit of effort depends on the relative strengths of two opposing forces: technical change and resource depletion. Technical change reduces the effort that is required to discover, develop, and produce oil, whereas resource depletion increases effort. If cost-reducing technical improvements dominate the effects of cost-increasing resource depletion, policies aimed at increasing effort by the domestic oil industry can reduce US dependence on imported oil with little costs to the economy and environment. On the other hand, if the cost-increasing effects of resource depletion dominate technical improvements, policies aimed at increasing effort by the domestic oil industry will not increase significantly US production and will inflict high costs on the economy and environment. There is a substantial and consistent body of empirical evidence that indicates that the cost-increasing effects of resource depletion have outweighed the cost-decreasing effects of technical change for several decades in every step of the oil supply process in the lower 48 states. Furthermore, there is no evidence that the relative strengths of these two forces will reverse anytime soon.

Even if the long-run balance between resource depletion and technical change does not deteriorate further, the efficacy of policy to increase production by increasing effort is limited by diminishing returns to effort. In the short run, the amount of oil discovered, developed, or produced per unit of effort declines as the rate of effort increases. Like all other sectors, the oil industry ranks investment opportunities in order of descending ROI. Increasing effort at a point in time forces the industry to explore formations with a smaller chance of containing oil and with smaller expected field size, and to develop fields with higher costs. As a result, there is an inverse relation between effort and the rate of return. Although tax policy can increase the ROI to higher effort and thereby lower the economic barriers to drilling marginal prospects, the empirical analyses described below indicate that increasing the rate of effort reduces significantly the quantity of oil discovered or developed per foot of well drilled. Such declines limit the viability of policies aimed at increasing domestic oil supply by increasing effort. These two factors, the predominance of resource depletion relative to technical change in the long run, and diminishing returns to the rate of effort in the short run, imply that policies that succeed in increasing effort by the oil industry will not increase substantially the supply of oil.

THE DISCOVERY OF NEW FIELDS   The discovery of new oil fields by exploratory drilling is the first step in the oil supply process. The ROI of drilling exploratory wells can be enhanced by factors such as higher oil prices and tax changes that accelerate the rate at which operators can depreciate oil-field equipment. The feasibility of policies that speed exploratory drilling depends, in part, on their costs and benefits. The social costs of the policy include the economic incentives needed to boost exploratory drilling such as accelerated tax write-offs and the benefits include the amount of oil discovered by the increased rate of drilling. The social costs of increasing drilling effort probably are quite high. The real cost of discovering a barrel of oil reached an all time high in the 1980s (3). Combined with low prices prior to the invasion of Kuwait, the high costs of discovery imply that the ROI of new-field wildcat exploratory drilling is low. The paucity of prospects with the potential for a ROI competitive with other investment opportunities probably is one reason why the rate of exploratory drilling in 1989 was at its lowest level since World War II. But even if policy generates economic incentives sufficient to spark a resurgence in exploratory drilling, the benefits of such a resurgence are likely to be small. An analysis by Cleveland & Kaufmann (4) indicates that resource depletion and diminishing returns to effort limit the quantity of oil discovered by an increase in exploratory drilling.

The ability of increased exploratory drilling to increase discoveries can be evaluated by quantifying the factors that determine the yield per effort (YPE) for exploration, which is the quantity of oil discovered per foot of exploratory well drilled. Three factors account for most of the variation in YPE for exploration in the lower 48 states: the cumulative footage of exploratory wells drilled, the annual rate of exploratory drilling, and real oil prices.[1] Cumulative drilling is a proxy for the net effect of resource depletion and technical advance on YPE (5, 6). We find that the steady decline in YPE since the late 1920s or early 1930s in the lower 48 states is due to the dominance of resource depletion over technical change (Figure 1). The overall decline in YPE however, is not monotonic. Periods of relative stability or increase in YPE are associated with systematic changes in yearly rates of drilling effort and real oil prices. If not properly accounted for, these short-run factors can mask temporarily the decline associated with depletion, and generate incomplete and/or misleading projections concerning oil discovery. Woods (7, 8) and Fisher (9–11) overlook changes in short-run factors and argue incorrectly that resource depletion slows or technical change accelerates during periods when short-run factors obscure the long-run effects of depletion. They conclude that the amount of oil awaiting discovery is greater than that predicted by a monotonic decline in YPE. For example, Woods argues that YPE stabilizes in the late 1960s and early 1970s. The stable rates of YPE during this period are due mainly to very low rates of drilling and to a lesser degree to technical improvements such as digital seismic exploration techniques. During periods of low drilling effort the industry drills only those prospects thought to have the highest ROI, and this temporarily increases YPE. Similarly, the large

---

[1]Cleveland & Kaufmann (4) estimate the following model for the lower 48 states with data from 1930 to 1988:

$$\ln (YPE_t) = 256.98 - 0.001 \, (h) + 0.057 \, (p) - 0.032 \, (d)$$
$$(40.1) \quad (-10.0) \quad (4.0) \quad (-5.1)$$

in which YPE is yield per effort for exploration (barrels per foot), h is cumulative exploratory drilling (million feet), p is the real wellhead price of oil (1982$/barrel), and d is the rate of exploratory drilling (million feet per year). The numbers in parentheses are t statistics. The adjusted R^2 of the equation is 0.894 and the Durbin Watson statistic is 1.6.

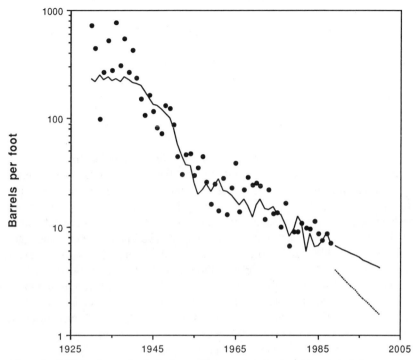

Figure 1   YPE for discoveries in the lower 48 states. Actual values (closed circles), regression model results from Cleveland & Kaufmann (4) (solid line), forecast for YPE based on Cleveland & Kaufmann holding effort and real prices at their 1989 levels (solid line), and forecast for YPE with a doubling of effort and real prices (dotted line).

increases in real oil prices account for the relative stability of YPE from the late 1970s through the mid-1980s. High oil prices make small fields economical, which allows producers to "discover" fields by completing wells that would have been considered dry holes in a low-price environment (12).

The model of YPE based on resource depletion, annual rates of drilling effort, and real oil prices can be used to project the amount of oil likely to be found by stimulating exploratory drilling effort. The historical relations among those factors suggest that a policy that increases drilling effort would discover little additional oil because the negative effects of high rates of drilling and continued resource depletion outweigh the positive effects of higher oil prices on YPE. If annual rates of exploratory drilling and real oil prices are held at their 1988 levels, a total of 2.7 billion barrels of oil will be found through 2000[2] (Figure 1). If the government implements policy that doubles effort by improving the economic conditions for exploration and improves the economic attractiveness of the oil discovered by doubling real oil prices to producers, a total of 4.8 billion barrels will be discovered.[2]

The model for exploratory YPE indicates that resource depletion and diminishing returns limit discoveries to amounts that are tiny relative to current rates of production and consumption. In 1989, rates of domestic crude oil production and consumption were 2.8 and 6.3 billion barrels per year, respectively. The small amounts of oil awaiting discovery in the lower 48 states cannot increase supply significantly, regardless of the rate at which exploratory wells are drilled.

THE DEVELOPMENT OF EXISTING FIELDS   New discoveries may not be a viable means for increasing US oil supplies because the industry may have discovered all of the large fields, but more intensive development of existing fields may boost oil supplies by enlarging the amount of oil recovered from fields that were discovered many years ago. Proved oil reserves are defined as "volumes of crude oil which geological and engineering information indicate, beyond reasonable doubt, to be recoverable in the future from an oil reservoir under economic and operating conditions . . ." (13). Because the phrase "beyond a reasonable doubt" limits reserves to volumes proved by the drill, new-field wildcat discoveries are estimated conservatively and therefore account for a small fraction of the oil ultimately added to proved reserves from a newly discovered field. Developmental drilling and certain types of exploratory drilling in years subsequent to discovery usually increase the initial estimate for the size of productive acreage, which increases proved reserves. Reserve additions resulting from such drilling are called extensions. Developmental drilling, production history, and the installation of improved recovery equipment can lead to upward revisions of recovery efficiencies within existing reservoirs. Reserve additions resulting from increased recovery factors are called revisions. Revisions and extensions increase the quantity of oil ultimately recovered from fields 3.2 to 9.4 times relative to the estimate made in the year of discovery (14).

Reserve growth in existing fields provides another strategy for boosting US production: increase the economic incentives to drill wells in existing fields or to adopt improved recovery techniques. This strategy would increase additions to reserves by increasing revisions, new pools in oil fields, and extensions. As for exploration, the feasibility of such a strategy depends on the size of the economic incentives needed to boost developmental drilling and the amount of oil added to proved reserves as a result of that effort. We find that the probable scenario for the future of developmental drilling parallels the

[2]The quantity of oil awaiting discovery is calculated using the model described in Note 1. The quantity of oil discovered in each year from 1990 to 2000 is projected by calculating the annual YPE for exploration and multiplying YPE by total exploratory footage drilled.

one described for exploratory drilling. Large economic incentives probably will be needed to spur a significant increase in developmental drilling because the real cost of developing a barrel rose to record levels in the 1980s (3). Combined with low prices, high costs of development imply that the profitability of developmental drilling is low. A paucity of prospects with ROIs competitive with other investment opportunities probably is one reason that current rates of drilling are near their post–World War II lows. Even if sufficient economic incentives are provided, an analysis by Cleveland & Pendleton (15) indicates that increasing developmental drilling will not increase significantly additions to proved reserves.

The effectiveness of a program to add oil to proved reserves by developmental drilling can be evaluated by the YPE for development. YPE for development is the ratio of total annual additions to oil reserves to the total oil footage drilled. Revisions and extensions dominate the numerator of this measure of YPE. Four factors account for most of the variation in the amount of oil added to proved reserves per foot of oil well drilled: the cumulative amount of oil footage drilled, the annual rate of drilling, the fraction of total reserve additions accounted for by revisions, and the ratio of new-field discoveries to the sum of new-field discoveries, new pools in old fields, and extensions.[3] Cleveland & Pendleton (15) find that resource depletion, as measured by cumulative drilling, is associated with a significant overall decline in the YPE for development since 1946 in the lower 48 states. As is the case for the YPE for new discoveries, short-run factors temporarily drive the YPE for development above or below the long-run trend. For example, Cleveland & Pendleton find that the increase in YPE in the 1960s is due to a decline in the drilling rate, an increase in the importance of revisions and new-field discoveries, and to a lesser degree due to the commercialization of digital seismic exploration technologies. They find no evidence to support the argument that the increase in YPE during the 1960s is either permanent or due primarily to the expansion of activity in the Gulf of Mexico (7, 16).

The efficacy of a policy to boost development effort can be evaluated by extrapolating the YPE for development based on a statistical analysis of the

---

[3]Cleveland & Pendleton (15) estimate the following model for the lower 48 states (onshore and offshore) with data from 1946 to 1988:

$$\ln (YPE_t) = 3.0 \quad - 0.003\ (h) \quad - 0.023\ (d) \quad + 1.9\ (r) \quad + 0.71\ (s)$$
$$ (18.7) \quad (-11.4) \quad (-3.5) \quad\quad (7.4) \quad\quad (2.2)$$

in which YPE is yield per effort for developmental drilling (barrels per foot), h is cumulative drilling (feet), d is the rate of oil exploration and developmental drilling (feet per year), r is the ratio of annual net revisions to total annual additions to oil reserves, and s is the ratio of annual new-field discoveries to the sum of new-field discoveries, new pools in old fields, and extensions. The numbers in parentheses are t statistics. The adjusted R^2 of the equation is 0.91 and the Durbin Watson statistic is 2.2.

historical record. As in exploration, the statistical analysis of the historical record for YPE suggests that the negative effects of resource depletion, diminishing returns to annual rates of effort, and the decreasing fraction of new-field discoveries outweigh the positive effects of the increasing importance of revisions. A policy that stimulates a large increase in developmental drilling is not likely to increase additions to proved reserves significantly. If annual rates of drilling, the importance of new-field discoveries, and the fraction of additions to proved reserves from revisions are held at their 1988 levels, 10.9 billion barrels of oil will be added to proved reserves in the 11 years from 1990 to 2000 (Figure 2). If annual rates of drilling are doubled and the importance of new-field discoveries and the contribution of revisions are held constant at their all-time highs, about 18.4 billion barrels will be added to proved reserves.

The model for development YPE indicates that resource depletion and diminishing returns limit additions to proved reserves to rates well below

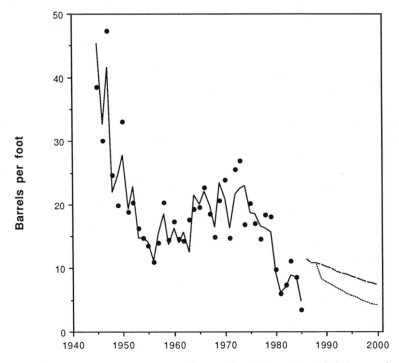

Figure 2   YPE for development in the lower 48 states. Actual values (closed circles), regression model results from Cleveland & Pendleton (15) (solid line), forecast for YPE based on Cleveland & Pendleton holding effort at 1986 levels (dashed line), and forecast for YPE with a doubling of effort (dashed line). The sharp rise between 1985 (last year of available data) and 1986 (the first year of the forecast) is associated with the large decline in drilling effort that accompanied the collapse in oil prices.

those for production and consumption. Under either of the previous scenarios for developmental drilling over the next decade, the low rates of additions to proved reserves relative to production imply a sharp reduction in proved oil reserves. If the current rate of oil production in the lower 48 states remains constant, an unrealistic assumption for reasons discussed below, 22.9 billion barrels of oil will be produced from 1990 to 2000. This result implies that additions to proved reserves will account for only 48% of production and that proved reserves in the lower 48 states will fall 12 billion barrels. This reduction represents 61% of the 19.8 billion barrels in proved reserves that existed in the lower 48 states as of December 31, 1989. Diminishing returns to annual rates of effort limit the effectiveness of policies that seek to avoid the decline in proved reserves by doubling drilling effort and increasing the other two determinants of YPE. Under these conditions, a 100% increase in drilling effort increases additions 69%, which represents 7.5 billion barrels. Thus, there is little reason to believe that increasing drilling will avoid a sharp decline in proved reserves.

PRODUCTION   The rapid rise in US dependence on imported oil is associated with a 20% production decline in the lower 48 states from 1985 to 1989. This suggests a third avenue of reducing import dependence: implement policies that increase the economic attractiveness of producing oil. As with policies designed to stimulate exploration and development, the feasibility of encouraging production depends on the cost of the economic incentives and the degree to which such incentives boost production. The historical record of oil production in the lower 48 states indicates that the efficacy of standard economic incentives to boost production, such as higher oil prices, is unclear. Seeming to defy standard economic principles, oil production in the lower 48 states moves in the direction opposite of real oil prices in the post–World War II period (Figure 3). Real oil prices declined slightly between 1947 and 1970, but production in the lower 48 states nearly doubled. Conversely, real oil prices tripled between 1970 and 1985, but production fell about 20%.

Oil production in the lower 48 states moves in the direction opposite real oil prices because annual rates of production are determined simultaneously by the effects of resource depletion, technology, and economics, not a simple relation between price and production. We find that attempts to boost production in the lower 48 states by increasing economic incentives may fall on the same sword as attempts to boost discoveries and additions to proved reserves. Empirical analyses indicate that the effects of resource depletion currently outweigh the effects of technical improvements in the production stage, and there is no indication of an imminent change in the relative strengths of those factors. As a result, economic incentives will not move significant quantities of oil from submarginal to economic status. Under such conditions, economic incentives will increase the ROI of pumping oil from fields that currently are economically viable but will not increase significantly annual rates of production.

The effects of resource depletion, economic changes, and political changes on oil production in the lower 48 states are quantified by Kaufmann (17). Kaufmann modifies the model for the production cycle developed by Hubbert (16), who posits that physical factors cause production to follow a symmetric bell-shaped curve. Kaufmann shows that the bell-shaped curve is a proxy for a U-shaped long-run cost curve, which embodies the net effect of resource depletion and technical change. In addition to these long-run forces, production is determined by real oil prices, the ratio of oil prices to natural gas prices, and the level of capacity allowed to operate by the Texas Railroad Commission (TRC).[4] Together, these variables account for most of the variation in the annual rate of oil production in the lower 48 states between 1947 and 1985 (Figure 3).

The long-run cost curve can account for the apparent contradiction between post-War movements in real prices and production. The U-shape for the long-run cost curve reflects changes in the relative strength of resource depletion and technical change (17). A decline in the cost of production may account for the doubling of oil production between 1947 and 1970 despite the slight decline in real prices. Using an independent source of data, Cleveland (3) demonstrates that the cost-reducing effect of technical change over-

---

[4]Kaufmann (17) estimates a two-step model for oil production in the lower 48 states. The bell-shaped production curve is given by the first difference of the following equation

$$\text{Cumulative production} = 172 * 10^9/[1 + 50.55 * e^{-0.08725 * (\text{Year} - 1925)}]$$

The econometric analysis of the residual is given by the following table:

	Beta	Standard error	t Statistic
Constant	$-5.36E-01$	$1.40E-02$	-22.94
Price (1–2)	$6.05E-03$	$1.43E-03$	4.24
Price (3–5)	$1.35E-02$	$1.41E-03$	9.61
Oil/gas price	$3.15E-02$	$9.41E-03$	3.34
Capacity	$3.55E-01$	$3.11E-02$	11.40
Symmetry test	$2.62E-09$	$3.31E-10$	7.91

in which Price (1–2) is a running average of oil prices lagged one and two years, Price (3–5) is a running average of oil prices lagged three, four, and five years, Oil/gas price is the price of a barrel of oil divided by the price for ten thousand cubic feet of natural gas, Capacity is the fraction of operable capacity allowed to operate by the Texas Railroad Commission, and Symmetry test is the first derivative of the production curve after the peak (prior to the peak, Symmetry test equals 0). The adjusted R^2 for the econometric equation is 0.97, the Durbin Watson statistic is 1.73, and the Cochrane Orcutt Rho is 0.4.

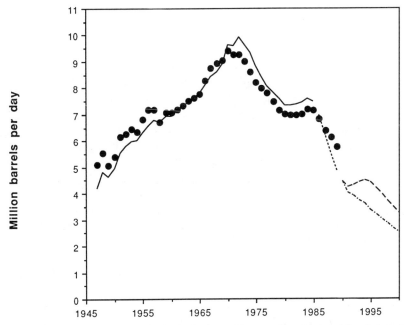

Figure 3   Annual rate of oil production in the lower 48 states. Actual values (closed circles), regression model results from (17) (solid line), ex ante out of sample forecast for 1986–1989 (dotted line), forecast for production at 1989 prices (dot-dash), and forecast for production at prices double their 1989 level (dashed line).

shadows the cost-increasing effect of resource depletion in the production stage between 1936 and the mid-1960s. A decline in the average cost of production implies that the marginal cost of production also declines. All else being equal, a decline in the marginal cost of production increases the profit-maximizing level of production. Conversely, an increase in the cost of production may lie behind the decline in production after 1970. Cleveland shows that the cost-increasing effects of resource depletion overshadow technical change after the mid-1960s. An increase in the average cost of production implies that the marginal cost of production also increases. All else being equal, an increase in the marginal costs of production decreases the profit-maximizing level of output. Kaufmann's model indicates that production in the lower 48 states follows the changes in long-run costs during the post-War period because movements along the U-shaped long-run cost curve overwhelm the effects of price and policy variables. Production in the lower 48 states increases between 1947 and 1970 because the long-run costs decline faster than real oil prices decline and faster than the rate at which the TRC shut in (closed) operable capacity. Conversely, production declines after 1970 because long-run costs increase faster than real oil prices and faster than the TRC could open spare capacity.

These results indicate that future rates of production in the lower 48 states, and the ability of public policy to alter that production path, depend on changes in long-run costs relative to changes in economic and policy variables. Kaufmann's regression results indicate that if real oil prices and the price of oil relative to natural gas remain at 1989 levels and if the Texas Railroad Commission allows producers to operate at capacity, the effects of resource depletion will reduce production in the lower 48 states over the next decade (Figure 3). If existing conditions do not change, production in the lower 48 states will be 2.5 million barrels per day (mbd) in 2000, down from 5.7 mbd in 1989. Furthermore, the effects of resource depletion probably will overwhelm economic incentives that the government could afford. Even if real oil prices double relative to their 1989 level, production falls to 3.2 mbd by 2000. The declines in production seem dramatic, but are consistent with the large decline since 1985 and the forecast for additions to proved reserves described above. For example, the reserve-to-production ratio in the lower 48 states is 8.0 in 2000 if the rates of production forecast by Kaufmann's model are combined with additions to proved reserves forecast by Cleveland & Pendleton (15).[5] A reserve-to-production ratio of 8 is near the historical low and the minimum level implied by reservoir engineering and implies that the low rates of production projected by Kaufmann's model are near the maximum that can be produced from the proved reserves projected by Cleveland & Pendleton (15).[6]

Two conclusions emerge from this forecast of oil production: US production will continue to shrink and there is little policy makers can do to stop it.

---

[5]The reserve-to-production ratio for crude oil in the lower 48 states is calculated by projecting proved reserves and dividing it by the projected level of production. Next year's reserves are projected by adding total additions (calculated from the model in Note 3) to current reserves and subtracting the rate of production projected by the model in Note 4. This process starts with reserves as of December 31, 1989 (19.8 billion barrels) and is repeated through 2000.

[6]The oil industry differs from most other sectors in that production is from inventories, i.e. proved reserves. As a result, the consistency of forecasts for oil production can be evaluated by the ratio of production to proved reserves (R/P). Proved reserves set a maximum rate of production, as described by the 8 : 1 minimum R/P ratio. A high R/P ratio indicates inefficient inventory behavior. The reserve to production (R/P) ratio of 10 : 1 in 2000 indicates that the forecast for production by Kaufmann (17) is consistent with the forecast for additions to reserves by Cleveland & Pendleton (15). If the forecast for production was low relative to the forecast for additions to proved reserves, proved reserves would accumulate, and the R/P would rise. A significant rise would not be consistent with the historical record of the US oil industry, whose R/P ratio in the lower 48 states has varied between 7 and 13.3 during the post-War period. On the other hand, if the forecast for production was too high relative to the forecast for additions to proved reserves, proved reserves would shrink, and the R/P ratio would drop below minimum levels. The consistency between the forecasts enhances the credibility of both models because they are estimated from independent sets of data using very different techniques.

The first conclusion indicates US dependence on imported oil will increase if there are no changes in the economic and political environment relative to 1989. Oil production in the lower 48 states will continue to decline because it is not cost effective to replace the giant, low-cost reservoirs from which they produce most of their oil with production from many smaller fields. Even if geological conditions permit such replacement, the huge increases in drilling effort would not be economically viable. The YPE for exploratory drilling declines two orders of magnitude between 1930 and 1988 (Figure 1) and the YPE for development declines one order of magnitude between 1946 and 1988 (Figure 2).

The second conclusion is that incentives aimed at increasing production will not increase domestic production significantly relative to the 1989 base-case scenario, and therefore will do little to ease US dependence on imported oil. Doubling real oil prices, for example, does not alter significantly the overall decline in production in the lower 48 states. Higher oil prices will make high-cost prospects profitable, but the models indicate that the amount of oil in these fields and the rate at which they can be produced are small.

Attempts to Boost Supply from a Depleted Resource Base Can Damage the Economy

Oil is one of many goods and services produced by the US economy. As such, arguments to increase production should be judged by the same criteria that are used to evaluate other productive activities. Economic efficiency in a competitive economy requires that, absent any externalities associated with domestic production, at the margin the ROI for the oil industry should be the same as the ROI for all sectors. If the ROI for oil is significantly lower than the ROI for other goods and services, the criterion for economic efficiency is violated. A lower ROI for oil production implies that some of the factors of production used by the oil industry could be reallocated to other sectors such that the total quantity of goods and services produced by the US economy could be increased. Any policy designed to boost domestic oil supply beyond levels dictated by an efficient market must evaluate the opportunity cost of these non-oil goods and services.

The feasibility of policies that aim to increase effort by the domestic oil industry can be evaluated by analyzing the effects of previous incentives. During the late 1970s and 1980s, higher prices and a series of tax incentives increased the ROI for many domestic oil projects. The increased ROI spurred activity, but the effects of resource depletion and diminishing returns to effort suggest that effort increased more rapidly than did supply. Data for gross domestic production (GDP) and gross fixed capital formation (GFCF) are consistent with this hypothesis, namely, that the US oil industry drew on factors of production faster than it added to national production at high rates of annual effort.[7] Over the past 15 years, the fraction of real GDP that originated in the crude petroleum and natural gas production sector moved in a direction opposite to the fraction of real GFCF that was used by the sector (Figure 4). In general, the fraction of GDP that originated in the crude petroleum and natural gas production sector declined steadily between 1973 and 1987, while the fraction of gross fixed capital formation increased and then decreased. The resultant gap between investment and output suggests that the industry used an increasing fraction of investment capital without a corresponding increase in its contribution to national production between 1975 and 1985.

The gap between output and investment in the crude petroleum and natural gas production sector strains the US economy in several ways. The gap represents capital that could be used by other sectors of the economy to increase production of non-oil goods and services. The amount of goods and services lost probably is significant. From 1975 to 1987, the difference between the investment and output curves totals $100.7 billion (1980 dollars).[8] This sum is 13% of the total GFCF by the entire manufacturing sector during the same period. The inefficient allocation of capital that is suggested by the gap adds to the US deficit in trade and capital accounts. For example, a portion of the decline in US exports is due to outdated production facilities, especially in manufacturing, some of which could have been re-placed by capital diverted to the oil and gas sector. Similarly, the "lost" capital represented by the gap is not available to finance the federal govern-ment deficit. This increases the government's dependence on foreign sources of capital and contributes to the shift in the US position from the world's largest creditor nation to the world's largest debtor nation in the past decade.

The gap between output and investment in the crude petroleum and natural gas production sector disappears in 1986, but there is good reason to believe the gap would reappear if the US implements policies designed to increase effort by the domestic oil industry. The gap in the late 1970s and early 1980s is caused by resource depletion in the long run and diminishing returns to annual rates of effort in the short run, which limit the ability of a substantial increase in effort to elicit a corresponding increase in output. Notice that the size of the output-investment gap rises and falls with drilling effort by the industry (Figure 4). There is little evidence that the dominance of depletion

[7]Data for gross fixed capital formation and GDP by kind of activity are measured in 1980 US dollars. Data are from (18).

[8]The investment gap is calculated by subtracting the fraction of GDP originating in the oil and gas industry from the fraction of GFCF and multiplying this difference by the real dollar value of GFCF. This calculation is based on the assumption that the oil and gas industry operates efficiently relative to other sectors when the fraction of GDP from the oil and gas industry is about equal to the fraction of GFCF undertaken by the oil and gas industry. This relative equality occurs before and after the large increase in drilling effort.

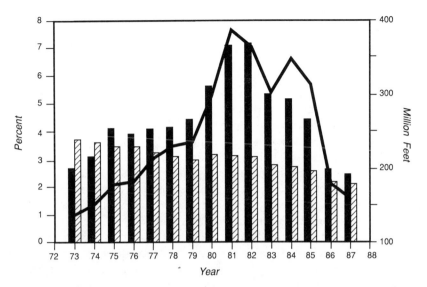

Figure 4   The percentage of gross fixed capital formation by the crude oil and natural gas production sector (solid bars), the percentage of GDP that originates in the crude oil and natural gas production sector (shaded bars), and total footage drilled in million feet (solid line).

over technical change will be reversed over the next 10 years or that the effects of diminishing returns to short-run increases in effort will be damped. This suggests that policies to increase effort will not bring forth a proportional increase in output and such a mismatch will reopen the output-investment gap in the oil and gas industry. The amount of non-oil productive capacity that is lost depends on the degree to which the policy succeeds in increasing effort.

Attempts to Boost Supply from a Depleted Resource Base Can Damage the Environment

The production and consumption of oil release many deleterious by-products into the environment, among which carbon dioxide (CO_2) has become the focus of increasing public and scientific attention. There is concern that the increasing concentration of CO_2 in the atmosphere will raise the mean temperature of the Earth and cause widespread environmental and economic disruptions (19). Whether the oil comes from Texas or Saudi Arabia has little effect on the amount of CO_2 released by combustion. But the source of oil has a large effect on the amount of carbon dioxide released in the process of exploration, development, and production. The technologies of oil exploration, development, and production use large quantities of refined petroleum products, natural gas, and electricity. By definition, resource depletion and diminishing returns increase the amount of energy and other natural resources used to produce a barrel of oil. This suggests that policies aimed at increasing domestic oil production will increase the amount of CO_2 released per barrel of oil produced and thereby speed global warming.

The amount of carbon dioxide released per barrel of oil produced can be evaluated using net energy analysis, which tracks the amount of energy society uses to deliver a unit of energy (20). The energy costs of energy supplies include both the direct costs, i.e. the energy used at the site of production, and indirect costs, such as the energy used to produce the raw materials and capital, and to support labor. The direct energy costs of oil production include purchased fuels and electricity, and fuels produced and burned on-site in the oil fields. The indirect energy costs of oil production include the energy used to manufacture drilling muds and rotary rigs, and to support roughnecks. The direct and indirect energy costs can be used in conjunction with fuel-specific emission factors to calculate the amount of CO_2 released in the process of producing a barrel of oil.[9]

A net energy analysis of US oil production indicates that resource depletion and high rates of effort increase the amount of energy used to produce a barrel of oil (4). These increases in energy costs increase the amount of carbon dioxide released per barrel of oil produced (Figure 5). The emissions of CO_2 per barrel decrease between 1919 and the mid-1960s, after which they increase rapidly. The level and rate of increase of CO_2 per barrel of oil production represents an important source of CO_2. In 1972, the CO_2 released in the process of producing a domestic barrel of oil is only 3.4% of the CO_2 emitted by the combustion of that barrel. By 1982, that percentage nearly doubles to 6.5%. These data underestimate the quantity of CO_2 released by the oil supply process because they do not include CO_2 released by exploration, development, transportation, and refining.

Conclusions

In 1990, the United States is more dependent on imported oil than at any time in its history. The human and economic costs of sending troops to the Middle East to ensure reliable supplies and low prices are some of the obvious costs of our import dependence. If a consensus emerges from current debate that the price for dependence is too high, policy makers will be charged with the difficult task of decreasing US dependence. Policy makers can increase the effort expended by the domestic oil industry by increasing the economic attractiveness of finding, developing, and producing oil. But the feasibility and desirability of such a policy hinge on the degree to which the increased

---

[9]Conversion factors for translating fuel use into carbon dioxide emissions are from (21). The factors are 210,000 kg/10^9 kcal for natural gas, 288,000 kg/10^9 kcal for oil, and 360,000 kg/10^9 kcal for coal.

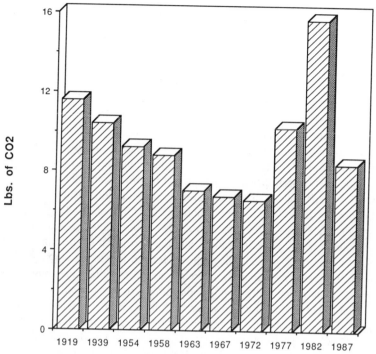

Figure 5 The quantity of carbon dioxide released directly and indirectly in the process of producing a million Btu of crude oil.

effort increases the supply of oil, contributes to economic well-being, and maintains environmental quality.

The results of our empirical models are robust and indicate that policy aimed at increasing US oil supply will not satisfy any of these goals. Costs are increasing at all stages of the oil supply process because resource depletion, manifest as smaller field size, deeper fields, and more remote environments such as offshore, overwhelms the positive impacts of technical change. Resource depletion began to predominate in exploration at least a half century ago, and in production a quarter century ago. Unless there is a dramatic change in the relationship between the factors that have generated the historical record of drilling and discovery over the past 50 years, a period characterized by significant technical improvements, there is little evidence that increased effort will increase supply. Policy based on the slim chance that such a change will occur has a high probability of incurring large economic and environmental costs. Large increases in effort directed at a waning resource base will reduce overall economic well-being and speed global warming.

Our analyses focus on the lower 48 states because it is the most thoroughly explored region in the world with respect to oil. Our predictions of future supply from conventional drilling are based on the historical record of drilling and discovery, and assume implicitly that the pace of technical change in the future will not be significantly different than in the past. Given the recent commercialization of new technologies such as horizontal drilling (22) and three-dimensional seismic surveys (23), one might question the validity of that assumption. Similarly, some argue that relaxing environmental restrictions to allow drilling in the Arctic National Wildlife Refuge (ANWR) and offshore regions will result in substantial new supplies. However, evidence from the past quarter century suggests that the oil resource base in the lower 48 states is so depleted that technical improvements and relaxed environmental controls will not significantly improve the supply of oil. Two of the most significant advances in the past quarter century are digital seismic, common-mid-point recording in exploration, and the expansion of drilling activity in offshore regions, especially the Gulf of Mexico. In concert with a decline in the drilling rate, the rapid commercialization of digital and common-mid-point data recording in the 1960s boosted YPE in that period (4, 15) (Figure 2). That effect is temporary however, and the YPE falls to record lows in the 1970s. The expansion of exploration and discovery in the Gulf of Mexico in the 1950s and 1960s does not alter significantly the decline in YPE for development in the lower 48 states (15). The YPE for drilling in the Gulf of Mexico declines sharply in the 1950s and stabilizes thereafter.

Once heralded as potentially prolific sources of oil, Alaska and other offshore regions now have dim prospects. Drilling results in Outer Continental Shelf (OCS) waters in the past decade have been particularly disappointing. The 46 exploratory wells drilled in the Atlantic OCS between 1978 and 1988 generated some of the most expensive dry holes in the history of the industry, but produced no commercially recoverable oil (24). Promising prospects in the OCS of Alaska such as Cook Inlet, the Gulf of Alaska, and the Navarin Basin have been expensive busts. Since the discovery of Prudhoe Bay in 1968, Alaska has contributed less than 10% of new-field discoveries and has produced the nation's most expensive dry hole, the Mukluk project. President Bush and Secretary of Interior Lujan are on record favoring exploration in ANWR. But the US Geological Survey's mean estimate of recoverable oil in ANWR is 3.45 billion barrels (25). At 1989 rates of oil use, this represents about 200 days' supply. Moreover, if found, oil from the ANWR would not be available for seven years. By then, the United States may import between 60 and 70% of its oil (14).

The high cost associated with a futile attempt to stimulate production suggests an alternative for reducing US dependence on imported oil: reducing US oil demand. Such a strategy has several potential advantages relative to increasing production. Fuel-efficient technologies could be implemented by redirecting the investment capital that would have been used by the oil industry to increase drilling and production effort. The current level of US dependence on imported oil would be lower, perhaps much lower, if the investment capital represented by the gap between the fraction of GDP and GFCF by the crude petroleum production sector had been invested in conservation technologies during the first two price run-ups. Moreover, it is inconsistent to subsidize drilling effort with public funds after public funds for renewable fuels and conservation technology were cut in the 1980s based on the argument that the market would direct the efficient level of investment towards these sources (26).

Reducing oil demand also would satisfy the political need to incorporate environmental factors into the decision-making process. Reduction of greenhouse gas emissions is near the top of that agenda. Lowering oil demand would accomplish this goal in two ways. First, lowering oil demand would reduce emissions per unit of economic activity. Second, it would shift activity away from a sector that will emit an ever-increasing amount of CO_2 per unit output. A policy of not subsidizing domestic effort also will prevent the possible irreversible degradation of fragile and irreplaceable ecosystems such as those in ANWR. Contrary to industry rhetoric (27), the environmental record of operations on the North Slope of Alaska is not laudable (28). Oil spills such as that from the *Exxon Valdez* grab the spotlight, but equally important are the little-publicized day-to-day impacts of routine exploration and production operations.

Finally, the United States could choose to neither increase supply nor decrease demand and let US dependence on imports grow. Dependence per se is not necessarily bad. Successful economies such as Germany, Japan, and the Republic of Korea depend almost entirely on imported oil. These economies succeed, in part, by selling non-oil goods and services to oil exporters. Increases in oil prices may hurt these economies less than during previous price spikes because members of OPEC are more integrated in the world economy and are able to absorb increased revenues. But the inability of US producers to compete in the world market implies that the United States would be unable to offset the effect of higher oil prices on its trade balance by increasing exports. If the United States cannot muster a consensus regarding policies to retard US dependence on imports, it may need to develop a trade policy to ensure that investment flows to those industries best able to compete in export markets. A consensus on trade policy, however, may be even more difficult to build than a policy to manage oil supply and demand.

Literature Cited

1. Schipper, L., Howarth, R. B., Geller, H. 1990. United States energy use from 1973 to 1987: the impacts of improved efficiency. *Annu. Rev. Energy* 15:455–504

2. Friedan, B. J., Baker, K. 1983. The market needs help: The disappointing record of home energy conservation. *J. Policy Anal. Manage.* 2:432–48

3. Cleveland, C. J. 1991. Physical and economic aspects of resource scarcity: the cost of oil supply in the lower 48 U.S., 1936–1987. *Resourc. Energy.* In press

4. Cleveland, C. J., Kaufmann, R. K. 1991. Forecasting ultimate oil recovery and its rate of production: Incorporating economic forces into the models of M. King Hubbert. *Energy J.* 12:17–46

5. Pindyck, R. 1978. The optimal exploration and production of nonrenewable resources. *J. Polit. Econ.* 86:841

6. Uhler, R. S. 1976. Costs and supply in petroleum exploration: the case of Alberta. *Can. J. Econ.* 9:72

7. Woods, T. J. 1985. Resource depletion and lower 48 oil and gas discovery rates. *Oil Gas J.* Oct. 28:pp. 140–45

8. Woods, T. J. 1985. Long run trends in oil and gas discovery rates in the lower 48 US. *Am. Assoc. Pet. Geol. Bull.* 69:1321–26

9. Fisher, W. L. 1987. Can the U.S. oil and gas resource base support sustained production? *Science* 236:1631–36

10. Fisher, W. L. 1989. *Changing Perceptions of the Oil and Gas Resource Base.* Paper presented in the Explor. Econ. Sess., Annu. Meet. Explor. Geophys., Dallas, Tex., Oct. 31

11. Fisher, W. L. 1990. *Oil Reserve Growth: Understanding Complex Reservoirs.* Paper presented at Am. Acad. Arts. Sci. Symp. Improved Recovery of Oil and Gas from Existing Fields, New Orleans, La., Feb. 17

12. Schuenmeyer, J. H., Drew, L. J. 1983. A procedure to estimate the parent population and the size of oil and gas fields as revealed by the study of economic trucation. *Math. Geol.* 15:145–61

13. Lovejoy, W. F., Homan, P. T. 1965. *Methods of Estimating Reserves of Crude Oil and Natural Gas, and Natural Gas Liquids.* Washington, DC: Resour. for Future

14. Kaufmann, R. K. 1988. *Higher oil price: Can OPEC raise prices by cutting production.* PhD thesis. Univ. Penn.

15. Cleveland, C. J., Pendleton, V. 1991. Are oil reserve additions declining?: Yield per effort for oil exploration and development in the lower 48 US and Gulf of Mexico, 1946–1988. *Am. Assoc. Pet. Geol. Bull.* In press

16. Hubbert, M. K. 1962. *Energy Resources.* Natl. Res. Counc., Natl. Acad. Sci. *Publication 1000-D.* Washington, DC

17. Kaufmann, R. K. 1991. Oil production in the lower 48 states: reconciling curve fitting and econometric models. *Resourc. Energy.* In press

18. Organ. Econ. Coop. Dev. 1988. *National Accounts Detailed Tables,* Vol. 2. 1975–1987. OECD Dept. Econ. Stat.

19. Flavin, C. 1989. *Slowing Global Warming: A Worldwide Strategy. Worldwatch Paper 91.* Washington, DC

20. Bullard, C. W., Penner, P. B., Pilati, D. A. 1978. Net energy analysis: Handbook for combining process and input-output analysis. *Resourc. Energy* 1:267–313

21. Hall, C. A. S., Cleveland, C. J., Kaufmann, R. 1986. *Energy and Resource Quality.* New York: Wiley

22. Fritz, M. 1989. Horizontal drilling sparks activity. *AAPG Explorer.* Nov: p. 1

23. Shirley, K. 1990. 3-D Seismic gives old field new life. *AAPG Explorer.* March: p. 1

24. Melancon, J., Brooke, J., Kinler, C., Knipmeyer, J. 1989. *Outer Continental Shelf Estimated Oil and Gas Reserves, Gulf of Mexico, December 31, 1988. OCS Report MMS 89-0074.* US Dept. Interior, Minerals Manage. Serv.

25. Energy Inf. Admin. 1987. *Potential Oil Production from the Coastal Plain of the Arctic National Wildlife Refuge. Document SR/RNGD/87-01.* Revised ed.

26. Sutherland, R. J. 1989. An analysis of the US Department of Energy's civilian R & D budget. *Energy J.* 10(1):35–54

27. *Oil Gas J.* Mar 20, 1989, p. 19; Mobil Corp. 1991. *New York Times* editorial, p. 27

28. Holing, D. 1991. America's energy plan. *The Amicus J.* 13:12–20

Annu. Rev. Energy Environ. 1991. 16:401–31

POTENTIAL IMPACTS OF BIOMASS PRODUCTION IN THE UNITED STATES ON BIOLOGICAL DIVERSITY

James H. Cook

Scully Science Center, National Audubon Society, 550 South Bay Avenue, Islip, New York 11751

Jan Beyea

National Audubon Society, 950 Third Avenue, New York, NY 10022

Kathleen H. Keeler

School of Biological Sciences, University of Nebraska-Lincoln, Lincoln, Nebraska 68588-0343

KEY WORDS: biofuel, wildlife, habitat, global climate change, energy.

ABSTRACT

Biomass could be a renewable source of energy and chemicals that would not add CO_2 to the atmosphere. It will become economically competitive as its cost decreases relative to energy costs, and biotechnology is expected to accelerate this trend by increasing biomass productivity. Pressure to slow global warming may also make biomass more attractive.

Substantial dependence on biomass would entail massive changes in land use, risking serious reductions in biodiversity through destruction of habitat for native species. Forests could be managed and harvested more intensively, and virtually all arable land unsuitable for high-value agriculture or silvicul-

401

1056-3466/91/1022-0401$02.00

ture might be used to grow energy crops. We estimate that it would require an area equal to that farmed in 1988, about 130 million hectares, just to supply the United States with transportation fuel.

Planning at micro to macro scales will be crucial to minimize the ecological impacts of producing biomass. Cropping and harvesting systems will need to provide the spatial and temporal diversity characteristic of natural ecosystems and successional sequences. To maximize habitat value for interior-dependent species, it will be essential to maintain the connectivity of the habitat network, both within biomass farms and to surrounding undisturbed areas.

Incorporation of these ecological values will be necessary to forestall costly environmental restoration, even at the cost of submaximal biomass productivity. Since it is doubtful that all managers will take the longer view, some sort of intervention will very likely be necessary. Given concerns about global warming, both bioenergy proponents and conservationists have an incentive to work together.

I—INTRODUCTION

Various types of biomass—municipal waste, farm and forest wastes, low-quality wood, and herbaceous and short-rotation woody crops—have been actively promoted as renewable sources of energy and chemical feedstocks (1) that do not contribute net carbon dioxide to the atmosphere (2–7). Indeed, the US Environmental Protection Agency has projected that biomass could become the world's largest single energy source following intervention to protect the climate (8). It is important to recall, however, that a biomass cycle would add net CO_2 to the atmosphere if carbon stored in standing trees, debris, and soil were released (9).

Even if biomass proved to be a renewable alternative to fossil feedstocks that did not hasten global climate change, increasing dependence on biomass would lead to more intensive harvesting of forests and other natural ecosystems, and a substantial demand for land to grow herbaceous and short-rotation woody crops (10, 11). The demand for biomass would compete for arable land with other human needs, such as food and fiber production, and would increase the pressure to convert "idle" land—land used primarily by other species—to human uses (10, 11).

This is an ominous prospect, since human activities, primarily the conversion of complex natural ecosystems to monoculture agroecosystems and the harvesting of natural ecosystems at unsustainable levels (12), are eliminatiag other species at thousands of times the pre-human rate (13). Indeed, one quarter of the world's biological density may be lost during the next 20–30 years (12). The economic implications of this loss are profound, yet, it is crucial to preserve global biodiversity for ethical, as well as economic, reasons (12). Although our focus in this paper is on US biodiversity, the issues raised and principles enunciated will be widely applicable to other temperate climates.

The extent of habitat destruction from biofuels development will depend on the intensity with which natural ecosystems are harvested and the amount of "idle" land brought into production. Competing human demands have primarily spared land in the United States that is marginal for current uses—both natural land and land recovering from previous human disturbance—to provide habitat for native species (14–18). Such marginal land will be attractive for biomass farms (19).

Although low-cost biofuels—biomass wastes and low-quality wood—are supplying an increasing percentage of US energy usage, energy crops are not economically competitive with fossil fuels (2). However, new technologies promise to relax some of the constraints inhibiting widespread implementation. For example, biotechnology may be used to improve energy crop production (20) and conversion to ethanol or other high-quality feedstocks (1), while new combustion technology (73) or gas turbines based on aerospace jet engines (7, 21) may increase the efficiency and lower the cost of generating electricity from biomass. Thus, while the developments that produce an economically sound biomass technology are not directly at issue in this paper, we strongly suspect that technological developments will make biomass more competitive with other alternatives to fossil feedstocks. In fact, we consider possible side effects of the economic success of chemical and energy industries based on genetically engineered biomass crops and processes to be much more serious than possible failures of biotechnology that are currently of concern in regulatory circles.

In reviewing the environmental implications of large-scale biomass harvesting and production, analysts have noted such potential impacts as competition for arable land with food production, water pollution, loss of soil fertility, and the spread of bioengineered organisms (10, 11, 22–25). Concern is largely focused on impacts affecting primary human needs for food and shelter. This article addresses the potential for increasingly intensive use of land to reduce biological diversity by eliminating habitat for native species and by destroying lands with special qualities. These concerns have not been widely enough addressed, given the large role that natural vegetation (primarily trees from existing forests) and dedicated biomass crops may play in meeting demands for organic feedstocks.

Since energy demands will very likely dominate biomass markets, we focus on supplying segments of the US energy economy. Although biomass can be used to produce organic chemicals and plastics currently made from oil (26–30), chemical feedstock markets are so much smaller than energy markets—2.5 Quads out of 72 Quads total US consumption in 1982 (1)—that concerns about demands for biofuels will probably dominate all other environmental concerns about biomass technology.

Given the prospects for a potentially large increase in biomass production, it is time to develop strategies for mitigating the loss of habitat for native species—and the resulting loss of biological diversity—that might follow intensive harvesting of natural vegetation and growing of dedicated biomass crops on a large scale in the United States. The preservation of natural biodiversity in the context of a growing demand for land to grow biomass will require the development and use of biomass production systems that provide both a sustainable yield of biomass and adequate habitat for native species (12).

Even though sustainable agricultural practices that maintain biodiversity have been identified and promoted, the push to maximize production and minimize cost has routinely led to monoculture agroecosystems (15) that virtually eliminate natural biological diversity, providing habitat for a limited range of plant and animal species (12). Implementation of biomass cropping systems that preserve naturally diverse ecosystems may require producers to accept submaximal biomass yields. Incentives or regulations may therefore be necessary to ensure economic viability and adequate implementation. The costs to the economy of limiting yields will be offset by the future benefits of maintaining our natural resources.

Although the prospects for biodiversity of large-scale use of biofuels are ominous, the threat to biodiversity from global warming is equally serious. Therefore, there are strong reasons for conservationists to work with bioenergy advocates to develop a technology and guidelines for use that are mutually acceptable. Incentives for biomass advocates to cooperate in negotiations are also strong: failure to address large-scale environmental problems will lead to public pressure that may limit the technology's development.

The next three sections consider the potential demand for biofuels from three perspectives—US biomass production, the amount of land required to meet various US fuel demands, and considerations regarding market penetration. Subsequent sections discuss some of the likely impacts on the natural environment and mechanisms for minimizing those impacts.

II—ESTIMATING BIOMASS PRODUCTION

There are at least three major sources of biofuels whose expanded use might reduce biodiversity: biomass from relatively natural ecosystems (primarily existing forests—including forest industry residues and wastes), agricultural residues and wastes, and dedicated energy crops. Many plant species have been suggested as suitable biomass crops, and alternatives for various geographic areas and ecosystems are being investigated (2, 19, 31). Examples include several species of fast-growing hardwood trees (3, 32), perennial grasses (19, 31), cattails (33, 34), water hyacinth (35), and algae (36, 37), as well as traditional crops such as sugar cane (38) and maize (39, 40).

Of particular concern from a biodiversity perspective are the potential impacts of managing forests more intensively for increased wood production and converting natural lands to dedicated energy farms. Table 1 shows a breakdown of current US land uses and estimates of annual biomass production under three scenarios: current land uses, more intensive forestry, and the use of land for energy farming. Our analysis, following that of Pimentel and coworkers (11), indicates that the current total net primary biomass production by all vegetation in the United States is about 47 EJ per year (yr^{-1}). This is a very approximate estimate, being based on annual biomass productivity estimates for broad classes of land. Even so, the estimated 47 EJ yr^{-1} of total current net US primary biomass production is less than the 76 EJ yr^{-1} of current US fossil fuel use, suggesting that great pressures could arise to manage our land more intensively.

The US Congress Office of Technology Assessment (OTA) has speculated that full stocking of US commercial forestland with highly productive tree species could raise productivity from 2–4 metric tons per hectare per year (t $ha^{-1} y^{-1}$) to approximately 5–10 t $ha^{-1} yr^{-1}$ (41). It is unlikely that all 200 million hectares of commercial forestland would be available for such intensive management, given public concerns and logistical constraints (41). If 80–200 million hectares of US commercial forestland were more intensively managed, as shown in Table 1, about 8–40 EJ yr^{-1} could be produced, increasing total forest production to 16–44 EJ yr^{-1} and total US biomass production to about 51–79 EJ yr^{-1}.

Wright et al have argued that, while energy crops require relatively even and fertile land—cropland or at least potential cropland—they can be grown on land that is subject to drought, erosion, or seasonal flooding (19). About 100 million hectares—40 million hectares of uncultivated cropland and about 60 million hectares of potential cropland—could be readily available for growing energy crops (Table 1). Yields of 8–17 dry t $ha^{-1} yr^{-1}$ have become common for short-rotation hardwoods in research trials (32), and yields for herbaceous energy crops have reached 12–40 dry t $ha^{-1} yr^{-1}$ (42). Assuming an average productivity of 10–28 t $ha^{-1} yr^{-1}$ dry biomass for a 1:1 mixture of herbaceous and short-rotation woody crops, this land could produce about 20–56 EJ of biofuels annually, increasing total US biomass production to about 59–96 EJ yr^{-1} (see Table 1).

To put these current and potential biomass productivity estimates into perspective, consider current human uses of biomass in the United States. Although biofuels supplied only 3 EJ of primary energy in the United States in 1987 (about 3.5% of the total) (43), Pimentel and coworkers have estimated

Table 1 Breakdown of current US land uses and estimates of annual biomass production under three scenarios: current land uses, more intensive forestry, and the use of land for energy farming

Category	Area[a] (10^6 ha)	Total current biomass production[b] (EJ)	Forestland available for more intensive forestry[a,c] (10^6 ha)	Potential wood production from more intensive forestry[d] (EJ)	Total biomass production with more intensive forestry[e] (EJ)	Prime land available for energy crops[a] (10^6 ha)	Potential energy crop production[f] (EJ)	Total biomass production with energy crops[g] (EJ)
Cropland (90)	170	20	—	—	20	40[h]	8–22	24–38
Pastureland (90)	50	3	—	—	3	24[i]	5–13	6–15
Rangeland (75, 89)	340	11	—	—	11	20[i]	4–11	14–21
Forestland (11, 41, 156)	290	12	80–200	8–40	16–44	17[i]	3–10	14–20
Other	70	<1	—	—	1	—	—	1
Total (11)	920	47	80–200	8–40	51–79	101	20–56	59–96

[a] These relative areas do not necessarily reflect general priorities from a biodiversity perspective at the local level (see Section VII).

[b] Biomass production was calculated by multiplying the area of each class of land by its estimated total biomass productivity. The following dry mass productivity estimates were taken from Table V of (11): 6 t ha^{-1} yr^{-1} for cropland, 3 t ha^{-1} yr^{-1} for pastureland, 2 t ha^{-1} yr^{-1} for forestland, and 0.5 t ha^{-1} yr^{-1} for other lands. The productivity estimate for rangeland of 1.6 t ha^{-1} yr^{-1} is a weighted average of productivity estimates for various classes of rangeland published by the USDA Forest Service (75). An energy content of 20 GJ t^{-1} was assumed (43).

[c] We assume that between 40% and 100% of US commercial forestland would be available for more intensive management (41).

[d] Biomass production was calculated for commercial forestland available for intensive management using a productivity estimate of 5–10 t ha^{-1} yr^{-1} (41), or 100–200 GJ ha^{-1} yr^{-1} based on an energy content of 20 GJ t^{-1} dry biomass (43).

[e] Biomass production was calculated as in footnote b, except for production from the 80–200 million hectares of intensively managed commercial forestland, which was calculated as in footnote d.

[f] We assume a 1:1 mix of herbaceous and short-rotation woody energy crops, which leads to a productivity range of 10–28 dry t ha^{-1} yr^{-1}, or 200–560 GJ ha^{-1} yr^{-1} based on an energy content of 20 GJ t^{-1} dry biomass (43). [Yields of 8–17 dry t ha^{-1} yr^{-1} have become common for short-rotation hardwoods in research trials (32), and yields for herbaceous energy crops have reached 12–40 dry t ha^{-1} yr^{-1} (42).]

[g] Biomass production potentials for each land class under a scenario of large-scale energy farming are the sum of potential energy crop production and biomass production on the land not used for energy crops, calculated as described in footnote b after deducting the land in each class used for energy crops.

[h] Total cropland less 130 million hectares used to grow crops in 1988 (19).

[i] Land considered by the US Department of Agriculture to have medium to high potential for conversion to cropland (88).

that the US human population directly and indirectly appropriates for its use about 26 EJ yr^{-1} of agricultural crops, livestock forage, and forest products (11)—about half of the 47 EJ total current net biomass production.

Both of the potential biomass production scenarios presented in Table 1 would be major perturbations to current biomass production and use. Either alternative might mean doubling the net primary biomass productivity of the United States, and perhaps tripling the amount of biomass taken from the land. In reality, some of the potential of both alternatives will probably be realized: some forestland will be managed more intensively and some arable land will be used for growing woody and herbaceous energy crops. This trade-off is discussed in Section VII.

III—ESTIMATING LAND REQUIREMENTS

Our estimates of the land required for growing biomass in alternative ways to meet various US energy demands are presented in Table 2. Separate land requirement estimates are given for wood from forests managed more intensively, but not so intensively as to destroy their forest character, and dedicated energy crops. These estimates are based on current energy demands and do not consider changes in demand, either projected increases or possible decreases through conservation, or competition from other energy sources that may limit the actual demand for biomass. The actual amount of biomass used will depend on a variety of factors. For example, biotechnological innovations that reduce the costs of energy crops will increase the demand for energy crops and arable land. Similar effects can be expected from developments that increase the costs of alternative energy sources. The ranges given reflect different assumptions regarding biomass productivity and the use of biomass residues and wastes.

Furthermore, these estimates must be considered in the context of other land requirements. Forces such as urbanization and demand for recreational land will limit the availability of land for producing biofuels. And forestland will also be needed to provide wood for traditional uses: the OTA has estimated that about 2–4 EJ yr^{-1} will be needed for final forest-industry products plus up to about 4 EJ yr^{-1} process energy in the forest products industry (41). Even so, this would require about 40–80 million hectares, less than most of the estimated forestland requirements in Table 2.

Even though our estimates in Table 2 are approximate, a few implications stand out. Increasing the production of existing forests, short of turning them into energy farms, would not provide sufficient biomass to displace all current US CO_2 emissions from fossil fuel. In fact, a land area at least twice the 200 million hectares of US commercial forestland would be required to replace fossil fuels with biomass. The results shown in Table 2 indicate that even a demand for biomass equivalent to only 20% of current fossil fuel use could

Table 2 Estimates of land required to replace various sectors of the US energy economy with wood from more intensive use of forests or with dedicated energy crops

Energy sector	Energy used (EJ)	Biomass required (EJ)	Biomass required in addition to wastes[d] (EJ)	Forestland required to grow needed wood[e,f] (× 10^6 ha)	Arable land required to grow needed energy crops[e,g] (× 10^6 ha)
Coal for electricity[a]	20	20	7–20	35[h]–200[i]	13[h]–100[i]
Transportation fuels[b]	23	46	33–46	165[h]–460[i]	60[h]–230[i]
All fossil fuels[c]	76	99	86–99	430[h]–990[i]	160[h]–500[i]

[a] Approximately 6.9×10^8 t of coal were used to generate electricity in 1989 (73), with an energy content of 29.3 GJ t^{-1} equivalent (157). Similar efficiencies are assumed for generating electricity from biomass and coal in plants of similar size (7, 21).

[b] The transportation sector currently uses about 23 EJ yr^{-1} of liquid fuels (73). We neglect refinery losses and assume the conversion of biomass to liquid fuels at 50% efficiency (2).

[c] Total energy consumption in 1990 was 86 EJ, about 76 EJ of which was supplied by fossil feedstocks (158). In all sectors other than transportation fuels, similar efficiencies are assumed for using biomass and fossil fuels.

[d] Assuming 0–13 EJ yr^{-1} of biomass wastes are used (42).

[e] These land categories do not necessarily reflect general priorities from a biodiversity perspective at the local level (see Section VII).

[f] Without considering energy farms; otherwise as described in footnote d to Table 1.

[g] Without considering intensive forest use; otherwise as described in footnote f to Table 1.

[h] Assuming 1. full use of biomass wastes, reducing the need for wood and energy crops, and 2. maximal forest or energy-crop productivity estimates.

[i] Assuming 1. no use of biomass wastes, and 2. minimal forest or energy-crop productivity estimates.

put tremendous pressure on forests, with impacts on biodiversity that are difficult to imagine.

There may be tremendous pressure to harvest most existing forests, or to increase the management intensity to the point that many are managed like crops. In particular, although the removal of "low-quality" and slowly growing wood from existing forests has been justified for improving stand quality (19, 41), it is likely that the demand for biofuels will drive the removal of wood at nonsustainable rates until adequate supplies of energy crops are available.

In contrast to the limited supply that could be provided by increasing the productivity of forests, Table 2 shows that there might be enough land to permit complete replacement of fossil fuels with biomass grown on energy farms, although the area required could be so vast that the environmental impacts would be potentially enormous.

Even replacement of the 23 EJ of current transportation fuels alone would require the transformation of immense amounts of land (Table 2). About 46 EJ of biomass would be needed, neglecting refinery losses and assuming the conversion of biomass to liquid fuels at 50% efficiency. A midrange estimate of the land needed would be 130 million hectares—as much land as that farmed in 1988 (19).

The above land-use estimates might be taken to suggest that biodiversity would be most efficiently protected by making forests off limits to biofuel harvesting, and focusing production on relatively small areas of very intensively managed energy crops. Aspects of this issue are discussed in Sections IV and VII.

IV—MARKET SHARE

Our analysis indicates that, for biomass cultivation to affect large areas of the United States, and thereby potentially threaten biodiversity, demand for biofuels must be so high that they displace a large share of the current energy market. This may happen in a number of ways. We distinguish between two main types of driving force: first, a market-force dynamic in which costs of biofuels are significantly lowered by new technologies, such as biotechnology and efficient gas turbines; and second, a policy-oriented driving force, in which concerns about global warming lead governments to override market forces. These two driving forces have vastly different implications for environmental regulation.

A. Market-Driven Penetration of Biofuels

Even in a world concerned with global climate disruption, the demand for biomass will be limited by competition with other energy sources and energy-conservation strategies that do not add carbon dioxide to the atmosphere. For example, electricity can be efficiently generated with photovoltaics. It is even possible that a new generation of "inherently safe" nuclear plants and innovative waste-disposal methods could allay public fears and provide economically competitive electricity (44, 45), although the development of politically acceptable breeder reactors will also be necessary to provide sufficient nuclear fuel to preclude biofuels from becoming a major energy source. Nevertheless, the environmental impacts of large-scale biomass use must be explored should its alternates fail to prove viable, either economically or politically.

Energy conservation is one potential source for reducing the future market share of biofuels. In a policy-driven path to biomass dominance, which will be based on concerns about global warming, gains from energy conservation will likely be applied to reducing CO_2 emissions, not biofuels use. In a market-driven world, the market share will be determined by relative costs. For this paper, we assume that energy-efficiency improvements will keep total US demand for primary energy constant over time, even with population and economic growth.

B. Biotechnology and Biomass

In a market-driven scenario for the large-scale use of biofuels, we expect that the application of various biotechnologies—classical genetics, fermentation, plant nutrition, cell culture, cloning, molecular biology, and genetic engineering—will increase the demand for biomass by reducing production and processing costs. Biomass productivity will also benefit by technology transfer from work on traditional crops to reduce losses caused by weeds and pests (20, 46–48). A variety of traditional crops with improved insect resistance, virus resistance, and herbicide tolerance have already been field tested and, pending regulatory approval, may reach commercial markets by the mid-decade (48–50).

Efforts are currently under way using the biotechnologies listed above to improve such qualities of potential biomass energy crops as productivity, feedstock value, pest resistance, and tolerance of marginal growth conditions (20, 48, 51–54). For example, hybrid poplar clones resistant to a broad-spectrum herbicide glyphosate (Roundup) have been produced by gene insertion (3). Similarly, work is under way on insect-resistant hybrid poplar clones through insertion of a DNA sequence coding for the active fragment of the *Bacillus thuringiensis* toxin (55). Other work has demonstrated that the optimized delivery of nutrients to trees during the growing season can dramatically increase productivity (56).

Commercially proven biochemical methods are available for making ethanol from starches and sugars in such traditional crops as wheat, corn, sorghum, and sugar cane (38). However, crop-derived ethanol is currently too expensive to compete with fossil fuels, unless subsidized, because the production of these crops is resource intensive, and because only the monosaccharides and starches are exploited (11, 22). For example, although approximately 6.5 t corn can be grown annually per hectare (11) and converted to roughly 2 t ethanol using current technology, 33–97% of the energy value of that ethanol was required to grow the corn (11, 40), leaving a maximum net ethanol yield of about 1.3 t ha^{-1} yr^{-1}. Indeed, when other energy inputs required for ethanol production are considered, the ratio of output to input energy can be less than one unless energy credits are taken for coproducts (40).

Such commercially proven biochemical techniques are not available for converting the cellulose and hemicellulose in plant fiber to high-quality feedstocks, such as methane, methanol, or ethanol, and efforts are under way to improve existing processes and create new ones (1, 2, 57, 58). Although existing biochemical methods permit the highly efficient conversion of cellulose to ethanol, the conversion is too slow for industrial use (59, 60), and there is much room for improvement through genetic methods (61, 62). Similarly, genetic methods are playing an important role in the improvement of organisms and enzymatic processes for converting hemicellulose to ethanol (2). In contrast to the situation for corn, the ratio of output to input energy for producing ethanol from cellulose appears to be about five (40).

Various implications of biotechnologically improved biomass crops and conversion processes for the quality of the natural environment have been considered. On the positive side, the use of crops engineered for insect resistance might reduce the demand for insecticides that damage nontarget species (48). Conversely, genetically engineered organisms might spread, becoming pests and destroying the integrity of natural ecosystems (63), including parks, wildlife refuges, and wilderness areas. The implications of other developments are likewise mixed. Planting a herbicide-resistant crop might reduce overall herbicide use, and resulting damage to nontarget species, by replacing several herbicides with one nonspecific, and perhaps less-persistent herbicide (48). Yet, the availability of crops resistant to a particular herbicide might encourage the use of this herbicide in new contexts (64).

Some potential failures, such as the possibility that genetically engineered biomass crops could escape their bounds and become noxious weeds, have received a great deal of attention and may be adequately addressed by the regulatory community. However, there may be a greater risk that new biomass industries will become so economically successful that the side effects of biomass production become a major problem for the environment.

Biotechnology-mediated improvements in biomass production and processing will probably foster the success of biomass-based chemical and energy industries. This success, in turn, may lead to the expansion of biomass production into natural areas currently marginal for agricultural use and reduce biological diversity by destroying habitat for native species (65).

In any case, the fact that different aspects of new biomass technologies may have opposite impacts on biodiversity suggests that opportunities exist to channel the technology in positive directions.

C. Policy-Driven Penetration of Biofuels

Bioenergy is but one of a number of promising, low- or zero-CO_2 emitting alternatives to which governments may look in a policy-driven scenario. In particular, as with biofuels, technological advances are reducing the costs of photovoltaics (PVs). Considering the relatively low efficiency with which biological photosynthesis collects incident solar energy, PVs may, in fact, have significant theoretical advantages for electricity generation. To make these advantages explicit, consider the following efficiency arguments.

Traditional agriculture generally stores less than 1% of the solar energy incident during the growing season (38, 66), although the practical maximum photosynthesis efficiency has been estimated to be 8–9% (67). Sugar cane, the most efficient plant known, has yielded up to 110 t ha^{-1} yr^{-1} of dry biomass when cultivated intensively on a year-round basis, representing the conversion of 3.3% of the incident solar energy (38). While photosynthetic efficiency will doubtless be improved someday by agricultural genetic engineers, a great deal more basic research on photosynthesis is needed, placing such breakthroughs rather far in the future (68–70).

Assuming seven months of growth per year, the conversion of 1–3% of solar radiation incident during the growing season to biomass (38), and the conversion of 33–34% of the energy content of that biomass to electricity (7, 21, 19), we estimate that a biomass cycle could convert solar energy to electricity with an efficiency of 0.2–0.6%. In contrast, conversion efficiencies for commercially available silicon PVs range from 9% for inexpensive amorphous silicon units to 15% for crystalline silicon units. Under laboratory conditions, conversion efficiencies for crystalline silicon PVs have been demonstrated to be as high as 31% (71), and efficiencies for gallium arsenide–gallium antimonide stacked junction cells have reached 35% (6). Furthermore, the costs of PVs are dropping so rapidly that some analysts predict utility reliance on them for peaking power in the late 1990s (71).

Thus, it would require about 15–45 times the land to generate electricity from solar energy using a biomass cycle as it would using amorphous silicon PVs. Indeed, it has been estimated that 3.4 million hectares of PVs operating at 12% efficiency could supply all electricity currently used in the United States (6). Even so, wood-fired electrical generation remains a regionally attractive option (19, 72), given an abundance of inexpensive low-quality wood in the northeast and southeast, and the development of new gas turbines capable of turning biomass into electricity with high efficiencies (7, 21).

Furthermore, the lack of an efficient way to store solar electricity (73) makes it unlikely that PVs will eliminate the potential market for biofuel-derived baseload electricity in the foreseeable future.

The long-term role for biomass in transportation is even clearer than its role in electricity. As it is the only source of renewable energy that yields liquid transportation fuels (19), demand for biomass is likely to be strong in the transportation market. Only conversion to electric or hydrogen-powered (using hydrogen from the electrolysis of water) vehicles would allow such energy sources as solar or nuclear power to compete in transportation markets (73). Although such a scenario is possible, it is not probable enough to justify ignoring the need to regulate the large-scale production of biofuels.

It is the combination of market forces and policy pressure that poses the most threat to biodiversity. With such a combination looking more and more plausible in recent years, the prospects for large-scale use of biofuels cannot be dismissed. Moreover, there is likely to be a synergistic effect from the two forces that will accelerate biofuels development: government pressure to manage global warming will lead to increased technological research and development in biofuels, and lowered costs will make legislators more willing to impose regulations favoring biofuels.

V—POTENTIAL LOSS OF BIODIVERSITY

Biodiversity is the existence of a variety of plants, animals, and other life forms living independently of humans. These species have inherent value, and contribute substantially to the aesthetic and recreational quality of our continent. Furthermore, their activities directly and indirectly sustain us, producing soil, feeding economically important plants and animals, and cleaning water and air. Any decrease in biodiversity will reduce the aesthetic and recreational value of our environment, and will have complex and largely irreversible consequences for human health and economic stability.

Habitat destruction is the most pervasive cause of biodiversity loss (12). While other causes contribute, habitat destruction removes more species faster, and irreversibly. Biomass production may destroy habitat through increasingly intensive forest management and the conversion of natural lands to energy farms. These issues are discussed below.

A. Intensified Use of Forests

Land that is currently forested is not farmed primarily because it is either too infertile or too rough (18, 74). However, these forests are highly diverse and provide habitat for thousands of plant and animal species (18). In particular, forestland includes many highly productive and increasingly rare wetland environments (75).

Even current demands for forest products have brought about the widespread clearcutting of natural, uneven-aged stands and their replacement with even-aged, often monocultural, stands (76). A number of adverse consequences have followed.

Clearcut harvesting destroys the forest-interior habitat required by many interior-dependent species (77). For example, although both mature aspen stands and aspen stands regenerating after clearcutting had similar avian species richness, avian species that are declining regionally or nationally generally predominated in mature stands (78). Clearcutting destroys forest-interior habitat both directly and through the fragmentation of large stands (77). Furthermore, the reduced structural diversity of the even-aged stands that develop after clearcutting may limit the diversity of the habitat they provide (79). Intensive management may also progressively destroy the long-term habitat potential of the land by depleting soil nutrients and increasing soil erosion (76, 79).

Unless regulated, the use of intensive, even-aged management seems likely to increase with increasing demands for biofuels from forests. Indeed, the OTA has speculated that fertilizing forests and stocking them with high-yielding hybrid trees might increase their yield from 2–4 dry t ha^{-1} yr^{-1} to as much as 6.3–13 dry t ha^{-1} yr^{-1} (41). If done intensively enough, this would amount to managing forests as dedicated energy crops.

B. Intensified Use of Cropland

The demand for biofuels may increase cropland management intensity, and may also result in the use of cropland to grow energy crops. The implications for biodiversity of growing energy crops on cropland will depend on how the land is currently used.

It has been proposed that crop residues be harvested for biomass (80). This removal of residues could reduce soil fertility and increase the rate of soil loss by reducing the quantity of organic material and micro-nutrients plowed back into the soil (11, 22), although some proposals to use residues take these concerns into account (80). Furthermore, many wildlife species exploit crop residues. For example, more than 400,000 sandhill cranes (80% of the US population) depend on Nebraska waste corn near the Platte River to store energy before migrating north: the corn provides 90% of their caloric requirement (81). Even so, it may be possible for people and other species to share this resource: although crop residues are critical to sandhill cranes, the birds use less than 20% of them (81).

C. Dedicated Energy Crops

Intensive management of trees or herbaceous crops for energy is necessary to obtain high yields (19, 82). In preparation for planting tree crops, existing vegetation is eliminated by herbicide treatments, and bare soil is exposed by plowing and disking (32). After fertilization and planting, competing vegetation is controlled for the first year or two by mowing, cultivation, and herbicide treatments (32). Maintenance for the rest of the rotation includes biennial fertilization and pesticide treatment (32). Management practices for producing herbaceous crops would probably be comparable to those for typical agricultural crops (19). Potential impacts of converting various lands to energy crops are discussed in the following sections.

1. CROPLAND USED FOR ENERGY FARMS   Replacement of food crops by energy crops would change the nature of the habitat provided, favoring some species and hurting others. The effect of a particular habitat change on overall biodiversity depends on the ecological role and rarity of the affected species (12).

Compared to urban and suburban areas, agricultural regions are rich in wild and native species. For example, the diversity of some arthropod species (insects and spiders) in cultivated fields is similar to the diversity of these species in forests; this is true for both soil species (83) and above-ground predatory species (84). On the other hand, cultivated areas clearly support fewer vertebrate and plant species than more natural areas (85). The frequency of disturbance is crucial to the quality of the habitat provided for vertebrates: e.g. frequent harvesting of hay fields destroys bird nests before the young have matured (86). Similarly, fruit and vegetable farming, which involves intensive management, does not support many vertebrates (18).

2. RANGELAND AND PASTURELAND USED FOR ENERGY FARMS   Grassland, characterized by low rainfall, periodic drought, and recurrent fire, once dominated the center of North America (87). Although the eastern prairies have been almost completely converted to cropland, some of the prairies further west remain, particularly where the soil is poor and adequate water not available. Much of these central and western grasslands are classified as pastureland and rangeland: while both are used to graze livestock, pastureland is more intensively managed than rangeland (88).

There are about 330 (75) to 350 (89) million hectares of rangeland in the United States, of which 52% is federally owned and 98% is located in the Great Plains, the Southwest, or Alaska (89). With the exception of some mountain ecosystems, much of the federal rangeland outside of Alaska is arid, with a relatively low primary productivity (89). About 53 million hectares of US nonfederal land is classified as pastureland (90).

Although the native grasslands of the Great Plains maintained their productivity under grazing for thousands of years (87), more than half of non-Alaskan rangeland is now in poor to very poor condition, having been damaged by overgrazing and/or regional climate change (17, 18, 89). Much of the Alaskan rangeland may still be in relatively good condition because it has not been grazed heavily by introduced livestock (89). About 72% of nonfederal pastureland is in fair to good condition (88).

Rangeland and pastureland provide habitat for at least some of the original native grassland species (89, 91). Intensive biofuels production on approximately 20 million hectares of nonfederal rangeland and 24 million hectares nonfederal pastureland with medium or high potential for conversion to cropland (88) (see Table 1) could eliminate habitat for many native grassland species. Intensive biofuels production could also reduce the potential habitat value of the land through soil erosion, salinization, groundwater depletion, or subsidence (89).

3. WETLAND HABITATS USED FOR ENERGY FARMS   Wetlands include prairie potholes; inland, delta, and coastal marshes; flood plains; and swamps (18, 92, 93). Water, nutrients, and exposure to full sunlight together make wetlands some of the most productive wildlife habitat, critical to surrounding ecosystems and the survival of a wide variety of native species (92). In Texas and Oklahoma, for instance, the density of birds in riparian habitats is seven times that in other habitats (94).

About 40% of nonfederal lands with excess water are classified as cropland, accounting for about 26% of the total (88). Indeed, some of the most productive US croplands were once too wet to crop (92). Furthermore, excess water is the main limitation inhibiting the cropping of about 20 million hectares of forestland, about 4 million hectares of rangeland, and about 10 million hectares of pastureland (88).

These potential croplands with wetness problems may be prime candidates for conversion to biofuels plantations (19). For example, workers are evaluating silver maple for short-rotation intensive culture on occasionally flooded bottomland in Iowa. As Wright et al (3) note, "The site is typical of Iowa bottomland that was cleared at one time for farming and later used for pasture or abandoned." Alternatively, Lakshman has proposed that marshes be developed as cattail biomass farms (34). Although this might be preferable to draining them, wetlands are integral to the survival of surrounding ecosystems and must be preserved.

D. Overall Implications for Biodiversity

Although existing forests, and energy crops on uncultivated cropland and lands with medium to high potential for conversion to cropland, could apparently supply a significant part of current US energy use, this would very likely require the increasingly intensive use of forestland and the conversion of the most fertile pastureland, rangeland, and forestland to energy cropland. The lands converted might include increasingly rare riparian and wetland habitats.

The intensive harvesting and management of such relatively natural ecosystems, culminating in their conversion to energy farms, would reduce biodiversity by eliminating habitat for native species—including many that are rare, threatened, or endangered (85). Such changes might eventually have an impact on the biodiversity of the United States equal to that of modern agriculture, which has dramatically reduced the ecological complexity of at least 170 million hectares in the five centuries of European settlement. For example, approximately 2100 wild vertebrate species occur in North America, including 650 species of birds; in contrast, there are only a few dozen domestic livestock species (95).

It is unknown today how intensively existing forests will be harvested and how much land will be used for energy farms, but the required expansion of biomass production and consequent loss of biodiversity are potentially so enormous that even skeptical environmental policy makers need to concern themselves now with putting into place regulations, guidelines, and incentives that will preserve biodiversity.

VI—TRULY SUSTAINABLE BIOMASS PRODUCTION

In response to ongoing pressures on biodiversity in the United States, including the potential of a growing demand for biofuels, policies are required that protect additional land (96), especially areas with a rich diversity of species (97), such as old-growth forests (79) and mature examples of other kinds of ecosystems. However, it will be difficult to reserve from commercial use enough land to protect biological diversity adequately. Policies are needed that preserve wildlife habitat and biodiversity on all lands, in a gradient from the most intensive energy farms to the most protected lands (98, 99).

Existing forests will need to be carefully managed to preserve habitat, biodiversity, and productivity in the face of an additional demand for wood. There is a rich literature on good forest management (100, 101). The challenge will be to get people to use good forest-management techniques.

There is no such extensive literature on good energy-farm management. Yet, large-scale biomass farming will seriously reduce biodiversity in the United States unless biomass farms provide adequate and appropriate habitat. Biomass production that destroys biodiversity should not be considered truly sustainable, even if otherwise renewable. Although it may be possible to design reasonably productive biomass cropping systems that are truly sustainable, being biologically diverse and providing habitat for a wide variety of native species, this may require growers to accept submaximal biomass yields. Determining the compromises in yield, if any, required to preserve biodiversity should be an ongoing goal of biomass research.

Determinants of agricultural and forestry practices that preserve biodiversity and are therefore sustainable on a long-term basis are emerging in several disciplines. Theories of sustainable agriculture (102–104), other alternatives to current agricultural practices (22, 105), and anthropological studies of land-management practices (106, 107) have emphasized the importance of protecting the soil and cropping multiple species. New forestry has emphasized the crucial importance of a complex web of interacting species in forest ecosystems (79, 100); landscape ecology has clarified the role that spatial relationships and transport processes play in the functioning of ecosystems (108–110).

Taken together, this work demonstrates that long-term sustainability entails the management of complex natural ecosystems and successional sequences, or the use of crop-management systems modeled on such natural processes. The contrast to modern farming and forestry, which have increasingly emphasized the simplification and intensive management of ecosystems, is profound.

Even though the broad outline of biologically diverse and long-term sustainable biomass production technology is apparent, more research is needed before full-scale implementation can responsibly begin. As an important step in this direction, the Oak Ridge National Laboratory has begun a study of the environmental impacts of biomass production (111).

A. Elements of Long-Term Sustainability

Biological, spatial, and temporal diversity are clearly central determinants of long-term sustainability at local to global scales (112). Spatial diversity includes both vertical and horizontal diversity: it is three dimensional. Natural ecosystems are normally very complicated, with many interacting and interdependent species (79). Loss of diversity tends to destabilize ecosystems, increasing their sensitivity to stress and disturbance (76, 113). Structure, the arrangement of these diverse elements, is crucial at all scales: natural ecosystems are mosaics (108).

For example, work in the Pacific Northwest has demonstrated that a wide variety of mycorrhizal fungi form attachments with tree roots and facilitate nutrient uptake. The spores of these fungi are spread by small fungus-eating rodents that live in rotting logs on the forest floor. Removal of rotting logs eliminates these rodents, thereby blocking dissemination of mycorrhizal fungi and reducing forest productivity (79, 100).

Another example of ecological complexity comes from the study of plant disease in agricultural ecosystems. Many plant pathogens are naturally suppressed by diverse populations of indigenous soil microorganisms (114). Soil sterilization with broad-spectrum pesticides destroys this natural community and therefore increases the risk of a subsequent severe infestation by these pathogens.

A diversity of plant communities is also crucial for wildlife habitat. Since species have different habitat requirements, the habitat value of an area for a range of wildlife depends both on the diversity of its plant communities and diversity within each community (115, 116). The spatial arrangement of these communities is also important. Old-growth areas, riparian zones, wetlands, open areas, and transitions between them are all crucial habitat elements (116–119).

Agricultural fields, grasslands, and initial stages of the forest succession— such as the herb, shrub, and open sapling-pole conditions—are exploited by many animal species for feeding or reproduction (81, 117, 120). These species include game animals as well as myriads of arthropods, birds, and small mammals (83, 84, 121).

Although such open areas were naturally produced by wildfires and severe storms, they are now primarily created on a much larger scale by human activities such as farming and clearcutting. Modern agriculture has become synonymous with the intensive management of monocultures; still, hedgerows and fallow fields do provide some habitat diversity. Moreover, research on the cropping of multiple species (122–124) and studies of traditional farming practices of indigenous peoples (125–127) confirm that highly productive agriculture and silviculture can include abundant species and structural diversity.

Old-growth forests are complex late-successional ecosystems with much internal horizontal and vertical diversity (16, 100, 116, 117, 120). These areas are used by a multitude of species, including many with very specific habitat requirements (121, 128). Snags (standing dead trees) and dead-and-down woody material are important ecosystem components that provide food and shelter for many species (129, 130). Crucial to the ecosystem are a numerous and very diverse community of arthropods, fungi, and microorganisms (79, 100). In particular, although it might be possible to recreate the old-growth forests of the Pacific Northwest using a 300–400-year rotation (117), these ancient forests are effectively irreplaceable and must not be harvested, if biodiversity is to be preserved (79).

Transitions between distinct plant communities, termed *edges,* provide valuable habitat for many species, especially those that use both of the adjoining community types (116, 121). On the other hand, creating edges fragments the landscape, reducing the size of communities and their potential habitat diversity, although the impact can be reduced by interconnecting corridors of mature forest (131, 132). Even so, if habitat blocks are made too small, they are dominated by edge effects and lose much of their central habitat value (100, 116). For example, loss of forest-interior nesting habitat through fragmentation appears to be primarily responsible for declines in populations of migratory songbird species (133, 134).

Wetlands, such as marshes, swamps, and bogs, are highly productive areas crucial to surrounding ecosystems (92). Riparian zones upland from streams, lakes, and wetlands are ecologically rich water-to-land transitions providing food, water, and cover for many wildlife species (118, 132). An especially valuable type of ecosystem, they serve as corridors connecting other habitats (118, 132). To preserve biodiversity, neither wetlands nor riparian zones should be harvested because of their value to surrounding ecosystems.

As noted above, it appears that productive cropping systems can include considerable within-field species diversity (122, 124, 126). Species diversity might actually increase productivity by minimizing competition between adjacent plants for sunlight, water, and soil nutrients (3). Additionally, living mulches can check soil erosion, smother weeds, aid in pest control, and supply nitrogen to crops (135). Both species diversity and structural diversity among fields can also be provided through crop rotation. In addition, these practices may maintain soil fertility and improve long-term site productivity (76, 79).

Other steps to maintain the health and fertility of the soil will also be crucial to long-term habitat value. It would be best to avoid the use of pesticides, which not only endanger wildlife but sterilize the soil and destroy natural resistance to pathogens (114).

Structural diversity within fields can also be provided by relatively minor management changes. For example, a wavy planting geometry has been developed that provides both jack pine for paper production and the clumped jack pine habitat needed by Kirtland's warblers (136). Small patches of older trees, including snags and down rotting logs, can be preserved during the conversion of secondary forests to biomass farms. These patches of mature habitat would act as reservoirs of organisms important to the health and productivity of the ecosystem (79, 100, 132, 137).

In managing an ecosystem for wildlife habitat, one can focus on providing habitat for a particular species of interest (*featured species*) or a diversity of species (117, 138). Featured species can include both game animals and threatened or endangered species. Both perspectives are essential, since native species may have conflicting habitat requirements. For example, edge between late-succession forest and grass-forb or shrub conditions is primary golden eagle habitat, while northern spotted owls require large unbroken stands (more than 120 ha) of old-growth forest (121, 128, 139). Creating more edge for the eagles would fragment the unbroken stands required by the owls.

B. Synthesis

Biological, spatial, and temporal diversity are clearly the central prerequisites of long-term sustainable biomass production. Landscape ecology provides a context for integrating these concepts in the design of biomass production facilities that provide diverse habitat for native species (108).

A network of mature habitat patches, connected by shelterbelts, hedgerows between fields, and corridors of secondary forest, could be retained or created (98, 99, 131, 140, 141). Riparian buffer zones and wetlands would be part of the network. The resulting mosaic would provide diverse habitat and reduce soil erosion. To maximize habitat value for interior-dependent species, it will be essential to minimize habitat fragmentation and maintain habitat connectivity, both within biomass farms and to surrounding undisturbed areas. Maintaining habitat connectivity through corridors may also be crucial to permit species movements in response to greenhouse warming and increasing sea level (140, 142).

Such a synthesis may permit the design of highly productive ecosystems that efficiently produce biomass, maintain soil fertility, and provide high-quality wildlife habitat. Biomass production technology could then be implemented while preserving irreplaceable natural resources.

VII-LAND USE PRIORITIES

How much intensively managed forestland and how large an area used for energy farms would be tolerable from the biodiversity perspective? Insufficient research has been done to allow this judgment to be made. Nevertheless, increasing human impact on biological diversity is an inescapable reality, given the threat of global warming. The use of appropriate management practices will reduce the negative impacts of biofuel production.

Prioritizing land uses will also be important, although it is unlikely that scientific consensus could be obtained at this early stage. It may turn out to be appropriate to try to channel biomass development away from forests towards cropland. This seems to be the suggestion from Tables 1 and 2. However, it is risky to make such a judgment at this state of our knowledge. Consider the fact that some of our best cropland is drained wetlands—areas that might be more valuable for biodiversity, if reclaimed as wetlands, than some dry forestland. In fact, recent farm bills have attempted to channel crop development away from environmentally important land types that have been cropped in the past. Just because land is classified as cropland does not mean that its optimal use is as cropland.

Although we do not doubt that it will eventually be possible, after further research, to build a scientific consensus on prioritizing land categories for biofuels development, we suspect that the categories will be more specific than simply forestland, cropland, pastureland, and rangeland. Even land that is usually low in biological significance may be very important locally, if a conjunction of factors are present that facilitate exploitation by wildlife.

Assuming that broad categories of land are identified as having highest priority for biofuels development from the perspective of biodiversity, the question arises of how intensively the land should be managed. Should energy farms or intensive forestry occur without joint management to improve wild-life habitat, thereby minimizing the land required to produce a given amount of energy? Or should some management for wildlife habitat take place even on lands devoted to energy crops? The answer has significant implications for the development of bioenergy technology. If no management for wildlife habitat needs to take place on land used to grow biofuels, then bioenergy advocates will have little need to modify their present plants for production and conversion technology. If, on the other hand, some management for wildlife habitat turns out to be optimal, then the bioenergy industry will need to work closely with wildlife agencies and experts in biodiversity preservation.

It is our contention that this latter situation is likely to be the case, since a small amount of wildlife management will probably pay large dividends in usable habitat. Although such management will likely decrease the biomass productivity, the habitat gain may be proportionately greater. To make such considerations precise, it is necessary to define curves of biomass productivity vs wildlife habitat for various categories of land (see Figure 1). For such curves to be useful, habitat quality must be quantified. For this paper, we

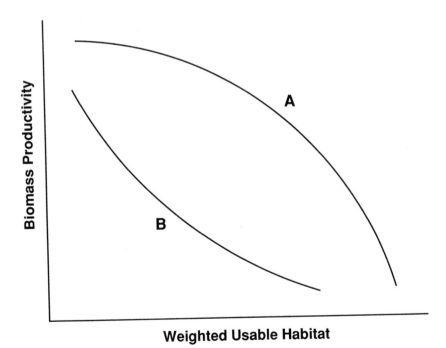

Figure 1   Possible relationships between harvestable biomass productivity and weighted useable wildlife habitat.

adopt the concept of "weighted usable area" used by the US Fish and Wildlife Service. In this method, biological experts weigh various qualities of lands in terms of the habitat requirements of target species. A hectare of land that provided little habitat, then, would be valued less than a hectare of land that provided much habitat. Specifically, two hectares of land with a rating of 0.5 would be considered equivalent in habitat value to one hectare of land with a rating of unity.

If there is a plateau region of the productivity-vs-habitat curve (Figure 1, curve A), it is clear that large gains in habitat value can be obtained for little loss in biomass productivity. Assuming that the cost of wildlife management in the plateau region is not great, it would make sense to operate there, jointly managing land for biofuel production and wildlife habitat. On the other hand, should there be no plateau (see Figure 1, curve B), it would make more sense to ignore habitat considerations on energy farms or in intensively managed forests, protecting habitat only on dedicated reserves.

Although our review of the literature leads us to believe that productivity-vs-habitat curves certainly have plateaus for some classes of land and target species, it is imperative that research on such curves be carried out for all land categories that are candidates for large-scale biofuels production.

Political, legal, and constitutional considerations will severely limit the options open to managing large classes of land. It is therefore naive to think that the mere existence of a solution acceptable in principle both to conservationists and biomass advocates will provide protection to the interests of either. Nevertheless a theoretical analysis can help to inform public policy.

VIII—RESEARCH AGENDA

As indicated above, research is needed to develop curves of biomass productivity vs wildlife habitat. Research is also needed on ways to flatten the curves, providing more habitat value with less loss in productivity. This is particularly true for energy farming, where little is known about management for biodiversity. Finally, classification of lands in terms of their significance for biodiversity is also an important research topic.

IX—MECHANISMS FOR PRESERVING BIODIVERSITY

Even though techniques for minimizing the negative impacts of biomass farming on habitat for native species are available, managers will not use them in the absence of relevant policy if doing so will reduce their short-term yields. Long-term productivity is not an effective determinant, since economic forces tend to emphasize immediate and personal benefits at the expense of future and social costs (143). Indeed, the values of biodiversity are largely external to economic markets, comprising services freely provided to society by natural ecosystems (144–146).

Since we cannot rely on the free market to protect biodiversity, some sort of intervention will be necessary. There are at least four ways to foster the preservation of biodiversity: encourage enlightened self interest through education and negotiation, enact regulatory legislation, internalize external costs, or give economic incentives (112, 144, 147–149). All four approaches will likely be necessary.

A. Negotiations

Negotiations among interested parties can be quite effective. In particular, since electrical production is a highly regulated industry, negotiation among producers, environmental advocates, and regulators can lead to mechanisms for mitigating environmental damage and/or internalizing nonmarket costs of biomass production.

The National Audubon Society, for example, is a party of interest to a proposal by the Vermont Department of Public Service to harvest low-quality wood for generating electricity. It is becoming clear from these negotiations that a model agreement between conservationists and biomass advocates may be possible. It might include the following points:

1. joint recognition that unwise management could damage forest health, diversity, and habitat value;
2. contracts between electricity-production facilities and their wood suppliers that would stipulate comprehensive forest management plans;
3. costs of forest management to be covered as part of the fuel cost of wood-fired electrical plants;
4. the state Fish and Wildlife Department to be the arbiter of wise forest management, and to be responsible for approving forest management and harvesting plans.

B. Regulations

Governments can use their police powers to outlaw or mandate particular activities. However, the success of this approach depends on careful design and effective enforcement. For example, although laws have been passed to preserve diversity and minimize habitat damage on public lands, they have not prevented significant damage from private uses such as mining, livestock grazing, and timber harvesting (150). Even National Wildlife Refuges, created expressly to provide habitat, are potentially vulnerable to commercial exploitation (151). The federal forestlands and rangelands have fared worse. Although the Forest Service and Bureau of Land Management have been mandated to manage for multiple uses, including wildlife habitat, substantial pressure to maximize commercial access has led to severe ecosystem damage (150).

C. Internalizing Nonmarket Costs

Several state utility commissions, most notably the New York Commission, have begun to require that utilities add environmental damage cost estimates to total electricity costs when calculating the cheapest form of future electricity generation (152). Incorporating estimated nonmarket values of biodiversity into these economic calculations is not yet required, and might foster the preservation of biodiversity on economic grounds.

Several approaches have been proposed for estimating such nonmarket values, most often in terms of their human utility (144–146). For example, one can sum the alternative economic values of an intact ecosystem, such as the nontimber values of a tropical forest (153). In the absence of significant economic values, one can poll a sample of the public, asking people how much they would pay to protect wildlife habitat or biodiversity (144–146).

It is more straightforward, though nontrivial, to gauge the direct costs of providing for biological diversity and wildlife habitat in existing forests and energy farms. Some measures may require the diversion of productivity from human uses. For example, it has been estimated that providing roughly three snags per hectare for cavity-excavating birds in ponderosa pine managed on a 150-year rotation would reduce productivity by about 6% (154). Similarly, if some of the available land were set aside as unharvested hedgerows and shelterbelts productivity might be reduced accordingly. Other measures might be economically neutral, and some might even increase long-term productivity by maintaining soil fertility.

D. Incentives

Experience with US cropland-management programs proves that economic incentives can profoundly change land use. For example, about 25 million hectares of productive farmland were kept out of cultivation by the Acreage Reduction Program in 1988 (155), while the Conservation Reserve Program kept an additional 10 million hectares of highly erodible land out of production (19). Only 130 million hectares were farmed (19). As incentives are developed to encourage the production of biofuels, they will need to require practices that enhance biodiversity.

Each of the mechanisms discussed—negotiation, regulations, internalization of nonmarket costs, and incentives—has a role in protecting biodiversity, and it is likely that all will be needed. More generally, only a consistent approach that fairly assigns all the costs of all potential energy sources, both conventional and alternative, will lead to an optimal mix of technologies and optimal versions of each technology (144).

X—A VISION FOR THE FUTURE

Consider highly productive land-management systems that work with natural forces, rather than against them, producing valuable energy and chemical feedstocks in a biologically diverse and long-term sustainable fashion. A naive vision? No, indeed. Prudent management of our natural resources is necessary to preserve our quality of life. Increasing human populations will require increasing amounts of food crops, timber, renewable energy, and opportunities for recreation. Only stewardship that manages for generations yet to come will allow our standard living to endure while preserving the natural world. The groundwork must be laid now by the combined efforts of ecologists, conservationists, foresters, range managers, and engineers.

ACKNOWLEDGMENTS

We thank the Joyce Foundation for supporting this work. We also thank Robert Socolow for his careful reading of this manuscript.

Literature Cited

1. Natl. Mater. Advis. Board. 1986. *Bioprocessing for the Energy-Efficient Prodution of Chemicals. NMAB-428*, pp. 1–8; Springfield, Va: US Dept. Energy; Natl. Tech. Inf. Serv.
2. Biofuels Munic. Waste Technol. Program. 1988. *Five Year Research Plan: 1988–1992*, pp. 1, 2, 8–13, 17–19, 24–27. Springfield, Va: US Dept. Energy; Natl. Tech. Inf. Serv.
3. Wright, L. L., Doyle, T. W., Layton, P. A., Ranney, J. W. 1989. *Short Rotation Woody Crops Program: Annual Progress Report for 1988. Environ. Sci. Div. Pub. No. 3373, ORNL-6594*, pp. 1–5, 31, 37, 41, 45. Oak Ridge, Tenn: Oak Ridge Natl. Lab.
4. Rader, N., Boley, K., Borson, D., Bossong, K., Saleska, S. 1989. *Power Surge: The Status and Near-Term Potential of Renewable Energy Technologies*, p. II-15. Washington, DC: Public Citizen
5. Schneider, C., Congressperson. 1989. Other options. Letter to the editor, *The New York Times*, Dec. 27
6. Weinberg, C. J., Williams, R. H. 1990. Energy from the sun. *Sci. Am.* 263:146–55
7. Hall, D. O., Mynick, H. E., Williams, R. H. 1990. *Carbon Sequestration vs. Fossil Fuel Substitution: Alternative Roles for Biomass in Coping with Greenhouse Warming, PU/CEES Report No. 255*. Princeton, NJ: Cent. Energy Environ. Stud., Princeton Univ.
8. US Environ. Protect. Agency. 1989. *Policy Options for Stabilizing Global Climate*. Draft report to Congress, p. 58, Figure 16. Washington, DC
9. Harmon, M. E., Ferrell, W. K., Franklin, J. F. 1990. Effects on carbon storage of conversion of old-growth forests to young forests. *Science* 247:699–702
10. Trimble, J. L., Van Hook, R. L., Folger, A. G. 1984. Biomass for energy: the environmental issues. *Biomass* 6:3–13
11. Pimentel, D., Warneke, A. F., Teel, W. S., Schwab, K. A., Simcox, N. J., et al. 1988. Food versus biomass fuel: socioeconomic and environmental impacts in the United States, Brazil, India and Kenya. *Adv. Food Res.* 32:185–238
12. McNeely, J. A., Miller, K. R., Reid, W. V., Mittermeier, R. A., Werner, T. B. 1990. *Conserving the World's Biological Diversity*, pp. 18–23, 25–35, 38–43, 55. Gland, Switzerland: Int. Union for Conserving Nature; Washing-

ton, DC: World Resourc. Inst., Conservation Int., World Wildlife Fund–US, and the World Bank

13. Wilson, E. O. 1988. The current state of biological diversity. In *Biodiversity*, p. 13. Washington, DC: Natl. Acad. Press

14. Brokaw, H. P., ed. 1978. *Wildlife and America: Contributions to an Understanding of American Wildlife and its Conservation*, p. 87. Washington, DC: Counc. Environ. Quality; US Govt. Printing Off.

15. Burger, G. V. 1978. Agriculture and wildlife. In *Wildlife and America: Contributions to an Understanding of American Wildlife and its Conservation*, ed. H. P. Brokaw, pp. 89–107. Washington, DC: Counc. Environ. Quality; US Govt. Printing Off.

16. Leopold, A. S. 1978. Wildlife and forest practice. See Ref. 15, pp. 108–20

17. Wagner, F. H. 1978. Livestock grazing and the livestock industry. See Ref. 15, pp. 121–45

18. Natl. Acad. Sci. 1982. *Impact of Emerging Agricultural Trends on Fish and Wildlife Habitat*, p. 83. Washington, DC: Natl. Acad. Press

19. Wright, L. L., Cushman, J. H., Layton, P. A. 1989. Expanding the market by improving the resource. *Biologue* 6:12–19

20. Ellis, D. D., McCown, B. H. 1988. *The Potential of Biotechnology in the Biomass Production Program*, Environ. Sci. Div. Pub. No. 3172, ORNL/M-606, pp. 1–10. Oak Ridge, Tenn: Oak Ridge Natl. Lab.

21. Larson, E. D., Williams, R. H. 1989. Biomass-fired steam-injected gas turbine cogeneration. *Biologue* 12/89–1/90:12–19

22. Jackson, W. 1980. *New Roots for Agriculture*, pp. 62–74. San Francisco, Calif: Friends of the Earth

23. McIntyre, T. C. 1987. An overview of the environmental impacts anticipated from large scale biomass energy systems. In *Biomass Conversion Technology*, ed. M. Moo-Young, p. 49. New York: Pergamon

24. Newman, D. R., Hall, D. O. 1990. Land-use impacts. In *Bioenergy and the Environment*, ed. J. Pasztor, L. Kristoferson, pp. 231–83. Nairobi, Kenya: United Nations Environ. Programme

25. Kvistgaard, M., Olsen, A. M. 1986. Biotechnology and environment: a technology assessment of environmental applications and impacts of biotechnology. *Trends Biotechnol.* 4:112

26. Lipinsky, E. S. 1981. Chemicals from biomass: petrochemical substitution options. *Science* 212:1465

27. Palsson, B. O., Fathi-Afshar, S., Rudd, D. F., Lightfoot, E. N. 1981. Biomass as a source of chemical feedstocks: an economic evaluation. *Science* 213:513

28. Bungay, H. R. 1982. Biomass refining. *Science* 218:643

29. MacQuitty, J. J. 1988. Impact of biotechnology on the chemical industry. In *The Impact of Chemistry on Biotechnology*, ed. M. Phillips, S. P. Shoemaker, R. D. Middlekauff, R. M. Ottenbrite, pp. 11–29. Washington, DC: Am. Chem. Soc.

30. Narayan, R. 1988. Preparation of bio-based polymers for materials applications. *Appl. Biochem. Biotechnol.* 17:7

31. Solar Tech. Inf. Program. 1989. *Biofuels: Technical Information Guide*, SERI/SP-220-3366, pp. 12–14. Washington, DC: Solar Energy Res. Inst.; US Dept. Energy

32. Wright, L. L., Ehrenshaft, A. R. 1990. *Short Rotation Woody Crops Program: Annual Progress Report for 1989*. Environ. Sci. Div. Pub. No. 3484, ORNL-6625, pp. 1–5, 18, 27, 35. Oak Ridge, Tenn: Oak Ridge Natl. Lab.

33. Pratt, D. C., Andrews, N. J. 1980. Cattails (*Typha* spp.) as an energy source. In *Energy from Biomass and Wastes IV*, ed. D. L. Klass, J. W. Weatherly III, p. 43. Chicago, Ill: Inst. Gas Technol.

34. Lakshman, G. 1988. Cattails. *Bio-joule* 10:9

35. Butner, R. S., Elliott, D. C., Sealock, L. J. Jr., Chynoweth, D. P. 1988. A process study of the biothermal conversion of water hyacinths to methane. *Appl. Biochem. Biotechnol.* 17:105

36. de la Noue, J., Proulx, D., Guay, R., Pouliot, Y., Turcotte, J. 1986. Algal biomass production from waste waters and swine manure: nutritional and safety aspects. In *Microbial Biomass Proteins*, ed. M. Moo-Young, K. F. Gregory, p. 141. London: Elsevier

37. Olguín, E. J. 1986. Appropriate biotechnical systems in the arid environment. In *Applied Microbiology*, ed. H. W. Doelle, C. G. Hedén, p. 111. Boston, Mass: Reidel

38. Elawad, S. H., Gascho, G. J., Shih, S. F. 1980. The energy potential of sugarcane and sweet sorghum. See Ref. 33, pp. 65–105

39. Scheller, W. A. 1981. An analysis of gasohol energetics. In *Biomass as a Nonfossil Fuel Source*, ACS Symp. Series 144, ed. D. L. Klass, p. 419. Washington, DC: Am. Chem. Soc.

40. Lynd, L. H., Cushman, J. H., Nichols, R. J., Wyman, C. E. 1991. Fuel ethanol from cellulosic biomass. *Science* 251:1318–23

41. Off. Technol. Assess. 1980. *Energy from Biological Processes: Volume II—Technical and Environmental Analyses,* pp. 23–24. Washington, DC: US Congress; US Govt. Printing Off.

42. Ranney, J. W., Turhollow, A. F., Donaldson, T. L. 1989. Biomass energy. In *Energy Technology R&D: What Could Make a Difference?, Volume 2—Part 2 of 3: Supply Technology, ORNL-6541/V2/P2,* ed. W. Fulkerson, D. B. Reister, J. T. Miller, pp. 84–85. Oak Ridge, Tenn: Oak Ridge Natl. Lab.

43. Energy Inf. Admin. 1989. *Estimates of Biofuels Consumption in the United States During 1987, CNEAF/NAFD 89-03,* pp. viii, 2. Washington, DC: US Dept. Energy

44. Haggin, J. 1986. New era of inherently safe nuclear reactor technology nears. *Chem. Eng. News.* June 30: 18–22

45. Beyea, J. 1991. The nuclear dilemma. *The Electricity J.* In press

46. Gardner, R. C. 1986. Biotechnology and gene transfer: the promise and the practice. *New Zealand Agronomy Society* 5: (spec. publ.) 333

47. Vasil, S. K. 1988. Progress in the regeneration and genetic manipulation of cereal crops. *Bio/technology* 6:397

48. Gasser, C. S., Fraley, R. T. 1989. Genetically engineering plants for crop improvement. *Science* 244:1293

49. Off. Technol. Assess. 1988. *New Developments in Biotechnology—Field-Testing Engineered Organisms: Genetic and Ecological Issues, OTA-BA-350,* p. 150. Washington, DC: US Congress; US Govt. Printing Off.

50. Monsanto Co. 1990. *Monsanto Company Field Trials: Genetically Engineered Plants and Microbes.* Jan.

51. Jeffries, R. L. 1980. The role of organic solutes in osmoregulation in halophytic higher plants. In *Genetic Engineering of Osmoregulation,* ed. D. W. Rains, R. C. Valentine, A. Hollaender, p. 135. New York: Plenum

52. Wang, D. I. C. 1982. Potential of the new bio-technology, including recombinant DNA, for production of biofuels and chemicals. In *The Increasing Use of Biomass for Energy and Chemicals,* p. 5. Aspen, Colo: The Bio-Energy Counc., Aspen Inst. For Humanistic Stud.

53. Pool, R. 1989. In search of the plastic potato. *Science* 245:1187–89

54. Williams, R. H. 1989. Invited testimony, p. 72. US Senate Comm. Energy Natural Resourc.; May 14

55. Stettler, R. F., Bassman, J. H., Bradshaw, H. D., Chastagner, G. A., Gordon, M. P., Gustafson, R. R., Heilman, P. E., Hinckley, T. M., McKean, W. T., Sarkanen, K. V., Smit, B. A., Van Volkenburgh, E., Whiteley, H. R. 1990. *Annual Progress Report on the Poplar Research Program.* Seattle, Wash: Univ. Wash.

56. Linder, S. 1989. Nutritional control of forest yield. In *Nutrition of Trees;* Marcus Wallenberg Found. Symp., Falun, Sweden. 6:62–87

57. Doelle, H. W. 1986. Microbial process development in biotechnology. See Ref. 37, p. 38

58. Keim, C. R., Venkatasubramanian, K. 1989. Economics of current biotechnological methods of producing ethanol. *Trends Biotechnol.* 7:22

59. Wright, J. D., Power, A. J., Douglas, L. J. 1987. Design and parametric evaluation of an enzymatic hydrolysis process (separate hydrolysis and fermentation). In *Biotechnol. Bioeng., Symp. No. 17,* ed. C. D. Scott, p. 285. New York: Wiley

60. Wright, J. D. 1988. *Economics of Enzymatic Hydrolysis Processes, SERI/TP-231-3310.* Golden, Colo: Solar Energy Res. Inst.

61. Forsberg, C. W., Taylor, K., Crosby, B., Thomas, D. Y. 1986. The characteristics and cloning of bacterial cellulases. In *Biotechnology and Renewable Energy,* ed. M. Moo-Young, J. Hasnain, J. Lamptey, p. 101. London: Elsevier

62. Pasternak, J. J., Glick, B. R. 1987. Cloning of cellulase genes: genetic engineering and the cellulose problem. See Ref. 23, p. 139

63. Tiedje, J. M., Colwell, R. K., Grossman, Y. L., Hodson, R. J., Lenski, R. E., et al. 1989. The planned introduction of genetically engineered organisms: ecological considerations and recommendations. *Ecology* 70:298–315

64. Goldberg, R., Rissler, J., Shand, H., Hassebrook, C. 1990. *Biotechnology's Bitter Harvest: Herbicide-Tolerant Crops and the Threat to Sustainable Agriculture,* p. 45. New York: Environ. Defense Fund

65. Beyea, J., Keeler, K. H. 1991. Biotechnological advances in biomass energy and chemical production: impacts on wildlife and habitat. *Crit. Rev. Biotechnol.* 10:305–19

66. Hall, D. O., Coombs, J., Scurlock, J. M. O. 1985. Biomass production and data, Appendix C of *Techniques in*

Bioproductivity and Photosynthesis, ed. J. Coombs, D. O. Hall, S. P. Long, J. M. O. Scurlock, pp. 274–87. Oxford, United Kingdom: Pergamon. 2nd ed.

67. Bolton, J. R. Hall, D. O. 1990. The maximum efficiency of photosynthesis. *Photochem. Photobiol.* 51. In press

68. Crowley, J. J., ed. 1986. *Research for Tomorrow,* p. 231. Washington, DC: US Dept. Agric.

69. Hardy, R. W. F. 1988. Biotechnology for agriculture and food in the future. In *The Impact of Chemistry on Biotechnology,* ed. M. Phillips, S. P. Shoemaker, R. D. Middlekauff, R. M. Ottenbrite, p. 316. Washington, DC: Am. Chem. Soc.

70. Gotsch, N., Rieder, P. 1989. Future importance of biotechnology in arable farming. *Trends Biotechnol.* 7:29

71. Hubbard, H. M. 1989. Photovoltaics today and tomorrow. *Science* 244:297

72. Ostlie, L. D. 1988. The whole tree burner: a new technology in power generation. *Biologue* Dec.–Jan.:7–9

73. US Dept. Energy. 1990. *National Energy Strategy: Interim Report, DOE/S-0066P,* pp. 12, 16, 68, 105. Washington, DC: Natl. Tech. Inf. Serv.

74. Wellman, J. D. 1987. *Wildland Recreation Policy,* p. 284. New York: Wiley

75. USDA Forest Serv. 1980. *An Assessment of the Forest and Range Land Situation in the United States, FS-345,* pp. 32, 40, 86. Washington, DC: US Govt. Printing Off.

76. Robinson, G. 1988. *The Forest and the Trees: A Guide to Excellent Forestry,* pp. 63, 66–81, 83–96. Washington, DC: Island

77. Franklin, J. F., Forman, R. T. T. 1987. Creating landscape patterns by forest cutting: ecological consequences and principles. *Landscape Ecol.* 1:5–18

78. Probst, J. R., Rakstad, D. S., Rugg, D. J. 1991. *Breeding Bird Communities in Lake States Aspen Stands.* In press

79. Maser, C. 1988. *The Redesigned Forest,* pp. 14–38, 96–98, 149–53. San Pedro, Calif: Miles

80. Risser, P. G. 1981. Agricultural and forestry residues. In *Biomass Conversion Processes for Energy and Fuels,* ed. S. S. Sofer, O. R. Zaborsky, p. 25. New York: Plenum

81. Krapu, G. L., Reinecke, K. J., Frith, C. R. 1982. Sandhill cranes and the Platte River. *Trans. N. Am. Wild. Nat. Res. Conf.* 47:542

82. Ranney, J. W., Wright, L. L., Layton, P. A. 1987. Hardwood energy crops: the technology of intensive culture. *J. Forestry* 85:17–28

83. Paoletti, M. G. 1988. Soil invertebrates in cultivated and uncultivated soils in Northeastern Italy. *Firenze* 71:501–63

84. Paoletti, M. G., Favretto, M. R., Ragusa, S., Strassen, R. Z. 1989. Animal and plant interactions in the agroecosystems: the case of the woodland remnants in Northeastern Italy. *Ecol. Int. Bull.* 17: 79–91

85. Berner, A. 1988. The 1985 Farm Act and its implications for wildlife. In *Audubon Wildlife Report, 1988/1989,* ed. W. J. Chandler, pp. 437–65. New York: Academic

86. Bollinger, E. K., Bollinger, P. B., Gavin, T. A. 1990. Effects of hay cropping on eastern populations of the bobolink. *Wildlife Soc. Bull.* 18:142–50

87. Risser, P. G., Birney, E. C., Blocker, H. D., May, S. W., Parton, W. J., Wiens, J. A. 1981. *The True Prairie Ecosystem,* p. 557. Stroudsburg, Penn: Hutchinson Ross

88. USDA Soil Conserv. Serv. 1987. *Basic Statistics: 1982 National Resources Inventory, Stat. Bull. No. 756,* pp. 11, 14, 24, 70, 105–106, 149, 150. Iowa State Univ. Stat. Lab.

89. Off. Technol. Assess. 1982. *Impacts of Technology on U.S. Cropland and Rangeland Productivity,* pp. 23–56, 59, 67, 68–70. Washington, DC: US Congress; US Govt. Printing Off.

90. USDA Soil Conserv. Serv. 1989. *Summary Report: 1987 National Resources Inventory, Stat. Bull. No. 790,* p. 10. Iowa State Univ. Stat. Lab.

91. Off. Technol. Assess. 1985. *Technologies to Benefit Agriculture and Wildlife: Workshop Proceedings,* p. 17. Washington, DC: US Congress; US Govt. Printing Off.

92. Counc. Environ. Quality. 1979. *Our Nation's Wetlands: An Interagency Task Force Report,* pp. 2, 5–14, 31–32. Washington, DC: US Govt. Printing Off.

93. Fed. Interagency Comm. Wetland Delineation. 1989. *Federal Manual for Identifying and Delineating Jurisdictional Wetlands.* Washington, DC: US Army Corps Eng., US Environ. Protection Agency, US Fish Wildlife Serv., US Dept. Agric. Soil Conserv. Serv.; US Govt. Printing Off. 76 pp. plus appendices.

94. Tubbs, A. A. 1980. Riparian bird communities of the Great Plains. In *Management of Western Forest Grasslands for Nongame Birds, Tech. Rep. INT 86,* p. 419. Washington, DC: US Forest Serv.

95. Evans, W. 1981. *Impacts of Grazing Intensity and Specialty Grazing Systems on Faunal Composition and Productiv-*

ity, Draft Report. Washington, DC: Comm. Developing Strategies for Rangeland Management; Natl. Res. Counc.; Natl. Acad. Sci.

96. Foreman, D., Wolke, H. 1989. *The Big Outside*, p. 4. Tucson, Ariz: Ludd

97. Scott, J. M., Davis, F., Csuti, B., Butterfield, B., Noss, R., et al. 1990. *Gap Analysis: Protecting Biodiversity Using Geographic Information Systems*, p. 22. Moscow, Ida: Univ. Idaho

98. Noss, R. F., Harris, L. D. 1986. Nodes, networks and MUMs: preserving diversity at all scales. *Environ. Manage.* 10:299–309

99. Noss, R. F. 1987. Protecting natural areas in fragmented landscapes. *Nat. Areas J.* 7:2–13

100. Franklin, J. 1989. Toward a new forestry. *Am. Forests* 95:37–44

101. Hunter, M. L. Jr. 1990. *Wildlife, Forests, and Forestry: Principles of Managing Forests for Biological Diversity.* Englewood Cliffs, NJ: Prentice Hall

102. Francis, C. 1986. *Multiple Cropping Systems.* New York: Macmillan

103. Francis, C. 1990. *Sustainable Agriculture in Temperate Zones.* New York: Wiley

104. Gliessman, S. 1990. Agroecology: researching the ecological basis for sustainable agriculture. In *Agroecology: Researching the Ecological Basis for Sustainable Agriculture*, ed. S. Gliessman, pp. 3–10. New York: Springer-Verlag

105. Mollison, B. 1988. *Permaculture: A Designers' Manual.* Tyalgum, Australia: Tagari Publications

106. Posey, D. A., Balee, W., eds. 1989. *Resource Management in Amazonia: Indigenous and Folk Strategies.* Bronx, NY: NY Bot. Garden

107. Alcorn, J. B. 1990. Ethanobotanical knowledge systems: resource for meeting rural development goals. In *Indigenous Knowledge Systems: The Cultural Dimensions of Development*, ed. D. M. Warren, D. Brokensha, L. J. Slikkerveer. London: Kegan Paul

108. Forman, R. T. T., Godron, M. 1986. *Landscape Ecology*, pp. 11–13, 495–530. New York: Wiley

109. Zonneveld, I. S. 1990. Scope and concepts of landscape ecology as an emerging science. In *Changing Landscapes: An Ecological Perspective*, ed. I. S. Zonneveld, R. T. T. Forman, pp. 3–20. New York: Springer-Verlag

110. Forman, R. T. T. 1990. Ecologically sustainable landscapes: the role of spatial configuration. See Ref. 109, pp. 261–78

111. Ranney, J. W. 1990. Personal communication, Oak Ridge Natl. Lab., Oak Ridge, Tenn.

112. Off. Technol. Assess. 1987. *Technologies to Maintain Biological Diversity, OTA-F-330*, pp. 37–59, 221–49. Washington, DC: US Congress; US Govt. Printing Off.

113. Dasmann, R. F. 1978. Wildlife and ecosystems. See Ref. 15, pp. 18–27

114. Baker, R. 1990. *Soil Microbial Diversity and the Biological Control of Plant Pathogens (Abstr.).* Feb. Am. Acad. Arts. Sci. Meet., New Orleans, La.

115. Bruce, C., Edwards, D., Mellen, K., McMillan, A., Owens, T., Sturgis, H. 1985. Wildlife relationships to plant communities and stand conditions. In *Management of Wildlife and Fish Habitats in Forests of Western Oregon and Washington; Part 1—Chapter Narratives*, ed. E. R. Brown, pp. 34–35. Portland, Ore: USDA Forest Serv.; Pacific Northwest Region

116. Logan, W., Brown, E. R., Longrie, D., Herb, G., Corthell, R. A. 1985. Edges. See Ref. 115, pp. 115–27

117. Hall, F. C., McComb, C., Ruediger, W. 1985. Silvicultural options. See Ref. 115, pp. 291–306

118. Oakey, A. L., Collins, J. A., Everson, L. B., Heller, D. A., Howerton, J. C., Vincent, R. E. 1985. Riparian zones and freshwater wetlands. See Ref. 115, pp. 57–80

119. Harris, L. D. 1984. *The Fragmented Forest*, pp. 19–23, 44–68. Chicago, Ill: Univ. Chicago Press

120. Hall, F. C., Brewer, L. W., Franklin, J. F., Werner, L. W. 1985. Plant communities and stand conditions. See Ref. 115, pp. 17–31

121. Brown, E. R., ed. 1985. See Ref. 115, Appendix 8: Occurrence and orientation of 414 wildlife species to Western Oregon and Washington plant communities, stand conditions and special or unique habitats.

122. Jackson, W., Piper, J. 1989. The necesssary marriage between ecology and agriculture. *Ecology* 70:1591–93

123. Gliessman, S. R. 1990. *Sustainable Agriculture and the Conservation of Biodiversity (Abstr.).* Feb. Am. Acad. Arts. Sci. Meet., New Orleans, La.

124. Amador, M. F., Gliessman, S. R. 1990. An ecological approach to reducing external inputs through the use of intercropping. See Ref. 104, pp. 147–59

125. Taylor, K. I. 1988. Deforestation and Indians in Brazilian Amazonia. See Ref. 13

126. Jimenez-Osornio, J. J., Gomez-Pompa,

A. 1989. *Human Role in Shaping of the Flora in a Wetland Community, the Chinampa*. Paper presented at the June 5–8 Int. Conf. Wetlands, Leiden, The Netherlands

127. Farrell, J. 1990. The influence of trees in selected agroecosystems in Mexico. See Ref. 104, pp. 169–83

128. Forsman, E. D., Horn, K. M., Mires, G. W. 1985. Northern spotted owls. See Ref. 115, pp. 259–67

129. Neitro, W. A., Binkley, V. W., Cline, S. P., Mannan, R. W., Marcot, B. G., Taylor, D., Wagner, F. F. 1985. Snags. See Ref. 115, pp. 129–69

130. Bartels, R., Dell, J. D., Knight, R. L., Schaefer, G. 1985. Dead and down woody material. See Ref. 115, pp. 171–86

131. MacClintock, L., Whitcomb, R. F., Whitcomb, B. L. 1977. Evidence for the value of corridors and minimization of isolation in preservation of biotic diversity. *Am. Birds* 31:6–16

132. Harris, L. D. 1984. See Ref. 119, pp. 141–65

133. Wilcove, D. S. 1985. Nest predation in forest tracts and the decline of migratory songbirds. *Ecology* 66:1211–14

134. Wilcove, D. 1990. Empty skies. *Nat. Conserv. Mag.* 40:4–13

135. Grossman, J. 1990. Mulch better. *Agrichemical Age*. Nov.: 4–5, 16–17

136. Taylor, S. 1990. Personal communication. Mich. Dept. Nat. Resourc., Mio, Mich.

137. Franklin, J. 1989. An ecologist's perspective on northwestern forests in 2020. *Forest Watch* 10:6–9

138. Brown, E. R., Curtis, A. B. 1985. Introduction. See Ref. 115, pp. 1–15

139. Stalmaster, M. V., Knight, R. L., Holder, B. L., Anderson, R. J. 1985. Bald eagles. See Ref. 115, pp. 269–90

140. Harris, L. D., Gallagher, P. B. 1989. New initiatives for wildlife conservation: the need for movement corridors. In *Preserving Communities and Corridors*, ed. G. Mackintosh, pp. 11–34. Washington, DC: Defenders of Wildlife

141. Gall, G. A. E., Stanton, M., eds. 1989. *Integrating Conservation Biology and Agricultural Production*, pp. 33–52. Davis, Calif: Publ. Serv. Res. Dissemination Program; Univ. Calif.

142. Lester, R. T., Myers, J. P. 1989. Global warming, climate disruption and biological diversity. In *Audubon Wildlife Report 1989/1990*, ed. W. J. Chandler, pp. 177–221. San Diego, Calif: Academic

143. Hardin, G. 1978. Political requirements for preserving our common heritage. See Ref. 15, pp. 310–17

144. Beyea, J. 1990. *Bringing Environmental Damage Costs into the Electricity Marketplace: Gains to Be Expected and Pitfalls to Be Avoided*. Natl. Conf. Environ. Externalities, Oct. 2, Jackson Hole, Wyo.

145. Costanza, R., Farber, S. C. 1984. Theories and methods of valuation of natural systems: a comparison of willingness-to-pay and energy-analysis based approaches. *Man, Environment, Space and Time* 4:1–38

146. Turner, M. G., Odum, E. P., Costanza, R., Springer, T. M. 1988. Market and nonmarket values of the Georgia landscape. *Environ. Manage.* 12:209–17

147. Brokaw, H. P., ed. 1978. See Ref. 15, pp. 275–309

148. Garrett, M. A. Jr. 1987. *Land Use Regulation: The Impacts of Alternative Land Use Rights*, pp. xv–xix. New York: Praeger

149. McNeely, J. A. 1988. *Economics and Biological Diversity: Developing and Using Economic Incentives to Conserve Biological Resources*, pp. ix–xiv. Gland, Switzerland: Int. Union for Conserv. Nature and Nat. Resourc.

150. Swanson, G. A. 1978. See Ref. 15, pp. 428–41

151. Greenwalt, L. A. 1978. See Ref. 15, pp. 399–412

152. Ottinger, R. L., Wooley, D. R., Robinson, N. A., Hodas, D. R., Babb, S. E. 1990. *Environmental Costs of Electricity*. New York: Oceana Publications

153. Peters, C. M., Gentry, A. H., Mendelsohn, R. O. 1989. Valuation of an Amazonian rainforest. *Nature* 339:655

154. Wick, H. L., Canutt, P. R. 1979. Impacts on wood production. In *Wildlife Habitats in Managed Forests: the Blue Mountains of Oregon and Washington: Part 1—Chapter Narratives*, ed. E. R. Brown, pp. 307–13. Portland, Ore: USDA Forest Serv.; Pacific Northwest Region

155. Armey, D. 1990. Farming for fun and profit. *Christian Science Monitor*. April 27, p. 19, col. 1

156. Wolfe, J., Mobley, M. 1989. How forests measure up. *Am. Forests*. Nov.–Dec.:33

157. Carbon Dioxide Inf. Anal. Cent. 1990. *Glossary: Carbon Dioxide and Climate, ORNL/CDIAC-39*, p. 44. Oak Ridge, Tenn: Oak Ridge Natl. Lab.

158. Energy Inf. Admin. 1991. *Monthly Energy Rev.* March, Table 1.1

Annu. Rev. Energy Environ. 1991. 16:433–531

LEAST-COST CLIMATIC STABILIZATION*

Amory B. Lovins and L. Hunter Lovins

Rocky Mountain Institute, 1739 Snowmass Creek Road, Old Snowmass, Colorado 81654–9199

KEY WORDS: global warming, energy efficiency, renewable energy, sustainable farming and forestry, least-cost energy policies.

INTRODUCTION: STABILIZING GLOBAL CLIMATE SAVES MONEY

The threat of serious, unpredictable, and probably irreversible changes in the earth's climate has moved from conjecture to suspicion to near-certainty (1). Denial is now confined to the uninformed (2). Yet the threat's cause continues to be widely misunderstood even by many experts on its mechanisms.

Global warming is not a natural result of normal, optimal economic activity. Rather, it is an artifact of the economically inefficient use of resources, especially energy. Advanced technologies for resource efficiency, and proven ways to implement them, can now support present or greatly expanded worldwide economic activity while stabilizing global climate—and saving money. New resource-saving techniques—chiefly in energy, farming, and forestry—generally work better and cost less than present methods that destabilize the earth's climate.

In short, even based on energy-efficiency assessments by such organizations as the Electric Power Research Institute (the utility industry's think-tank) and leading US national laboratories, most of the best ways known today to abate climatic change ("stabilize" the climate) are

---

1. not costly but profitable;
2. not hostile but vital to global equity, development, prosperity, and security; and
3. reliant not on dirigiste regulatory intervention but on the intelligent application of market forces.

Newer analyses summarized here reveal even bigger and cheaper energy-saving potential, as well as innovative ways to abate other sources of global warming at unexpectedly low costs. These findings imply that most global warming can be abated not at roughly zero net cost, as two government-sponsored analyses have recently found, but at negative net cost—without ascribing any value to the abatement itself.

However, technical and implementation options, obstacles, and strategies vary widely with culture and geography. This paper therefore identifies opportunities to use resources more efficiently, describes concrete successes in capturing those opportunities, and outlines an agenda for systematically harnessing them in three main types of societies: the industrialized Organization for Economic Cooperation and Development (OECD) countries, the USSR and its former satellites, and developing countries. The discussion highlights needed interactions (e.g. issues of trade, technology transfer, and emulation) and possible synergisms among these three regions.

MAJOR ABATEMENT TERMS

Over a century's time horizon, about half of global warming in the 1980s[1] was caused (1) by burning fossil fuel, which produces carbon dioxide (CO_2) and monoxide (CO), nitrous oxide (N_2O), fugitive methane (CH_4), and ozone (O_3). Another one fourth or more was driven by unsustainable farming and forestry practices, which produce biotic CO_2 and CH_4 and nitrogen-cycle N_2O. Virtually all the rest was driven by the release of halons and chlorofluorocarbons (CFCs), whose production (but not use) will be phased out during the 1990s to protect the stratospheric ozone layer. Among the ~57% of current worldwide contributions to global warming estimated by the US Environmental Protection Agency (EPA) to arise directly from energy use (Ref. 5, p. VII-28), 20% is ascribed to transportation, 22% to industry, and 15% to buildings. The sizeable uncertainties in all these figures are immaterial for the purposes of this paper, since, as will be shown, major abatements are available and cost-effective for each gas and in each application.

---

[1]Counting contributions by gases projected to be released in 2025 rather than in the 1980s (Ref. 3, Figures 2, 3), or integrating their effects over a longer period (4), would strengthen this paper's conclusions, which are most detailed for energy efficiency. These conclusions hold independently of such assumptions, since they rest on the demonstrated potential to reduce dramatically all three source terms (energy, agroforestry, and chlorofluorocarbons).

Stabilizing atmospheric concentrations of heat-trapping gases at current levels will require that their present rates of emission be reduced by different amounts, because different gases remain in the air for different periods and react to form different products. These required reductions are believed (1) to be more than 60%[2] for CO_2, 15–20% for CH_4, 70–80% for N_2O, 70–85% for the most important CFCs,[3] and 40–50% for HCFC-22. These reductions, too, are not as interchangeable in practice as they might seem mathematically (Ref. 6, Ch. 2). Yet reductions generally larger than these—large enough to return the atmosphere to a composition likely to entail no climatic change (Ref. 6, Ch. 2)—are shown below to be very cheap, free, or better than free.[4]

Specifically, this paper shows that demonstrated technologies and implementation methods can

1. save most of the fossil fuel now burned, at a cost below that of the fuel itself, making the abatement cost less than zero;
2. change soil from a carbon source to a carbon sink (and incidentally reduce related emissions of CH_4 and N_2O too) at a net cost around zero or less; and
3. displace CFCs (and often their proposed hydrohalocarbon substitutes) at a net cost close to zero—though this cost is irrelevant, since the substitution is already required by international treaty to abate stratospheric ozone depletion.

The cost of the main global-warming abatements therefore ranges, broadly speaking, from strongly negative to roughly zero to irrelevant—with policy implications discussed in the CONCLUSIONS section.

TECHNOLOGICAL AND ECONOMIC OPTIONS

Methodological Note

Throughout this paper, unless otherwise noted, costs of saved energy are expressed in 1986 US dollars, levelized at a 5%/y real discount rate using the Lawrence Berkeley Laboratory methodology (Ref. 7, pp. 11ff; Ref. 8, "Conventions"), whereby the marginal cost of buying, installing, and maintaining the efficient device is divided by its discounted stream of lifetime energy

---

[2]Simulations suggest 50–80% (5).

[3]Ref. 5 gives 75–100% for CFC-11 and -12.

[4]This paper does not consider several recently proposed innovations that could allegedly capture and sequester CO_2 from electric generation and other combustion processes at relatively low costs (6a, 6b). If such processes turn out to work, that is "icing on the cake." Early data suggest, however, that these processes, while probably cheaper than global warming, abate CO_2 at a cost several to many times that of the proven energy-efficiency options described below.

savings. Specifically, the levelized cost ($) of saving, say, 1 kWh equals $Ci/S[1 - (1+i)^{-n}]$, where C is installed capital cost ($), i is the annual real discount rate (in this case 0.05), S is the energy saved by the device (kWh/y), and n is its operating life (y). Thus a $10 device that saved 100 kWh/y and lasted 20 y would have a cost of saved energy of 0.80¢/kWh. Against a 5¢/kWh electricity price, a 20-y device with a one-year simple payback saves energy at a cost of 0.4¢/kWh. Similar accounting is used for the cost of saving direct fuel. Cost of saved energy is methodologically equivalent to cost of supplied energy (e.g. from a power plant): the price of the energy saved is not part of the calculation, and whether the saving is cost-effective depends on comparing the cost of achieving it with the avoided cost of the energy saved. If that avoided cost is P, for example, then the simple payback is C/SP.

The installed capital cost C is marginal cost for devices installed in the course of new construction or routine replacement—usually assumed for devices, like household appliances or commercial chillers, that are normally replaced anyhow every 10–20 years and can simply be replaced by better models. C is total cost for immediate retrofit—usually assumed for options whose lives are relatively short, like lights or motors, or many decades, like building shells. Unless otherwise noted, costs and savings used in this paper are empirical, based on purchasing and operating conditions typical of the application described. Financing costs are not normally included (because devices with paybacks of about a year, like most of those described, should normally be expensed rather than capitalized), but could be with a minor effect on the results.

If the new device has a different lifetime, required population, or maintenance cost than the original device, then C is corrected for the change in present-valued capital and operating cost to achieve the same cycle life. In some cases—as with the compact fluorescent lamp mentioned below, which lasts ~13× as long as the incandescent lamp it replaces—the saved present-valued maintenance cost can exceed the total cost of the device, resulting in a negative value for both C and the cost of saved energy. Naturally, this means that the present-valued dollar cost of the efficient device (lifetime kWh saving times ¢/kWh cost of that saving) is negative and can be used to offset the positive costs of other kinds of savings.

In general, this paper does not explicitly count the transaction costs of implementing the technologies described, since these costs are highly sensitive to implementation methods and marketing skill. In mature utility programs, however, transaction costs (program design, marketing, physical delivery, monitoring, evaluation, research, regulatory support, etc) are very small. For example, Southern California Edison Company in 1984 reported transaction costs totalling only hundredths of a cent per kWh. Details are in (Ref. 7, pp. 105–106). The 1984 transaction costs reported to the California PUC (8a) were of two kinds. Those directly allocated to programs included outreach, research, evaluation, and technical and regulatory support; levelized over the annualized savings during an assumed 20-y average measure life, they cost 0.0051¢/kWh in the commercial, industrial, and agricultural sectors and 0.039¢/kWh in the residential sector, or respectively 3.5% and 2.0% of direct program costs. In addition, general demand-side overheads for administration, public awareness and education, advertising, measurement, and evaluation— reduced 25% pro rata on the direct-cost share of nonefficiency programs (solar, independent power production, and load management)—cost 0.026¢/kWh. These latter costs may have been partly or wholly offset by the avoidance of similar costs to support the supply-side resources displaced. As noted by Nadel (8b), however, transaction costs can be one or two orders of magnitude higher in less mature or less well-designed programs, especially those on a small scale, in startup, or with heavy residential emphasis.

This paper does not try to assess the macroeconomic consequences of energy efficiency (nor, similarly, of other global warming abatements such as changes in farming and forestry practices). In general, saving energy and capital should lower their prices and hence increase the net dollar saving to society. At the levels of savings described here, this indirect benefit might exceed the direct dollar savings. The cheaper energy and capital would also be expected to boost economic output, and this too would conventionally be considered a benefit. Whether the increased output in turn decreased the net energy savings would depend on the composition of the marginal output, and on whether, for example, people used their extra income to buy snowmobiles and motorboats, symphony tickets, or additional energy-saving measures. Such differences between the engineering-economics approach used here and the econometric approach are discussed elsewhere (9–12); perhaps the most important difference is that econometric models generally omit any potential for negative-cost savings (i.e. saving fuel more cheaply than buying and burning it) because, in the economic paradigm, such cost-effective savings are assumed to have been achieved already.

Energy Efficiency

Removing a 75-watt incandescent lamp and screwing into the same socket a 15-watt compact fluorescent lamp[4a] will provide the same amount of light for 13 times as long, yet save enough coal-fired electricity over its lifetime to keep about a ton[5] of CO_2 out of the air (plus 8 kg of sulfur oxides and various

---

[4a]An "integral" model, in which the adapter base, ballast, lamp, and envelope (if any) are a single throwaway unit, may replace a 75-W incandescent lamp with 14–18 W using an electronic or 16–20 W using an electromagnetic ballast. With a reflector-equipped "modular" model, lower wattages (down to ~8 W) may deliver adequate, equivalent light in a directional beam.

[5]All tons (t) in this paper are metric; all miles (mi) and gallons (gal) are US; and standard metric prefixes are used (M = 10^6, G = 10^9, T = 10^{12}, P = 10^{15}, E = 10^{18}), along with mostly metric units (e.g. 1 ha = 2.4 acre).

other pollutants). If the quintupled-efficiency lamp saves oil-fired electricity, as it would in many developing countries, it can save more than enough oil to drive a standard 20-mi/gal car for a thousand miles, or to drive the most efficient prototype car across the United States and back. If it saves nuclear electricity, it avoids making a half-curie of long-lived wastes and two-fifths of a ton-TNT-equivalent of plutonium. Yet far from costing extra, the lamp generates tens of dollars' net wealth—it saves tens of dollars more than it costs—because it displaces replacement lamps, installation labor, and utility fuel. It also defers hundreds of dollars' utility investment (13).[6]

This is an example—one of the costlier ones that could be given—of the proposition that today it is generally cheaper to save than to burn fuel. The CO_2 and other pollution avoided by substituting efficiency for fuel are thus avoided not at a cost but at a profit.

Most of the best energy-saving technologies—especially the superefficient lights, motors, appliances, and other end-use devices that save electricity—are supplanted by still-better models approximately annually. However, this is no reason to wait, as the loss from deferring the benefit usually exceeds the benefit of the new model's often modest incremental improvement. Saving electricity gives the most climatic leverage, because it takes 3–4 units of fuel (in socialist and developing countries, often 5–6 units) to generate and deliver a single unit of electricity, so saving that unit displaces many units of fuel, mainly coal, at the power plant. Power plants burn one third of the world's fuel and emit one third of the resulting CO_2, as well as one third of the nitrogen oxides (NO_x) and two thirds of the sulfur oxides (SO_x), both of which also contribute to global warming (Ref. 6, Ch. 2)—a little directly, and more by degrading forests and other ecosystems that otherwise store carbon. Electricity is also by far the costliest form of energy,[7] so it is the most lucrative kind of energy to save. Saving electricity saves much capital: the United States in the mid-1980s spent as much private capital and public subsidy (15) expanding its electric supply, about $60 billion per year, as it invested in all durable-goods manufacturing industries. Moreover, one fourth of the world's development capital goes to electrification, and about five times as much such capital is projected to be needed in the 1990s as is likely to be available. This ~$80 billion annual shortfall (16, 17) may imperil proposed development. For these reasons, this discussion emphasizes the frequently undervalued opportunities to save electricity.

---

[6]These calculations (13) assume that each kWh saved directly by the lamp also saves 0.36 kWh of net space-conditioning energy, as it would in a typical US commercial-sector building (8, 14).

[7]Average 1989 US retail electricity at 6.44¢/kWh (1989$) is equivalent in heat content to oil at $110/bbl.

ELECTRICITY   Many utilities still think that only ~10–20% of the electricity used can be cost-effectively saved. However, a recent reassessment by the Electric Power Research Institute found a potential, mainly cost-effective, to save 24–44% of US electricity within this decade, not counting a further 8.6% already expected to occur spontaneously or another 6.5% likely to be saved by utilities' planned efficiency programs (18, 19). The California Energy Commission has similarly identified a potential to save electricity 2.5%/y faster than projected load growth (Ref. 20, Figures 3–1 and C-4). As shown below, such electrical savings, and analogous non-electrical energy savings, can save enough money to pay for most non-energy kinds of global-warming abatement. Most of this electricity-saving potential is untapped: for the non-Communist world during 1973–1987, oil intensity fell by 32%, but non-oil intensity by only 1%.

Analyzing technologies even a few years old, however, can make potential savings seem much smaller and costlier than they now are (21). Today's best electricity-saving technologies can save twice as much as five years ago, but at only one-third the real cost. Still more detailed assessments[8] of these new opportunities, based on measured cost and performance data, thus reveal that full retrofit of US buildings and equipment with today's most efficient commercially available end-use technologies would deliver unchanged or improved services while saving far more electricity, and at far lower cost, than previously supposed. This makes it possible to abate a large fraction of global warming—enough, it appears, to stabilize the earth's climate (i.e. stop perturbing it)—at negative net cost.

The modern US electric-efficiency potential, compared to "frozen" 1986 efficiency levels, includes saving

1. half of motor-system (or one fourth of total) electricity through 35 motor, control, drivetrain, and electric-supply improvements collectively paying back in ~16 months against 5¢/kWh electricity (22, 19)[9]—a key opportunity in reindustrializing countries like the USSR, whose motors already use 61% of all its electricity (26, 27);
2. 80–92% of lighting electricity (or one fourth of total electricity including net space-conditioning effects) at a net cost less than zero, because much

---

[8]These are presented chiefly in the COMPETITEK Hardware Reports cited below. The first three such reports (8, 22, 14) total 1268 single-spaced pages documented with more than 3000 notes, chiefly citing primary sources. The findings of the three remaining Hardware Reports are already known in sufficient detail through earlier analyses (e.g. 23, 24, 7) and from many other authors' studies, some of the most important of which are cited below.

[9]Improvements to or beyond the machine driven by the motor are not included here, but can often save about half the remaining energy (e.g. 25, 25a). Each unit of energy saved downstream, e.g. by reducing friction in pipes exiting a pump, can save nine units of fuel at the power plant.

of the lighting equipment more than pays for itself by costing less to maintain (8, 28);

3. one sixth of total electricity through numerous design improvements to household appliances, commercial refrigeration and cooking, and office equipment—where the potential savings exceeds 90% at roughly zero or negative cost (14);

4. two thirds of water-heating electricity through eight simple improvements (insulation, high-performance showerheads, etc) (23);

5. most of the electricity used for space-heating and -cooling, through both mechanical-equipment retrofits and improved building shells[10]—including "superwindows" that can now insulate 2–4 times as well as triple glazing but cost about the same (32, 33, 24, 7);

6. three fourths of all electricity used in typical US houses and commercial buildings at respective retrofit costs of 1.6¢/kWh and -0.3¢/kWh (7);

7. about three fourths of total US electricity at a net cost averaging about 0.6¢/kWh (Figure 1)—several times cheaper than just operating a typical coal or nuclear power plant, even if building it cost nothing. Of course, considerably more could be saved at less than long-run marginal cost, which is at least tenfold higher, and higher still when externalities are included.[11]

Two examples—industrial drivepower (22, 31a) and commercial fluorescent lighting (8)—illustrate how such large, cheap savings can be achieved by, and only by, whole-system engineering with meticulous attention to detail.

Most analysts emphasize only two drivesystem improvements: high-efficiency induction (asynchronous) motors, and adjustable-speed drives (ASDs) using variable-frequency electronic inverters. The motors gain several percentage points' efficiency by using more and lower-loss materials plus better design and manufacturing. This is worthwhile because a large motor typically consumes electricity worth its entire capital cost every few weeks. Further, the output of many pumps, blowers, and fans is controlled by running them at full speed against mechanical obstruction. Yet their power

---

[10]And through lighter-colored paving and building surfaces and smarter landscaping, especially urban forestry: direct shading, evapotranspiration, and reduction in the mesoscale urban "heat island" effect will enable a half-million trees in Sacramento, for example, to save 15 peak MW and 35 GWh/y in the tenth year of a program costing 2¢/kWh (29, 30). In Sacramento, simulations show savings up to 14% in peak and 19% in annual cooling energy just from whitewashing buildings, and up to 35% and 62% from all measures to modify urban albedo (to average values of 40% overall and 90% for houses) (31).

[11]An exhaustive compilation (34) found that external costs due mainly to SO_x, NO_x, CO_2, and particulate emissions total about 5.8¢ for coal-fired generation, 2.7¢ oil-fired, 1.0¢ gas-fired, and 2.5¢ nuclear, all per busbar kWh.

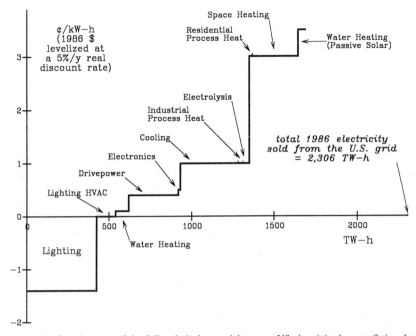

Figure 1 Supply curve of the full technical potential to save US electricity by retrofitting the best commercially available 1989 end-use technologies wherever they fit in the 1986 stock of buildings and equipment. The vertical axis is levelized marginal cost (1986 ¢/kWh delivered, 5%/y real discount rate). Costs are negative if the efficient equipment's saved maintenance costs exceed its installed capital costs. The horizontal axis is cumulative potential saving corrected for interactions. Measured cost and performance data are summarized for about a thousand technologies, condensed into end-use blocks. Fuel-switching, life-style changes, load management, further technological progress, and some technical options are excluded. How much of the potential shown is actually captured is a policy variable, but many utilities have in fact captured 70–90+% of particular efficiency markets in a few months or years through skillful marketing, suggesting that most of the potential shown could actually be captured over a few decades. Note that savings totalling around 50% have a net internal average cost of zero, and that new-construction savings would be larger and much cheaper (often negative-cost) than the retrofit savings shown. For comparison, utilities such as Bonneville Power Administration and Wisconsin Power & Light Co. report empirical total costs of 0.5¢/kWh for saving business customers' electricity. The "electronics" saving has turned out in more recent analyses to be larger and cheaper than shown (Ref. 14, Ch. 6), and the drivepower saving, to be probably twice as large as shown (22, 19), although saving more drivepower energy leaves less cooling energy available to be saved.

consumption varies roughly as the cube of their flow rate, so if only half the full flow were needed, seven-eighths of the full input power (less minor ASD circuit losses) could be saved by removing the obstruction and halving the speed. ASDs' full use could thus save ~20% (18) or ~14–27% (22) of all US motor energy, with typical paybacks estimated at ~1 to ~2½ y respectively.

So far so good. But adding 33 further drivesystem improvements—in the choice, sizing, maintenance, and life of motors, in control systems of three further kinds, and in upstream electrical supplies and downstream mechanical drivetrains—can at least double the savings from these two measures. It can also cut total retrofit cost by perhaps fivefold (19), because of the 35 combined measures, 28 are free by-products of the seven that must be paid for, yielding greater savings at no extra cost.

For example, immediately retrofitting an in-service standard induction motor to a high-efficiency model, without waiting for it to burn out, is commonly assumed to incur an unattractively long (~10–20 y) payback. Yet many US motors are so grossly oversized that probably half never exceed 60%, and a third never exceed 50%, of their rated load. This oversizing often makes actual at-load efficiency lower than the nameplate rating implies, and may enable the replacement motor to be smaller, hence cheaper. Making the new motor the right size reduces the payback of immediate retrofit to ~3 y. Also counting the new motor's longer life (because it runs cooler and has higher-quality bearings) makes the immediate-retrofit cost negative, averaging about −$13/kW.

In addition, the new motor automatically eliminates any increased magnetic losses that may have been caused by improper repair of the old motor. This plus proper motor sizing yields direct electrical savings roughly twice as big as would be expected from the new motor's better nameplate efficiency alone. The high-efficiency motor also has better power factor and greater harmonic tolerance (hence better ASD operation). Thus it provides a half-dozen important operational advantages—but need be paid for only once.

Many of these savings, however, depend on others. For example, not only efficiency but also motor life depends on other energy-saving improvements: reducing voltage imbalance between the phases, improving shaft alignment and lubrication practice, reducing overhung loads (sideways pulls) on the shaft (e.g. by substituting toothed, non-stretch, low-tension "synchronous" belts for V-belts), and improving housekeeping—not siting motors in the sun or next to steam pipes, not smothering them beneath multiple coats of paint, etc. Motor choice, life, sizing, controls, maintenance, and associated electrical and mechanical elements all interact intricately. A few interactions are unfavorable, but most make the savings of the whole drivepower package far larger and cheaper than would appear from considering just a few fragmented measures, as most analyses do.

Or consider commercial fluorescent lighting: say, four 40-W lamps driven by two 16-W electromagnetic ballasts in a ~65%-efficient enclosed luminaire.[11a] An imaging specular reflector—a very shiny, computer-

---

[11a]Surprisingly, starting with three 34-W lamps, two "high-efficiency" electromagnetic ballasts, and a modern louvered parabolic troffer yields the same percentage savings within a few percent, partly because of more favorable thermal interactions within the luminaire. Either way, there are ~1½ billion such fixtures in US service.

designed, specially shaped piece of sheet-metal—inserted above the lamps nearly doubles optical efficiency, because each exit ray bounces barely more than once rather than nearly three times, yielding almost the cube of the reflectance benefit. Half the lamps can then be removed, the rest relocated, and approximately the same delivered light obtained as before. (The removed lamps appear to be still there, but they are only virtual images, and virtual lamps require no electricity or maintenance. The avoided maintenance costs end up paying for half the retrofit package.) While being relocated, the lamps can also be replaced with new ones whose "tristimulus" phosphors—tuned to red, green, and blue retinal cones—emit up to 18% more light per watt, with more pleasant and accurate color.

The two two-lamp ballasts can then be replaced with one four-lamp high-frequency electronic ballast shared between two adjacent luminaires. The ballast and its control systems save electricity in at least 15 ways. These include lower ballast dissipation and higher lamp output at ~20 kHz than at line frequency (together boosting system lumen/W by >40%); reduced sensitivity to abnormal supply voltage or lampwall temperature (cutting design margins by an eighth); continuous dimming of the lamps to match available daylight (often saving >50% in perimeter zones); brightening the lamps as they dim with age and dirt, so they need not be too bright when young, fresh, and clean in order to provide enough light when old, tired, and dirty (saving a seventh of the energy over each group relamping cycle); and facilitating automatic control by occupancy sensors and timers. Together, these ballast and control mechanisms typically save about half the energy per unit of delivered light in the center of a large building, and 70–80+% in a typical mix of core and perimeter zones. The better lamp phosphors and reflector optics cut W/lux by a further ~15% and ~35+% respectively—a cumulative total saving of ~83–91%.

Such large savings are not unusual even in awkward cases, because further opportunities are available too: reducing endemic overlighting, concentrating local light on the visual task, making the light more visually effective (e.g. by radial polarizers that reduce veiling glare), using lighter-colored surfaces to bounce light around better, bouncing daylight several times as far into the room (via lightshelves, top-silvered blinds, glass-topped partitions, etc), and improving maintenance. In all, 70–90+% savings on electricity used for lighting are typically available at a cost of ~0.6¢/kWh (or twice that if the maintenance costs saved by the customer were ignored), with no reduction in illuminance and greatly improved lighting quality.

As with motors, however, achieving both such large savings and higher-quality service depends on harnessing complex interactions—thermal, optical, and electrical—between all the components. It requires including all the right parts, and combining them into something greater than their sum. This requires not just new technology but also new thinking, and new ways to deliver integrated packages of modern hardware plus managerial and cultural changes.[12] That is not easy; but neither is expanding electric supplies.

Illustrating the power of properly combined packages of technologies, five engineering firms' conceptual designs recently found (34a) a cost-effective retrofit energy-saving potential of 67–87% in a 1900-m^2 Pacific Gas & Electric Company research office that was one-third more efficient than a typical US office to start with. One firm found 86% potential retrofit electrical savings, at ~27% of PG&E's long-run marginal supply cost, by combining daylighting, advanced lighting components and controls (delivering 500 lux with 3.1 W/m^2), ~20 W/workstation office equipment, ~0.7 W/m^{2}K super-windows with spectral response "tuned" to each elevation of the building, and a ~65%-downsized mechanical system with a coefficient of performance (COP) of >10.

Potential savings appear to be only slightly smaller and costlier in the most efficient countries than in the United States. Detailed studies have found a potential to save half of Swedish electricity at an average cost of 1.3¢/kWh (35), half the electricity in Danish buildings at 0.6¢/kWh or three fourths at 1.3¢/kWh (36), and 80% (including fuel switching) in West German house-holds with a 2.6-year payback (37). To be sure, Europeans do (for example) light their offices less intensively, and turn the lights off more, than Americans do, but that does not affect the percentage savings available in the lighting energy that is used—a function only of the lighting technology itself, which is quite similar in both places.

Abundant observational evidence confirms that the potential savings in socialist and developing countries (38) are much larger and (at world equipment prices) cheaper than in OECD. Differences in what electricity is used for between industrialized and developing or capitalist and socialist countries are surprisingly small (38, 17, 39), and major savings are available in essentially every significant application. The feasibility of major electric savings is confirmed by comparisons at all scales: the micro-scale of individual technologies (e.g. 14), sectoral intensities[13] and aggregate intensities.[14] It therefore seems reasonable, and probably conservative, to treat the US values as a surrogate for the global average of potential electrical savings' fractional quantity and average cost.

---

[12]With motors, for example, an important cultural need is to change lubrication from a low-caste, dirty-hands occupation to a high-caste, white-lab-coat occupation.

[13]For example, Kahane (40) found that in the car, paper, and cement industries, electricity per ton of product was falling in Japan but rising in the United States.

[14]For example, International Energy Agency statistics show that Japan's GNP in 1986 was 36% less electricity-intensive than the US GNP, and this gap was projected by those governments to widen to 45% in 2000.

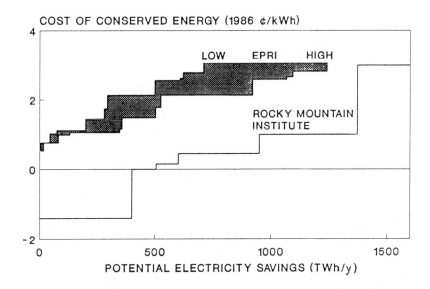

Figure 2 Supply curves for potential US savings of electricity, compared with ~1986–1988 frozen efficiencies, from analyses by EPRI (18) and RMI (19). This comparison, slightly modified from Figure 12 of Hirst (41), shows EPRI's uncertainty range; the RMI curve is a midcase. Important methodological differences between the two curves are summarized in the text. The RMI curve is now considered by the authors to be technically conservative, and will be updated in early 1992.

A recent comparison (41) of the Rocky Mountain Institute (RMI) electric-efficiency supply curve in Figure 1 with the Electric Power Research Institute's version (19) shows substantial similarities (Figure 2). But nearly all the remaining difference is from the EPRI curve's showing a drivepower saving three times smaller and five times costlier than EPRI agreed to in the same article (19), and from methodological differences:

1. the EPRI supply curve shows only potential savings by the year 2000, excluding a further 9–15% saving that EPRI believes will occur by then automatically (through price response, government standards, and present utility programs), while the RMI suppply curve shows full long-term savings potential, and
2. EPRI excludes but RMI includes credit for maintenance costs saved by customers, so commercial lighting savings cost 1.2¢/kWh in the EPRI but −1.4¢/kWh in the RMI supply curve.

Normalizing for these nonsubstantive differences would make the two curves nearly identical. The remaining differences—believed to be due to the modernity, thoroughness of characterization, and disaggregation[14a] of the measures analyzed—are less important than the EPRI/RMI consensus that cost-effective potential savings are many times larger than utilities currently plan to capture.

OIL   The potential for saving oil with today's best demonstrated technologies is also large and cheap. Unlike electricity-saving options, about half of the needed technologies are not yet on the market, though they could be within a few years. There are large potential savings in transportation (~44% of world oil use), industrial heat (~12%), building heat (~14%), electric generation (~10%), and feedstocks (~14%). The rest of the oil is used or lost in refineries (42).

Transportation   Personal mobility, the most familiar and pervasive use of oil, accounts for about two thirds of OECD transportation energy use (Ref. 5, p. VII-37) and offers some of the most dramatic savings. To start with, a US Department of Energy (DOE) study (43) describes 15 proven, readily available improvements in car design. These, plus two more equally straightforward improvements (44), can maintain average 1987 US new-car size, ride, and acceleration at 33.6 actual mi/gal (7.0 l/100 km). That is 35% less fuel-intensive than the average new 1987 US car. The measures' average cost is 53¢/gal saved (14¢/l). This result is conservative[15]: at least seven attractive cars with actual performance of more than 42 mi/gal (55 EPA-rated mi/gal), albeit with non-1987-average size or acceleration, are already on the US market. Savings ~72% as large are achievable in light trucks at about half the unit cost (Ref. 47, Table 10). Contrary to widespread belief, such improvements need not entail downsizing: only 4% of the improvement (by twofold) in new US-made cars' fuel economy during 1976–1987 came from making them smaller, while the other 96% came from making them smarter (48).

A further doubling or tripling of car efficiency has been demonstrated by 10 automakers[16] whose prototypes have achieved average on-road efficiencies of

[14a]Disaggregation alone—counting many small savings as well as a few big ones—can roughly double the quantity of savings.

[15]E.g. a drag coefficient of 0.3 is assumed, but < 0.1 is readily achievable and consistent with attractive appearance (~0.15 for vans); the Ford Sable and several other models already get 0.3, Renault's Vesta 2 prototype, 0.186, and the experimental Ford Probe v, 0.137. A curb weight of 2800 lb (1273 kg), only 12% below the 1991 US average of ~2180 lb, is also assumed, but is ~2.0–2.7 times as heavy as some four- to five-passenger prototypes. Indeed, a US car fleet averaging 2000 lb and hence 50 mi/gal could be achieved by materials substitution alone (45); each 200-lb reduction improves fuel economy by ~5% (46).

[16]Ten examples from eight companies are discussed by Bleviss (49, 50) and by Goldemberg (38). More recent varieties reported in the trade press, from Audi, Citroen, General Motors, Fiat, Honda, Nissan, Volkswagen, and others, support the same conclusions.

67–120 mi/gal (2.0–3.5 l/100 km)—~4–6 times the present OECD fleet or ~2½–4 times the low-powered USSR/Eastern European fleet (51). The Toyota AXV, for example, carries four to five passengers at EPA ratings of 89 mi/gal city and 110 highway, while Renault's four-passenger Vesta 2 prototype was tested in 1987 at 64 and 138 mi/gal. Two prototypes—a 71-mpg Volvo and a 92-mpg Peugeot—should cost about the same to mass produce as ordinarily inefficient cars of comparable size.[17] Thus such radical efficiency improvements appear to be far cheaper than paper studies had predicted by simply extrapolating the cost of far smaller incremental savings. Though each prototype has individual design peculiarities, collectively they prove that cars more than three times as efficient as the world fleet can be at least as comfortable, peppy, safe,[18] and low in emissions as today's typical new OECD cars. Comparable opportunities apply also to light trucks (46).

Even this dramatic efficiency gain results only from incremental progress, not basic redesign. For example, one could choose crushable-metal-foam bodies [which would be extremely crashworthy at a curb weight of ~750 lb (~350 kg)], series-hybrid drives, switched-reluctance motors integrated into wheel hubs at zero marginal weight, power-electronic regenerative braking (eliminating the hydraulic system), variable-selectivity windows, miniature sodium headlamps, and other innovations. Systematically combined, such features appear to be able (51a) to yield a very safe, peppy station wagon averaging around 150 mi/gal (1.6 l/100 km). Other major innovations show additional promise, e.g. membrane oxygen enrichment of air intake, variable-geometry turbochargers, direct-injection diesels, ceramic engines, new two-stroke engines, monolithic solid-oxide fuel cells, etc (50).

---

[17]This is largely because their ~1200–1555 lb curb weight (~550–707 kg) requires extensive use of plastics and composites, so large, complex assemblies can then be molded as a unit and snapped together. Not having to make and assemble many small parts, cheaper tooling, and being better able to design for easier assembly save more money than the more exotic materials cost. The leftover money pays for better aerodynamics, smarter chip controls, etc, and the total net marginal cost is about zero. This negative cost of weight reduction is consistent with consultancy data recently provided to Mark Ledbetter (personal communication), and with Chrysler's finding (52) that a largely plastic/composite car could cut a steel car's part count by 75% and plant cost by up to 60%. Earlier incremental calculations (e.g. 53) of a $500 marginal cost to achieve 71 mpg have therefore proven, as hoped, to be overly conservative. Moreover, plastics and composites can improve safety and cut maintenance costs, although they require careful design for recyclability.

[18]Today, some light cars are among the safest and some heavy ones are among the most dangerous. This proves that for safe momentum transfer in a crash, design and materials are far more important than mass. Some of the prototypes, such as the Volvo LCP, are designed for survival in a 35-mph head-on crash—a 36% higher energy-dissipation capability than the US 30-mph standard. Energy-absorbing materials and body designs can thus ensure that ultralight cars are safer than today's—without taking credit for their greater maneuverability, faster acceleration (because they are so light, despite their ~25–50 hp engines), and shorter stopping distance.

There is also the option of a more diverse mix of car sizes. Minicars (no more than 4½ × 10½ ft [1.4 × 2.3 m] and 0.55 l displacement) recently held one fifth of the Japanese domestic car market. General Motors, too, has invested more than $50 million in successfully developing a one- or two-passenger, ~0.75-liter "Lean Machine" rated at ~150–200 mi/gal (1.2–1.6 l/100 km). Said to be safer than a normal car (because of its "bouncy" composite materials and its maneuverability), and occupying less than half the driving or parking width of a normal car (54), it is licensed to Opel and Suzuki but stalled by regulatory ambiguities.

Similarly, the minivehicles popular in many developing countries can be much improved. Taiwan is now doing this with motorcycles and scooters. An unwelcome development, in contrast, is the widespread improvisation of carrying passengers on adapted tractors, especially in China and India: a typical Chinese tractor carrying one ton is estimated by the World Bank to use 75% more fuel per km than a fully loaded four-ton truck, and such tractors used 27% of China's diesel fuel around the mid-1980s (Ref. 5, p. VII-58). Efficient alternatives are clearly important.

The world's half-billion motor vehicles—12 times the 1945 fleet, and now consuming nearly half the world's oil—has grown by about 5%/y for two decades (46). If this growth continues, we will run out of land and air long before we run out of oil. Transportation system alternatives are thus essential to supplement light-vehicle efficiency, because urban highway congestion handicaps even the most efficient cars. California's congestion, for example, is frequently severe on the urban roads that carry 40% of traffic. It costs $17 billion/y in wasted fuel and lost time (360 million person-h/y), and is expected to triple in the 1990s (Ref. 20, p. 7–13). Congestion is even worse in many developing countries. The main technical options for alleviating it include:

1. Improved road design, signalling, signage, and controls (nearside turn on red, computer controlled traffic lights, freeway entry-ramp flow controls, etc). In-car computers linked to computer-driven transmitters that suggest time-minimizing route changes are already under test in Berlin. Automatic proximity-control systems could safely pack 2–3 times as many cars per lane-mile and "save up to 20% of the fuel consumed" (46).

2. Symmetrical treatment of competing transportation modes. A thorough study of 32 cities on four continents (55) found that after controlling for per-capita income and other variables, vehicle-miles travelled per capita in Australian and Western European cities and in Tokyo are respectively 85%, ~45%, and ~25% of the US average—correlated with mass-transit shares of 8%, 32%, 63%, and 4% respectively. The US cities' high gasoline intensity was fundamentally due to their overprovision of roads and of downtown parking as apparently free goods, while other modes had to pay far more of their own costs. In California, for example, cars pay only ~10% of their full

costs through taxes and tolls, while mass transit pays ~20–25% through fares (Ref. 20, p. 7–9).

3. Coordinated land-use/transport development. A mile of travel by mass transit is commonly assumed to displace a mile of travel by car, implying that mass transit has limited effect and uninviting economics (though almost always severalfold cheaper than roadbuilding: Ref. 56, Appendix B). Recent studies, however, indicate that the one-for-one assumption is misleading, because "the availability and usage of transit services also changes the location of trip origins and destinations in a way that reduces the need to travel by car, and reduces the distance of travel required by [most] . . . people who will continue to drive their cars" (56).[19] That study found a nearly twofold difference in vehicle-miles per comparable, income-normalized household in two nearby California communities, one with and one without light-rail service. Each mile of mass-transit travel displaced 10 miles of car travel. Less thorough studies elsewhere have found a 1:4–5 ratio, essentially independent of cultural factors (Ref. 55, pp. 77–100). These results are empirical, reflecting actual transit ridership, not market potential. Because doubling residential density reduces vehicle-miles per household by 25–30% (55), doubling a region's population through infill would increase car travel by only 40–50%, rather than the 100% expected from sprawl. The effect of transit corridors on commercial density, too, is even greater than on residential density, enabling many more errands to be done per trip. Capturing these benefits requires zoning that encourages infill and highly mixed land-use but discourages sprawl.[20] It also requires careful coordination of land-use and of parking with transport planning, and frequent, fast, safe transit. Such variables are far more important to per-capita gasoline use than gasoline prices, incomes, and vehicle efficiencies (55).

4. Ridesharing in private vehicles, and vanpooling, long organized by employers but paid for by riders, work well in some US cities, encouraged by dedicated carpool lanes and other incentives. In the United States, where the average car carries only 1.7 people, full four-passenger carpooling would save 45% of all gasoline (46). During 1973–1988 alone, California's vanpool

[19]This should come as no surprise, since the US Interstate Highway System has proven to be the most important determinant of land-use in this century: as night satellite photographs reveal, an estimated ~95% of US residents live in counties on or adjacent to interstates. Of course, the roads were built between cities and through or near towns to start with, but they have since accreted most new "greenfield" developments too.

[20]Newman & Kenworthy (55) found that fuel use rises steeply and nonlinearly as density drops too low to support satisfactory transit service. Urban per-capita gasoline use worldwide falls into roughly three classes (57): car cities (<30 persons/ha, typically North American), public transport cities (30–130, typically European), and walking/cycling cities (>130, typically Asian). Australian residents have claimed to be as satisfied with their life-styles in high- as in low-density suburbs even if they might not originally have preferred the former (57a).

and rideshare programs saved 2.5 billion vehicle-miles and 156 million gallons of fuel, or more than 5 million tons of carbon (58). Another option, voluntary carsharing among 4–6 households, is proving quite successful in West Germany. And a Belgian innovation—a national hitchhikers' club with a variety of cost- and risk-reducing features—could be widely emulated.

5. Telecommuting via electronic media (Ref. 14, pp. 432–36) is now the main work-style of some 10 million Americans and growing rapidly. It saves money, time, stress, unhappiness, and pollution (59). A related Swedish experiment is fitting one car per subway train with computers, modems, FAXes, etc to make commuting time more productive.

6. Offering safe and convenient bicycle (and pedestrian) lanes or paths, and coordinating with public transit (so bikes can be taken on trains and buses or rented at stations), enables bikes to carry 9% of all Dutch commuter traffic, and "in some cities, they account for more than 40% of all passenger trips" (46). In contrast, although 54% of working US residents work within five miles of home, only 3% bike to work. The scope for overcoming obstacles to biking is enormous: all US biking is currently estimated to displace more than 14 billion car trips per year, saving the marginal portion of ~$6 billion in differential costs, but the gasoline displaced is currently less than 1% of total usage (Ref. 60, p. 26).

7. Some capital-short developing countries have devised cheap, highly effective transit designs. Jaime Lerner, for example, developed unsubsidized, 10¢/ride commuter bus systems in Rio de Janeiro and Curitiba, Brazil, some with one-minute intervals. His onstreet "boarding pods" and special door designs nearly trebled density to 12,000–18,000 passengers per corridor-hour (J. Lerner, current mayor of Curitiba, personal communication). Coordinated land-use policy and a three-tier bus system (including dedicated radial express lines) gave Curitiba "one of the highest rates of motor vehicle ownership and one of the lowest rates of fuel consumption per vehicle in Brazil"—because most car-owners prefer mass transit for routine city travel (46).

Heavy transportation can save considerable energy too. For example, commercial jet-aircraft efficiency is about twice as high in today's 757/767/MD9-80 aircraft as in older aircraft in the US fleet (81), and that fleet in turn is about twice as efficient as that of, say, Aeroflot. Further savings of ~50% have been demonstrated in the Boeing 7J7 and of ~40% in the McDonnell Douglas MD-91/02.[21] New methods of drag reduction (63) can save 20–40% of fuel. By such methods, Boeing's recently released 777 achieves almost twice the fuel efficiency of a 727—~94.5 seat-mile/gal, vs the 727-200's

---

[21]The latter, described by Henne (62), meets or exceeds all current or anticipated noise rules, sharply reduces pollution, and has less interior noise than any other commercial airliner. Both aircraft were fully developed but shelved in 1989 for lack of a market in those cheap-fuel, cash-short days. The 777, brought to market in November 1990, combines some of their features with conventional engines.

50—and 35% less fuel per seat-mile than a typical DC-10: ~185 lb/2000 mi, vs 285 for a DC-10–30, 210 for a 747–400, and 205 for a 767–300ER (D. Mikov, T. Cole, Boeing Corporation, personal communication).

Yet this 10%-better-than-767 performance is being achieved in today's most efficient commercial airliner without using (at least initially) General Electric's unducted propfan engine, with a bypass ratio of 36.[22] That engine uses 40% less fuel than the 727's JD8D-17 engines (64), at a 25% ($1 million/engine) price premium. Against its nearest competitor, its saved energy[23] costs only 19¢/gal or 5¢/l (a 14–21%/y rate of return @ $1/gal); against the fleet, it looks about twice that good. It is therefore slightly too costly for cash-strapped airlines, but extremely attractive at social discount rates, or at long-run replacement fuel costs, or counting externalities like global warming.[24]

Aircraft can also benefit from further improvements (still not begun in many countries) in computerized operations management, fuel-load minimization, idle reduction, flightpath optimization with improved weather monitoring, more frequent aircraft washing, weight paring,[25] etc. In all, EPA (Ref. 5, p. VII-52) has identified improvements that "could reduce fuel use per passenger mile to less than one-third of the current [US] average. . . ."

Even in the United States with its relatively modern railway equipment, potential energy savings of ~25% were estimated (66) before Caterpillar introduced a diesel locomotive with doubled efficiency (67). Substituting Japanese- or French-style (or more advanced) high-speed trains for long car or short air trips could likewise save ~80% and ~87% of travel energy, respectively, at current European efficiencies (68), and similar fractions of

[22]Compared to ~6 for conventional engines and 10 for GE's latest large conventional engine, the GE90.

[23]Levelized at 5%/y real over a 15-y operating life, assuming a 16.8% weighted saving from initial fuel consumption of 1.424 million gal/engine-y, typical of a narrow-body two-engine aircraft flying 3200 h/y (Ref. 64, p. 8). Unfortunately, the >40% drop in fuel use per passenger-mile in the United States during 1970–1980 and the >50% decline in real jet-fuel price during 1980–1987 greatly reduced airlines' economic incentive to save more fuel (Ref. 5, p. VII-52). The 1990–1991 Mideast crisis's doubling of fuel prices increased US airlines' average fuel share of total costs from ~15% to nearly 25%, but that price spike proved short-lived.

[24]The contribution to Ref. 34 by P. Chernick & E. Caverhill found the local and regional CO_2 externality alone to be worth 1.1¢/lb, equivalent to 7.5¢/gal or about one eighth of 1989 jet-fuel prices; adding SO_x, NO_x, and particulates increased this by fivefold. Soviet estimates excluding CO_2 are even higher, at more than 60% of the 1989 US jet-fuel price (65). Chernick & Caverhill further estimate the external cost of US oil imports at $2.26/million Btu, or 52% of the 1989 jet-fuel price. Thus counting the externalities would more than double the price, making new aircraft like the 7J7 or MD91/92, or engines like GE's unducted fan, immediately attractive even at the airlines' high discount rates.

[25]At mid-October 1990 prices, according to a 12–13 October 1990 CNN Headline News "Dollars and Cents" feature, it cost the average US airline ~ $22.50/y in fuel to carry one extra can of soda onboard.

travel cost. New proprietary developments in simple and cost-effective magnetic-levitation trains show particular promise. Simply electrifying main rail lines, in countries with efficient utility systems, can save tens of percent of primary fuel compared with typical OECD diesels (Ref. 5, p. VII-61).

Freight-energy savings estimated at ~80% (Ref. 69, p. 73; Ref. 54) are also available by substituting rail for long freight hauls by truck, using e.g. GM's Roadrailer, which converts in seconds between a semitrailer and a railcar, and pays back in a few years just from lower demurrage and enabling just-in-time inventorying.

Heavy trucks can directly save ~60% of their energy, with paybacks of probably a few years, through turbocharged and adiabatic (uncooled) low-friction engines, improved controls and transmissions, better tires and aerodynamics, exhaust heat recovery, regenerative braking, improved payloads and payload-to-capacity matching, and reduced empty backhauls through better shipping management (69a, 70). Savings of 40% have already been prototyped (71), and savings of 50% have been found to have reasonable cost with present technology (38). Most of these techniques also apply to buses, and many apply to agricultural traction, the need for which can be further reduced by improved cultural practices (see below).

Even after the doubling of ships' energy efficiency per ton-mile during 1973–1983, considerable further savings remain available (72) from improved propellers, engines, and hydrodynamics, antifouling paints, heat recovery, and (in some cases) modern versions of sails. The same engine innovations identified for trucks may, by themselves, achieve 30–40% savings (Ref. 5, p. VII-50).

Transportation equipment stays efficient only with proper maintenance and operation. Much of China's truck energy intensity [twice that of the United States (51)], and that of many developing countries, is due to poor maintenance. Nor is OECD adequate in this respect: about 1 mi/gal, or ~5% of the fuel used by the US car fleet, could be saved simply by proper tire inflation. Road quality, too, is a key determinant of vehicle efficiency and life, especially in the USSR and Eastern Europe. Poor roads, while not discouraging vehicle ownership (Ref. 5, p. VII-59), subtly nudge designers toward tank-like designs—in effect, substituting extra fuel (to haul around the extra weight, drive more slowly, and stop often) for pothole repairs.

Low-temperature heat   Buildings can use the same improvements that save electricity in water- and space-heating to save oil (or gas fungible for oil[26]).

---

[26]This discussion assumes such fungibility. This is reasonable on a timescale of several decades—sufficient to achieve flexibility in refinery product-slate allocation—but in the short term, saved gas may mainly displace residual oil currently in surplus, rather than scarcer light products.

New options include furnaces up to 97% efficient (while also saving >90% of fan energy), superwindows that gain net winter heat even facing away from the Equator, ventilation heat recovery, and cost-effective ways to insulate or "outsulate" a wide range of existing buildings. A major government study found that even with 1979 technologies, careful retrofits could save 50% or 75% of US space heat at average costs of $10/bbl and $20/bbl respectively—severalfold cheaper than heating oil (73). Technological progress since (32, 33, 14) has cut these costs by probably half.[27] EPA considers a 75% reduction in households' total energy intensity achievable by 2025 (Ref. 5, p. VII-6).

A compilation (159) prepared for a National Academy of Sciences study, which is discussed below, adopted a midrange finding that presently commercial technologies could save 45% of the electricity (18) and 50% of the direct fuel used by US buildings in 1989. Those savings would respectively save $37 billion and $20 billion per year more than they would cost. An additional $4.3 billion per year could be saved by cost-effective fuel-switching. The carbon avoided would thus total 232 million tons per year, or one sixth of total US emissions, at a net cost of minus $61 billion per year (or −$263 per ton of carbon). The paper also summarized eight other studies, several of which documented much larger and cheaper savings than those agreed upon by the Academy subpanel.

High-temperature heat   Most of the oil used for industrial process heat, being fungible for gas, could be replaced by less carbon-intensive natural gas saved in buildings—and by far less of it. US industry reduced its primary energy intensity, nearly all by saving process heat, by 30% during 1977–1985. Similar savings continue today, chiefly through improved insulation, heat recovery, controls, and process design (75): computerized process simulation and controls, and substitution of membrane and other nonthermal processes for distillation, offer especially important opportunities still largely untapped. Process redesign for waste minimization often provides an apt opportunity.

Numerous conversations with industrial energy managers confirm that many tens of percent more industrial energy remain to be saved. The typical paybacks are often around two years, even in the most efficient countries, where many firms have already cut energy intensity in half since 1973. Swedish industry in the mid-1970s, for example, was one third more energy-efficient than US industry despite having a more energy-intensive product mix (76). Even so, ~50% of its 1975 energy intensity could still be cost-effectively saved by using the best ~1980 technologies, or ~60–65% by

---

[27]Even against low gas prices, and with a relatively new building stock much of which was built under modern standards, the California Energy Commission has found it cost-effective to save half of the natural gas used in existing households (Ref. 20, Figure 3-1).

using the best technologies entering the market around 1982 (25). Both opportunities continue to expand: some leading European chemical firms that have already halved their fuel intensity since 1973 are privately reporting typical additional savings around 70% from pinch technology (thermodynamic process-design optimization) and better catalysts, with two-year paybacks.

Further large savings are available from long-term redesign and coordination of industrial systems to cascade industrial process heat through successively lower temperatures on a regional scale. Using heat pumps, cogeneration (with heat transmission up to 50 km), and heat-exchangers (up to 25 km), ~25% of industrial energy could thereby be saved in West Germany, 30% in the United States, and 45% in The Netherlands and Japan. Much of this potential appears cost-effective. Probably all of it is attractive at long-run marginal social cost (77).

These are technical improvements only. But the rapid "dematerialization" of the industrial economies (78, 79) has reduced industrial energy intensity in the United States and Western Europe nearly as much as improved energy efficiency has (69a). US steel consumption per real dollar of GNP, for example, is now below its 1860 level and falling. Worldwide, raw-material use per unit of industrial output has fallen by at least 60% since 1900, and this decline is accelerating so quickly that Japan's intensity fell by 40% just during 1973–1984 (80). As will be noted below, the scope for future compositional energy savings is especially large in the USSR and similar economies distorted by excessive output of primary materials that are largely wasted (51).

Furthermore, reductions in the throughput of resources needed to maintain a given stock of material goods represent an additional revolution just beginning (69a). These reductions involve recycling, reuse, remanufacture, scrap recovery, minimum-materials design (often by computer), near-net-shape processing, increased product lifetime, and substituting elegantly frugal materials [such as optical fibers for copper cables, reducing their tonnage by 97½% and their manufacturing energy by 95% (80)] or processes (such as ambient-temperature biological enzymatic catalysis for chemical-engineering pressure-cookers[28]). Recycling alone typically saves about half of materials-processing energy, so "[t]he potential energy savings are staggering" (Ref. 75, p. 96). This is especially true in a garbage-rich, landfill-poor country like the United States, which, for example, throws away enough aluminum to rebuild its commercial aircraft fleet every three months, even though recycling aluminum takes only ~5% as much energy as making it from virgin ore. Collectively, these materials-policy options can probably reduce long-run

---

[28]For example, if we were as smart as chickens (as Ernie Robertson points out), we would know how to make eggshell at ambient temperature, rather than calcining limestone at ~1250°C into Portland cement that is several times weaker.

industrial energy intensity per unit of maintained stock (not throughput) by an order of magnitude. If this were combined with technical gains in process efficiency, surprisingly little industrial energy use would remain: industrial energy use smaller than today's could support a worldwide Western European material standard of living (69a, 38).

Electric-utility hydrocarbon fuels   Another kind of oil savings comes via electricity.[29] The potential to save electric utilities' small remaining oil input by substituting other forms of generation has been exaggerated (Ref. 81, pp. 108–110). Yet that oil and the large amount of gas still used in thermal power plants were together equivalent to 13% of all oil burned in the United States in 1989. Nearly twice that much electricity could be saved by lighting retrofits alone, at negative net cost (28, 8). Globally, oil-and-gas use in power plants is equivalent to ~22% of total oil use (42), but lighting retrofits plus other cheap electrical savings can clearly displace far more than 22% of electricity at negative net cost (Figure 1).

Miscellaneous oil uses   The oil and gas used as feedstocks (10% and 3% of their respective total US consumption in 1986, 14% and 7% globally) are subject to unknown but probably substantial savings. These mainly involve more efficient petrochemical processes (75), internalization of solid-waste disposal costs [leading, as in Europe, to high plastics recycling rates, improved product design and longevity, agricultural reform (see below), and lower use of disposable packaging], and reduced but more durable highway construction leading to lower asphalt requirements (and less fuel burned to make cement). In addition, at least half of the ~2% of oil used to propel the other 98% through pipelines would be saved by the above measures, and analogously for refinery fuel and losses (~6%) and for gas compressor energy (~3%).

Total oil-saving potential   The combined potential to save oil by these means in the United States, shown in Figure 3, is ~80% at an average cost below $3/bbl (plus a further 20% of leftover saved gas at ~$10/bbl-equivalent). Qualitative evidence that potential oil savings and costs are comparable in other OECD countries includes:

---

[29]Besides the scope for saving power-plant fuel by saving electricity, small but useful amounts of oil and gas can be directly replaced by electrotechnologies that are cheaper and/or better in certain applications, but they will increase industrial electric use only a few percent as much as improved electric efficiency decreases it (18).

1. the similar efficiencies of new light vehicles (nearly 30 mi/gal for cars) throughout OECD[30];
2. the cost effectiveness of large additional industrial and building heat savings even in such efficient-in-aggregate countries as Sweden (25, 35), West Germany (69a, 37), and Denmark (81a), and hence even more so in less efficient countries; and
3. the virtual irrelevance of differences in oil end-use structure, because such large savings are available in each end-use.

These considerations apply a fortiori in the other two world regions, since they are even less efficient than OECD.

Aggregate energy intensities per unit of economic output are typically 2–3 times as high in socialist and developing countries as in OECD (51, 38). Both this fact and the field observations reported universally in the literature suggest that if all countries became as energy-efficient as OECD countries should be, the potential percentage savings would be even larger in socialist[31] and developing countries than in OECD, and the costs correspondingly lower (38). If such countries rapidly build or rebuild their infrastructure, too, the opportunities will arise more in new construction than in retrofit, as would be the case in most OECD countries. This will further increase savings and reduce costs. One can therefore conclude that most of the oil now used in the world can be saved at an average cost far below mid-1991 world oil prices— perhaps an order of magnitude below.

OTHER ENERGY   Natural gas, natural-gas liquids (NGL), and coal used for process or building heat or for feedstocks are subject to the same categories of savings just described, and can be saved with similar effectiveness and cost. This is especially true in heating applications, where oil, gas, and NGL are used essentially interchangeably and in nearly identical technologies. In

---

[30]New cars were until recently a few mi/gal less efficient in the United States than in Western Europe and Japan, but in the past few years new German and Japanese cars sold in those countries have been less efficient than new US-made cars sold in the United States. In any case, such differences are immaterial compared with the potential improvements, partly because only half as much of Europe's and Japan's oil use is for transportation as in the United States; rather, they use more oil in industry and buildings. The thermal efficiency of buildings in such countries as Germany, Britain, and Japan is particularly low: German houses, for example, have worse average shell thermal integrity than US houses.

[31]Their efficiency analyses so far (e.g. 51) show savings of only about one sixth for the Soviet Union by 2030 (or one third including changes in output structure). Extensive discussions there support our and other observers' belief that this is not because of a lower actual potential—quite the contrary—but only because Soviet analysts have not yet become familiar enough with disaggregated analyses, modern Western technology, and market mechanisms to apply these opportunities to their own difficult situation.

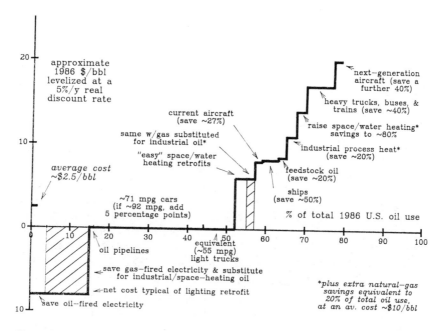

Figure 3 Supply curve of the full technical potential to save US oil use by retrofitting or substituting the best demonstrated 1988 end-use technologies, at least half of which are on the 1991 US market. The vertical axis is levelized marginal cost (1986 $/barrel delivered, 5%/y real discount rate). The horizontal axis is cumulative potential saving (% of total 1986 US end-use) corrected for interactions. Shaded areas represent savings of natural gas that then displaces oil used to heat buildings or industrial processes. The cost and performance data are empirical (81): costs above $10/bbl are quite uncertain, but this has little effect on the result. No life-style changes or intermodal transport shifts are assumed. The curve reflects many conservatisms: e.g. omission of any light-vehicle improvements (51a) whose marginal cost exceeds zero, omitting new industrial and aircraft developments, and translating the negative-cost lighting retrofits (which save the oil- and gas-fired electricity) directly into equivalent $/bbl without taking credit for the value of the fuel displaced. The overall uncertainty appears to be ~10 percentage points in total quantity and <2× in average cost.

broad outline, therefore, no additional treatment of these other fuels is necessary. There are two exceptions: 1. Seven eighths of Chinese household energy comes from coal, nominally for cooking. Although its use is officially forbidden for space heating, despite indoor winter temperatures often below freezing in many provinces (51), cooking coal nonetheless contributes precious heat. Better-insulated houses would thus improve comfort more than they would save coal in that instance: only in combination with gas (e.g. biogas) cooking will they displace much coal. 2. Much Soviet and Eastern European steel is still produced in open hearths, which are half as efficient as basic-oxygen plants that are themselves no longer state-of-the-art. The USSR is by far the world's largest steelmaker but consumes at least two-thirds more energy than Japan to make each ton of steel (Ref. 5, p. VII-107). It can thus save large amounts of coal by this improvement alone. Continuous casting and more advanced processes can save even more (82).

For some two billion people, most fuels are noncommercial, but that does not mean they are renewable. Unsustainably harvested wood and dung burned for fuel release biotic carbon from soil to air: in effect, they mine carbon fixed in soil organisms by reducing their populations and diversity. This contributes ~23% of global CO_2 emissions and hence increases poor countries' 1987 share of those emissions from 19% (from fossil fuels) to 42% (17). The use of such fuels also causes erosion, deforestation, loss of soil fertility, blindness among women and children, and many other social ills. A combination of energy efficiency,[32] reforestation, cascaded fuel-switching, and social changes (chiefly related to the role of women) is needed to address this complex problem (38). Among the desirable substitutions here is substituting the use of liquefied petroleum gas (LPG) and biogas for electric and kerosene lighting and cooking (38); a good gas stove can also be 5–8 times as efficient as a traditional <10%-efficient woodstove (38). Similarly, a compact fluorescent lamp driven by a 25%-efficient electric generator and grid is some 200 times as fuel-efficient as the kerosene lamps used in ~80% of Indian households (38).

Such substitutions are only part of a complex chain of successive fuel-switching (Ref. 38, pp. 255–73) needed to address simultaneously the fuelwood and oil problems of countries like India. The main steps, devised by A. K. N. Reddy, are: replace household kerosene, wood, and dung with biogas (which also produces more and better fertilizer); use sustainably grown gasified fuelwood and a little biogas to run old diesel pumpsets, new pumpsets (if not photovoltaic), and short-haul freight; use the electricity saved from pumpsets to electrify all homes and replace kerosene for lighting; desubsidize kerosene and diesel fuel; and shift long-haul freight from trucks back to revitalized railways. End-use efficiency is the key throughout.

CONVERSION AND DISTRIBUTION EFFICIENCY   More efficient use of delivered energy is only part of the energy savings available. Major savings are also available in converting and distributing primary fuels. Some salient opportunities from around the world that could reduce global fossil-carbon emissions by one third or more at negative net cost include:

---

[32]Comprising not only efficient stoves (83, 84) but probably also efficient pots, perhaps with double walls and lids, enhanced heat transfer, and internal hot-gas paths—a concept that has as yet received no serious engineering attention for developing-country use.

1. improving maintenance and operational techniques at most developing countries' thermal power plants, where efficiencies below 20%, output several times lower than nameplate ratings, and poor availability are endemic;

2. substituting 42+%-efficient combined-cycle gas turbines, or better still, >50%-efficient intercooled steam-injected gas turbines (85), for 25–35%-efficient classical thermal power plants (a ~50% CO_2 reduction), at one-fourth the marginal capital cost and lead time of scrubbed coal plants—an especially attractive opportunity in the gas-rich Soviet Union (51) and when integrated with efficient biomass gasifiers (86, 87);

3. substituting low- for high-carbon fuels (e.g. natural gas emits half as much carbon per unit energy as coal);

4. expanding economic power wheeling (transmitting bulk power from areas with higher to those with lower generating costs) by using Japanese advances in power electronics and mainly Soviet advances in control theory to raise grid capacity;

5. saving several percent of electric energy in OECD and up to 10 times that in socialist and developing countries through advanced distribution management and metering at negative cost;

6. reducing the major losses of natural gas (~8%, Arbatov (88)[33]) and district heating energy (half to two thirds, Ref. 91) in the Soviet grids, and the even more dismaying loss of much of the delivered Soviet district-heating energy (which heats about three fourths of all buildings) owing to the lack of operable controls, let alone meters, for each office or apartment, making open windows serve as thermostats[34];

7. extending advanced Scandinavian district-heating technology to cold countries not yet taking advantage of it (i.e. most of them) wherever superinsulation retrofits are not cheaper;

8. displacing electric space and water heating with gas or passive-solar

[33]Makarov & Bashmakov (89) put the losses at ~2%, but Arbatov's higher estimate, or something close to it, has been informally confirmed by other knowledgeable Soviet experts. Arbatov found 50 Gm^3/y of losses (mostly CH_4) due to leakage and ruptures, excluding 17 Gm^3/y lost in extraction, out of 800 Gm^3/y of total production, 80 burned for compression, and 30 of associated gas burned (not vented). The losses are very nonuniform in time and space. In contrast (90), US production losses are ~0.13%, ~0.54% is "lost and unaccounted for" in interstate pipelines, and corresponding losses in retail distribution are a highly variable 1–6%, averaging 2–3%. These figures are all too high, and should be greatly reduced to save both money and fire-danger, as well as to reduce global warming. Abrahamson (90) further cites direct CH_4 measurements in ambient urban air consistent with urban leakage of ~2.9–5.9% of US natural gas consumption, some perhaps from storing and using coal.

[34]Similarly, Makarov (51) states that "metering and individual control of residential heating and hot-water systems, coupled with repairs and improved management of district heating systems, could permit [Polish] housing space to double with only 15% growth in energy consumption." This implies a 43% efficiency gain. Jászay (92) estimates a corresponding 30% sectoral saving potential in Hungarian buildings.

techniques, as some North American electric utilities already pay customers to do;

9. recovering gas-pipeline compression energy with a city-gate turboexpander/generator (a US opportunity in the multi-GW range);
10. eliminating gas flaring, which accounted for nearly 1% of 1986 fossil-carbon releases (93), by using the gas as a feedstock or to fuel steam-injected gas turbines;
11. making industrial cogeneration—a common and lucrative practice in West Germany—the universal practice in all countries that can use both the electricity and its process-heat coproduct (examples are cited below); and
12. using simple kiln improvements to double, or more, the efficiency of traditional charcoal-making in developing countries (Ref. 5, p. VII-140).

RENEWABLE ENERGY SOURCES   Many renewable energy sources are already, or are rapidly becoming, competitive without subsidy. Those sources that do not quite compete yet usually have a smaller margin of disadvantage than their pollution reduction would justify if counted (34, 94). The speed of progress and the variety of options showing great promise of further improvement are most encouraging (95, 94).[35] Integrating different types of renewable power sources, in different places, tends to provide needed storage automatically (94), or at least more cheaply than with nonrenewable central stations (Ref. 97, pp. 268–70 & Apps. 1–3; Ref. 98). Even the costlier electric technologies, such as photovoltaics, are already competitive in remote sites (95), for certain substations' support, or in uses requiring high power quality and reliability.[36] Progress is similarly rapid in bringing down the costs and raising the conversion efficiency of both thermal and biochemical processes for converting biomass to liquid fuels (95, 94). And new developments in solar process heat[37] promise to provide it economically even in unfavorable climates.

[35]For example, apparently competitive solar-thermal-electric technologies now include not only the Luz trough-concentrator technology mentioned in Ref. 95, but also the Sunpower/Cummins combination of an Ericsson engine with a solar dish, now believed to cost ~5¢/busbar kWh with cheap dishes (96) and perhaps less with very cheap ones like Solar Steam's. Yet only a few years ago, solar-thermal-electric technologies looked unpromising.

[36]The Federal Aviation Administration, for example, is converting hundreds of ground avionics stations to PV power even where there is already grid power to the site, because cleaning up and backing up the grid power costs more than starting with an isolated source.

[37]Especially David Mills's development, at the University of Sydney, of semiconductor-sandwich surfaces that should soon be able to absorb visible light at least 85–90 times as well as they emit infrared (personal communications, December 1989 & October 1990). Such a surface in a hard vacuum, if it is sufficiently heat-resistant, can be calculated to yield heat at high enough temperatures for most industrial processes, even on cloudy winter days at high latitudes. More recently (personal communications, April 1991), even higher selectivity ratios have been achieved, although the applications currently envisaged are at only a few hundred °C.

The menu of renewable sources is very large and rich (94) and many sources have the potential to grow rapidly. A contrary impression can be created by dividing renewable energy into many small pieces and discussing only one at a time. However, in the most detailed official study to date in the United States, five National Laboratories (94) found that either fair competition plus restored research, development, and demonstration (RD&D) priority,[38] or proper counting of avoided environmental costs, could increase competitive renewable electric output from 363 TWh/y in 1988[39] to ~1573–1895 TWh/y in 2020, which equals 60–72% of all 1989 US electricity sold. Renewable generation at that level could run an expanded but efficient economy without fossil-fueled or nuclear power stations. All the technologies assumed would compete with assumed 2030 prices of 6¢/kWh baseload (1988 $), 9¢ intermediate, and 15¢ peaking.

The same study also found that in 2030—about the retirement date of a standard power station ordered today—cost-effective electric and nonelectric renewable sources could together supply 44–60 EJ/y (41–57 quadrillion Btu/y), equivalent to 48–67% of present total US energy demand, including 72–148% as much electricity as the United States uses today.[39a] A vibrant and much expanded economy, if it used energy in a way that saves money, would need no more than that.[39b] Even these impressive renewable outputs are conservative: they are based on midrange expectations of economic performance and artificial constraints on intermittent sources such as windpower (99), and do not represent "an upper limit on the potential contribution of renewables" (Ref. 94, p. ix).

Electric generation is not the only role where renewables plus efficiency can do the job. Studies a decade ago showed that a combination of renewables (including sustainably grown biofuels[40]) with efficiency (chiefly doubled light-vehicle efficiency) could cost-effectively eliminate the need for fossil

---

[38]Real federal renewable-energy RD&D funding fell by 89% during 1979–1989. SERI (94) assumed an increase equivalent to a total of $3 billion over the next 20 years—about the cost of one 1-GW nuclear plant. This is assumed in the following paragraph.

[39]Excluding nonelectric sources, such as at least 3.3 EJ/y of direct biofuels. Renewable supply of all kinds was probably not the cited ~8% but rather ~10–12% of total US primary supply, and the fastest-growing part, outpaced only by savings (81).

[39a]The lower figures, here and in the previous paragraph, assume that intermittent electricity is artificially constrained not to exceed 20% of regional generation. Circumstances requiring this now appear rare (97, 99).

[39b]E.g. improving energy productivity 1%/y faster than GNP growth would cut total energy use in 2030 to 68% of 1991 use—similar to the 64–70% of unconstrained renewable supply projected to be cost-effectively available in 2030.

[40]I.e. those whose production can be indefinitely repeated because it depletes nothing. If such biomass were not burned, it would rot or be eaten by respiring animals and release its carbon anyhow; the issue is only whether that carbon release is taken up again, promptly and in equal measure, by new photosynthesis.

fuels both for light vehicles[41] and for power plants in each region of the globe. Indeed, one such study, done for the German government, found that such sources could provide essentially all the energy needed, at levels of end-use and energy-system efficiency available and cost-effective in 1980, to sustain a 1975 West German standard of living throughout a long-run world with a population of eight billion (69a). Another study, though twofold more conservative, reached broadly similar conclusions about the combined potential of efficiency and renewables (38), consistent with a decade's shorter-term analyses for the United States (73, 94).

Four further investigations of efficiency-plus-renewables strategies also merit emphasis:

1. A detailed analysis by the Swedish State Power Board (35) found that doubled electric end-use efficiency (costing 78% less than marginal supply), plus fuel-switching to natural gas and wood, plus environmental dispatch,[42] could together support 54% growth in Swedish real GNP during 1987–2010 and handle the voter-mandated phaseout by 2010 of the nuclear half of the country's electric generation, yet at the same time reduce the heat and power sector's CO_2 output by one third and reduce the cost of electrical services by nearly $1 billion per year. (The latter reduction arises because efficiency would save more money than fuel-switching and environmental dispatch would cost.) This result is especially striking because Sweden is arguably the world's most energy-efficient country (in aggregate or in many details: e.g. 76) to start with, with a heavily industrialized economy and a severe climate. Any other country should therefore be able to do better.

2. At the same time, a study for the Indian state of Karnataka analyzed the combination of several end-use efficiency measures with small hydro dams, bagasse cogeneration, biogas/producer gas, a small amount of natural gas, and solar water heaters. This far-from-comprehensive combination would achieve far greater and earlier development progress than the fossil-fueled plan advanced by the state utility (but later rejected by the government). The efficiency-plus-renewables combination would also have used three-fifths less electricity, cost one-third as much, and emitted only 1/200th as much fossil-

---

[41]Pure-electric cars are not considered here. They appear unlikely in principle to compete in cost, range, and performance with efficient fueled cars (including those that convert the fuel to electricity with an onboard motor-generator or fuel cell (51a), the "series hybrid" concept). They also do not reduce global warming if powered by anything like the present utility fuel mix (100). For the same reasons of economics and an actual worsening of global warming, coal synfuels are not considered either (100). Compressed or liquefied natural gas can modestly reduce CO_2/ vehicle-mile, but far less than efficiency and biofuel options, and with some drawbacks, so they are best considered transitional niche fuels.

[42]I.e. operating most the power plants that emit the least carbon, and vice versa, by including externalities in economic dispatch.

fuel CO_2 (17). This is encouraging too, since India already emits ~5% of global carbon (17) and projects that this fraction, assuming the traditional coal-based strategy, will increase enormously.

These two analyses are especially interesting when considered together, because between them they span essentially the full global range of energy intensity and efficiency, technology, climate, wealth, income distribution disparities, and social conditions. Yet both find that the money saved by efficiency more than pays for the renewables, yielding a net profit on the whole carbon-displacement package in the energy sector. In addition:

3. An analysis of British, Dutch, German, French, and Italian potential for carbon reduction from 1985 to ~2015 (100a) found that if electric demand grew 82% as officially forecast, then optimizing the electric supply system— by emphasizing gas cogeneration and renewables—could nonetheless cut its carbon emissions by 54%, from 293 to 135 MtC/y, at zero net cost, despite the higher gas prices that the increased gas demand was assumed to cause. Such optimization plus end-use efficiency (assumed to keep demand constant and, generously, to cost just as much as the coal and nuclear plants it displaced) could cut emissions to 50–75 MtC/y—a 75–83% reduction from the base case—while reducing electric service costs by 10%.

4. A similar Lawrence Berkeley Laboratory analysis for the New England power sector (100b) found that by using 75% of the identified cost-effective potential for each of three resource groups—efficiency (including a little fuel-switching), wind and biomass, and gas and biomass cogeneration—the region could meet 67% of the forecast 160 TWh/y of electric demand in 2005. This would halve carbon emissions from a sector that is already one-third nuclear. Yet even though this falls well short of long-term potential, using 75% of each option set's potential is not least-cost, and all plants were assumed to retire at the end of original book life (thereby removing 3.2 GW or 44% of the 7.4 GW of regional nuclear capacity), production-cost redispatch modelling found only a 4.3% rise in the cost of electrical services. A least-cost resource mix would instead decrease this cost.

National analyses commissioned by the governments of Australia (101) and Canada (102) similarly found that national CO_2 emission reductions of ~20% via energy efficiency would be highly profitable. A 36% Australian energy saving from projected 2005 levels, reducing forecast fossil-fuel CO_2 emissions by 19%, would produce net internal-cost savings, in today's Australian dollars, of $6.5 billion per year by 2005. Each $5 invested in efficiency could save $15 worth of new energy supplies and 1 ton of CO_2—an average abatement cost of −A$37/t. Similarly, the Canadian report found a cumulative net saving of $100 billion through 2005 (present-valued Canadian dollars) from a 20% CO_2 cut compared to present emissions. In addition, a private analysis for California has detailed potential CO_2 emission reductions

of 26% from projected levels in 2000 and of 54% in 2010, both at negative net
cost (60).

Farming and Forestry

As noted earlier, nearly all greenhouse gas emissions not related to energy use
or CFCs arise from unsustainable farming and forestry practices. These
emissions include:

1. ~46% of anthropogenic CH_4, which comes from livestock-gut fermentation and rice-paddies, rising to 68% if biomass burning[43] is included;
2. ~57% of CO, from forest clearing, chiefly for agriculture (103) and
fuelwood;
3. ~52% of anthropogenic NO_x and ~15% of anthropogenic N_2O, emitted
by biomass burning;
4. a further ~33% and ~18% of anthropogenic N_2O, derived from cultivat-
ing and fertilizing natural soils,[44] respectively; and
5. about one fourth of all CO_2, from deforestation, desertification, and
simplification of terrestrial ecosystems including farmland (Ref. 6, Ch. 3).

Ecological simplification in its myriad forms is less visible, but no less
important to global warming, than the clearcutting of American forests or the
burning of Brazilian forests. Just the loss of above-ground biomass and
diversity, assuming no loss of soil carbon, means that replacing an old-growth
Pacific Northwest forest with a young one reduces its total carbon inventory
by two- to threefold (104).[45] But typically at least as much terrestrial biomass
is belowground as aboveground—and in temperate farmland, some 20–30
times as much is in the soil as in the plants above it (Ref. 6, p. I.3–32[46]). This
invisible but enormous carbon stock, typically upwards of 100 metric tons of

---

[43]Much biomass is burned, often unsustainably, for fuelwood, slash-and-burn shifting cultivation, disposing of crop residues, and clearing forests to extend cultivation. The last three of these
terms probably release ~1.4–2.9 GtC/y, vs ~1.6 from burning wood and dung for fuel (Ref. 6,
p. I.3–15n8). Burning inevitably emits non-CO_2 trace gases whether the carbon is recycled or
not—i.e. independently of whether the carbon release is compensated by re- or afforestation or by
other biotic carbon sinks. How the carbon is harvested will of course affect the ecosystem's
ability to sustain such compensation (Ref. 6, p. I.3–15).

[44]Forcing the nitrogen cycle boosts the yield from side-reactions whereby denitrifying bacteria
in the soil produce N_2O from nitrate and nitrifying bacteria produce N_2O from ammonium.
Cultivation also appears to increase microbial N_2O emissions even without fertilizer, and nitrate
runoff into surface- and groundwaters appears to result in increased N_2O emissions (Ref. 6, p.
I.3–20).

[45]Hence Oregon has estimated that cutting of old-growth forest is responsible for ~17% of the
state's total carbon emissions (105).

[46]Contrary to the conservative assumption made by Harmon et al (104), the data cited by
Krause et al (6) do show a 10% soil-carbon loss when the natural forest becomes managed.

carbon per hectare (tC/ha), is at risk of mobilization into the air if insensitive practices defeat living systems' ability to fix carbon into soil biota. In essence, turning (for example) prairie into corn and beans, and substituting synthetic for natural nutrient cycles, puts a huge standing biomass of soil bacteria, fungi, and other biota out of work. They then tend to lose interest, die, oxidize or rot, and return their carbon to the air.

At the same time, soil erosion, still endemic throughout most farmlands, transports soil organisms and other soil organic constituents ("finely pulverized young coal") into riverbeds and deltas, where they decay into CH_4—a greenhouse gas many times as potent as CO_2 (1). Reduced soil fertility from erosion, biotic simplification, compaction, or the use of poisons requires ever greater inputs of agrichemicals, notably nitrogen fertilizers, whose production consumes ~2% of industrial energy (75) and whose use increases N_2O emissions from the soil. Other well-known agricultural problems include:

1. burgeoning pest resistance—the world loses more of its crops to pests now than before the pesticide revolution—and pesticide-caused health problems, especially among fieldworkers;
2. rapidly growing OECD demand for food free of chemical contamination (105a)[47];
3. crops' narrowing genetic base as diverse native stocks are inexorably lost to habitat destruction and seed-bank neglect;
4. problems of water quality and quantity;
5. many farmers' marginal profitability as their revenues immediately flow back to input suppliers; and
6. the distressing spectacle of simultaneous food surpluses and famines.

These trends indicate the need for "a major overhaul of current agricultural production methods" (Ref. 6, p. I.3–14). Such an overhaul appears necessary to achieve adequate, acceptable, and sustainable food and fiber output even if global warming were not of concern. Achieving it may be difficult because of the rapid loss of rural culture and traditional ecological knowledge as farmers vanish into cities (106): every year's delay adds to the loss of those irreplaceable human resources. However, as with the changes to the energy system described above, changes in the agricultural system needed to reduce global warming will increasingly be seen as attractive for a wide range of other reasons, including economics. And of these changes, those with the highest climatic leverage involve livestock.

---

[47]For example, a 1986 National Institutes of Health study found that every US-registered fungicide is a known carcinogen (E. N. Davies, Ecogen Corp., personal communication). The Dutch Parliament is shortly expected to pass a law requiring 25% biocide reductions by 1993 and 50% by 2000 (E. N. Davies, personal communication).

LIVESTOCK   Just as saving electricity reduces CO_2 emissions disproportionately by displacing severalfold or manyfold more fuel, so can affecting the numbers and rearing of livestock—which convert $\sim$3–20+ units of grain to one unit of meat—disproportionately help to protect existing forest-, farm-, and rangeland while reducing emissions of CO_2, N_2O, and CH_4. High-priority actions include (6, 5):

1. reducing OECD dairy output to match demand[48];
2. desubsidizing livestock production, especially for cattle, which emit $\sim$72% of all livestock CH_4 (107): many dairy and beef cattle would not be grown without large subsidies, especially in OECD (108);
3. reforming beef grading and distribution, particularly in the United States, to reduce the inefficient conversion of costly, topsoil-intensive grains to produce fat that is then largely discarded (109);
4. regulating or taxing methane emissions from manure so as to encourage its conversion to biogas for useful combustion;
5. improving livestock breeding,[49] especially in developing countries, to increase meat or milk output per animal, consistent with other important qualities and with humane practices;
6. shifting meat consumption to less feed- and methane-intensive animals[50] and to aquaculture (preferably integrated with agriculture, a highly flexible and productive approach that may also help cut rice-paddy CH_4); and
7. developing, if possible, alternative feed, fodder, and rumen flora that minimize CH_4 output.[51]

Many of these livestock options would have important side-benefits. For example, many OECD cattle herds are fed, at conversion ratios of 8:1 or worse, with grain from developing countries. The Western European herd consumes two thirds of the domestic grain crop, yet still imports >40% of its feed grain from developing countries (Ref. 6, p. I.3–19). OECD consumption

[48]Dairy cows produce extra methane because they are fed at about three times maintenance level (Ref. 6, p. I.3–16).

[49]This does not mean using such biotechnological innovations as bovine growth hormone, which is not an improvement in the herds but rather an artificial way to make existing cows produce, probably briefly, much more milk than they were meant to.

[50]For example, shifting half of beef consumption to pork and poultry would maintain dairy output and total meat consumption while reducing methane emissions by $\sim$40%—about twice the stabilizing CH_4 reduction (Ref. 6, p. I.3–18). In OECD, the market is already shifting in this way, largely because of health concerns. Ultralean, organic range beef (which grazes only on natural grasslands and is not grain-fed), which alleviates those concerns and can cost less, may also produce less methane than equivalent feedlot beef (Ref. 5, p. VII-270).

[51]EPA is encouraged about this option and believes that the resulting productivity increases would often yield a significant net profit: as J. S. Hoffman of EPA put it, "This creates a very economic picture for methane [abatement]" (109a). IPCC (3) apparently concurs. Validating field experiments are now under way.

of this large amount of feedlot beef is thus "directly related to starvation in the poor countries of the world." If OECD countries replaced part of their feedlot beef consumption with range beef and lamb, white meats, aquaculture, marine fish, or vegetable proteins, then Central America might feel less pressure to convert rainforest to pasture. Many developing countries could free up arable land. There could be less displacement of the rural poor onto marginal land, and renewed emphasis on traditional food crops rather than on export cash crops. Above all, this one action could save enough grain, if properly distributed, to feed the world's half-billion hungry people (Ref. 6, p. I.3–18).

LOW-INPUT SUSTAINABLE AGRICULTURE   Organic farming techniques that are already rapidly spreading in OECD for economic, health, and environmental reasons (105a) can simultaneously reduce biotic CO_2, N_2O, and CH_4 emissions directly from farmland, and indeed may reverse the CO_2 emissions. These techniques can and often do use standard farm machinery, but require it less often,[52] and can work well on any scale. They substitute natural for synthetic nutrients (e.g. legumes for synthetic nitrogen), mulches and cover crops for bare ground, natural predators and rotations in a polyculture for biocides in a monoculture, and nature's wisdom for humans' cleverness. They integrate livestock with crops, and garden and tree crops (see below) with field crops. They maintain often tens and sometimes hundreds of cultivars instead of just one or a few. In Asia, they draw on a particularly rich tradition of integrating many kinds of production—vegetables, fish, rice, pigs, ducks, etc—in a sophisticated quasi-ecosystem that efficiently recycles its own nutrients.

Green Revolution seeds and artificial fertilizers are often assumed to be essential to grow enough food in land-short developing countries. Yet diverse African field studies have demonstrated that "ecoagriculture," which substitutes good husbandry and local seed for otherwise purchased inputs, yields nearly as much maize, sorghum, etc in the short term. The small yield difference probably narrows with time "[i]n view of the accelerated degradation of soils that usually accompanies chemical agriculture." Such results "suggest that regenerative farming could be greatly expanded both in industrialized and developing countries without negative consequences for the goal of increasing Third World agricultural yields. On the contrary, without

---

[52]In California (20), ~3% of energy is used directly in agriculture. Of the one third of that used for irrigation, ~40% can readily be saved through simple and highly cost-effective water-efficiency measures—thereby saving electricity (coal-fired on the margin), since pumping water is the largest single use of electricity in the state. The further ~22% of agricultural energy use for synthetic pesticides and fertilizers would be virtually eliminated by organic techniques, while the ~24% for traction could be cut about in half (20) through the reduced need for field operations. In all, therefore, organic farming would save nearly half of California's agricultural energy use (60). To the extent it were less centralized, it would also reduce transportation needs for both inputs and outputs.

[such] a conversion . . . the loss of arable land, notably in the tropics, threatens to accelerate out of control . . ." (Ref. 6, pp. I.3–23 & –24).

In both OECD and developing countries, ordinary organic farming practices modelled on complex ecosystems generally produce comparable or slightly lower yields than chemical farming but at much lower costs. They therefore produce comparable or higher farm profits (110)—without counting the considerable premium many buyers are willing to pay for food free of unwelcome biocide, hormone, and antibiotic residues (105a). The organic practices' economic advantage has been demonstrated in large commercial operations over a wide range of crops, climates, and soil types (110). That advantage tends to increase at family-farm scale, which brings further social benefits (111). Similar economic benefits have been found in many hundreds of diverse US and West German farms (112–114).

Little is yet known about CH_4 and N_2O cycles, so it is only a plausible hypothesis, not yet a certainty, that the reduced tillage and fertilization that accompany profitable organic farming, together with reductions in the burning of biomass and fossil fuels, will suffice to eliminate most of the $\sim$35% of total N_2O that is released by human activity. Yet even if N_2O reductions from organic fertilizers turned out to be less than hoped, the CO_2 benefits would still be large, because CO_2 can be absorbed by building up organic matter in soil humus through the gradual accumulation of a richly diverse soil biota. Today, in both OECD and developing countries, and reportedly in Eastern Europe and the USSR too, soil loss, and especially the physical loss or biological impoverishment (hence carbon depletion) of humus, is far outpacing soil and humus formation and enrichment. But successful conversions to organic practices, chiefly in the United States and West Germany, have demonstrated that after a few year's reequilibration, these carbon losses can be not only eliminated but reversed.

Not just forests, then, but also farmland can be changed from carbon sources to carbon sinks. For example, an ordinarily impoverished soil in the US cornbelt could plausibly start at 2% organic content or $\sim$1% C. A decade of organic practices—rotating corn with alfalfa or clover, using manure and green manure, and integrated pest management—could raise the organic content by a conservative 0.02%/y,[53] thereby adding $\sim$4.8 tC/ha over the

---

[53]For example, Holmberg (115) cites the late Herman Warsaw, a successful (370 bushels/acre in 1985) organic corn farmer who in the past 30 y increased the organic content of his soil from $\sim$3½% to $\sim$8% in the top 3" and from 1% to 3% at 1' depth. The average increase in the top foot of topsoil was $\sim$4% ($\sim$2% in carbon terms). He believed he learned how to achieve the same improvement in five years. We assume half that speed. This is probably quite conservative: for example, Holmberg also cites Steve Pavich's 1% carbon gain profitably achieved in 12 y and probably reproducible in half the time in dry Arizona and California soils, and US Department of Agriculture/Beltsville test plots' achievement of 0.2–0.6%/y carbon gains through light (40 t/ha-y) applications of compost and manure. The carbon uptake assumed here is 45$\times$ slower than Beltsville's two-year achievement at a 160 t/ha-y manure application rate.

decade (115). Doing this on the United States's 50 million hectares of farmland would offset the annual combustion of ~10% of current US gasoline use per year. Thus a very efficient US car fleet (~5–7 times as efficient as now), getting a substantial part of its fuel from sustainably grown biomass, would emit only as much fossil carbon as the farmland would reabsorb into soil humus (115). Based on organic-farming comparisons by the National Research Council (110) and others [e.g. McKinney (116)], this carbon sink could be achieved at zero or negative net internal cost. Specifically, each hectare of sustainably grown corn or other fuel feedstock could, for example, produce 600 gal of anhydrous ethanol, fix enough soil carbon to offset the combustion of 200 gal of gasoline (115),[54] and increase the farmer's profits.

There are also many techniques for substantially reducing the use of nitrogen fertilizer (Ref. 6, p. I.3–20) within the context of conventional OECD farming practice or of lower-input but still not truly organic modifications. Most of these techniques are cost effective because they reduce chemical and application costs and nitrate-runoff pollution without cutting yields. In many developing countries, too, additional measures to reduce CH_4 emissions are available and desirable: e.g. biogas-digester preconditioning of rice-paddy fertilizer, improved dryland rice options, and reducing in-paddy anaerobic fermentation of rice residues.[55] Substantial reduction in N_2O and CH_4 releases can undoubtedly be obtained by the simpler of these management techniques at costs on the order of \$3–30 per ton of carbon-in-CO_2-equivalent. On closer examination, side-benefits, such as saving fertilizer and reducing runoff through more precise application, may well turn out to pay for some important abatement measures.

SUSTAINABLE FORESTRY   The needs and opportunities in forestry are strikingly analogous to those in farming. Often the two are directly linked, chiefly by ways to reduce agriculture's pressure on forests and by opportunities for agroforestry—applying agricultural traditions to tree crops. The former options include (5, 6):

1. reducing the area and increasing the fallow period of slash-and-burn to sustainable levels;
2. replacing slash-and-burn with sustainable techniques (often proven by indigenous cultures), including agroforestry using native or cultivated tree-crops or both;

---

[54]This is not to say that corn is necessarily the best feedstock nor ethanol the best biofuel, but the conclusion holds for other examples too.

[55]Perhaps by frequent pond-switching between rice and aquaculture, a phenomenally productive traditional South Asian technique in which fish graze the rice stubble, or by using the rice straw more widely for roofing or as a biogas feedstock.

3. using trees felled during land-clearing for timber and biofuel;
4. using crop wastes not as direct fuel but rather for composting and mulch, and for efficiently burned, low-leakage biogas or gasification to run steam-injected-gas-turbine cogeneration (87);
5. controlling artificial burns;
6. adopting low-input/organic farming and forestry techniques (the former because they reduce pressure for land-clearing);
7. improving developing countries' farm productivity in order to reduce land needs per person; and
8. reducing feedlot if not total beef consumption (see above).

Agroforestry (117) is especially applicable to developing countries. Projects in many African conditions have demonstrated profitable 40–90% crop-yield gains while providing surplus woodfuel (117). Such practices can also permit the beneficial substitution of organic for artificial fertilizer and other agrichemicals (Ref. 6, p. I.3–23ff). Some tree crops can yield not only wood and food but also oils, resins, or terpenes (Ref. 6, p. I.3–41) that are directly usable as fuel, especially in diesel tractors and pumpsets. The oils can also be combined with dirty, wet alcohols in a simple solar catalytic reactor to yield superior diesel fuels such as methyl and ethyl esters.

Nonagricultural ways to relieve pressure on existing forests (Ref. 6, p. I.3–40) involve energy and materials policy. This includes recycling forest products[56] and "stretching" their effect (as by using honeycomb structures, see below), substituting electric efficiency for tropical-forest hydroelectric projects, designing frame structures so as to minimize the waste of timber, improving the protection and hence the lifetime of outdoor structural wood, and wringing far more work from biofuels (e.g. 83, 87, 38).

Additional forest-protection measures include taking better care of existing forests, harvesting more thoughtfully, planting shelterbelts, excluding livestock (e.g. with photovoltaic-powered electric fences), regenerating degraded forests, and promoting recreation and ecotourism to create a supportive constituency. These actions should supplement silvi/agri/aquacultural integration, the reforestation of surplus OECD farmland and degraded drylands, the planting of fuelwood around developing-country cities and along roads, and urban forestry. Together, such forestry practices could probably sequester a maximum of ~1.3 billion tons of carbon per year (GtC/y)—enough, with the concomitant stabilization of existing carbon pools, to offset

[56]"The average OECD person consumes about as much wood in the form of paper as the average Third World person consumes in the form of fuelwood" (Ref. 6, p. I.3–43). Producing a ton of paper takes ~¾ ton oil-equivalent of energy (79). If that energy were biomass, as the majority of it is in the US forest products industry, it could still be used instead to make liquid biofuels for sale to replace oil.

one-fourth of worldwide 1985 fossil carbon releases (Ref. 6, p. I.3–49). This broader menu of options nearly doubles the 0.7 GtC/y net carbon sink calculated by EPA for conventional forestry (Ref. 5, VII-7). Yet much of the extra carbon capture comes from practices like agroforestry that are generally profitable without counting their environmental benefits.

Some of the most promising and profitable new forests, too, can be in cities. Urban tree-planting programs are an especially cheap carbon sink because one tree planted in a typical US city sequesters or avoids ~10–14 times as much carbon release as if it were planted in a forest where it could not also save space-cooling energy (30). Urban forests and woodbelts can go far to relieve developing countries' fuelwood shortages and urban sprawl. The biomass produced by urban trees, even if not systematically and densely planted, can be substantial: Los Angeles County alone sends ~3600–7300 t/d of pure, separated tree material to landfills, not counting mixed truckloads. That ~1 GW of currently wasted (and costly-to-dispose-of) thermal energy is equivalent, at 70% conversion efficiency, to 0.5 million gal/d of gasoline—enough to drive a 60-mi/gal car >10 mi/d for every household in the County.

Urban forestry is also consistent with urban agriculture—long practiced in Western Europe and in China, where it provides 85+% of urban vegetables (100% in Beijing and Shanghai), plus large amounts of meat and treecrops (118). Urban farming in turn further reduces greenhouse-gas emissions from centralized agriculture, saves energy otherwise needed to process and transport food, and improves nutrition, aesthetics, community structure, and urban culture. Even the most crowded cities can farm on rooftops. With superwindows and air-to-air heat-exchangers, climate is no obstacle: Rocky Mountain Institute's 99+%-passive-solar headquarters grows bananas with no heating system despite outdoor temperatures as low as $-44°C$ ($-47°F$).[57]

In considering these supplements to conventional forestry, however, it could be overly sanguine to suppose that forestry in its present form will be able to sustain its vital carbon-sequestering role. Most discussions of CO_2 abatement through fiber-production forestry emphasize planting trees, often of specific, fast-growing, genetically engineered kinds. But planting trees is very different—often by severalfold in carbon inventory (104)—from maintaining a diverse forest that changes at its own pace. Much "modern" forestry repeats the ecological errors of monocultural, chemical-driven agriculture, treating trees like rows of giant corn—short-rotation (annual) monoculture

---

[57]The 372-m² building, at 2165 m in a 4900 C°-d/y climate, also saves half of normal water use, 99–100% of water-heating energy, and 90+% of household electricity (reducing the lights-and-appliances bill to $5 a month @ 7¢/kWh). Its marginal cost for all these savings, $16/m², paid back in 10 months with 1983 technology, and would pay back faster today, mainly because windows can now insulate twice as well (<0.5 W/m²K or >R-11) at nearly the same cost.

instead of long-rotation (perennial) polyculture (111). This is as true with *Pinus radiata* in New Zealand or southern pine in the United States as with eucalyptus in developing countries, where >40% of all new hardwood plantings are now eucalypts (Ref. 6, p. I.3–46).

Much modern forestry rests on mechanistic assumptions that appear from historic evidence to be ecologically unsound and unsustainable (119–121). Clearly what is needed is to sustain and increase both the quantity and the ecological quality (diversity, health, cycle time, resilience) of existing and new forests. This will require forest managers—just like farmers—to think like ecologists, not accountants.

Currently, forestry economics is being questioned chiefly on an accounting basis (the US Forest Service, for example, is the world's largest socialized roadbuilder), not on fundamental grounds. But new questions are emerging, and new answers will follow. In this decade, researchers may discover whether in forestry, as in farming, ecologically sound practices are also the most profitable kind, and whether, as some forest scientists are starting to suspect (121), a forest is worth more as a going (growing) concern than in liquidation. Unfortunately, the kinds of re- and afforestation that might prove not to be very productive or sustainable in the long run are the kinds currently considered in global-warming economic analyses. By normal forestry-economics standards, the fast-rotation plantings now dominating forestry practice appear profitable: that is why they are so widely practiced. But they could be increasing carbon inventories aboveground at the expense of the larger, invisible inventories belowground.

However it is practiced, planting trees is certainly a cheap way to sequester carbon (at least aboveground). Massive tree-planting programs have been found in several analyses, including a US Forest Service/ex-Council of Economic Advisors report to EPA, to sequester carbon at low costs, typically on the order of $10/tC (Ref. 122, p. 39), despite taking no credit for the discounted revenues from ultimate timber harvest, nor for other benefits meanwhile.[58] Cost estimates used by a National Academy of Sciences study (discussed below) are in a similar range. A study for Pacific Power suggests that timber revenues could in fact repay the cost of the reforestation about twice over (123). If this proves true, as the apparent profitability of current forestry activities suggests, then these canonical ~$10/tC abatement costs are too high: profitable tree planting abates global warming at negative cost.

Consistent with this, EPA (Ref. 5, p. VII-7) considers reforestation "one of the most cost-effective technical options for reducing CO_2 and other gases." If the planting programs are based on agroforestry, urban forestry, and other

---

[58]These include e.g. erosion and flood control, groundwater recharge, runoff holdup (stretching reservoir capacity), fish and wildlife enhancement, aesthetic value, and recreation.

ecologically sensitive techniques, then the average cost of sequestering carbon in trees should be even lower.

CFC Substitution

The projected cost of the CFC/halon production phaseout now required by the London Amendments to the Montréal Protocol has declined by roughly half over the past two years through closer scrutiny and ingenious technological innovations. In some instances, including refrigeration with reoptimized design, the substitute may actually improve performance (Ref. 14, p. 62). In others, notably the cleaning of printed-circuit boards with terpene derived from orange peels, or new low-flux soldering techniques using aqueous or no cleaning, the substitute not only works better but also costs less. Although ~$135 billion worth of equipment in the United States alone uses CFCs, the need is expected to be met by a combination of ~29% efficiency/maintenance/recovery/recycling/reclamation, 39% "drop-in" replacements (those that directly replace CFCs with little or no modification of equipment), and 32% substitution by replacements requiring different or redesigned equipment (124).

According to EPA analyses (125, 122), late-1989 data indicated that a US phaseout of CFCs by 2000 would cost ~$1.3 billion, or ~$2.4/tC-equivalent, at a 6%/y real discount rate. This figure is the sum of many disparate terms, all subject to technological change that tends to reduce costs. New technologies, like the advanced thermal insulation mentioned next, are reducing the total cost quickly enough that it may before long become slightly negative.

There is no obvious reason why abatement should cost more in other countries, given access to OECD technologies; quite the contrary, since one third of global CFC use (Ref. 6, p. I.3–3) is for aerosol propellants long ago cheaply displaced in the United States. The United States accounts for ~29% of global CFC consumption (126). The global abatement cost appears, therefore, to be ~$4½ billion, and is continuing to decline with further technological development. Since CFCs in the 1980s accounted for ~24% of global warming (1), $4½ billion is equivalent to <$2 per ton of carbon-equivalent in CO_2.

Multigas Abatements

Important opportunities exist to abate two or more greenhouse gas emissions simultaneously, both at a net profit. A few examples illustrate the diversity of such options.

1. Five advanced classes of thermal insulation now under development or in early commercialization, along with other design refinements, could save 90+% of the electricity used by refrigerators and freezers (Ref. 14, pp. 44–60)—the biggest users of electricity in households lacking electric space

and water heating, in countries as diverse as the United States and Brazil (17). This saving alone can avoid burning roughly enough coal each year to fill up the refrigerator. (Savings >80% are available with conventional insulation.) The emerging insulations also substitute a vacuum for the CFCs normally used to fill plastic foam; eliminate most of the refrigerant inventory (currently CFCs); and can accommodate the modest efficiency losses, if any, caused by switching to non-CFC refrigerants. Best of all, certain advanced insulations can make the appliance's walls thinner, because they insulate up to 12 times as well as CFC-filled plastic foam. The resulting increase in interior volume may be worth about enough to pay for the insulation (Ref. 14, pp. 59–60).

2. Landfills emit ~30–70 Mt/y of uncontrolled methane, about one sixth of total methane from human activities (Ref. 6, p. I.3-11). Capturing this gas and using it as fuel—ideally for cogeneration—both prevents its emission and displaces a larger amount of CO_2 otherwise emitted by a coal- or oil-fired power plant or boiler. ICF (Ref. 122, pp. 43–45) calculates that 75% recovery just from US landfills holding >0.9 million tons of waste would burn ~55% of US landfill methane at a cost roughly half the market value of the electricity. At least 123 US landfills already recover methane as fuel (Ref. 5, p. VII-191), but far more do not yet. Capturing and burning coal-bed methane (Ref. 5, p. VII-131) looks similarly profitable (or at least breakeven) and makes mines safer. Converting livestock or human manure into biogas, whether on a commercial or a village scale, and using the biogas to displace fossil fuels or unsustainably grown firewood (38), also appears economic with modern technologies available at many scales. Interestingly, since ~20% of US methane emissions can be captured just from landfill, coal, and natural-gas leakage, and there are apparently attractive agricultural abatement options too, there is a growing consensus that the ~15–20% CH_4 abatement needed to stabilize this gas's heat-trapping will prove costless or profitable.

3. Recycling paper, or composting food and garden wastes, reduces land-fill CH_4 output; saves the CO_2 and NO_x otherwise emitted when fuel is burned to produce and transport those materials; and saves money. Compost can also displace synthetic fertilizer, whose manufacture releases CO_2 and whose use releases N_2O. By improving tilth, compost can also help the soil to retain water, saving irrigation pumping energy and hence CO_2. If part of a locally based agriculture, compost can further help to substitute fresh food for perishable food refrigerated with CFCs and fossil-fueled power plants (~9 GW in the United States: Ref. 14, p. 115). Local agriculture can also save oil otherwise burned to transport food; the average molecule of US food has been estimated to travel ~1200 miles before being eaten.

4. Native building materials such as adobe, caliche, mudbricks, rammed earth, etc can displace CO_2-intensive production and transportation of cement. Sustainably grown timber or bamboo incorporated into buildings can

also temporarily sequester carbon (Ref. 6, p. I.3–14). Where timber is scarce, however, pressed-wood/paper-honeycomb materials can reduce tree use per building by up to ~90% at negative net cost, incidentally increasing the building's energy efficiency too (D. Hartwell, Bellcomb Corporation, personal communication). Improved energy infrastructure, especially for more efficient use of fuelwood, can also divert large amounts of wood from fuel to fiber use (114, 38, 83).

5. Efficient motor vehicles can cost-effectively and simultaneously reduce emissions of CO_2, CO, O_3, N_2O, NO_x, SO_x, hydrocarbons, and other radiatively active gases or photochemical products such as peroxyacetyl nitrate (6). The reduced air pollution, especially O_3 (103), can also reduce forest death and other vegetative damage, maintaining more and healthier trees as carbon sinks. On a microdesign scale, more efficient car air conditioners, or reduced cooling loads due to such improvements as lighter-colored paint or spectrally selective glass, can reduce the inventories and leakage of CFCs, save compressor operating energy (hence CO_2), and save CO_2 in all driving by transporting a smaller, lighter-weight compressor (51a).

6. Electrical savings that displace new hydroelectric dams can achieve their fuel-saving goal but preserve the carbon inventories in the impoundment area's above- and below-ground biota, rather than emitting them both as CO_2 when the area is cleared and, even worse, as CH_4 after flooding converts it to an anaerobic swamp. Since the electric savings are cost-effective (cheaper than the dam or a thermal power station), both these abatements cost less than zero.

Though not thoroughly catalogued or characterized, such multigas abatements will generally reduce the total cost of abating global warming by providing multiple benefits for single expenditures. Omitting them from supply-curve analyses is thus a conservatism. It may well be a significant one.

Supply Curves for Abatement

In 1989, the Amsterdam office of McKinsey & Co. prepared for the Dutch government one of the first attempts at a supply curve for global-warming abatements (D. Six, personal communications). It was explicitly illustrative and incomplete, but heuristically valuable. Since then, an increasing body of ever more detailed and empirically grounded evidence has taught two lessons: that using supply curves to relate the marginal quantities and marginal costs of abatement is a useful way to gain understanding of policy options, and that closer scrutiny tends to raise the quantities and lower the costs. For example, the data from one such compilation (122)[59] of diverse government and

---

[59]Plotted from Tables 10–15; some utility-sector data in Table 14 differ from those in Appendix H, although their totals differ little.

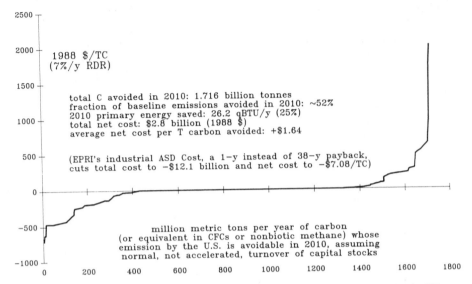

Figure 4 Supply curve, from data prepared by ICF Resources Inc. (a contractor for the US Environmental Protection Agency), of potential abatements of US emissions of CO_2, nonbiotic CH_4, and CFCs in 2010 (122).

industry studies of the US potential for abating emissions of CO_2, nonbiotic CH_4, and CFCs in 2010 are plotted in Figure 4.

This curve's basic structure is arrestingly simple: a long flat section in the middle at roughly zero cost (the cheap CFC abatements, landfill and coal-bed methane capture, and reforestation), plus "tails" of essentially equal area at both ends. These "tails" comprise chiefly ~25% energy savings at negative net cost on the left, and the costlier kinds of renewable energy and industrial fuel-switching at positive cost on the right (i.e. costing respectively less and more than competing fossil fuels). The calculated potential reduction from this quite incomplete list of measures[60] totals 1.72 billion metric tons of carbon per year (GtC/y), or ~52% of the base-case emissions projected for that year. This large abatement is significant for two reasons:

1. some major options, such as sustainable farming, forestry other than standard reforestation, most industrial and heavy-transport savings, and other trace gases, were not counted; yet

[60]As a small example, only half of the industrial motor-system retrofit potential was considered, and that half—all from adjustable-speed drives (ASDs)—was assigned a cost ~38 times EPRI's value (19) or ~15 times RMI's (22). Correcting this apparent error changes the net cost of abating a ton of carbon-equivalent emission from +$1.6 to −$7.

2. the net private internal cost of halving the US contribution to global warming is roughly zero.

A similarly conservative linear-programming optimization for The Netherlands (127) found a very similarly shaped supply curve (Figure 5). It counted only reductions in fossil-fuel CO_2, not other gases, and hence is akin to combining the left- and right-hand ends of the ICF supply curve. But this Dutch analysis found a 35% CO_2-saving potential in 2000 from measures all costing less than zero, whereas for the ICF study the corresponding figure was roughly 20%. The Dutch study found a net cost of ~$90/tC for CO_2 reductions totalling up to ~65%. As can be seen by comparison with Figure 4, this is a relatively modest cost for such a large fossil-fuel saving in a comparatively efficient country. The quantity of potential savings would of course be even larger if any credit were taken for the essentially free non-fossil-fuel abatements included in the middle portion of the ICF curve.

These large, approximately costless potential savings are qualitatively consistent with the findings of the Mitigation Subpanel of a 1991 National

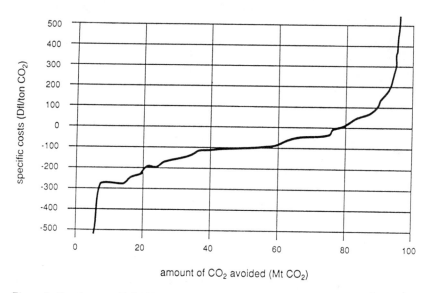

Figure 5 Supply curve (Ref. 127a, cited in Ref. 127, Figure 3.13, reproduced by authors' kind permission) of Dutch fossil-fuel-CO_2 abatements from saving fossil fuel through end-use efficiency, small-scale cogeneration, and renewables compared with a midcase projected emission of 224 million tons of CO_2 (excluding feedstick uses) in the year 2000. Of the 97 $MtCO_2$/y reduction potential shown, 80% is from end-use efficiency, of which 80% was found to be cheaper than the fuel it saved. Most of the latter measures were therefore included in the base-case projection. The list of measures considered was less complete than that evaluated in this paper and was technically conservative (c.f. 100a).

Academy of Sciences (NAS) study (9).[61] The Subpanel (9a) analyzed, on very conservative assumptions,[62] a much wider range of technical measures (some of which the Subpanel felt might entail modest life-style changes). The Subpanel therefore found a larger abatement potential. The measures on which the Subpanel reached consensus could, if fully implemented, collectively abate ~61–64%[63] of 1988 US contributions to global warming at an average net cost of −$6/tC—again, slightly profitable rather than costly. (The range of average costs found was −$24 to +$9/tC.) The mean net cost would thus equal zero at a global-warming abatement somewhat larger than 61–64%. The report, citing The Intergovernmental Panel on Climate Change (3), states that at least a 60% abatement is likely to be needed to stabilize the climate.

The point is not whether any of these three sets of figures is exactly right—they all inevitably reflect many uncertainties—but rather that sound public policy requires an open process to identify all the parts of such supply curves and harness the public's imagination and ingenuity in refining and achieving each. It should be possible to reach near-agreement about the numbers, or at least to understand the origins of residual disagreements.

As a small initial contribution to that goal, how might the additional opportunities described earlier in this paper change the conclusions of the ICF or NAS analyses? Qualitatively, these extra options would clearly have two effects: reducing global warming even more, and converting those studies'

[61]The Panel on Policy Implications of Greenhouse Warming of the Academy's Committee on Science, Engineering, and Public Policy, whose Synthesis Panel report was published 10 April 1991 (9). The Chairman was Washington Governor Dan Evans, a distinguished engineer and former Chair of the Northwest Power Planning Council. The report will be a valuable contribution to public policy formation. Its main shortcomings include outdated energy-efficiency potential, inadequate attention to less conventional forestry options and to organic agriculture, omission of multigas abatements, and an excessively restrictive and short-term view of renewable energy potential (94). Improving the analysis in any of these respects would raise the calculated abatement and lower its cost (see below).

[62]The mean potential savings assumed included 20% in household refrigerators and dishwashers (compared to a cost-effective potential with best present technology—see citations in the first part of this paper—of >90% in prototypes vs the present stock), zero in other household appliances (typically two-thirds or more in models entering production), 25% in commercial cooking and commercial/industrial space-heating (>50% and ~100% in commercial equipment), "up to 30%" for all industrial electricity and fuel (>50%), 40% for commercial ventilation (>70%), 45% for commercial lighting (~70–90+%), 50% for residential space-heating and commercial water-heating (~100%) and for commercial space-cooling (~80+%), and 55% for residential water-heating (~65–100%). Light vehicles were taken only to 25 on-road mi/gal without, or 36 with, reductions in size or performance, and similarly with other transportation modes. The Subpanel did not examine the RMI/COMPETITEK analyses cited earlier.

[63]The quantities of savings cited add up, corrected for interactions, to 61%; their cited percentage savings, to 64%. It is not clear which set of figures is more accurate.

low net cost to a substantially negative net cost, whether for the United States or worldwide. The rough magnitude of some of these changes can be estimated as follows.

1. The Academy apparently assumed industrial electricity-savings potential equivalent to about one-third of the savings that RMI has demonstrated (22) and EPRI concurred with (19) for motor systems alone. The RMI/EPRI-agreed cost of such savings is also an order of magnitude lower. Ignoring the substantial nonmotor electricity-saving possibilities, just substituting the RMI/EPRI motor findings would therefore triple the saving and cut its cost (i.e. increase its net present-valued financial saving) by about $60 billion. Considering industrial fuel savings more fully, and taking full credit for industrial energy savings from the leaner materials flows described earlier, would also yield major gains. So would substituting the gas saved in industrial and building heat for more carbon-intensive fuels.

2. EPRI's assumed potential for electricity savings in buildings was presented to the NAS panel as 45% at an average cost of 2.5¢/kWh. Yet detailed, empirically based retrofit analyses for Arkansas, which has a slightly more difficult climate than the US average, found a retrofit potential to save 77% in houses at 1.6¢/kWh and 74% in commercial buildings at −0.3¢/kWh (128). The residential retrofit, costing ~$5500 in a typical 127-m^2 frame house, includes improved insulation, superwindows added over the existing single glazing, reduced infiltration, light-colored wall paint, lights and appliances comparable to the best European models, a hot-water-saving package, and an air conditioner whose more than doubled efficiency (to COP = 4.54) was nearly paid for by making it two-thirds smaller. The resulting package— carefully chosen after rejecting nearly 100 other options—was simulated to save 77% of annual and 83% of peak electric use (and, fortuitously, 60% of the gas without improving the gas appliances) with a 2.9 year retrofit payback at measure prices determined from local quotations, Arkansas Energy Office field experience, and standard R. S. Means construction-cost data. The commercial-sector retrofit included ~100 measures: the lighting package discussed earlier, standard improvements to space-cooling and air-handling systems, some shell improvements (anti-gain window films, lighter roof color, etc), and better internal equipment (refrigeration, computers, etc) and building controls. Interactions were explicitly accounted for. The cost of nearly doubling the efficiency of replacement chillers was negative because it was less than the cost saved by downsizing to accommodate the reduced cooling loads. The ~76% calculated electric savings cost slightly less than zero because the savings in lighting maintenance costs (chiefly from replacing incandescent with modular compact fluorescent lamps) paid for the lighting equipment with more than enough money left over to pay for all the nonlighting improvements too. More recent developments (14) would support even

more favorable results, but just the Arkansas results would raise the EPRI/NAS building-electricity carbon saving by two-thirds and nearly double its net financial saving.

3. The 50% fuel saving assumed in buildings is smaller and costlier than the potential found by a major federal study a decade ago (73), and improvements in the building stock meanwhile do not seem to account fully for the difference. Such key developments as superwindows, whose simpler versions have captured a large share of the US insulated-glass market in the past few years, were not considered. Interestingly, the Arkansas house-retrofit analysis just cited found that gas savings of 60% would result at no extra cost as a free by-product of the 77% electric savings—or far more, at modest cost, if the gas appliances were also made more efficient. In an era when skilled practitioners can retrofit superinsulation that saves most, or in some cases nearly all, space-heating, and retrofit savings of two-thirds of water-heating fuel are straightforward, a 50% fuel-saving potential is clearly outdated.

4. US cars and trucks alone released ~0.32 GtC in 1988 (129), or 14% of the US contribution to global warming (122); all transport modes, 20%. Since most of that consumption can be saved at negative cost, as was documented earlier, today's most efficient transportation technologies (51a) (let alone further system-design improvements) could probably increase the total abatement potential by ~15+ percentage points—several times the assumption in the Academy's less complete survey.

5. The renewable energy potential considered economically competitive in the recent Interlaboratory White Paper (94)—i.e. cheaper than the internal cost of carbon-emitting alternatives—is officially projected to be adequate, in concert with even cheaper energy efficiency, to displace most or all of the US fossil-fueled power stations now operating or planned for the next few decades, plus much direct fuel.[64] Since most of the world, especially the most populous parts, has a renewable-energy potential broadly comparable or superior to that of the United States (99, 69a), the same conclusion should hold generally. Any serious long-term energy scenario must therefore consider carefully the opportunity to squeeze down fossil-fuel use between efficiency and renewables.[65]

6. It is not clear that zero- or negative-cost forestry options, such as urban forestry and maintenance of old-growth forests, have been properly taken into account. Similarly, using organic agriculture to change 50 million hectares of US farmland to a carbon sink, on the assumptions described earlier, would

---

[64]The total cost-effective renewable energy output in 2030 was projected to be 37–67% as large as current US total primary energy use.

[65]Assistant Secretary of Energy J. M. Davis has recently presented such a "jaws" scenario—conceptually an updated version of a "soft path" scenario published 14 years earlier (130), and driven by identical economic logic.

reduce the US contribution to global warming by 1%,[66] nearly as large as the Academy's assumed savings of oil and gas in commercial buildings. The cost of this extra 1% abatement would be negative, because such practices are probably more profitable for the farmer (110, 112, 114, 113, 116). But as noted above, Holmberg's (115) assumptions about carbon uptake may well be conservative by at least severalfold.

7. Adding synergisms between measures, described earlier as "multigas abatements," should further improve the total abatement potential and its economics.

8. Like most such analyses, the Academy study assumed that saving electricity would displace the average kWh generated. However, "environmental dispatch" in which the most carbon-intensive (or otherwise polluting) power stations were backed out first would save ~59% more carbon per kWh (131). This would be a natural consequence of internalizing external costs, such as the 5.8¢/kWh authoritatively estimated for coal-fired electricity in the United States (34). Since ~17 state utility regulators across the United States have already adopted this practice for planning in some degree, another 20+ are currently doing so, and all OECD member nations officially accept the principle of internalization, it is probably only a matter of time before economic dispatch of power stations starts taking avoided pollution-, including carbon-, abatement costs fully into account.

A simple thought-experiment illustrates the importance of giving the energy-efficiency potential the most searching and up-to-date possible scrutiny:

1. In 1989, Americans paid $453 billion at retail for commercial fuels and power (Ref. 74, 1989 $).
2. The US energy system is quite competitive in most respects.[67] Competitively providing the world's commercial energy would therefore probably cost, shorn of taxes and subsidies, not far from the same amount per joule.
3. Scaling up for consumption yields an internal shadow expenditure for global retail energy on the order of one and two-thirds trillion dollars per year. (We use this very rough estimate because actually quoted prices often do not reflect actual costs.)
4. Just the potential described earlier for directly saving electricity and oil would save upwards of three-fourths of that energy, and at typical late-1980s energy prices, would repay its cost in a few years.
5. The savings would be slightly smaller and costlier in the most efficient

[66]Its global significance would be far greater—on the order of 0.7 GtC/y if successfully applied throughout the world's 1.5 billion hectares of cultivated land (Ref. 6, p. I.3–33).

[67]Direct federal subsidies reduced the apparent total energy bill by ~10% in FY1984 (15), but became smaller in 1986, so the overall distortion is probably lower now, though it may be more unevenly allocated between competing options.

OECD countries, but substantially bigger and cheaper in other countries. The US potential would thus be a reasonable-to-conservative global average.

6. The present-valued cost of such large savings was shown above to be on the order of one tenth of present energy prices.

7. Subtracting that tenth from the gross savings leaves a potential long-run net monetary saving[68] of at least $1 trillion a year from fully implementing the efficiency opportunities described.

That is about as big as the global military budget. The money now wasted on inefficiently used energy is certainly needed for more productive purposes than wasting energy. But more importantly for the purposes of this paper, the negative-cost energy savings that yield this ~$1-trillion-a-year of net wealth creation represent probably the cheapest increment of global-warming abatement. Cheap energy efficiency can reduce global warming by about one third—and free about a trillion saved energy dollars per year for other abatements, with money left over. Recent official assessments show abatements in the ~50–64% range at roughly zero net cost despite assuming energy savings several times smaller and costlier than those demonstrated above. But those studies slighted the modern energy-efficiency potential (and some other opportunities too). Properly counting that potential is thus bound to abate even more global warming, and to change the total net internal cost of abating global warming from about zero to a value robustly less than zero.

It is also important to note that a trillion saved dollars per year is so much money that it can buy a great deal of additional abatement even at relatively high unit costs. For example, many of the costlier ways to achieve the last pieces of abatement required for climatic stabilization might cost (say) $165 per ton of carbon. If so, then the net money saved by the energy efficiency program could buy six billion tons of carbon abatement per year—about equivalent to today's entire global output of CO_2 from all combustion processes. Since most of that fossil-fuel CO_2 would already have been abated by the energy efficiency itself, and since most other kinds of abatement (biotic carbon, CFCs, and most CH_4 and N_2O) appear, as noted above, to be relatively inexpensive, it is not easy to construct an optimized menu of abatements that could stabilize the climate without having a considerable amount of money left over. In other words, although the amount of money that energy efficiency can save is not precisely known, it is clearly large enough to ensure a financial surplus despite the considerable uncertainties about how much the non-energy abatements may cost.

---

[68]Long-run because the sunk capital costs of the energy system cannot be saved in the short run, but reincurring those costs—rolling over retiring capacity into replacement capacity—can still be avoided by achieving durable, reliable, long-term efficiency gains.

Opportunity Cost Requires "Best Buys First"

There is, however, one caveat requiring emphasis. To abate global warming promptly with finite resources, it is vital to choose the best buys first. This is because of "opportunity cost"—the impossibility of using the same money to buy two different things at the same time.

If, for example, you spend a dollar on a costly source of electricity, such as nuclear power or photovoltaics, then you will get relatively little electricity for that dollar—that is what "expensive" means. You will therefore be able by such means to displace little coal-burning in power stations. But if you use the dollar to buy a very cheap option instead, such as superefficient lights or motors, then the resulting bounty of electricity could displace a lot of coal. Therefore, whenever you spend the dollar on a costly option instead of on a cheap one, you will unnecessarily release into the air the extra carbon that would not have been released had you bought the cheapest option first. That is why, for example, nuclear power makes global warming worse[69]: it emits less carbon per dollar than coal-fired plants, but in this opportunity-cost sense, many times more than efficiency (132–134).

This is not an academic point; it is at the crux of essential policy choices. Investors must understand which options are cheapest, hence most profitable, and policymakers must avoid or amend regulatory structures that divorce these two attributes. Even in societies where capital is allocated by planning rather than by markets, the planners must be able and eager to determine the best buys and then buy them. Any other sequence of investments prolongs and enlarges climatic risk. We therefore turn next to how the least-cost investment sequence can actually be discovered, bought, and successfully marketed and delivered in diverse societies.

IMPLEMENTATION TECHNIQUES

The foregoing discussion has highlighted both commonalities and differences among the technical options available to the world's three main regions. In all

[69]Algebraically, if K is the carbon intensity of existing coal-fired plants (in tC/kWh), C_n is the levelized cost of a marginal nuclear kWh, and C_e is the levelized cost of a marginal saved kWh (both in \$/kWh), then $K([C_n/C_e]- 1)$ tons of extra carbon are released per kWh made in a new nuclear plant instead of saved by improved end-use efficiency. This assumes simple fungibility of dollars between the two investments. Recent experience of the US utility industry, however, suggests that matters are actually worse: when utilities overinvested in capacity by ~\$200 billion, many, seeking to recover sunk costs, turned their efficiency departments into surplus-power marketers, making power-plant dollars not just a neutral complement to but a direct enemy of efficiency dollars. Thus EPRI estimated a few years ago that today's "strategic marketing" programs will directly result, by 2000, in some 35 GW of new onpeak demand—about two-thirds of the savings projected from the industry's efficiency programs. Whenever nuclear investments, therefore, mean not only foregoing an efficiency investment but deliberately seeking to boost electric demand, more carbon will be released than C_n/C_e would indicate.

regions, for example, energy efficiency is an urgent priority, though for somewhat different reasons: for example, high energy efficiency in industrialized countries is vital to global climatic protection and is economically valuable for those countries, while high efficiency in the other two regions is currently less important for the world but economically vital to their own development (which would in turn raise their CO_2 output to the majority of the global total). Moreover, non-OECD countries now building infrastructure and stocks of consumer goods have a chance to build in efficiency from scratch, often as a natural and inexpensive feature of investments they are making anyhow, whereas OECD faces the daunting task of retrofitting trillions of dollars' worth of obsolete buildings and equipment.

Most instructive, however, are the differences in and synergisms between implementation strategy between the regions. Many of these have been extensively treated in a huge literature, so we seek here only to summarize main points that may previously have been overlooked or underrated.

OECD Countries

The OECD countries emitted nearly three-fifths of 1950–1986 fossil CO_2 (Ref. 6, p. I.1–9) and cause the lion's share of global warming today. They have such large economies largely because they have vigorous markets[70] with high innovation rates and rapid discovery, application, learning, and corrective feedback in most sectors. Most OECD countries also have a powerful public consensus for environmental protection. That is why, for example, government policy is committed to major CO_2 reductions in West Germany, Denmark, and the Australian State of Victoria,[71] to CO_2 stabilization (var-

---

[70]Centrally planned exceptions, such as the French electric sector, are likely to prove short-lived when 1992 European economic integration unleashes new competitive pressures and transparent-pricing requirements. Increasing North American integration under Canadian and (soon) Mexican free-trade agreements will similarly hone competition. Some economies, such as Sweden's and Japan's, and in some respects the European Economic Community itself, are more properly regarded as mixed, with an overlay of sophisticated public-sector planning and coordinating apparatus. Yet the vitality of the market sector, the diversity of the public institutions within OECD, and the forces of market and political accountability in most OECD countries have permitted as impressive a range of initiatives to flower in the mixed as in the most laissez-faire economies.

[71]The German proposal, whose implementation is being worked out, is to reduce CO_2 emissions by 25% of 1987 levels by 2005. A stricter proposal, with very large long-term reduction targets, was overwhelmingly recommended by the Enquête-Kommission of the Bundestag, which has produced two excellent reports on the subject. Victoria has also adopted the 25%-reduction Toronto goal. Denmark's plan calls for 20% reductions by 2005 and 50% by 2030.

iously defined) in Britain, Sweden, Canada, and Japan, and to stabilization followed by reduction in The Netherlands and Norway.[72]

This region is ideal for adopting and adapting the best worldwide experience of market mechanisms for capturing profitable global-warming abatements. Favorable conditions include widespread acceptance of the polluter-pays principle;[73] availability of generally sound statistical data and reporting systems; widespread sophistication in industrial organization to meet financial objectives; a large population of skilled entrepreneurs; and the world's highest mobility of labor, capital, and (most importantly) information.

Although most Western European "Greens" traditionally distrust markets and the private sector, there are clear signs that many are starting to appreciate the power of properly structured and informed markets,[74] hence the importance of making markets in avoided depletion and abated pollution (128). The basic concept is simple: economic glasnost—prices that tell the truth—can scarcely achieve efficient behavior without a market where the buyers and sellers of technological solutions can meet and do business. My potential loss from a carbon tax must be convertible to your potential profit from selling me a more efficient lamp. If I cannot buy that lamp, my only response is a behavioral change (using a dimmer lamp or using it less often), which is relatively weak and impermanent.

Ways to make markets in saved energy and water, developed at Rocky Mountain Institute and elsewhere, are now entering widespread use both in the United States and abroad. Based on successful early experience, they show promise of accelerating energy efficiency and appear applicable also to other key ways to abate global warming, as described next.

ENERGY   *Electric utilities*   More than half of US residents can already get financing from electric utilities for electricity-saving equipment, in the form of concessionary loans, gifts,[75] rebates, or leases (19, 135). Such financing is

---

[72]There is no analogous US policy, apart from some commendable state initiatives (Ref. 60, p. 12); on the contrary, in both global warming and ozone protection, a strong consensus in the rest of OECD was openly thwarted by US representatives during 1989–1990. At this writing, the United States and Turkey are the only OECD countries lacking CO_2 targets.

[73]Finland and The Netherlands introduced carbon taxes in early 1990, Sweden did so in January 1991, and Germany is weighing one.

[74]Acknowledging this trend, in Royal Dutch/Shell's latest Group Planning scenarios, the "Sustainable World" scenario was somewhat dirigiste as first drafted, but was changed to rely far more on market forces—logically enough, since its options cost less.

[75]Southern California Edison Company, for example, has given away more than a million compact fluorescent lamps, because that was cheaper than operating the company's existing power plants.

essential because customers typically want to get their energy-saving investments back within a couple of years, whereas if they do not become efficient and the utility builds a power station instead, its technical and financial strengths enable it to accept a payback closer to 20 years (135). This roughly tenfold "payback gap" is rational to both parties, but societally, it causes a severe misallocation (~$60 billion a year in the United States in the mid-1980s) by effectively diluting price signals tenfold. Utility financing— ~$2 billion in 1991— helps to close this gap by reducing customers' implicit >60%/y real discount rate approximately to utilities' ~5–6%/y (135).

Utility financing of efficiency enables all supply- and demand-side options to compete on a "level playing field." Such competition, through either a planning or a market process, is now mandatory (to varying practical degrees) in the ~43 states with a "least-cost utility planning," or best-buys-first, policy (136). Such financing works so well that if all US residents saved electricity at the same speed and costs at which the ~10 million people served by Southern California Edison Company actually saved electricity during 1983–1985, chiefly financed by SCE's rebates, then national forecast needs for power supplies a decade ahead would decrease by ~40 GW/y, at a total cost to the utility of ~1% of the cost of new power stations (19; 8a; Ref. 136a, pp. 180–83; Ref. 7, p. 171).

This sort of service-delivery, engineering-driven model of how utilities promote customers' efficiency gains retains an important place in dealing with specific market failures, such as the split incentives between builders and buyers or landlords and tenants. (Government performance standards and labels are also key parts of the policy toolkit in such situations.[76]) But a complementary approach is now starting to supplement and might ultimately supplant the "we-will-wrap-your-water-heater" philosophy. Rather than merely marketing "negawatts," many utilities are also starting to make markets in negawatts: to make saved electricity into a fungible commodity subject to competitive bidding, arbitrage, derivative instruments, secondary markets, etc. For example, some utilities are

1. buying back savings from customers by paying "generic rebates" per kWh or peak kW saved—including rebates for beating government standards or for scrapping old equipment;
2. in eight states, operating "all-source bidding" in which all ways to make or

---

[76]California's Title 24 building standards alone, now saving the state's citizens ~$1 billion worth of electricity every year, are a good model of how to combine performance standards with prescriptive options to reduce hassle to the builder.

save electricity compete in open auction and the utility takes the low bids, which are generally for efficiency;[77]

3. starting to buy saved electricity from other utilities—a form of arbitrage on the difference between the cost of supply and efficiency;

4. considering making spot, futures, and options markets in saved electricity (an electricity futures market was in fact launched in Britain in spring 1991);

5. exploring ways to allocate saved electricity among customers, rewarding customers who go "bounty-hunting" by correcting inefficiencies anywhere in the system;[78]

6. selling electric efficiency in other utilities' territories (Puget Power Company, for example, sells electricity in one state and efficiency in nine states); and

7. in seven states, considering or experimenting with sliding-scale hookup fees for new buildings—"feebates" whereby the builder either pays a fee or gets a rebate when the building is connected to the grid (which and how big depends on the building's efficiency).[79] Feebates can offer major economic advantages to all the parties, can generate tens of thousands of dollars' net wealth per US house so built, and are readily coupled with efficiency labelling.

In addition, gas utilities can make money selling electric efficiency, thereby changing the behavior of buildings in ways that also help them open up new gas markets (7). Electric utilities can also sell gas efficiency, and both should be rewarded for selling either. Wisconsin's utility regulators are even considering ordering that state's utilities to help customers switch to any competing energy form that costs less.

Rapid experimentation in these and other market-making methods has been facilitated by the great diversity of the US electric utility system: ~3500 utilities of all shapes and sizes in ~50 major and hundreds of minor regulatory

---

[77]Maine, largely through such bidding, also raised its private, mainly renewable, share of power generation from 2% in 1984 to 20% in 1989 to 30% in 1991 (based on 1990 construction), according to former Maine Public Utilities Commission Chairman David Moskovitz.

[78]This has already been done with saved water (Ref. 138, p. 21). All the other mechanisms described here are also being applied to water efficiency, and many appear useful for other resources or for services such as transport (128).

[79]Formally, the "feebate" should satisfy three boundary conditions: revenue-neutrality (the fees pay for the rebates); fees for inefficient buildings are based on long-run marginal costs including externalities; and rebates exceed the builder's marginal cost of achieving the efficiency. The slopes and intercepts can be adjusted annually as needed to maintain these conditions. When construction of inefficient buildings has been driven off the market, one can declare victory and stop.

jurisdictions. Although learning is often painfully slow, with many utilities still reporting relatively small, slow, and costly savings (8b), new results are encouraging. A few years ago, some utilities had captured ~70–90% of particular efficiency micromarkets, mainly difficult ones (residential shell retrofits), in only one or two years.[80] In 1990, greater marketing experience (139) has enabled, for example, New England Electric System to capture 90% of a small-commercial pilot retrofit market in two months, and Pacific Gas & Electric Company to capture 25% of its entire new-commercial-construction market—150% of the year's target—in three months. (PG&E then raised its 1991 target, and captured all of it in the first nine days of January.)

Such entrepreneurship is being encouraged by a nationwide agreement in principle among US utility regulators[82] to change the rules of price formation so as to ensure that utilities' cheapest options are also their most profitable ones. Although many ways to do this are available (136), the most common is to decouple utilities' profits from their sales and then let them keep as extra profit part of what they save their customers. Under a new policy approved in summer 1990, for example, Pacific Gas & Electric will be allowed to keep 15% of certain savings—adding ~$40–50 million to its 1991 profits. But the customers are better off getting 85% of an actual, prompt saving than getting all of nothing. At this writing, five states have approved such reforms and another 20+ are doing so. The previous regulatory scheme rewarded utilities for selling more electricity (or gas) and penalized them for selling less. Similar perverse incentives still exist in many countries, despite the diversity of utility structures, and can be corrected by similar means.

More than a dozen states' least-cost planning comparisons, too, already credit efficiency and renewables, or penalize their competitors, with shadow prices reflecting externalities (34); more than half the states are expected to do so by the end of 1991. The correct externality values are not exactly known, but they certainly exceed zero, so in adopting a nominal figure, the New York Public Service Commission's Chairman, Peter Bradford, notes that it is better to be approximately right than precisely wrong. This approach, too, is attracting considerable interest in Europe.

These regulatory moves toward simulating efficient market outcomes have accelerated the already rapid shift in US utilities' culture and mission, away from selling more kilowatt-hours and toward the profitable production of customer satisfaction (19). About a third of US utilities have already made this transition. Selling more efficiency may reduce their electric sales and

---

[80] E.g. the Hood River County experiment in Oregon, and several Iowa municipal utilities' load-management promotions. In at least one case—water-heating wraps in Osage, Iowa— reported market capture was 100%.

[82] Unanimously approved by the National Association of Regulatory Utility Commissioners in November 1989.

revenues, but their costs go down more; and under the new rules, they can keep part of the difference, making money on margin instead of volume. Such a utility can indeed make money in six ways: it saves operating, construction, and replacement costs, plus associated risks and externalities; under the new US Clean Air Act its fuel savings will be able to generate tradeable emission rights (currently for acid gas, but extendible by future legislation to fossil carbon); as soon as its regulators reform their ratemaking rules, it will be specifically rewarded for efficient behavior; and it can earn a spread on financing customers' efficiency improvements.[83]

Oil efficiency   Analogous concepts are starting to enter the oil-efficiency market. Most importantly, on 30 August 1990, the "Drive+" feebate proposal passed the California Legislature with overwhelming bipartisan support[84] (140). This bill would enact a revenue-neutral, open-ended sales-tax adjustment based on both fuel efficiency (measured as CO_2 emissions per mile) and smog-forming emissions. Buyers of dirty, inefficient cars would pay two fees; buyers of clean, efficient cars would get two rebates. By influencing car choice directly, feebates could overcome the ~6:1[85] dilution of gasoline prices by the other costs of owning and running a car. Governor Deukmejian vetoed Drive+ on 30 September 1990, but it seems likely to become law in 1992, perhaps volume-normalized. Drive+ may then launch a national trend. Similar proposals are pending in Iowa and Massachusetts, are being drafted in other states, and have been proposed nationally.

This rapidly spreading idea has gotten several of the makers of superefficient prototype cars seriously interested in entering the US market immediately, in order to maximize their share of the early adopters' market. Interestingly, General Motors did not oppose Drive+ in 1990, reportedly because the company prefers market-oriented feebates to standards or other direct regulations.[86] Nor did the White House's generally anti-efficiency National Energy Strategy (February 1991) explicitly exclude feebates.

---

[83]Arbitrageurs get rich on spreads of a fraction of a percent, but the difference in discount rates between utilities and customers is more like a thousand percent.

[84]By 61–11 votes in the Assembly and 31–2 in the Senate.

[85]In the United States at pre- or post-Gulf-crisis prices; typically the ratio is nearer ~3–4 : 1 in other OECD countries with higher motor-fuel taxes, and 8:1 in some US cases.

[86]This is understandable: the Corporate Average Fuel Economy (CAFE) standards passed by Congress in 1975 were " at least twice as important as market trends in fuel prices, and may have completely replaced fuel price trends as a base for long-range planning about [new-car efficiency]" (141), but improved standards would be well complemented by feebates, which reward manufacturers for bringing the best technologies to market soonest and reward buyers for choosing the best from the mix offered.

A useful early refinement to basic feebates would be "accelerated scrappage": basing rebates for efficient new cars on the difference in efficiency between the new car and the old one that is scrapped (if it is worse than a certain level). Drivers who scrapped a functioning car and did not replace it would get a bounty on presenting a death certificate that it had been duly recycled. By offering a far higher price for scrappage than dealers offer for tradeins, the state would put a premium on getting the least efficient, most polluting cars off the road soonest. This incentive would greatly accelerate the energy and pollution savings. It would also help poor people, to whom the worst cars tend to trickle down,[87] to afford to buy a highly efficient new car that they could then afford to run.

Such feebates have wide potential application. They could spread from cars, light trucks, and buildings to appliances, aircraft, heavy road vehicles, etc.[88] In each case, they would transfer wealth from those whose inefficient choices impose large external costs on society (global warming, acid rain, oil-import dependence, etc) to those who save such costs. Their self-financing makes them politically attractive, as the California Legislature's vote confirms. They entail three straightforward steps: set a target level of efficiency; charge an open-ended variable fee for all new devices that are worse; and rebate the fees (less administrative costs), also on an open-ended variable basis, to all new devices that are better (74).

Feebates could be usefully supplemented by three further policy innovations[89]:

1. In many European Economic Community (EEC) countries, and some others like New Zealand, much urban commuting—often half or more—is in company-owned cars, provided as a perquisite because this form of compensation is taxed less than equivalent salary (or not at all). This tax dodge is often meant to inflate domestic car sales for the benefit of domestic automakers. It contributes disproportionately, however, to congestion and hence to fuel waste and pollution. Since removing even a small number of cars from a crowded highway can markedly relieve its congestion, eliminating tax breaks for company cars should be a high priority.

2. Greater symmetry between modes of transport requires that cars pay more of the cost of providing their infrastructure. Singapore, for example,

---

[87]And whose inefficient cars, despite their fewer miles driven per car, were disproportionately responsible for imports of the same Gulf oil that those same poor people were sent in disproportionate numbers to defend.

[88]Draft recommendations by the California Legislature's Joint Committee on Energy Regulation and the Environment (12 September 1990) include feebates for buildings and, on a pilot basis, appliances.

[89]A possible fourth, a novel "pay-as-you-drive" way to charge for car insurance, is starting to receive scrutiny too (142).

limits traffic by a no-exceptions daily tax on downtown driving. Oslo and Bergen are doing the same. In an even more interesting system being introduced in Stockholm, downtown residents who wish to drive their cars during a given month must buy a permit which also serves as their free pass to the regional public transit system for that month. Then they have it, so they might as well use it.

3. "Golden carrot" rebates are designed to elicit the production of specific energy-saving products that are cost effective but are not yet brought to market because cautious or undercapitalized manufacturers are unwilling to risk retooling costs for uncertain sales. This concept could be adapted to saving oil from its current uses in saving electricity. For example, utilities from San Diego to Vancouver may soon join together to pay, say, a $300 rebate for each of the first 10,000 or 100,000 refrigerators sold in their territories that beat the 1993 federal efficiency standard by at least 50%.[90] Such incremental efficiency would require one or more of the advanced insulating materials mentioned above to be put into mass production. If the refrigerators are not sold, no rebate is paid, so the utility is at no risk of not getting the desired savings; and once that many are sold, the manufacturing hurdle has been leapt, the special rebates can be discontinued, and continuing sales will then yield far larger savings. A larger "platinum carrot" can then be offered for the next incremental advance, and so on until cost-effective opportunities are exhausted. This proposal was originally developed for North American electric utilities and appliances, and has in fact been successfully applied by the Swedish National Energy Commission (Statens Energiverket) to refrigerator design. It is now being considered by other European and South Pacific electric utilities. There is no obvious reason why it could not be offered by governments and applied to other products such as light vehicles, in addition to or in lieu of feebates.

So far, less progress has been made in fundamentally changing mission and culture in free-market oil companies than in regulated, franchise-monopoly utility companies. But there are strong reasons for private oil companies, too, to promote customers' energy efficiency (81):

1. Under long-run competitive equilibrium conditions (ignoring fluctuations as war or peace breaks out in the Middle East), the major oil companies expect fairly flat long-term real prices, implying little prospect for large upstream or downstream rents. The major rent still largely uncaptured is the spread between the cost of extracting and of saving barrels.

---

[90]This scheme is promoted by A. H. Rosenfeld (Lawrence Berkeley Laboratory) and R. Cavanagh (Natural Resources Defense Council, San Francisco), and has already been accepted by the main utilities involved.

2. With long-run real oil prices fluctuating between ~$12/bbl and ~$25/bbl, it is easier to make one's margins selling "negabarrels" that cost $5 to produce than barrels costing $15.
3. Selling a variable mixture of fuel and efficiency can be used to hedge risks in supply-side markets (81).
4. Efficiency can fundamentally reduce long-term price volatility (81), benefiting a capital-intensive industry.
5. Efficiency promotes global development, which is good for the oil business, and deflects Persian Gulf wars, acid rain, and global warming, which are bad for it.

For these and similar reasons, several major and independent oil companies are starting to express interest in significant commitments of effort and resources to marketing efficiency. Some of these firms are, in essence, very large, technically oriented banks, and are starting to realize that they can get better at selling financial products, such as leasing efficient end-use technologies.

Generic issues   The main cultural obstacles to this transition in oil companies, as in many utilities, are changes in mindsets: selling services rather than commodities, working on both sides of the customer's meter or the vendor's pump, and getting used to doing fewer big things and more smaller things. This difference of unit scale is perhaps the most uncomfortable, because traditional energy-supply systems are ~5–8 orders of magnitude larger than end-uses. But such large scale is not technically or logically necessary, and often it is not economic either. The well-known economies of scale in engineering and manufacturing certain energy systems can easily be swamped by even a small subset of the dozens of known diseconomies of scale.[91] There is now abundant empirical evidence that minimizing whole-system cost generally entails matching scale, at least roughly, between supply and end-use. This does not mean that everything should be small: it would be nearly as silly to run a huge smelter with thousands of little wind machines as to heat millions of houses with a fast breeder reactor. But it does mean that making supply systems the right size for the job usually makes them cheaper.

Oil and gas companies are starting to compete in electricity markets by promoting smaller-scale technologies such as packaged gas-fired cogeneration plants, steam-injected gas turbines, and combined-cycle retrofits of

---

[91]These arise from, e.g., higher distribution costs and losses, reduced unit availability, increased reserve requirements, longer lead times (hence greater risks of cost escalation, technical or political obsolescence, or mistimed demand forecasts), higher ratio of onsite fabrication to factory mass-production, more difficult maintenance, greater awkwardness of using waste heat, etc—~50 identified mechanisms in all (Ref. 97, App. 1).

classical combustion turbines. By some estimates, private additions to US generating capacity exceeded utility additions starting in 1990 (143). The 27 states that recently ran supply-side auctions were offered, on average, eight times as much private generation as they wanted (144).[92] Fuel vendors are becoming significant players in this competition. Gas companies diversifying downstream now routinely note that building and running a gas/combined-cycle power plant undercuts the running cost alone of a typical nuclear plant. As oil and gas companies increasingly bundle both electric and gas efficiency with downstream applications of their fuels, they become involved with customer scale and start to think more like customers.

Another obstacle being slowly overcome is the pervasive asymmetry in public policy, long dominated by special interests subject to scant performance accountability. Public policy in almost all countries has for decades been overwhelmingly biased toward supply over efficiency, depletables over renewables, electricity over heat and liquids, and centralization over appropriate scale. There are reasons for this, but they are certainly not economic reasons. They are the same reasons that efficiency got 2.8% and renewables 3.1% of US DOE's FY1988 proposed civilian RD&D budget, vs 12.4% for fusion, 16.6% for fossil fuels, and 11.1% directly for fission; or that US fission, after decades' devoted effort, continues to receive strong policy support despite having missed its cost target by an order of magnitude, while renewables, which quickly met or bettered their cost targets and now deliver twice as much energy (81), continue to be dismissed as futuristic or impractical. Too often, the balance of official effort between competing options is like the old recipe for elephant-and-rabbit stew: one elephant, one rabbit. But in the United States, the EEC, and many developing countries, the pendulum is starting to swing towards economic rationality, if only because there is no longer the capital to misallocate.

These frustrating, though gradually resolving, problems must not obscure the major gains already made. During 1979–1986, for example, the United States got more than seven times as much new energy from savings[93] than from all net expansions of supply, and more new supply from sun, wind, water, and wood than from oil, gas, coal, and uranium (81). By 1986, US CO_2 emissions were one-third lower than they would have been at 1973

[92]Many, but not all, of these proposals could be relied upon to yield actual, reliable capacity if accepted. Opinions differ on the "real" fraction. Many utilities have also found private cogeneration to be more reliable than their own central plants.

[93]As indicated in aggregate by improvements in primary energy consumption per unit real GNP—a crude and sometimes misleading measure, but useful shorthand in most cases. About 65–75% of that improvement is generally considered to be due to technical gains in energy efficiency, nearly all the rest to changes in composition of output, and only a few percent to behavioral change.

efficiency levels; the average new car alone expelled almost a ton less carbon per year; annual energy bills were ~$150 billion lower; and annual oil-and-gas savings were three-fifths as large as OPEC's capacity (Ref. 74, p. 4).

During 1977–1985, the United States increased its oil productivity four-fifths faster than it had to in order to match both economic growth and declining domestic oil output. By 1986, the annual savings, chiefly in oil and gas, were providing two-fifths more energy than the entire domestic oil industry, which had taken a century to build (81). Oil, however, has dwindling reserves, rising costs, and falling output, whereas efficiency has expanding reserves, falling costs, and rising output.

By 1989, the United States was getting 91% as much annual primary energy from post-1973 savings as from all oil, domestic and imported. Some other OECD countries have done even better. During 1973–1988, while energy intensity declined 26% in the United States, it fell 30% in Japan. In the 1980s, the countries with the highest energy efficiencies, such as Japan and West Germany, have proven among the toughest economic competitors—and are now redoubling their efficiency efforts as they start to discover the new technological opportunities.

The task now for OECD is twofold: to accelerate these historic efficiency gains by harnessing today's far more powerful and cost-effective technologies and delivery methods, e.g. by promoting superefficient cars through feebates with accelerated scrappage; and to extend to electricity the rapid, consistent efficiency gains obtained during 1973–1986 for direct fuels. In principle, it should be possible to save electricity as least as quickly as oil, because

1. far more electricity than oil is used in highly concentrated applications[94] and in standardized commodity-like devices installed in relatively few places;[95]
2. there is far more economic and environmental incentive to save electricity than oil;
3. electric applications do not have psychological complications analogous to those of cars, especially in the United States;
4. service quality is more likely to be markedly improved by saving electricity than by saving oil; and
5. for electricity, unlike oil, a skilled engineering and financial institution, with a relationship with every customer, stands ready to deliver efficiency programs; and
6. existing regulation can readily reward such delivery.

---

[94]About a million industrial motors >125 hp use ~12% of all US electricity (Ref. 22, pp. 28–29 & 39); of the ~53–60% of all US electricity used by motors, half is used by about a million, and three fourths by about three million, large motors.

[95]For example, the ~1.5 billion 2×4′ fluorescent lamp fixtures installed mainly in large US office buildings.

The best utility programs confirm this thesis: Southern California Edison's 1983–1985 program mentioned above (plus slightly more important concurrent state actions) reduced the decade-ahead forecast of peak load by the equivalent of 8½% of the then-current peak load per year, at average program costs of a few tenths of a cent per kWh saved (7, 19). Taken together, many US and some foreign utilities' experience, especially during 1989–1991, now extensively confirms that such rapid, reliable, cost-effective electric savings are possible. As usual, the limiting factor in rapidly propagating such success is the number of skilled practitioners. Major initiatives to expand recruitment, training, cross-pollination, and career development opportunities are therefore emerging, and merit reinforcement. Internal technology transfer, and new tools such as expert systems, also require far greater emphasis if utilities and design professionals are to learn as quickly as computer companies—a formidable cultural challenge. And few utilities are yet good at maximizing all three elements of efficiency success: participation, savings per participant, and competition in who saves and how.

One last category of policy initiative requires emphasis: incentives within the public sector. Washington State, for example, splits the money saved by energy-efficiency improvements to government buildings into three unequal parts: returns to the General Fund, rewards to everyone responsible for achieving the saving (in the form of supplements to institutional budgets or personal cash bonuses), and further efficiency purchases, which bootstrap successively longer-payback investments without requiring recourse to the capital budget. California already returns half its dollar savings to the institution achieving the efficiency, and proposes to earmark the other half for a revolving loan fund to achieve more savings throughout state government. Such mechanisms, plus careful tracking of energy costs so that responsibility for reducing them can be assigned, are the beginnings of sound energy management. Without such exemplary leadership by governments at all levels, citizens will take calls for efficiency less seriously.

FARMING AND FORESTRY   More is known about the mechanisms of transition to sustainable energy systems than to sustainable farming and forestry, but the latter seem analogous in principle and seem to offer similar scope for such market mechanisms as feebates. Sweden, for example, has long taxed agrichemicals and rebated the proceeds to help farmers make the transition to organic techniques. Iowa has a similar agrichemical tax to finance groundwater protection. Fees on synthetic nitrogen fertilizers could be rebated to users of manures, green manures, or legumes. Fees on logging could be rebated to tree-planting or (better) to forest protection, especially of old-growth stands.

Many agrichemicals are already costly enough, and engender enough health anxiety among farmers, that little more incentive is needed not to use them. What is lacking is transitional advice, reassurance, risk-sharing, and financing: most farmers do not have practical knowledge of sound alternatives, and have too little financial safety-margin (or lender flexibility) to undertake the perceived risk of trying anything new. But in the United States, Canada, Germany, Japan, and other OECD countries, lively networks of farmers are emerging to help match successful conversion case-studies with farmers facing similar challenges. Expanding and endorsing such clearing-house activities should be a high government priority.

Alert agrichemical companies are already starting to plan their own transitions. One of us (ABL) has recently been told by senior planners at three major agrichemical manufacturers that they all plan to get out of the business; the only question is how soon, how gracefully, and what to do instead. Like utilities that have revised their product from electricity to end-use services, these agrichemical firms have revised their mission from selling chemical commodities to helping farmers grow nourishing food. Like utilities, too, such firms often have financial resources, marketing skills, and technical capabilities that will be important for the agricultural transition. They are becoming interested in providing such assets as sophisticated soil-test kits (conveniently testing the health and diversity of soil biota is still in its infancy), mineral amendments, targeted predators and other integrated-pest-management tools, adaptable native seed, technical advice, farm-management software, and transitional financing and risk insurance. Together, such elements could make an attractive package for a harried farmer who wants to change but is deterred by "hassle factor," novelty, and perceived risk.

It is also important to make markets in carbon sequestered or not emitted ("negatons" of carbon). Applied Energy Services, Inc., an Arlington, Virginia, firm, is planting trees in Guatemala to offset the output of its new coal-fired generation plant, and is funding the planting by a voluntary <5% surcharge on the plant's output. As utilities or private market-makers start to trade such carbon offsets to and from utilities and industrial fuel users, farmers and foresters should be able to bid to provide carbon-absorbing services. For example, it should be straightforward to make a market in forest- or prairie-preservation rights, which would certify that at least a certain carbon inventory or density will be maintained for a given period. Such a market would, for example, add value to farmers' decisions to enhance humus through organic practices, or to foresters' decisions to lengthen cycle times or preserve old-growth forest. Analogies already exist: in Southern California, cogeneration deals were made in 1989 in which one of the partners' in-kind equity contributions was reduced smog formation, assessed at the day's market value as quoted by the local pollution-reduction broker (a new profession created by EPA's "bubble concept" for air-pollution offsets).

In fact, a market in carbon offsets could in effect transfer some of the large amounts of money saved by electric efficiency into financing the transition to sustainable farming and forestry. Electric utilities, especially those burning coal, would have to reduce their carbon emissions, or purchase "offset rights" representing extra carbon sequestered in trees or soil biota. (Utilities in the parts of the United States, Germany, and Britain that burn the most coal tend to be the least interested in energy efficiency, so they would need to buy the most offsets.) Until the utilities got tired of this income transfer, it would provide a timely injection of capital to help launch a broad farming and forestry transition—and thereafter, a spur to the utilities to get serious about their own demand-side opportunities. Such a capital flow could easily exceed $10 billion per year in the United States alone. That is not large compared with the $175-billion-a-year electric bill, but would be a godsend to cash-short farmers and small-scale foresters unable to finance fundamental changes through traditional lenders.

It is hard to estimate the attainable speed of reforms to make farming and forestry more sustainable and carbon-conserving. It is not even known how quickly organic farming is spreading in the United States. Anecdotal evidence suggests it is far faster than anyone had expected (105a), and informal reports from many regions in 1989–1990 have indicated that demand for organic produce is often tens of times larger than supply. Now that organic farming has finally received an official economic endorsement (110), formal definitions and standards are emerging, and most areas and types of operations can find a successful example of conversion relatively close by, many previously skeptical farmers are starting to consider a transition as a more serious near-term option. Extension agents from many parts of the United States report overwhelming demand for transitional counsel. In time, the supply of the needed information and risk capital will catch up. Bundling the global-warming benefits with the other, more familiar benefits of sustainable practices will tend to attract more capital, reduce perceived risk, and hence speed the transition.

Some encouragement may be drawn from the speed with which other farming changes have lately been adopted: low- or no-till herbicide-based cultivation, land set-aside programs, and hedging in commodity futures markets. US farmers responded eagerly enough to financial set-aside incentives to cause spot shortages of some crops. Similar incentives for conversion—analogous to electric utilities' loans, gifts, rebates, and leases for efficiency—could bear similar fruit.[96] If government agricultural departments assigned a

---

[96]Just the major energy benefits in reducing heat-island effects could motivate utilities to fund urban forestry, including agroforestry, on a large scale, as Sacramento's municipal utility is already starting to do (see note 10 above).

tenth as high a priority to helping free farms of their chemical dependence as law-enforcement departments do for citizens, and mounted a major campaign to renew the old arts of soil conservation and tilth improvement, there is every reason to think that the time is ripe and many farmers could make surprisingly rapid changes.

CFC SUBSTITUTION   Mandatory production phaseouts and rapidly increasing taxes are already a fact of life for CFCs. Less mature, however, are mechanisms to recover, store, and destroy the large existing inventories of CFCs. A few utilities that already pay customers to scrap old, inefficient refrigerators and freezers are starting to integrate CFC recovery, usually by hiring specialist appliance-recycling firms (Ref. 14, pp. 95–96). The City of Palo Alto, California, is also considering collecting all CFC-containing products found in local landfills and recovering the CFC for reuse or disposal (Ref. 126, p. 197). CFC recovery from air conditioners in both in-service and scrapped cars is also a rapidly growing business.

The recoverable CFC inventories currently in circulation, or sitting in (and leaking from) scrapped cars and appliances, are important for both global warming and ozone depletion. A typical 18-ft^3 US refrigerator/freezer contains ~0.9 kg of CFC-11 in the foam insulation and ~0.23 kg of CFC-12 refrigerant.[97] These CFCs are ~14,000–20,000 times as heat-trapping per molecule as CO_2 (Ref. 6, p. I.1–10), so they add significantly to the direct effect of CO_2 from the power plant. Their potency, and the high policy priority therefore accorded to their replacement, suggests that "offsets" for CFCs would also be worth marketing—permitting their continued use (even though their production may meanwhile have been phased out) so long as an equally potent quantity of CFCs is removed from the environment (Ref. 126, pp. 197–198). Properly done, such "tradeable use rights," analogous to the EPA's "bubble concept" for conventional air pollution, could cap the effective prices of CFC substitutes, reduce energy demand, and smooth the transition to CFC substitutes (126).

Formerly Centrally Planned Economies (FCPEs)

The Soviet Union emits 15% of the world's fossil carbon from ~12% of world economic output (calculated by Soviet methodology: Ref. 89; actual economic output by OECD definitions appears far lower). The USSR and its former satellites, which are similarly or more carbon-intensive, are engaged in an historic transition of extraordinary dimensions. These countries have great opportunities to help abate global warming. How much they will

---

[97]Technology to recover the former may be available in Germany, but information seems scarce and R&D appears to be a high priority.

do so, however, can be neither assessed nor achieved without wrenching structural, political, and economic changes that are only just beginning (126a,b). These changes include:

1. Major reallocation of national resources, especially in the USSR, from military to civilian production within the context of lessening tensions, European partial demilitarization, increased popular control over governmental military adventures (this is needed in the United States as well as in the USSR), and a widespread commitment to a global security regime that makes others more secure, not less (145).
2. Radical economic reform based largely on market principles, truthful prices, and integration into the world economy, including convertible currency and fair opportunities for foreign ownership and joint ventures under reliable legal arrangements.
3. Equally radical social and political reform reinvigorating productivity, initiative, and personal responsibility, and hence requiring comprehensive educational renewal.
4. Economic restructuring markedly reducing the relative role of primary materials production, enhancing the nascent service sector, favoring smaller enterprises, decentralizing much overcentralized production, introducing competition to monopolies (including electric utilities), and creating a working distribution system that now scarcely exists at all.

This is a tall order; yet there is no way out but through. As the economist P. G. Bunich (now an advisor to President Yeltsin of the Russian Republic) remarked in early 1989,

> Here we are, 280 million Soviets gathered together on a vast beach and wading together into the surf. We're at that awkward point where it's too deep to wade but not yet deep enough to swim, so we're losing our footing—and anyway, none of us know how to swim. But, by God, some of us will figure out how to swim, and those who figure out first will teach the rest. Of course, we'll have to make some lifejackets, but only a few million not 280 million.

That historic transition comes at just the time when the extremely energy-intensive Soviet/Eastern European economies are broken and need fixing. Poland, for example, has about the same per-capita energy use and carbon emissions as Austria, but only one-fourth the per-capita GNP (146). Hungary has about the same per-capita energy use as Japan, but one-fifth to one-fourth the per-capita GDP (92). The Soviet Union is one-third to two-thirds more energy-intensive than the United States, and 2–4 times more than the most efficient countries such as Sweden (with a similar climate) and Japan.

However, significant progress is being made. Hungary got 38% of its

1970–1988 increase in energy services from efficiency (31%) and structural change (7%), and can both save much more and considerably expand biomass growth and use (92). Although Poland has so far sustained substantial efficiency gains only in transportation, further cost-effective structural and efficiency improvements could even hold long-term per-capita energy use constant despite 2–3%/y income growth (146). The Soviet Union's heavy use of natural gas contributed to the 22% drop (about the same as in Western Europe) in its CO_2/primary-energy ratio during 1970–1984, far above the 1% drop in the United States (89).

These high intensities have four especially important aspects:

1. Energy prices are heavily subsidized (51). A senior Soviet colleague who set up a cooperative selling energy-efficiency services in Moscow said, "You might suppose this activity is three times less cost-effective than in the West, because in the Soviet Union we sell energy for about a third of its production cost. But we're also about three times less efficient, so it works out the same." Energy subsidies in Poland in 1987 (before the "big bang" price reforms) were ~49% of the delivered price of coal, 83% of gas, and 27% of electricity (146). For Eastern Europe as a whole, "Until recently, electricity, coal and natural gas were priced at one fourth, one half and four fifths of world market levels" respectively (51). Yet "it is not enough simply to get the prices right: prices must also matter. Making prices matter means not permitting enterprises to pass the cost of energy waste on to consumers" without risking bankruptcy for their uncompetitiveness (51). Absent such fundamental reform, based on understanding that prices are information, not a social entitlement, the Eastern economies are simply, as one Soviet scholar put it, "vast machines for eating resources."

2. Decades of central planning by empire-building bureaucracies rewarded for setting ever-larger quantitative targets have left the Eastern economies exceptionally overbuilt in primary materials industries. An astonishing 70% of Hungary's energy consumption goes to raw materials production, which provides only 15% of national output (92). Similarly, by producing less unnecessary material and using it more efficiently, market reforms could save one sixth of Soviet energy and half of Czech steel production—even more in chemicals and nonferrous metals (51). This is especially good news because more than two-thirds of Soviet fossil-carbon output is from the industrial and energy sectors (89).

3. Distortions of production, blockage and compartmentalization of information, administered prices, and mandatory allocation of goods and services have led to serious weaknesses in some sectors, notably electronics, that are vital to energy efficiency. That is why "only 25–30% of the cost of [Soviet] energy efficiency . . . [is the cost of] implementing measures at the point of use. The remaining 70–75% results from the expense and difficulty of

expanding domestic production of energy-efficient equipment and materials" (89). Price distortions make inferior domestic equipment look only half as costly as foreign versions, but those cannot be imported without a convertible ruble, soft credits from abroad, or less restricted and less risky joint ventures.

4. As both Western and Soviet analyses independently showed in 1983,[98] energy efficiency is by far the most critical technical measure for the success of perestroika, because it produces a double benefit: it frees scarce resources (capital,[99] hard currency, technical skills, etc) to modernize industry and agriculture, and it frees the saved oil and gas for export to earn hard currency to buy technologies for the same purpose. For these reasons, they estimated, each 1% improvement in aggregate energy productivity (especially in electricity) may increase national output by several percent, and bring even more important qualitative improvements. Because saved energy is fungible for hydrocarbons that can be sold for hard currency, Soviet energy savings are properly denominated not in kopecks per gigajoule but in dollars per barrel.

These circumstances require a combination of three approaches: economic reforms, structural shifts in the composition of output, and strong improvement in end-use efficiency. The alternative, on standard projections, is to double Soviet fossil-carbon output by the 2020s (89), and Polish a decade earlier (146). Such projections may be too high because the growth is self-limiting: Makarov & Bashmakov (89) state flatly that "The Soviet economy can no longer sustain continued growth in energy consumption and the corresponding demand for increasing energy production. If current trends continue, capital and other resources will be required in amounts so large as to preclude the possibility of realizing any but the Pessimistic Variation of the Base Case Scenario." But the prospect conjured up by this kind of involuntary grinding to a halt is not pleasant either.

Besides economic reforms, structural changes, and end-use efficiency, some Eastern European countries would benefit greatly from raising fuel quality by substituting mainly natural gas. This is most true for Poland: the world's fourth largest extractor of coal, 75% coal-fired throughout its economy, and the user of one third of Eastern Europe's 20 EJ/y of primary energy (51). Most Eastern European countries have major energy supply problems.

[98]By Royal Dutch/Shell's Group Planning in London and Corresponding Member V. A. Gelovani et al at the USSR Academy of Sciences' Institute for Systems Studies in Moscow.
[99]Nearly one fifth of Soviet investment goes to fossil-fuel extraction (not counting the electric sector) (51). The same fraction went to Polish coal-mining alone in 1980, almost double the 1970 share, while gas and oil extraction's share of investment went from <1% of total industrial investment in 1980 to 39% in 1986. These trends squeezed out non-energy industrial investment, which fell from 74% of total industrial investment in 1970 to 61% in 1986 (146). That is partly why Sitnícki (146) advocates halting all energy-supply investment until the demand side is set in order. In Hungary, likewise, "investment in energy supply grew to 40% of all industrial investment by 1986" (92).

Hungary imports half its energy; Soviet fuel reserves are shifting inexorably from high-grade western to remote, costly, and often low-grade Siberian and Far Eastern sites,[100] etc. But Poland epitomizes the most acute supply problems. Severe air and water pollution are contributing to economic shrinkage (51), coal-mining consumes one fifth of all steel (up >150% since 1978) and nearly one tenth of grid electricity, the average depth of mines is increasing 2–4%/y, more difficult mining conditions are cutting labor productivity in the worst mines to one sixth of the British or West German norm, social and administrative costs are high and rising, and land is so scarce that some mines "transport waste rock and coal washing refuse as far as 80 km for disposal" (146). Coal exports for hard currency must cease in the 1990s in order to fill domestic needs as coal quality and accessibility decline. In any event, the economic benefits of the exports have been illusory because they greatly speeded the shift from high- to low-quality coal (high sulfur, high ash, high cost, more global warming).

Such conditions offer unusual leverage in abating global warming, because the costs and the environmental impacts of burning such poor fuels are inflated in four ways: more fuel must be burned to yield a given amount of primary energy, and most of that energy is then wasted by inefficient conversion, delivery, and end-use. Conversely, end-use efficiency improvements bring benefits that are amplified manyfold by avoiding system losses. Any such improvements, in a country like Poland, will push the most intractable coal-mining problems further into the future, buying more time to substitute gas which (efficiently used) might bridge to renewables.

In principle, the Soviet Union has three advantages in promoting energy efficiency. One is its gifted scientists and technologists. Soviet achievements in such areas as control theory, materials coproduction, materials science (diamond films, ceramics, supermagnets, composites, etc), and mathematical physics can not only help meet domestic needs, but also compete in world markets. It is now becoming common, with good reason, for the computer literature to speculate that later in this decade, the combination of Taiwanese hardware and Soviet software may make a strong market showing. Thus the USSR has far more to sell than its unparalleled storehouse of raw materials. So far, however, Soviet technical prowess has not been mobilized to advance energy efficiency in any fundamental way, and indeed no sufficiently detailed and modern (by Western standards) efficiency research or analysis yet exists in the country, though there is plenty of talent to apply to it.

Second, Soviet energy-using hardware is highly standardized. Lighting

---

[100]In recent years, Soviet coal-mining's shift to the east has resulted in mining more coal but getting less energy out of it, because grade is falling faster than tonnage rises. Some eastern coals are of such poor quality that no one has figured out a reliable way to burn them (147).

retrofits in the United States or Western Europe are complicated by a fragmented industry distributing thousands of types of fixtures from hundreds of manufacturers. In the Soviet Union, only 10–20 types of fixtures are in general use. For that matter, somewhere in the files of GOSPLAN (if one knows which numbers are real and which are fake) one can ostensibly find a complete record of the entire Soviet capital stock, because its production and shipment were planned. No Western country has a comparable "paper trail": American analysts, for example, know less about their stock of electric motors, which use more primary energy than highway vehicles, than they know about moonrocks.

Third, what is left of the Soviet command economy—which has been largely retained in the strictly monopolistic utility/electrotechnical sector—may still be useful in shifting production towards end-use efficiency. The design of every electricity-using device manufactured by the State, for example, must be approved by a single engineer in St. Petersburg everything from toasters to giant motors, lamps to computers. He likes efficiency, but his authority is circumscribed. For example, a small lighting institute on the Volga has been trying for years to get permission to pilot-produce some improved compact fluorescent lamps (none are made in the Soviet Union). But the Ministry of Energy and Electrification has never approved the request: the Ministry's job is to build and run power plants, and there is no demand-side institution with any force as a counterweight. There are recent signs that the Ministry is starting to think more about efficiency—having only 3½–5% more generating capacity than expected 1990 peak load concentrates the mind wonderfully—but such change of mission will probably be slow. Happily, the view of the leading Soviet climatologist and many of his Academy colleagues—that global warming would probably, on the whole, be good for the Soviet Union—is apparently giving way to a realization that the risks of mispredicting are too great (M. Budyko, Yu Izrael, G. Golitsyn, personal communications). Yet there is still a very limited understanding of energy efficiency within the old Ministries.[101]

The Soviet/Eastern European energy problem is part of a nearly infinite onion containing layer upon layer of challenging social, political, and economic problems. Peeling that onion will be slow and difficult. Soviet experts have suggested, however, some of the specific initiatives that are most needed. Paraphrasing Makarov & Bashmakov (89), these include:

---

[101]For example, Budyko and Izrael have misinterpreted analyses by Makarov et al as indicating that major gains in Soviet energy efficiency would cost the entire capital investment available. What the studies said was that such gains could accompany a complete, from-the-ground-up reconstruction of the whole Soviet economy and infrastructure—which of course would cost that much, but would yield far more joint benefits than just reducing global warming.

1. International lenders' soft credits for importing energy-saving equipment.
2. Joint ventures to make such equipment in the Soviet Union.[102]
3. Joint exploration of efficiency opportunities and implementation methods.
4. Public education stressing "that energy savings is the principal, and most economically effective, means of solving many global problems of world energy development."
5. Collaboration to make nuclear plants safer (a major concern for Western utilities whose operations are hostage to the next Soviet accident) and cheaper. (In our view, and that of many Soviet colleagues, such cooperation may be worthwhile but is very unlikely to lead to a Soviet nuclear revival: as in most other countries, the economic and political obstacles are too daunting and the alternatives too attractive.)
6. Systematic reduction of the natural gas system's methane leaks, which increase the Soviet contribution to global warming by anywhere from ~8% (89) to ~26% (88).
7. Expanded cooperation in renewable energy development—an area of much ingenious Soviet design—and in trapping and isolating CO_2.

Our own experience suggests the need to add seven major items to that list.

1. A high initial priority should be given to superwindows—the coating technology exists in the Soviet military sector but has not yet been transferred to the civilian sector—and to superinsulated modular house/apartment construction. It would be silly to try to relieve the housing shortages, let alone meet emergency housing needs in e.g. Armenian-earthquake reconstruction, using the same poorly insulated, shoddily built, seismically hazardous technologies whose use is still so widespread.

2. Soviet analysts are prone to suppose that building up consumer goods and the service sector will require rapid growth in the electric share of end-use energy, accounting for half the projected rise in CO_2 output (Ref. 89, p. 5). Instead, there is good reason to believe that the electricity required could come from larger-than-expected savings both in the new equipment itself and elsewhere, especially in industrial drivepower. But those efficiency opportunities are revealed only by very detailed and up-to-date analysis, so they are not yet well understood by most Soviet experts.

3. As is now starting, it is important that skilled Western and Eastern Asian firms feel able to participate for mutual benefit in major Soviet oil and gas projects. Potentially very large reserves of both will require OECD technology to find and extract. Such new reserves plus high end-use and conversion efficiency could probably "completely cover" Soviet domestic

---

[102]Poland now provides a three-year tax holiday for such ventures, and has introduced measures to encourage reinvestment domestically. Those measures include low-interest credits and 50% taxable-income deductibility for investments in modernization—100% if they are for environmental protection (146).

energy needs from high-grade resources—hydrocarbons could cover ~80% of energy supply—for many decades to come (51). Such new reserves would only be squandered in the USSR, however, and unaffordable to Eastern Europe, without major reforms in price, output structure, conversion efficiency, and end-use efficiency (51).

4. To help achieve those reforms, neighboring countries in both Asia and Western Europe could and should provide a massive infusion of capital and technology specifically focused on Soviet and Eastern European end-use efficiency. That is among the cheapest ways in the world to abate global warming, because equipment is so inefficient to start with and the fuel displaced on the margin is mainly low-grade coal (or high-grade, low-carbon gas that can in turn displace more coal). The improved efficiency, preferably using equipment made by joint ventures in the USSR, would free up oil and gas for resale to the efficiency providers. Those fuels, especially the gas, would displace German coal, reducing global warming even more, and earn hard currency to pay for the efficiency technologies and financing. That payback would be very rapid, leaving most of the oil-and-gas revenue stream available to buy other technologies for modernization outside the energy sector. Similarly, Western aeroturbine manufacturers could joint-venture with their Soviet counterparts in producing steam-injected and combined-cycle gas turbines—a good use of the military aircraft-engine production capacity now being partly demobilized. Those turbines, with their modularity and very short lead times, could help quickly to relieve Soviet and Eastern European power shortages while also yielding major long-term gas and coal savings.

5. A formal mechanism to provide up-front, pump-priming credits for Western investments in Eastern energy efficiency would be through the sale of carbon offsets from East to West. Since, for example, West Germany, a carbon-intensive country, is considering introducing a carbon tax, the introduction of German offsets without territorial limits could readily lead to a flow of hard currency eastwards to pay for carbon savings that can initially be achieved more cheaply in the East than in the West. That hard currency would then be recycled westwards to pay for German or other world-market efficiency technology installed in the East to fulfill the carbon-abatement contract previously sold. Sale of the subsequently saved gas to the West, and its use to offset still more carbon, could be part of the same deal.[103]

---

[103]This would be timely, since much German coal will become uncompetitive after 1992 anyway, when its "Kohlepfennig" subsidy becomes legally unsupportable. This could offer an opportunity to make lemons into lemonade. Ideally, such offset arrangements should be traded in markets that include derivative instruments with a range of maturities, so that rather than locking themselves into contracts lasting far longer than the timescales of installation and technological evolution, the parties would retain some flexibility to keep hunting for better buys as factor costs and gas and carbon values all shift. As noted below, Krause et al (6) suggest such periodic carbon auctions.

6. Japanese industry's remarkable training programs for energy-efficiency managers should be exported to the Soviet Union. Those managers' attention to detail is unrivalled elsewhere in OECD, and has much to teach all other countries. In principle, Japanese industry might obtain carbon offsets through fuel savings achieved in Soviet industry, yielding another mutually beneficial currency flow to be coordinated with increasing oil-and-gas collaboration in the Soviet Far East.

7. A requirement for Soviet restructuring is continued cooperation by the North Atlantic Treaty Organization (NATO) countries, especially the United States, in rapid disarmament and demilitarization of both sides' economies (89). This will require redefining not only military force structures and missions but also strategic doctrine and the whole concept of what security is and how to obtain it at least cost (145). Recent US-Soviet cooperation in the Persian Gulf crisis is an encouraging step in this direction. Rapid, effective, comprehensive Western help with Eastern energy efficiency is a critical component of Western security interests: the alternative is not only an unstable meteorological climate but also political instability, stagnation, or worse. NATO has for many years sponsored leading international seminars on technical aspects of energy efficiency, and may be able and willing, given high-level encouragement, to turn its energies in this direction.

In short, the United States and the Soviet Union, which between them release about half the world's fossil carbon, are not only natural technical and financial partners in helping to abate global warming; their intricate security embrace demands such cooperation, mobilizing the best talent in both countries.

This partnership should not stop with energy. Both countries have much to learn from each other's experience of sustainable farming and forestry—rather than continuing misguided efforts to sell US-style chemical farming to a country whose internal obstacles have so far happily hampered, at least outside Kazakhstan, its efforts at chemical self-contamination of its vast lands. Early citizen-organized farmer-to-farmer exchanges have proven such eye-openers to both sides that they clearly merit expansion: it will be difficult to elicit enough entrepreneurship to break up the state farms unless more Soviet farmers realize what is possible. Further, a large proportion of Soviet vegetables, fruit, meat, and dairy products come from the tiny fraction of farmland privately farmed with some semblance of real economic incentives. Yet the tools used on those plots are often medieval. Important productivity gains could probably be had by as simple a means as transferring to Soviet artisans and cooperatives the technologies of advanced gardening tools (from such firms as Smith & Hawken) that could do so much to ease the work and expand the output of those private plots. Superwindows could in time further provide more fresh winter produce by making indoor microfarming possible in major cities like Moscow.

Carbon offsets are especially important also to preserving the fragile carbon inventories of the Soviet Union's boreal forests, now being plundered by overbuilt pulp-mills at home and in Scandinavia. Similar opportunities exist elsewhere in the parts of Eastern Europe not yet severely damaged by acid rain: Hungary, for example, "possesses rich biomass production potential and could serve as a testbed for new biomass utilization and sequestration technology," involving e.g. "increasing the humus in soil from 2 to 3 percent and . . . forest area from 17 to 24%" (Ref. 92, p. 9). Among the comparative advantages of the Eastern cultures in this context is their often intact rural culture and knowledge of how the land works—assets all but destroyed in the overurbanized West (111). There may be more ecological wisdom to be learned from Byelorussian or Bulgarian countrypeople than from Western agronomists.

Developing Countries

For all their historic handicaps, most of the peoples of the Soviet Union and Eastern Europe represent industrial cultures with long and distinguished histories of sophisticated technical achievements. Comparable native talents, however, have had a very different historic pattern of expression in developing countries. While their achievements in many arts, sciences, and ways of living have been extraordinary, often predating analogous Western progress by many centuries (Arab and Chinese astronomy, Chinese medicine, Sri Lankan irrigation, and Polynesian navigation come immediately to mind), they have not in general had the opportunity to exploit cheap resources, usually taken from poorer countries or bought at competitively depressed prices, with Western manipulative technologies to produce "modern" infrastructure, sell manufactures at monopoly rents, and create, at least for themselves, widespread material wealth. Therein lies their great opportunity today not to seek to tread that same development path, but to proceed more wisely.

If developing countries try to repeat the mistakes of OECD, they will never develop. The cheap resources are dwindling; the force-fed monopolistic markets are going fast; the direction of the global economy no longer supports colonists' former ability to buy cheap and sell dear. Rather, one now increasingly buys raw materials at monopoly rents and sells manufactured goods at competitively depressed prices.

The collision between old cultures, new technologies, and perennial aspirations is perilous, both for the little peoples for whom a major mistake may be the end of the line, and for the great peoples whose ultimate weight in the world is so ponderous: a billion Chinese times anything is a big number, and before long, on present trends, India will have more people than China. This is not to say there are no hopeful signs: in the 1980s, the Chinese economy did apparently grow twice as fast as coal consumption (which represents three-fourths of energy use)—a commendable ~4.7%/y annual decline in aggregate energy intensity (51). Yet China is still at least three times as energy-intensive as Japan. A plausible scenario for 1990–2025 (51) is widely assumed to end up with at least 1.4 billion Chinese, with quadrupled per-capita income (rising to one third of the current US level), and with total energy and coal consumption tripling to nearly the present US level. If anything like that actually happens (let alone whatever happens after that), then OECD will need to pursue efficiency very vigorously to help stabilize the earth's climate.

Preventing Chinese energy use from tripling under such a 5%/y-GNP-growth scenario requires far more detailed and comprehensive efficiency efforts. Indeed, without such efforts, the growth may not occur, because under business-as-usual projections, investment in electrification alone will consume approximately the entire economic growth of the developing world.

Consider, for example, the sad story of Chinese refrigerators. When the government decided people should have them, more than a hundred factories were built, and the fraction of Beijing households owning one rose from 2% to 62% in six years. But through inattention, they were built to an inefficient design. China now needs a billion unavailable dollars' worth of new power plants to run those inefficient refrigerators. An effort to promote development instead created crippling shortages of both power and capital. The officials to whom this was pointed out said the error would not, if they could help it, be repeated: it had taught them that China can afford to develop only by making energy, water, and other kinds of resource efficiency not just an add-on program but the very cornerstone of the development process. Otherwise, as is already true even in such a fundamentally wealthy country as the USSR (see above), the waste of resources will require so much and so costly supply-side infrastructure that too little money will be left to build the things that were to use those resources. As a wise homebuilder once put it, "If you can't afford to do it right the first time, how come you can afford to do it twice?"

Recall, too, that the average poor country[104] derives nearly three times less economic output per unit of primary energy used (commercial and traditional) than the average rich country—which in turn can cost-effectively at least quadruple its energy productivity. These two facts together imply that if poor countries leapfrogged over the mistakes of the rich countries, they could in principle expand their economies roughly tenfold without increasing their

[104]Most poor countries are in fact " 'dual societies,' consisting of small islands of affluence in vast oceans of poverty. The elite minorities and the poor masses differ so much in their incomes, needs, aspirations and ways of life that, for all practical purposes, they live in two separate worlds" (17). For that reason, this discussion focuses mainly on the needs of poor people within poor countries.

energy use at all—while the rich countries could in principle sustain or improve their standard of living while using several times less energy than now. (Principle and practice probably differ here. Because of practical constraints in getting organized, the tenfold pace of savings seems less plausible than the fourfold.) Ultimately, both groups would meet in the middle, and there would be enough energy for all.

Lack of energy or fuel, however, is not the problem. In Reddy & Goldemberg's masterly summary (17):

> If current trends persist, in about 20 years the developing countries will consume as much energy as the industrialized countries do now. Yet their standard of living will lag even farther behind than it does today. This failure of development is not the result of a simple lack of energy, as is widely supposed.
>
> Rather, the problem is that the energy is neither efficiently nor equitably consumed. If today's most energy-efficient technologies were adopted in developing countries, then only about one kilowatt per capita used continuously—roughly 10 percent more than is consumed now—would be sufficient to raise the average standard of living to the level enjoyed by Western Europeans in the 1970s.

This discussion therefore seeks to supplement the large literature on energy and development—much of it compactly summarized elsewhere (38, 69a)—with a few observations on how it may be possible to "do it right the first time" and improve the prospects for a "leapfrog strategy" that largely jumps over both inefficiency and fossil fuels.

MARKETING FOR DIVERSITY   First, action requires understanding of choices, and "Consumers who do not obtain their energy efficiency fall into three categories: the ignorant, the poor and the indifferent" (17). Reddy & Goldemberg continue:

> The first [group] consists of people who do not know, for example, that cooking with LPG is more efficient than cooking with kerosene. They can be educated to become more energy efficient.
>
> The second category consists of those who do not have the capital to buy more efficient appliances. . . . An Indian maid may know that her employers spend less money cooking with LPG than she does with kerosene, yet she may not be able to switch because LPG stoves are about 20 times more expensive than kerosene stoves.

This is the position, albeit in a far starker context, of the Western householder who is deterred by the high cost of a compact fluorescent lamp. The remedy is analogous:

> . . . utilities or other agencies should help finance the purchase of efficient equipment with a loan that can be recovered through monthly energy payments. Alternatively, a utility can lease energy-efficient equipment. A consumer's savings in energy expenditures can exceed

the expenses of loan repayments and new energy bills. In principle, this method of converting initial costs into operating expenses can be extended to commercial and industrial customers as well, thereby improving efficiency and modernizing equipment at the same time.

This is what an RMI-cosponsored pilot project seeks to do with compact fluorescent lamps in Bombay (148). The lamps would be leased with an always-favorable cashflow to the user. The lamps would save the utility at least six times their total cost. There are two bonuses. First, the power saved in the heavily subsidized household sector, where it is sold at a loss, can then be resold to businesses which pay full rates, so the utility converts a loss to a profit. Second, in a city where 37+% of the evening peak load is from lighting, the saving should help to prevent the evening crashing of the grid, as well as improve reliability in some adjacent Western-grid states now short of capacity. Even such simple measures can have a profound economic effect: by one expert estimate, giving away such lamps throughout a very poor country like Haiti might raise the average household's disposable income by as much as one fifth, because so much of the sparse cash economy goes directly or indirectly for electricity, mainly for lighting.

> The third category of consumers consists of those with little incentive to raise their energy efficiency because their energy costs are so small or because the costs are almost unaffected by efficiency changes. . . . Enticing those customers . . . will depend on intervention at higher levels [e.g., via government] . . . efficiency standards.

Bombay has such people too—especially affluent householders who can afford tubular fluorescent lamps three times as efficient as the poor user's incandescents. Marketing to the affluent group may require emphasizing the compact fluorescent lamp's longer life and aesthetic superiority (better color, no flicker, no hum, nicer shape).

INSTITUTIONAL PARALLELS   These three categories and remedies are not unique to developing countries: they have exact parallels in OECD, and for that matter in the USSR and Eastern Europe. Parallel, too, is the problem that most "utilities, financial institutions and governments" lack "methods for converting the initial cost of efficient systems into an operating expense." Further parallel is the issue of scale:

> Spending $2 billion on end-use improvements is much more complicated [than spending $2 billion on a single power plant]. If each efficiency measure costs between $2,000 and $20 million, then between 100 and a million subprojects are involved. Organizing so many diverse activities is difficult. . . .

This is especially true in a nonmarket economy. And equally parallel is the almost universal bad habit of using energy consumption, or (worse) its rate of growth, as a measure of progress rather than as an indicator of inefficiency in meeting social goals with elegant frugality. Until energy planners start by asking what the energy is for, and how much of it, of what kind, at what scale, from what source, will do that task in the cheapest way, the outcomes will be far from rational or even affordable. Until energy planners appreciate that it is 10 times as important to eliminate the "payback gap" as to get the prices right, their exhortation simply to desubsidize energy prices will continue to prove inadequate.

In developing and ex-socialist countries, while seeking to harness or mimic market forces, it is especially important to remember that markets do not in general produce justice, equity, or sustainability. They were not meant to. Equity requires political and ethical instruments, and an appreciation that, as Reddy's field experience has taught him, "If you look after the needs of the poorest, everything else will look after itself."

LENDING INSTITUTIONS   One of the chief obstacles to sound energy-for-development policies is the World Bank's and other multinational lenders' persistent lack of interest and insight (149). This is partly because while reorganizing a few years ago, the Bank laid off most of its technical staff and retained chiefly economists who do not understand the engineering and tend toward not-invented-here reflexes.[105] But it is also because the system of rewards within such institutions, as in commercial banks, creates incentives to make bigger loans, not smaller ones, and to maximize loan volume in a way that a small, centralized staff can do only in large chunks. The Bank and its peer organizations also lack "templates" for successful Third World energy-efficiency projects. These structural adjustments within the lending organizations will be made only when their major supporters insist that lending follow the same cost-minimizing, all-things-considered rules that their own utilities are already institutionalizing—i.e. that energy financing be allocated only to the least-cost options (149). The Bank's ~$2 billion a year worth of power-plant investments would practically all fail such a test.

To encourage a major shift of emphasis, the Bank and its regional counterparts may wish to experiment with becoming carbon-offset brokers. For example, US, German, and Japanese coal-fired utilities and industries might choose to fund exceptionally cheap Third World carbon-abatement projects—agroforestry, lighter-colored buildings and pavements, Curitiba-style bus systems, lighting and motor retrofits, etc—through the Bank via carbon-offset contracts. (They could even train the relevant Bank and host-country staff in the implementation techniques those modern utilities have developed, and

---

[105]In spring 1991, however, this began to change, partly through the intervention of the Bank's own Facilities engineers, who became interested in making the Bank's new headquarters building cost-effectively efficient.

perhaps learn too about the often smarter ideas that resource-short hosts tend to develop out of necessity.) This is at first blush an adventurous financial concept, but in structure it is no odder than a debt-for-nature swap. Another alternative, suggested by Reddy & Goldemberg (17), is that carbon taxes, collected chiefly from industrialized countries under the OECD "polluter pays" principle, be earmarked for energy- and other carbon-saving projects in developing countries. Certainly OECD's historic responsibility for much of the cause of global warming justifies no less.

It has been argued (Ref. 6, pp. I.5–7 & –8) that even if tradeable emission rights have an initial allocation based on an equitable per-capita formula rather than one that "grandfathers" OECD's historically high emissions, such melancholy experiences as the Third World debt crisis suggest "that approaches based mainly on market exchange will not work among nations as structurally different as developing and industrialized countries." In today's political climate, that reservation may be valid (although, as with the Berlin Wall, the unlikely can happen). But as a limited component of a far broader implementation strategy, tradeable rights do make good sense (Ref. 6, pp. I.5–17). After all, in societal terms they are allocating not costs but benefits, and the only question is how to allocate the benefits to give all parties the necessary incentives to act. Since abating global warming is better than free, the object is not to figure out how to share sacrifices for the common good, but rather how to help individuals, firms, and nations behave in their economic self-interest. To this end, Krause et al (Ref. 6, p. I.5–18ff) have further proposed an ingenious mechanism with some potential advantages—a Super-fund-like international Climate Protection Fund combined with carbon-reduction auctions.

International lending, both bi- and multilateral, badly needs to be restructured "from support for specific projects (such as building dams) to support for goal-oriented programs (such as lighting more homes)" (17). Another worthy goal would be to support, as in Eastern Europe and the USSR, joint ventures to produce efficient equipment locally. Many utility officials in developing countries, for example, say that their countries have refrigerator factories—but are unaware that those factories could joint-venture with, or license from, a Danish firm to make refrigerators that look and cost about the same but use ~80–90% less electricity (14). It appears that such industrial marriage-brokerage is not on lenders' agenda. It should be. It is indeed extraordinary that some developing countries, like Brazil and Mexico, manufacture for export to rich countries certain appliances more efficient than are available for sale to their own people.

In some instances, it may be possible to package an advantageous three-way swap: e.g. using a Western European compact-fluorescent lamp technology, plus rare-earth phosphors from Soviet minerals, manufactured in an

Indian free-trade zone (using rupees as a bridge between guilders and rubles), for sale in both socialist and developing countries if not to the West as well. Currently, however, such high-leverage opportunities apparently are not on lenders' radar screens at all.

Critical to these shifts of emphasis are education, training, and institution-building. Interesting programs are starting to emerge in countries like Brazil, Ghana, India, Thailand, and Tunisia. Their university and government efforts could be nurtured into regional centers of excellence in energy-efficiency technology, field implementation, and policy. A significant private initiative to build such centers' capabilities for regional outreach, research, and training is under serious discussion. Trained people from within each culture are essential: only they can understand behavioral issues and novel kinds of mistakes that would baffle a Western expert.[106] A "Negawatt University" network that induced a few smart graduate students to devote their careers to energy efficiency could make a big difference. Today, a few such nascent centers exist (e.g. Bangalore, Bangkok, Berkeley, Genève, Grenoble, Lund, Lyngby, Princeton, Santiago, São Paulo, Sydney), but they need support.

Overarching all these issues, even education, is the power of example. US citizens, for example, cannot preach that others should protect their forests as we clearcut our own in Alaska, Hawaii, and the other 48 states. We cannot preach the virtues of population control, or energy efficiency, or sustainable rural economies, as we erode their foundations at home. But the power of a positive example can be even stronger than the power of a negative example. Acting from our highest traditions has moved the world before, and can again.

The almost-conventional agenda mentioned so far is essential. But several more steps, seldom discussed, seem warranted too. The central one is a major effort to inventory, then shift, the efficiency of energy-using equipment in international commerce.

TECHNOLOGY TRANSFER—POSITIVE AND NEGATIVE   In Osage, Iowa (population 3800), the municipal utility has worked with local vendors to make good equipment easy to get and bad equipment hard to get. The hardware stores have a good selection of compact fluorescents, but may keep incandescent lamps out back where you have to ask for them. The lumber yards do not stock 2×4" studs for poorly insulated frame walls, nor ordinary

---

[106]The opposite can also be true. We recently heard of a major tourist hotel, near the Equator, that installed a 250-ton chiller to attract affluent visitors. It never worked: the building seemed to get hotter in the summer and colder in the winter. No expert even from the capital could figure out what was wrong with it. The manager started asking every engineer who came to stay to take a look. Finally, one Dutch engineer instantly spotted the problem: the chiller's chilled-water output was being pumped straight into the cooling tower. Nobody else had thought to check something so obvious.

double glazing—"Sorry, sir, that's obsolete—special order, it'll take a month"—but instead sell wider lumber and superwindows, and explain their superiority to help make the sale.[107]

Nobody does that in global commerce. From Bangkok to Cairo, apparently cheap Taiwanese ballasts are being installed in fluorescent lighting fixtures. Those ballasts are often made from unrefined recycled-scrap-copper wire with such high resistivity that its overheating causes most of Taipei's house fires. Seemingly cheap Taiwanese, Czech, and other motors, gearboxes, and the like are also common, and often so inefficient that if run at more than a modest fraction of their rated power, they burn up. Although apparently no one has yet done a formal survey of this component market, nor of the equipment that incorporates such components, many anecdotes leave a strong impression that the world is awash in very inefficient end-use devices marketed to people with limited technical understanding, no independent information sources, and almost infinite discount rates.

Making and selling those devices, however, is not such an innocent pastime. The precious capital, especially foreign exchange, consumed in trying to build and fuel power plants and other energy-supply systems to run such inefficient equipment cannot be used to buy vaccines, to provide clean drinking water and family planning, to plant trees, to teach women to read. The overloaded power supply is unreliable and cannot run a productive industry, imposing huge backup-generator costs. Too little electricity is available for such basic needs as lighting (further hampering literacy) and pumping water. Tracing back such interlinked opportunity costs suggests that those inefficient ballasts and motors may ultimately create about as much human misery as the drug traffic. This "negative technology transfer" needs to be put squarely on the international agenda.

We do not know what it would take to get bad equipment off the world market, or at least to stigmatize it as the menace to development that it represents; but some countries have ideas. In Tunisia, for example, the national energy office has achieved an exemplary but little-known decoupling of economic activity from energy use, chiefly by developing nearly two dozen national minimum performance standards for basic appliances, lights, motors, and vehicles (M. Ben Abdallah, personal communication). You can buy a variety of cars, but they have all been bid, bought, and imported in bulk by

---

[107]This is a small part of a series of initiatives whereby Osage's residential energy savings have enabled the utility to repay all its debt, build up a large interest-bearing surplus, cut its rates five times in five years (to half the average Iowa level), thereby attract two big factories to town, and keep recirculating in the local economy more than $1000 per household per year. This plugging of unnecessary dollar leaks has made the local economy noticeably more prosperous than in neighboring towns. The same principle, elaborated in RMI's Economic Renewal Project, applies to villages, states, and countries.

the government on the basis of their fuel efficiency. Inefficient models are deliberately excluded. The saved capital is then available for what people really need—a prerequisite to the sensible investment approach that has paid off so richly in Costa Rica.[108]

Another possible analogy is the recent United Nations Environment Programme (UNEP) convention on international trade in hazardous wastes, requiring the "informed consent" of the recipient. Recipients of electricity-wasting equipment could surely be better informed—say, by labelling with life-cycle costs, and a list of alternatives, perhaps put out by the United Nations itself.

An international convention banning trade in energy-guzzlers, as in endangered species, may be too much to hope for, but perhaps Customs authorities, in countries from Bermuda to India where Customs plays a key role in trade regulation,[109] could be encouraged to attach tariffs varying inversely with energy efficiency (or even positive-or-negative duties structured like feebates). Devoting to import regulation of energy efficiency a small fraction of the engineering skill that now goes into designing power plants could pay big dividends. One can even imagine multicountry consortia that bulk-buy energy-using commodities (lights, motors, windows, etc), and maintain a common secretariat that tracks and helps to get the best buys. One way or another, it is important to make it cheaper and more politically attractive for both buyers and sellers to deal in efficient than in inefficient equipment. The industrialized countries, too, have a major but unacknowledged responsibility in this matter: they typically set a bad example in what they buy, and they often export their most inefficient and obsolete equipment to boot—like deregistered pesticides all over again. This occurs even with inefficient motors, etc, removed under present US utility rebate programs: if not scrapped, such items may end up being reinstalled down the street or in Mexico.

Another issue meriting exploration—though its importance is often exaggerated—is the ways in which intellectual property rights may collide with transfer of energy-saving technology to countries that need them but cannot afford the royalties. Where such problems occur, carbon-tax revenues or

---

[108]Research at Shell and by the late human-rights barrister Paul Sieghart has confirmed the basic premise of basic-human-needs strategies by showing three nearly perfect correlates and predictors of conventionally measured development success: absence of subsidies to basic commodities, adherence to basic human rights, and the health and education of the people 10 years earlier. (The Costa Rican electric utility, incidentally, runs a lottery for everyone who uses less electricity this month than last month; but their opportunities to use less are currently limited to behavioral change, since they cannot yet buy very efficient equipment and receive no financial help in doing so.)

[109]But often a perverse one, e.g. in India, where they charge far higher duties on energy-saving than on energy-supplying imports.

carbon-offset sales revenues might be used to pay those royalties—a sort of pro bono fund. It may also be possible to organize a form of nonmonetary recognition, perhaps by UNEP, for inventors and firms who waive royalties in such circumstances. There may be helpful analogies, too, with the ways in which intellectual property rights were handled when Green Revolution crop strains were developed for use by developing countries.

The need for further research, demonstration, development, and outreach—for there is already much good news to report—hardly needs further emphasis. But it is important to stress a major hole in most current research agendas: the integrative design of future energy systems that integrate efficiency with renewables. This includes the redesign of industrial processes to match solar heat more conveniently, as those processes were in the past successively redesigned to fit the characteristics of wood, coal, oil, gas, and electricity. It also embraces efforts to anticipate bottlenecks, problems, and technical gaps as renewables and efficiency come to play major roles in energy-service supply in each sector.

Reddy & Goldemberg (17), asking whether the needed transformation of energy policies can take place in the next few decades, conclude that aside from actions by the international bodies charged with, but largely ignoring, this responsibility, "the best hope for change lies in a convergence of interests." Industrialized countries need to protect the environment and ensure sustainable development; the environmental movement is joining hands worldwide; advocates for the poor, and increasingly the disenfranchised themselves, are crying out for policies that work better and cost less. The realization is spreading around the world that high levels of energy efficiency, and farming and forestry practices that treat nature as model and mentor, are not hostile but vital to a stable climate, a healthful environment, sustainable development, social justice, and a liveable world.

This convergence will be seen and used more clearly for what it is when everyone, especially we in the industrialized world, better understand what so often obscures our view of poor countries' energy and development needs. It is not our regions' differences, important though they are, but their similarities.

ADDITIONAL BENEFITS

Energy efficiency and sustainable farming and forestry practices, and the other ways described above to abate global warming and make money simultaneously, all have additional benefits which presumably could be expressed as monetary values. Counting those values would make their sources look even cheaper. The following list of omitted joint benefits is illustrative, not exhaustive.

Environmental Protection

Energy efficiency and renewable energy sources largely eliminate the environmental impacts of mining, transporting, and burning the fuel they displace. Reduced CO_2 is only one such benefit. Others include avoided SO_x, NO_x, O_3, hydrocarbons, particulates, and other air pollutants; despoiled land; acid mine drainage; oil spills; and impaired aesthetic, wilderness, and wildlife values.

A simple example suffices. Rather than raising electric bills to clean up dirty coal-fired power stations, a utility can help its customers to get superefficient lights, motors, appliances, etc, so they need less electricity to do the same tasks. The utility will then burn somewhat less coal and emit less sulfur (ideally backing out the dirtiest plants first), but mainly the utility will save a lot of money, because efficiency costs less than coal. That saved operating cost can then be used partly to clean up the remaining plants, partly to make electricity cheaper, and partly to reward the utility's shareholders. Similar principles permit the negative-cost abatement of urban smog by such measures as superefficient cars (150). In either case, not only the health of people and other living things (and the longevity of cultural monuments and natural artifacts) will benefit; forests and other ecosystems, being less degraded, will also be better able to sequester carbon.

Sustainable agriculture and forestry, too, do not just reduce biotic CO_2, CH_4, and N_2O emissions; they also help to preserve topsoil, genes (biodiversity), water, fuels, farms, and farmers. They control floods, reducing the siltation of dams and of navigable waterways. (Dams that last longer displace more fossil-fuel CO_2 from thermal power plants.) Sustainable practices increase the habitat and population of land, aquatic, and marine wildlife. They preserve or create diverse and beautiful landscapes. They protect rural culture—an important cultural "anchor"—and help to reverse rural depopulation, a key contributor to urban problems. They reduce or eliminate dependence on biocides, whose manufacture accounts for a substantial fraction of all toxic-waste generation, and reduce associated problems ranging from occupational exposure and drift in application to runoff and groundwater contamination after application. They similarly reduce fertilizer contamination of ground- and surface waters, hence eutrophication, nitrate toxicity in drinking water, etc. They also restore the wholesomeness of food now contaminated by biocides, hormones, and antibiotics, probably thereby benefiting public health.

Moreover, such sustainable practices reduce crop losses to pests by rebuilding predator stocks and diversity; may modestly raise net farm or forest income, and make that income stream less vulnerable to weather, pests, crop prices, and other uncontrollable variables; and should reduce dependence on government subsidies. They keep farming, in short, as profitable as now or (in most cases) more so (110), and far more consistently so. They thereby reduce strains on the rural banking system, reliance on commodity brokers, and risks arising from commodity speculation. They set a good example for agricultural evolution abroad. They foster a land ethic and the practice of stewardship. And they make farming more diverse, interesting, and appealing to the young.

In sum, informal estimates at EPA's Pollution Prevention Office suggest that most—perhaps around 90%—of the problems EPA deals with could be displaced, at negative cost, just by energy efficiency and by sustainable farming and forestry. That is a pleasant by-product of abating global warming at a profit.

Security

Security—freedom from fear or privation or attack—may be achieved by other and cheaper means than armed might and threats of violence (145). Security comprises access to reliable and affordable necessities of life (water, food, shelter, health, a healthful environment, a sustainable economy, a legitimate system of government, certain cultural and spiritual assets). It also requires some combination of conflict prevention, conflict resolution, and defense—preferably nonprovocative—that is predictably able to defeat aggression (145). Resource efficiency in all its forms is an essential element of both providing needed resources and forestalling conflict over resources.

Energy/security links take many forms, and the list grows with awareness that climatic stability and biotic productivity are essential elements of global security. High on anyone's security list are equitably providing the energy needed for sustainable development; reducing domestic energy vulnerability (97); abating the spread of nuclear weapons (61); and avoiding conflict over fuel-rich areas like the Persian Gulf.

The United States and other nations recently put their youths in 0.56-mile-per-gallon tanks and 17-feet-per-gallon aircraft carriers in the Middle East because the same youths were not put in 32-mile-per-gallon cars. A three-mile-per-gallon improvement in the ~20-mi/gal US household vehicle fleet would displace the mid-1990 rate of oil imports from Iraq and Kuwait; a 12 mi/gal improvement would displace all US oil imports from the Persian Gulf. If the military cost of the Gulf war were somehow internalized in the oil price, rather than socialized through taxes and deficits, it would be interesting to see how well the oil would sell at more than $100 per barrel (151, 152).

This paper has described cost-effective efficiency improvements in US oil use that could potentially displace the United States' oil imports from the Persian Gulf roughly seven times over. Similar opportunities, differing only in detail, are available to other oil-importing countries, even the most efficient. It is bad enough to pay in dollars the cost of continuing to ignore such opportunities. It is tragic to pay it in blood. Given that the Middle East is uniquely rich in oil, and full of diverse peoples who have fought each other for millennia, there is wisdom in at least making the oil under their disputed territories as irrelevant to the world's continued peace and prosperity as modern technology permits.

Equity

An original and pragmatic treatment of global warming (6) has proposed that the long-term fossil-fuel "budget" of ~300 billion tons of carbon (GtC) believed to be consistent with a probably tolerable rate of climatic change be split evenly between developing and industrialized countries. This would "push industrialized countries to fully mobilize their technological, financial, and organizational capacities for phasing out fossil fuels without creating infeasible goals" (Ref. 6, p. I-6.9). It would also leave developing countries more leeway to cope with their presumably lower adaptive capacity, greater needs, and greater population momentum. All regions together, under such a budget, might find it a reasonable milestone to return global carbon emissions to ~1985 release rates by ~2005—by which time the 20% reduction target set at Toronto should have been achieved by the industrialized countries. This seems a reasonable trajectory, and is certainly fairer than most. But one must also consider the "micro-equity" features of the specific tools proposed. In that respect, the strategy proposed both by Krause et al (6) and here seems attractive.

One of the best features of the efficiency-and-renewables energy strategy is that many of the technologies are "vernacular," able to be locally made with fairly common skills, and the renewable energy itself is equitably available to all. Sunlight is indeed most abundant where most of the world's poorest people live (99). In no part of the world between the polar circles is freely delivered renewable energy inadequate to support a good life indefinitely and economically using present technologies (99, 38, 69a).

Much the same is true of sustainable farming and forestry practices. By relying chiefly on natural processes and assets, these practices minimize dependence on inputs that must be bought or brought from afar. In both cases, future generations' rights seem to be far better protected than now.

Resilience

Life is full of surprises. Energy analysts who owe their careers to singular events in 1973 and 1979 cheerfully go on to assume a surprise-free future. It will not be like that at all. There will even be surprises of kinds nobody has thought of yet (just as some people unfamiliar with the history of climatic science suppose global warming is). However, the fundamental principles of resilient design, borrowed chiefly from biology and engineering (97), are completely consistent with the energy and agri/silvicultural strategies proposed here. More diverse, dispersed, renewable, and above all efficient energy systems can make major interruptions virtually impossible in principle (97). Growing green things in a way duly respectful of several billion years' design experience is the best way anyone knows to ensure that the earth will keep on handing down that experience.

Buying Time

In an earlier analysis of least-cost climatic stabilization (69a), two colleagues and we noted that if the terrible exponential arithmetic of burning more and more fuel, faster and faster, were simply reversed—if the amount of carbon released each year steadily shrank—then the "tail" of "global warming commitment" would soon become so slender that its length would be unimportant. A very long period would then be safely available for displacing the last remnants of fossil-fuel use. That simple idea remains valid.[110] In a more subtle sense, however, the time-buying value of techniques like energy efficiency is greater than meets the eye. Efficiency buys not just money and avoided pollution but also time—the most precious and least substitutable resource.

An anecdote is useful here. Around 1984, Royal Dutch/Shell Group planners foresaw the 1986 oil-price crash, and warned that a deepwater North Sea oilfield called Kittiwake would have to be brought in at 40% below the planned cost, because by the time the field opened, it would be possible to sell the oil only for $12 a barrel, not $20. The engineers, who had been sweating over one-percent cost cuts, were aghast. But offered the alternative of being fired and leaving the oil where it was, they cut the cost by 40% in about a year.

It turned out they had previously been asked the wrong question: how to bring on fields as fast as possible with cost no object, rather than how to do it cheaply even if it took longer. Asked the new question, they came up with

[110]Perhaps by coincidence, the rough and illustrative estimates presented in 1981, at a time when the role of trace gases other than CO_2 was poorly understood, are surprisingly close to those now emerging from today's far more sophisticated analyses. Indeed, Krause et al's rough timeline for a 300-GtC global carbon budget (6)—a 20% reduction in global carbon emissions between about 2005 and 2015, 50% around 2025–2035, and 75% before the middle of the next century—is "roughly equivalent to the efficiency-plus-renewables scenario of Lovins et al [Ref. 69a]" (152a). The 450-GtC budget associated with more sanguine climatological assumptions is close to the Goldemberg et al (38) scenario, "modified to include significant renewables penetration" (152a). Krause concludes (152a): "The former scenario can be seen as representing the limit of practical logistic feasibility, while the latter scenario can be seen as marking the limit of climatic acceptability. . . . The range of climate-stabilizing energy scenarios is circumscribed by" those two scenarios.

completely different technological answers. (In how many other situations have we gotten the wrong answer by asking that same wrong question?) But the key result was that the new technology made oil that used to cost $30 to extract into oil that costs only ~$18 to extract: the whole oil supply curve therefore flattened out. This in turn postponed depletion, and this in turn bought time in which to develop and deploy still better technologies, on both the demand side and the supply side, which broaden the range of choice and which reinforce each other by buying still more time, together pushing depletion far into the future and facilitating a graceful transition to renewables.

Time-buying is a sound principle worldwide. The startup obstacles to achieving major efficiency gains in the non-OECD regions strengthen the case for strong and rapid efficiency gains in the countries (OECD) best equipped to achieve them. If that might be "overachievement" relative to some theoretical goal of equitable sharing, nobody should mind: on the evidence presented above, maximizing the size and speed of energy savings is likely only to bring larger and earlier economic benefits.

CONCLUSIONS

This paper has rebutted ten prevalent myths (in italics) about abating global warming:

1. *Greater scientific certainty should precede action.* The uncertainties about global warming and its potential consequences are substantial, interesting, and likely to cut both ways. But they are also irrelevant to policy, because virtually all the actions needed to abate global warming (if it does turn out to be a real problem) should be taken anyway to save money. These "no-regrets" actions are about enough to solve the problem if it does exist, and are highly advantageous even if it does not. The problem with global warming is not decision-making under uncertainty; it is realizing that in this instance, uncertainty is unimportant.

2. *The issue is whether to buy a "climatic insurance policy" analogous to fire insurance or to defense expenditures (a major investment mobilizing most of the country's scientific and technological resources, and meant to forestall or respond to unlikely but potentially catastrophic threats to national security).* The "insurance" analogy is partly valid, because delaying action until obvious climatic changes are unambiguously under way makes abatement too little, too late, and too costly—just like trying to install a sprinkler system in a hotel that is currently on fire, or build military forces while you are already under attack, or buy collision insurance after you have crashed your car. Abating global warming will require significant efforts affecting large numbers of people and stocks of capital over long periods and with long lead

times, so waiting too long will certainly raise cost, difficulty, and risk of failure (153). But the analogy breaks down if, as was shown above, the real choice is not balancing uncertain future benefits against daunting present costs, but rather making the investment as wisely and quickly as possible in order to achieve both the uncertain future benefits and the guaranteed financial savings. Any insurance "premium" is actually negative: the actions that can stabilize global climate will save money anyway, without counting the avoided costs of trying, or failing, to adapt to possible climatic change. This "insurance" is unquestionably a good buy.

3. *Abating global warming would be costly.* Distinguished econometricians have claimed that just achieving the Toronto interim target of cutting CO_2 emissions by 20%—roughly a third of the reduction probably required for climatic stabilization (1)—would cost the United States alone on the order of $200 billion per year (153a, 154–156). Such calculations are wrong by at least an order of magnitude (e.g. 157, 158). Worse, their high-cost conclusion is a bald assumption masquerading as a fact (W. Nordhaus, personal communication). The econometric analysis merely asks how high energy prices would need to be, based on historic price elasticities of demand (typically from decades ago), to reduce fossil-fuel use by x%, then counts those higher prices (or their equilibrium econometric effects) as the cost of abatement. This approach ignores the compelling empirical evidence that saving most of the fuel now used is cheaper than even its short-run marginal cost, and hence is profitable rather than costly. The econometricians thus have the amount about right but the sign wrong; using modern energy-efficient techniques to achieve the Toronto target would not cost but save the United States on the order of $200 billion a year.[111] These techniques did not exist at the time of the behavior described by the historic price elasticities: those elasticities summarize how people used to behave under conditions that no longer hold. Indeed, cost-minimizing energy policy—if not derailed by the blunder of treating future energy needs as fate instead of choice—will seek to change those conditions as much as possible.

4. *Abating global warming would drastically curtail US and similar lifestyles, and would mean less comfort, mobility, etc.* Nothing could be further

---

[111]Rosenfeld et al (159) note that commercial direct fuels cost Americans $283 billion and electricity $175 billion in 1989. Saving (for illustration) a fifth of each at average prices would save $92 billion a year, but the costliest sources would in fact be displaced first, and long-run marginal costs generally exceed present prices. Together, these effects probably at least double the value of the savings. A similar result could be obtained by a conservative method using longer-term savings potentials: saving about two-thirds of the direct fuels (an understated and rough composite of the potential for all sectors) and, in the short run, only one-fourth (utilities' average fuel-cost share) of 75% of the electricity, would total ~$221 billion. More sophisticated calculations are of course possible, disaggregated by sector, fuel, region, timing, etc, but not very useful.

from the truth. The fuel-saving technologies that can stabilize global climate while saving money actually provide unchanged services: showers as hot and tingly as now, beer as cold, rooms as brightly lit, torque as strong and reliable, homes as cozy in the winter and cool in the summer, cars as peppy, safe, and comfortable, etc. The quality of these and other services can often be not just sustained but substantially improved by substituting superior engineering for brute force, brains for therms: e.g. efficient lighting equipment provides the same amount of light, but it looks better and one can see better. This same is broadly true of sustainable agriculture and silviculture, which provide comparable yields with superior quality, resilience, human health, and (generally) profitability.

5. *If such cost-effective abatements were available, they would already have been bought.* This is reminiscent of the econometrician who, asked by his mannerly granddaughter whether she could pick up a $20 bill she had just noticed lying on the sidewalk, replied, "No, my dear, don't bother: if it were real, someone would have picked it up already" (M. Gell-Mann, personal communication). The striking disequilibrium between how much energy efficiency is now available and worth buying and how much has already been bought arises from distinctive, well-understood market failures that leave cheap efficiency seriously underbought at present prices. (For example, consumers have poor access to information and to mature mechanisms for conveniently delivering integrated packages of modern technologies. Discount rates are about tenfold higher for buying efficiency than supply, severely diluting price signals. Many energy utilities misunderstand their business and want to increase their sales—even though reducing their sales would increase their profits by decreasing their costs even more. Perverse regulatory signals often reward inefficient and penalize efficient behavior. Markets in saved energy are sparse or absent. And present market signals, omitting externalities that may be as big as the apparent fuel prices, make consumers indifferent to whether they buy, for instance, a 20- or a 60-mile-per-gallon car, since both cost about the same per mile to own and drive.) Solutions exist for each of these market failures. These solutions have been proven in market economies and are rapidly emerging in a wide range of other societies, so there is an ample range of effective policy instruments to choose from. The technical and implementation options—the everyday work of energy-efficiency practitioners—are mostly unknown, however, to those econometricians who lie awake nights worrying about whether what works in practice can possibly work in theory.

6. *Abatements would be so costly and disagreeable that they could only be achieved by draconian, authoritarian government mandates incompatible with democracy.* On the contrary, the abatements described above are so profitable and attractive that they can be largely if not wholly achieved by

existing institutions, within the present framework of free choice and free enterprise. Planners unaware of market-driven alternatives seem anxious to set up new bureaucracies to tell people how to live. Many bizarre schemes have been suggested for substituting dirigisme for markets, penury for development, risks for rewards, and costs for profits. This paper seeks to provide an antidote to this perversion of economic rationality.

7. *Combating global warming requires tough trade-offs—swapping one kind of pollution or risk for another.* Abating global warming by resource efficiency can simultaneously reduce or eliminate many other hazards too—oil-security risks, nuclear proliferation, utilities' planning and financial risks, declining farm and forest yields, etc—without creating new ones.

8. *Available means of abatement, singly or combined, will be too small and too slow, so global warming is inevitable and we must start trying to adapt to it.* This counsel of despair is misguided. To be sure, some significant degree of climatic change or increased climatic volatility in some places may already be unavoidable if the more sensitive models prove valid (1, 6), or if greater climatological or ecological understanding continues to bring unpleasant surprises. A modest degree of adaptation may therefore be prudent if not inevitable (153): e.g. planning coastal developments to accommodate some sea-level rise and water projects to tolerate shifts in rainfall, or reversing the narrowing of crops' and forests' genetic bases. Nonetheless, the techniques described here, if their benefits are properly understood, show promise of such rapid and widespread deployment that most of the harm projected in today's best models could almost certainly be avoided. Many abatement measures also have the valuable side-effect of increasing resilience in the face of whatever climatic change may nonetheless occur.

9. *Abating global warming would lock developing countries into abject poverty, or at least prevent their achieving their legitimate aspirations—even though most global warming so far has been caused by the industrialized countries.* On the contrary, the abatement options discussed above are not merely compatible with but essential to affordable and sustainable global development and increased equity.

10. *Policymakers already know what their options are and have not chosen those described here, so either the policymakers are stupid or the options do not work.* Many policymakers suppose that abatement must be slow, small, costly, inconvenient, and nasty—not because that is true, but simply because they do not know any better. The difficulty, we suspect, may be the one economist Ken Boulding described: that a hierarchy is "an ordered arrangement of wastebaskets designed to prevent information from reaching the executive." The options described above are available, demonstrated, and often in widespread and successful use. Many, however, are so new that they are not widely known even to technical experts, and will take many years to

filter up to decisionmakers through normal channels. What is needed, therefore, is better and faster technology transfer to the policymakers. We hope
this paper contributes to that effort.

ACKNOWLEDGMENTS

The research and logistical support of the RMI staff, especially Rick Heede
and Catherine Henze, valuable comments by Robert Socolow and Howard
Geller, and the essential contributions of many colleagues around the world
are gratefully acknowledged.

Literature Cited

1. Intergov. Panel Climate Change. 1990.
   Report from Working Group I. June. UK
   Meteorol. Off, Bracknell: World
   Meteorol. Org./United Nations Environ.
   Prog. Summarized in 1a.
1a. *Science,* 3 Aug. 1990, p. 481
2. Brookes, W. T. 1989. The global warm-
   ing panic. *Forbes.* 25 Dec: pp. 96–100
3. Intergov. Panel Climate Change. 1990.
   *Report from Working Group III.
   Policymaker's Summary.* June. World
   Meteorol. Org./United Nations Environ.
   Prog.
4. Lashof, D. A., Ahuja, D. R. 1990. Relative global warming potentials of
   greenhouse gas emissions. *Nature*
   334:529–31 (5 Apr.)
5. Environ. Protect. Agency. 1989. *Policy
   Options for Stabilizing Global Climate.*
   Draft Report to Congress, ed. D. A.
   Lashof, D. Tirpak. 2 vols. Feb.
6. Krause, F., Bach, W., Koomey, J.
   1989. *Energy Policy in the Greenhouse.*
   Report to Dutch Ministry of Housing,
   Physical Planning and Environment, International Project for Sustainable Energy Paths, Vol. 1. El Cerrito, Calif. Sept.
   Summarized in Ref. 152a
6a. Hendriks, C. A., Blok, K., Turkenburg, W. C. 1990. Technology and the
   cost of recovery and storage of carbon
   dioxide from an integrated combined cycle plant. *Appl. Energy.* In press
6b. Wolsky, A. M., Brooks, C. 1989.
   Recovering CO_2 from large stationary
   combusters. See Ref. 152a, pp. 179–85
7. Lovins, A. B. 1988. *Negawatts for
   Arkansas,* Vol. 1. *RMI Publication
   #U88-30.* 30 June
8. Lovins, A. B., Sardinsky, R. 1988. *The
   State of the Art: Lighting.* Old Snowmass, Colo: Rocky Mountain Inst./
   COMPETITEK
8a. Southern Calif. Edison Co. 1985. *1984
   Conservation/Load Management Results.* Rosemead, Calif. 31 March

8b. Nadel, S. 1990. *Lessons Learned: A
   Review of Utility Experience with Con-
   servation and Load Management Programs for Commercial and Industrial
   Customers. New York State Energy Res.
   & Dev. Auth. #90-8.* Albany. Available
   from Am. Counc. Energy-Efficient
   Econ., Washington, DC. April
9. Evans, D. J., et al. 1991. *Policy Implications of Greenhouse Warming.*
   Washington, DC: Natl. Acad. Sci.
9a. Natl. Acad. Sci. 1991. Report of the
   Mitigation Subpanel, Panel on Policy
   Implications of Greenhouse Warming.
   Washington, DC. 16 June
10. Cherfas, J. 1991. Skeptics and visionar-
   ies examine energy saving. *Science*
   251:154–56 (11 Jan.)
11. Lovins, A. B. 1991. Energy conservation. *Science* 251:1296–98. 15 March
12. Lovins, A. B. 1991. Energy savings.
   Science 252:763 (10 May)
13. Lovins, A. B. 1990. *Factsheet: How a
   Compact Fluorescent Lamp Saves a Ton
   of CO_2. RMI Publication #E90–5.* See
   also 13a
13a. Gadgil, A. J., Rosenfeld, A. H. 1990.
   *Conserving Energy with Compact
   Fluorescent Lamps.* Lawrence Berkeley
   Lab. 1 May
14. Shepard, M., Lovins, A. B., Neymark,
   J., Houghton, D. J., Heede, H. R. 1990.
   The State of the Art: Appliances. Old
   Snowmass, Colo: Rocky Mountain Inst./
   COMPETITEK
15. Heede, H. R., Morgan, R. E., Ridley,
   S. 1985. *The Hidden Costs of Energy:
   How Taxpayers Subsidize Energy
   Development.* Washington, DC: Cent.
   Renew. Resourc. Summarized in
   15a
15a. Heede, H. R., Lovins, A. B. 1985.
   Hiding the true costs of energy sources.
   Wall Street Journal. 17 Sept, p. 28
16. Churchill, A., World Bank. 1989. Dis-
   cussion in *Digest of the 14th World En-*

ergy Conference, p. 120. Montreal, 17–22 Sept.

17. Reddy, A. K. N., Goldemberg, J. 1990. Energy for the developing world. *Sci. Am.* Sept: pp. 110–18

18. Electr. Power Res. Inst. 1990. *Efficient Energy Use: Estimates of Maximum Energy Savings. EPRI CU-6746.* Palo Alto, Calif.

19. Fickett, A. P., Gellings, C. W., Lovins, A. B. 1990. Efficient use of electricity. *Sci. Am.* Sept: pp. 64–74

20. Calif. Energy Comm. 1990. *1990 Conservation Report.* Staff draft. Sacramento. Aug.

21. Lovins, A. B. 1980. Economically efficient energy futures. In *Interactions of Energy and Climate,* ed. W. Bach et al. Dordrecht: Reidel

22. Lovins, A. B., et al. 1989. *The State of the Art: Drivepower.* Old Snowmass, Colo: Rocky Mountain Inst./COMPETITEK

23. Lovins, A. B. 1986. *The State of the Art: Water Heating.* Old Snowmass, Colo; Rocky Mountain Inst./COMPETITEK

24. Lovins, A. B. 1986. *The State of the Art: Space Cooling.* Old Snowmass, Colo: Rocky Mountain Inst./COMPETITEK. Preliminary ed.

25. Johansson, T. B., Steen, P. 1983. *I Stället för Kärnkraft: Energi År 2000. DsI 1983:18.* Industridepartementet, Stockholm (In Swedish). Summarized in *Science* 219:355–61

25a. Larson, E. D., Nilsson, L. J., Johansson, T. B. 1989. *The Technology Menu for Efficient End-Use of Energy: Vol. I, Movement of Material.* Environ. & Energy Stud., Lund Univ.

26. Orlov, R. V. 1988. Inst. Energy Econ., USSR Acad. Sci., Moscow. Personal communication. Nov.

27. Makarov, A. A., et al. *Otsyenka Parkov Dvigatyeleiy v Narodnom Khozyaistvye SSSR i SShA.* Moscow: Inst. Energy Econ., USSR Acad. Sci. (In Russian)

28. Piette, M., Krause, F., Verderber, R. 1989. *Technology Assessment: Energy-Efficient Commercial Lighting. LBL-27032.* Berkeley, Calif: Lawrence Berkeley Lab. Apr.

29. Sacramento Municip. Util. Dist. 1990. *Shade Tree Program ("Trees are Cool").* With Sacramento Tree Found.

30. Akbari, H., Martien, P., Ranier, L., Rosenfeld, A. H., Taha, H. 1988. *The Impact of Summer Heat Islands on Cooling Energy Consumption and Global CO_2 Concentration. LBL-25179.* Berkeley, Calif: Lawrence Berkeley Lab. Apr.

31. Taha, H., Akbari, H., Rosenfeld, A. H., Huang, J. 1988. Residential cooling loads and the urban heat island—the effects of albedo. *Building Environ.* 23(4):271–93

31a. Nadel, S., Shepard, M., Greenberg, S., Katz, G., de Almeida, A. 1991. *Energy-Efficient Motor Systems: A Handbook on Technology, Programs, and Policy Opportunities.* Washington, DC: Am. Counc. Energy-Efficient Econ.

32. Rosenfeld, A. H., Hafemeister, D. 1988. Energy-efficient buildings. *Sci. Am.* Apr:pp. 78–85

33. Bevington, R., Rosenfeld, A. H. 1990. Energy for buildings and homes. *Sci. Am.* Sept:pp. 76–86

34. Ottinger, R. et al. 1990. *Environmental Costs of Electricity.* Pace Univ. Law Sch. Report to US DOE and NYSER-DA. June

34a. ACT². 1991. *Newsletter* of, and technical publications available from, the Advanced Customer Technology Test for Maximum Energy Efficiency, Pacific Gas & Electric Co., Research Dept., San Ramon, Calif. 94583

35. Bodlund, B., Mills, E., Karlsson, T., Johansson, T. B. 1989. The challenge of choices: technology options for the Swedish electricity sector. See Ref. 160, pp. 883–947

36. Nørgård, J. S. 1989. Low electricity appliances—options for the future. See Ref. 160, pp. 125–72

37. Feist, W. 1987. *Stromsparpotentiale bei den privaten Haushalten in der Bundesrepublik Deutschland.* Darmstadt: Inst. Wohnen und Umwelt (In German)

38. Goldemberg, J. et al. 1988. *Energy for a Sustainable World.* New Delhi: Wiley Eastern

39. Lovins, A. B. 1979. Re-examining the nature of the ECE energy problem. *Energy Policy* 7:178. 1978 United Nations Econ. Comm. Europe paper.

40. Kahane, A. 1986. *Industrial Electrification: Case Studies of Four Industries. Steel, Paper, Cement, and Motor Vehicles in the United States, Japan, and France.* Lawrence Berkeley Lab. report to Electr. Power Res. Inst. Available from author at PL/12, Shell Centre, London SE1 7NA

41. Hirst, E. 1991. *Possible Effects of Electric-Utility DSM Programs, 1990 to 2010. ORNL/CON-312.* Oak Ridge, Tenn: Oak Ridge Natl. Lab. Feb.

42. Davis, G. 1990. Energy for planet Earth. *Sci. Am.* Sept: pp. 54–62

43. Difiglio, C., Duleep, K. G., Greene, D. L. 1989. Cost effectiveness of future fuel economy improvements. *J. Energy* 2(1):65–83

44. Ledbetter, M., Ross, M. 1990. A supply curve of conserved energy for automobiles. In *Proc. 25th Intersoc. Energy Convers. Eng. Conf., Reno, Nev., 12–17 Aug.* New York, NY: Am. Inst. Chem. Eng.
45. Flemings, M. C. et al. 1980. *Materials Substitution and Development for the Light Weight, Energy Efficient Automobile.* Report to US Congress's Off. Technol. Assess., 8 Feb. Cambridge, Mass: Mater. Sci. & Eng. Dept., Mass. Inst. Technol.
46. Bleviss, D. S., Walzer, P. 1990. Energy for motor vehicles. *Sci. Am.* Sept: pp. 102–9
47. Ledbetter, M., Ross, M. 1990. Supply curves of conserved energy for automobiles. See Ref. 122, Att. B
48. Patterson, P. 1987. *Periodic Transportation Energy Report 1. DOE CE-15.* Washington, DC: US Dept. Energy. 16 Nov.
49. Bleviss, D. L. 1988. Saving fuel. *Technol. Rev.* Nov/Dec: pp. 47–54
50. Bleviss, D. L. 1988. *The New Oil Crisis and Fuel Economy Technologies.* Westport, Conn: Quorum Books
51. Chandler, W. U., Makarov, A. A., Zhou, D. 1990. Energy for the Soviet Union, Eastern Europe, and China. *Sci. Am.* Sept: pp. 120–27
51a. Lovins, A. B. 1991. *Advanced Light Vehicle Concepts.* Presentation to Natl. Acad. Sci. panel. *RMI Publication #T91-20.* 9 July
52. Chrysler Genesis Project studies composite vehicles. 1986. *Automot. News* 5 May:p.36
53. von Hippel, F., Levi, B. V. 1983. Automotive fuel efficiency: the opportunity and weakness of existing market incentives. *Resource. Conserv.* 10:103–24
54. Sobey, A. J. 1988. Energy use in transportation: 2000 and beyond. In *Summary of Presentations at a Workshop on Energy Efficiency and Structural Change: Implications for the Greenhouse Problem. Oakland, Calif., 1–3 May,* ed. S. Meyers. Berkeley, Calif: Lawrence Berkeley Lab.
55. Newman, P. W. G., Kenworthy, J. A. 1989. *Cities and Automobile Dependence: A Sourcebook.* Aldershot, Hants, England: Gower Technical
56. Goldstein, D. B., Holtzclaw, J. W., Davis, W. B. 1990. *NRDC/Sierra Club Testimony for Conservation Report Hearing on Transportation Issues. Calif. Energy Comm. Docket #89-CR-90.* 23 Apr. San Francisco, Calif: Nat. Resour. Defense Counc.
57. Newman, P. W. G., Hogan, T. L. F. 1987. *Urban Density and Transport: A Simple Model Based on Three City Types. Transport Research Paper 1/87.* Dept. Environ. Sci., Murdoch Univ., Perth, Western Australia
57a. Duxbury, M. L., Neville, S. D., Campbell, R., Newman, P. W. G. 1988. *Mixed Land Use and Residential Satisfaction: An Evaluation. Transport Research Paper 2/88.* Dept. Environ. Sci., Murdoch Univ., Perth, Western Australia
58. Calif. Energy Comm. 1988. *1988 Conservation Report.* Sacramento. Oct.
59. Wash. State Energy Off. 1990. *Telecommuting: An Alternative Route to Work.* Olympia
60. Calwell, C., Edwards, A., Gladstein, C., Lee, L. 1990. *Clearing the Air.* San Francisco, Calif: Nat. Resourc. Defense Counc.
61. Lovins, A. B., Lovins, L. H. 1981. *Energy/War: Breaking the Nuclear Link.* New York: Harper & Row Colophon. Expanding Lovins, A. B., Lovins, L. H., Ross, L. 1980. Nuclear power and nuclear bombs. *Foreign Affairs* 58:1136ff, 59:172ff
62. Henne, P. A. 1989. *MD-90 Transport Aircraft Design.* Douglas Aircraft Co., McDonnell Douglas Corp. (Long Beach, Calif.), Am. Inst. Aeronaut. Astronaut. Aircraft Design, Systems & Operations Conf., Seattle, Wash., 31 July–2 Aug.
63. Vaughan, C. 1988. Saving fuel in flight. *Sci. News* 134:266–69 (22 Oct.)
64. Kavanaugh, M. 1990. Fuel economies available from ultrahigh bypass (UHB) jet engines. Memo to P. Schwengels, EPA, 10 March. See Ref. 122, Att. D
65. Chizhov, N., Styrekovich, M. A. 1985. Ecological advantages of natural gas over other fossil fuels. In *The Methane Age,* ed. T. H. Lee et al, pp. 155–61. Dordrecht: Kluwer and Laxenburg, Austria: Int. Inst. Appl. Systems Anal.
66. Ephraim, M. Jr. 1984. Locomotive energy-related technology: recent improvements and future prospects. In *Rail Vehicle Energy Design Considerations,* ed. B. C. Houser. Amsterdam: Elsevier
67. Updating the diesel locomotive. 1986. *Fortune.* 3 March: p. 52
68. Parson, E. 1990. *The Transport Sector and Global Warming,* p. 80. Global Environmental Policy Project Discussion Paper, Harvard Univ.
69. Natl. Res. Counc. 1990. *Confronting Climate Change.* Washington, DC: Natl. Acad. Press
69a. Lovins, A. B., Lovins, L. H., Krause,

F., Bach, W. 1981. *Least-Cost Energy: Solving the CO_2 Problem*. Andover, Mass: Brick House. 2nd printing by Rocky Mountain Inst., Old Snowmass, Colo. 1989. Summarized in 69b.

69b. Lovins, A. B., Lovins, L. H. 1982. Energy, economics, and climate. *Climate Change* 4:217–20; 1983, 5:105ff

70. Samuels, G. 1981. *Transportation Energy Requirements to the Year 2010*. *ORNL-5745*. Oak Ridge, Tenn: Oak Ridge Natl. Lab.

71. Cat's 3306B makes it big in the real world. 1983. *Automot. News*. 7 Nov.

72. Spyrou, A. 1988. *Energy and Ships: The Effects of the Energy Problem on the Design and Efficient Operation of Future Merchant Ships*, pp. 34–41. London: Lloyds of London Press

73. Solar Energy Res. Inst. 1981. *A New Prosperity: Building a Sustainable Energy Future*. Andover, Mass: Brick House

74. Rosenfeld, A. H., Mowris, R. J., Koomey, J. 1990. *Policies to Improve Energy Efficiency, Budget Greenhouse Gases, & Answer the Threat of Global Climate Change*. Am. Assoc. Adv. Sci. Meet., 15–20 Feb., New Orleans, 26 Jan.

75. Ross, M. H., Steinmeyer, D. 1990. Energy for industry. *Sci. Am.* Sept: pp. 88–98

76. Schipper, L., Lichtenberg, A. J. 1976. Efficient energy use and well-being: The Swedish example. *Science* 194:1001–13

77. Groscurth, H.-M., Kümmel, R. 1989. *The Cost of Energy Optimization: A Thermoeconomic Analysis of National Energy Systems*. Physikalisches Inst. der Univ. Würzburg. 16 May

78. Larson, E. D., Ross, M. H., Williams, R. H. 1986. Beyond the era of materials. *Sci. Am.* June: pp. 34–41

79. Herman, R., Ardekani, S. A., Ausubel, J. H. 1989. Dematerialization. In *Technology and Environment*, pp. 50–69. Washington, DC: Natl. Acad. Press

80. Colombo, U. 1988. The technology revolution and the restructuring of the global economy. In *Globalization of Technology: International Perspectives*, ed. J. H. Muroyama, H. G. Stever, pp. 23–31. Washington, DC: Natl. Acad. Press

81. Lovins, A. B., Lovins, L. H. 1989. Drill rigs and battleships are the answer! (But what was the question?) In *The Oil Market in the 1990s*, ed. R. G. Reed, F. Fesharaki, pp. 83–138. Boulder, Colo: Westview

81a. Nørgård, J. S. 1979. *Husholdninger og Energi*. København: Polyteknisk Forlag. (In Danish)

82. Eketorp, S. 1989. Electrotechnologies and steelmaking. See Ref. 160, pp. 291–96

83. Baldwin, S. F., Geller, H., Dutt, G., Ravindranath, N. H. 1985. Improved woodburning stoves: signs of success. *Ambio* 14(4–5):280–87

84. Baldwin, S. F. 1987. *Biomass Stoves: Engineering Design, Development and Dissemination*. Arlington, Va: Volunt. Tech. Assist.

85. Williams, R. H., Larson, E. D. 1989. Expanding roles for gas turbines in power generation. See Ref. 160, pp. 503–54

86. Larson, E. D., Williams, R. H. 1988. Biomass-fired steam-injected gas turbine cogeneration. In *Proc. 1988 ASME Cogen-Turbo Symp.* Montreaux, 30 Aug.–1 Sept.

87. Larson, E. D., Svenningsson, P., Bjerle, I. 1989. Biomass gasification for gas turbine power generation. See Ref. 160, pp. 697–739

88. Arbatov, A. A. 1990. Deputy Chairman, Comm. Stud. Production Forces and Nat. Resourc., USSR Acad. Sci., Moscow, personal communication

89. Makarov, A. A., Bashmakov, I. A. 1990. *The Soviet Union: A Strategy of Energy Development with Minimum Emission of Greenhouse Gases*. *PNL-SA-18094*. Richland, Wash: Battelle Pacific Northwest Lab. Apr.

90. Abrahamson, D. 1989. *Relative Greenhouse Heating from the Use of Fuel Oil and the Use of Natural Gas*. Minneapolis, Minn: Hubert H. Humphrey Inst. Public Affairs, Univ. Minn.

91. Demirchyan, K. S. 1989. Deputy Director, Div. Physical-Technical Problems of Energetics, USSR Acad. Sci., personal communication

92. Jászay, T. 1990. *Carbon Dioxide Emissions Control in Hungary: Case Study to the Year 2030*. *PNL-SA-18248*. Richland, Wash: Battelle Pacific Northwest Lab. May

93. Marland, G., et al (5 coauthors). 1989. *Estimates of CO_2 Emissions from Fossil Fuel Burning and Cement Manufacturing*. *ORNL/CDIAC-25*. Oak Ridge, Tenn: Oak Ridge Natl. Lab.

94. Solar Energy Res. Inst. 1990. *The Potential of Renewable Energy*. Interlaboratory White Paper. *SERI/TP-260-3674*. Golden, Colo. March

95. Weinberg, C. J., Williams, R. H. 1990. Energy from the sun. *Sci. Am.* Sept: pp.147–55

96. Beale, W. 1990. Address to Global Forum energy section, Moscow, Jan.

97. Lovins, A. B., Lovins, L. H. 1982.

Brittle Power: Energy Strategy for National Security. Report to US DOD Defense Civil Preparedness Agency/FEMA/CEQ. Andover, Mass: Brick House. Summarized in 97a, 97b

97a. Reducing vulnerability: the energy jugular. 1983. In *Nuclear Arms: Ethics, Strategy, Politics*, ed. R. J. Woolsey. San Francisco, Calif: Inst. for Contemporary Studies

97b. The fragility of domestic energy. 1983. *Atlantic*. Nov: pp. 118–26

98. Lovins, A. B. 1978. Soft energy technologies. *Annu. Rev. Energy* 3:477–517

99. Sørensen, B. 1979. *Renewable Energy*. New York: Academic

100. DeLuchi, M. A., Johnston, R. A., Sperling, D. 1988. Transportation fuels and the greenhouse effect. *Transport. Res. Rec.* 1175:33–44

100a. Krause, F., Koomey, J., Olivier, D., Radanne, P. 1991. *Energy Policy in the Greenhouse*, Vol. 2: *The Cost of Carbon Reductions—A Case Study of Western Europe*. Report to Dutch Ministry of Housing, Physical Planning and Environ., Int. Project for Sustainable Energy Paths. El Cerrito, Calif. Forthcoming

100b. Krause, F., Busch, J., Koomey, J. 1991. *Internalizing Global Warming Risks in Power Sector Planning: A Case Study of Carbon Reduction Costs for the New England Region*. *LBL-30797*. Berkeley, Calif: Lawrence Berkeley Lab. In press

101. Greene, D. 1990. *A Greenhouse Energy Strategy: Sustainable Energy Development for Australia*. Report to DASETT, Canberra, by Deni Greene Consulting Services, Melbourne

102. DPA Group. 1989. *Study on the Reduction of Energy-Related Greenhouse Gas Emissions*. Report to Federal-Provincial Task Force on Global Warming, Toronto, March

103. Newell, R. E., Reichle, H. G. Jr., Seiler, W. 1989. Carbon monoxide and the burning Earth. *Sci. Am.* Oct: pp. 82–88

104. Harmon, M. E., Ferrell, W. K., Franklin, J. F. 1990. Effects on carbon storage of conversion of old-growth forests to young forests. *Science* 247:699–702 (9 Feb.)

105. Ore. Dept. Energy. 1990. *Oregon Task Force on Global Warming Report to the Governor and Legislature*, p. 1–9. June

105a. Big firms get high on organic farming. 1989. *Wall Street Journal*. 21 Mar., p. B1; Back to the future: A movement to farm without chemicals makes surprising strides. 1989. *Wall Street Journal*. 11 May, p. A1

106. Jackson, W. 1980. *New Roots for Agriculture*. San Francisco, Calif: Friends of the Earth; Salina, Kans: The Land Inst.

107. Crutzen, P. J., Aselmann, I., Seiler, W. 1986. Methane production by domestic animals, wild ruminants, other herbivorous fauna and humans. *Tellus* 38B:271–84

108. Soden, K. 1988. *U.S. Farm Subsidies*. *RMI Publication #A88-22*

109. Browning, W. 1990. Steaks and mistakes. Unpublished manuscript in course of revision for publication by Rocky Mountain Inst.

109a. Stevens, W. K. 1989. Methane from guts of livestock is new focus in global warming. *New York Times*, 21 Nov., p. C4

110. Natl. Res. Counc. 1989. *Alternative Agriculture*. Washington, DC: Natl. Acad. Press

111. Jackson, W., Berry, W., Colman, B. 1984. *Meeting the Expectations of the Land*. San Francisco, Calif: North Point

112. Brody, J. E. 1985. Organic farming moves into the mainstream. *New York Times*, 8 Oct., p. 20

113. Bechmann, A. 1987. *Landbau-Wende: Gesunde Landwirtschaft—Gesunde Ernährung*. S. Fischer. Frankfurt/M. (In German)

114. Bossell, H., Kramer, P., Schaffner, J., Weiernantel, H., Zerger, U. 1986. *Technologiefolgeabschätzungen für die Landwirtschaftliche Produktion*. Hanover: Inst. Systemanalyse Prognose (In German)

115. Holmberg, W. 1988. *Farm Ethanol—Time for a National Policy*. Washington, DC: Inf. Resourc. Inc. 16 Sept.

116. McKinney, T. R. 1987. *Comparison of Organic and Conventional Agriculture: A Literature Review*. *RMI Publication # A87–28*

117. Leach, G., Mearns, R. 1988. *Beyond the Woodfuel Crisis: People, Land, and Trees in Africa*. London: Earthscan

118. Wade, E. 1981. Fertile cities. *Dev. Forum*. Sept./Dec.

119. Plochmann, R. 1968. *Forestry in the Federal Republic of Germany*. Hill Family Foundation Series. Corvallis: Ore. State Univ. Sch. Forestry

120. Cramer, H. H. 1984. On the predisposition to disorders of middle European forests. *Pflanzenschutz-Nachrichten* 37:98–207

121. Maser, C. 1988. *The Redesigned Forest*. San Pedro, Calif: Miles

122. ICF Resourc. Inc. 1990. *Preliminary Technology Cost Estimates of Measures Available to Reduce U.S. Greenhouse Gas Emissions by 2010*. Draft report to

Environ. Protect. Agency. Fairfax, Va. Aug.

123. Reichmuth, H., Robison, D. 1989. *Carbon Dioxide Offsets*. Portland, Ore: Lambert Engineering/Pacific Power. March

124. Manzer, L. E. 1990. The CFC-ozone issue: Progress on the development of alternatives to CFCs. *Science* 249:31–35 (6 Jul.)

125. Environ. Protect. Agency. 1989. *Costs and Benefits of Phasing Out Production of CFCs and Halons in the United States*. Off. Air and Radiation. 3 Nov review draft

126. Turiel, I., Levine, M. D. 1989. Energy-efficient refrigeration and the reduction of chlorofluorocarbon use. *Annu. Rev. Energy* 14:173–204

126a. Chandler, W. U. 1989. *Control of Carbon Emissions in the Soviet Union, Poland, and Hungary*. IPPC draft expert report. 15 Jan.

126b. Chandler, W. U., ed. 1990. *Carbon Emissions Control Strategies: Case Studies in International Cooperation*. Washington, DC: Conservation Found. In press

127. Okken, P. A., Ybema, J. R., Germers, D., Kram, T., Lako, P. 1991. *The Challenge of Drastic CO2 Reduction*. *ECN-C-009*. Petten: Energieonderzoek Centrum Nederland. Mar.

127a. Blok, K., Worrell, E., Albers, R. A. W., Cuelenaere, R. F. A., Turkenburg, W. C. 1990. The cost-effectiveness of energy conservation from the view of carbon dioxide reduction. In *Proc. 13th World Energy Eng. Congr., Atlanta, 10–12 Oct.*

128. Lovins, A. B. 1989. *Making Markets in Resource Efficiency*. RMI Publication #E89–27

129. Rowberg, R. E. 1990. *Energy Demand and Carbon Dioxide Production. 90-204 SPR*. Washington, DC: US Congr. Res. Serv., Lib. Congr. 17 Apr.

130. Lovins, A. B. 1976. Energy strategy: the road not taken? *Foreign Affairs* 55(1):65–96 (Oct)

131. Rosenfeld, A. H., Meier, A. 1990. *Energy Efficiency vs Global Warming: A No-Regrets Policy*. Berkeley, Calif: Lawrence Berkeley Lab. Apr.

132. Keepin, W. N., Kats, G. 1988. Greenhouse warming: Comparative analysis of nuclear and efficiency abatement strategies. *Energy Policy* 16(6):538–61 (Dec.), and calculational suppl., *Greenhouse Warming: A Rationale for Nuclear Power?* Available from Rocky Mountain Inst.

133. Keepin, W. N., Kats, G. 1988. Global warning. *Science* 241:1027 (26 Aug.)

134. Lovins, A. G. 1989. Energy options. *Science* 243:12 (6 Jan.)

134a. Electr. Power Res. Inst. 1986. *Impact of Demand-Side Management on Future Customer Electricity Demand*. EM-4815-SR. Palo Alto, Calif.

135. Lovins, A. B., Shepard, M. 1988. *Financing Electric End-Use Efficiency*. *RMI/COMPETITEK Implementation Paper #1; and Update*. 1989

136. Moskovitz, D. 1989. *Profits and Progress Through Least-Cost Planning*. Washington, DC: Natl. Assoc. Regul. Util. Comm.

136a. Lovins, A. B. 1985. *Least-Cost Electricity Strategies for Wisconsin*. Exh. 33. *Docket #05-EP-4*, Wisc. Public Serv. Comm. *RMI Publication #U85-23*. 9 Sept.

137. Lovins, A. B. 1989. *End-Use/Least-Cost Investment Strategies*. Paper #2.3.1, World Energy Conf., Montreal, Sept.

138. Menke, K., Woodwell, J. 1990. *Water Productivity and Development: Strategies for More Efficient Use*. *RMI Publication #W90-10*. See also the fuller EPA-sponsored RMI water-efficiency implantation handbook, in press, 1991

139. Lovins, A. B., Shepard, M. 1988. *Customer Behavior and Information Programs*. *RMI/COMPETITEK Implementation Paper #2*

140. Levenson, L., Gordon, D. 1990. Drive+: Promoting cleaner and more fuel efficient motor vehicles through a self-financing system of state sales tax incentives. *J. Policy Anal. Manage.* 9(3):409–15

141. Greene, D. L. 1989. *CAFE or Price? An Analysis of the Effects of Federal Fuel Economy Regulations and Gasoline Prices on New Car MPG, 1978–89*. Oak Ridge Natl. Lab./US DOE Off. Policy, Planning, & Anal., 10 May draft

142. El-Gasseir, M. M. 1990. *The Potential Benefits and Workability of Pay-As-You Drive Automobile Insurance*. 8 Jun. testimony to Calif. Energy Comm., *Docket #89-CR-90*

143. Borré, P. 1990. Presentation to Aspen Inst. for Humanistic Stud. Energy Comm. Aug

144. Blair, P. 1990. Presentation to Aspen Inst. for Humanistic Stud. Energy Comm. Aug.

145. Harvey, T. H., Shuman, M., Arbess, D. 1991. *Reclaiming Security: Beyond the Controlled Arms Race*. Old Snowmass, Colo: Rocky Mountain Inst. In press

146. Sitnícki, S., Budzinski, K., Juda, J., Michna, J., Szpilewicz, A. 1990. *Poland: Opportunities for Carbon Emission Control*. Richland, Wash: Battelle Pacific Northwest Lab. May

147. Off. Technol. Assess. 1981. *Technology and Soviet Energy Availability*. Washington, DC

148. Gadgil, A. J., Sastry, M. A. 1990. *The BELLE Project: Bombay Efficient Lighting Large-Scale Experiment. Overview: An Innovative Method for Reducing Peak Electric Demand*. Lawrence Berkeley Lab./Rocky Mountain Inst. Sept.

149. Van Domelen, J. 1989. *Power to Spare*. Washington, DC: Conservation Found.

150. Lovins, A. B. 1989. Abating air pollution at negative cost via energy efficiency. *J. Air & Waste Manage. Assoc.* 39(11):1432–35

151. Lovins, A. B., Lovins, L. H. 1990. Make fuel efficiency our Gulf strategy. *New York Times*. 3 Dec., p. A15. Available in annotated edition from Rocky Mountain Inst., *Publication #S90-26*

152. Lovins, A. B., Lovins, L. H. 1991. The energy saboteurs are in the White House. *Los Angeles Times*. 21 Jan. *RMI Publication #S91-4*

152a. Krause, F. 1990. Required speeds of fossil fuel phase-out under a 2 degree C global warming limit. In *Energy Technologies for Reducing Emissions of Greenhouse Gases*, pp. 2:102–109. OECD/IEA Experts' Semin., Paris, 12–14 Apr 1989. Paris: Organ. Econ. Coop. Dev.

153. Schneider, S. 1989. *Global Warming: Are We Entering the Greenhouse Century?* New York: Vintage

153a. Davis, B. 1990. Bid to slow global warming could cost U.S. $200 billion a year, Bush aide says. *Wall Street Journal*, 16 Apr., p. B4

154. Nordhaus, W. 1989. *The Economics of the Greenhouse Effect*. New Haven, Conn: Yale Univ., and presentation at Am. Assn. Adv. Sci. Meet., New Orleans, 19 Feb. 1990

155. Manne, A. S., Richels, R. G. 1990. Global CO_2 emission reductions—the impacts of rising energy costs. *Energy J.* and EPRI, Aug.; see also CO_2 emission limits: An economic analysis for the U.S.A. *Energy J.* and EPRI, Nov. 1989

156. Passell, P. 1990. Cure for greenhouse effect: the costs will be staggering. *New York Times*. 19 Nov., pp. 1, 10

157. Williams, R. H. 1990. Low-cost strategies for coping with CO_2 emission limits. *Energy J.* 11(3): In press

158. Zimmerman, M. B. 1990. *Assessing the Costs of Climate Change Policies: The Uses and Limits of Models*. Washington, DC: Alliance to Save Energy. 10 Apr.

159. Rosenfeld, A. H., Atkinson, C., Koomey, J., Meier, A., Mowris, R. A. 1990. *A Compilation of Supply Curves of Conserved Energy Use for U.S. Buildings*. Paper for Mitigation Subpanel, Panel on Policy Implications of Greenhouse Warming, US Natl. Acad. Sci. 20 Sept

160. Johansson, T. B., Bodlund, B., Williams, R. H., eds. 1989. *Electricity*. Lund Univ. Press. Apr.

Annu. Rev. Energy Environ. 1991. 16:533–55

ENERGY APPLICATIONS OF SUPERCONDUCTIVITY

Thomas R. Schneider

Electric Power Research Institute, Palo Alto, California 94303

KEY WORDS: superconductors, magnets, electrical equipment, electricity.

INTRODUCTION

Background

Energy use in society underwent a major revolution in the late 19th century when electricity was transformed from a laboratory curiosity into a backbone of modern society (1).

Further developments in the first half of the 20th century built on the early discoveries, and much of the shift from an agricultural economy to an industrial society was assisted by further improvements in the ability to deliver work at the point of use with precision and flexibility through electrical machines and devices. The developments in electronics and microelectronics of the second half of this century have provided even more precision in energy use; indeed microelectronics are the brains to electrical equipment's brawn. The underlying technology has, however, remained largely unchanged over the past 100 years. Although individual components have evolved, we are not that far from the technology of hanging copper wires on pine trees with iron nails!

The discovery of superconductivity—the remarkable ability of some materials to conduct electrical current with no resistance when cooled to extremely low temperature—in 1911 by Onnes (2) had no effect on the emerging electrical technology. The discovery did, however, spark scientists' imaginations about the possibility of lossless electricity transmission.

From 1911 until 1986, research efforts failed to raise superconducting

1056-3466/91/1022-0533$02.00

temperatures above 23 K ($-418°$ F). The focus of research shifted in 1986 to new classes of superconducting ceramic compounds, which are normally insulators, rather than metals, as a result of the remarkable discovery of superconductivity in the 30-K range by Bednorz & Muller (3). The result was a dramatic increase in critical temperatures. The real leap came in early 1987, when research teams led by Ching-Wu (Paul) Chu of the University of Houston and Maw-Kuen Wu of the University of Alabama reported a transition temperature of 94 K ($-290°$ F) in a new ceramic compound, yttrium barium copper oxide (4). These discoveries were followed in 1988 by the announcement of compounds containing bismuth (5) and thallium (6) with superconducting transition temperatures as high as 125 K. This progress in temperature is illustrated in Figure 1.

The exciting discoveries of the new superconductors initially led many to speculate on the possibility of rapid progress in the application of these new materials to a wide range of uses—from supercomputers to magnetic levitated trains. This article puts many of those opportunities into perspective and describes the broad range of possible energy applications. The possibilities are numerous, and the enthusiasm generated for the new materials has also revitalized interest in the older low-temperature superconductors.

The excitement over the new superconductors repeats the enthusiasm following the initial discovery in 1911 as well as that following the discovery of more practical metallic materials in the early 1960s. Higher temperature and the rapid progression from 35 K to 95 K and then to 125 K in a period of less than two years increased the normal enthusiasm to what could be described whimsically as hyperbole. Such excessive optimism led to some backlash and various expressions of pessimism in the media as the difficulty of the technical challenge of development was recognized. Now nearly five years after the initial discoveries and with fairly steady progress in both the science and the technology, the hype has been reduced and a more rational atmosphere prevails.

Realistically, it will take the balance of the decade to advance the materials science and develop the technology to commercialize the new materials. At the same time, progress on the older lower-temperature materials will continue. The prognosis for the practical application of superconductivity to energy technology is excellent.

Nevertheless, high-temperature superconductivity generally promises an evolutionary rather than a revolutionary impact on energy applications. In other words, it could improve efficiency and reliability, but not make large power plants or petroleum obsolete.

Superconductor Behavior

Although new materials have been discovered, and the understanding of the structure and chemistry of the new superconductors has been greatly ad-

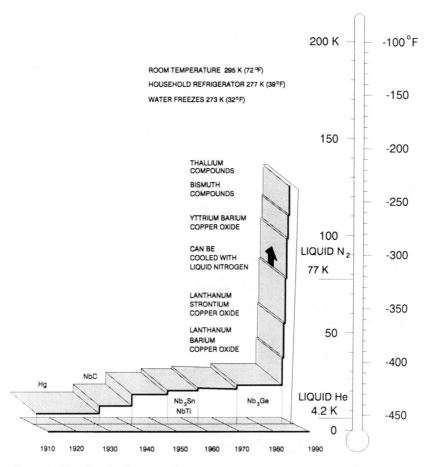

Figure 1   Time line for discovery of superconductors

vanced, a comprehensive theory has not yet emerged, and a physical picture
of the nature and behavior of the new materials is still evolving. For the
purposes of this article, however, the "classical" description of these new
materials is sufficient (7).

Superconductivity applications are affected by three interrelated physical
properties of superconducting materials:

1. *Critical temperature* (T_C) is the maximum temperature at which
   superconductivity can be maintained.
2. *Critical magnetic field* (H_{C2}) is the maximum magnetic field strength at
   which superconductivity can be maintained. H_{C1}, a lower critical field, is
   the magnetic field strength at which a material makes the transition from a

state that excludes a magnetic field to a state that a magnetic field can penetrate while the material remains superconducting.

3. *Critical current density* (J_C) is the maximum current density that a superconductor can carry before it loses its superconducting properties.

The three-dimensional graph shown in Figure 2 illustrates how these properties are related. In order for superconductivity to be maintained, the values of temperature, magnetic field, and current density must be within the envelope shown. The location and shape of the envelope depend on the specific superconducting material represented.

The increase in the first parameter, T_C, for the new "high-temperature" superconducting materials caused great excitement, as these materials superconduct well above the boiling point of liquid nitrogen (77 K) and allow relatively inexpensive, easy-to-handle liquid nitrogen to be used as the coolant rather than liquid helium. As actual performance requires operation within the envelope illustrated in Figure 2, actual operation is at only 2/3 to 3/4 of T_C. The ideal of room temperature operation would require T_C to be roughly 400 K.

The second parameter, critical magnetic field, does not appear to pose any serious technical limitations, as the measured values of H_C are higher than the levels required for currently conceivable applications. However, the magnetic properties are highly anisotropic, and it is not yet clear how to manage this characteristic in actual engineering applications.

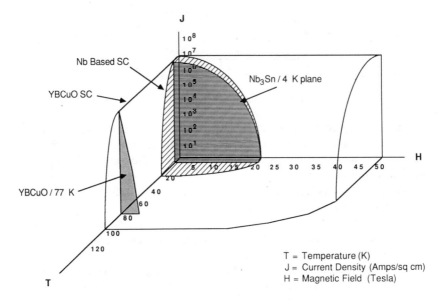

Figure 2 Critical surfaces for superconductivity

The third parameter, critical current density, is the most challenging technological hurdle for many energy applications of superconductivity. Although the new superconductors have achieved very high—and technologically exciting—critical current density values ($>10^7$ A/cm^2) in the form of small single-crystal thin films (up to a few microns thick), the results for polycrystalline "bulk" forms, which need to be developed for use in energy applications, have been less encouraging. Polycrystalline strands or tapes are the most logical route to fabricating a wire. Very recently, the highest-performing superconductors in wire or tape form have achieved 10^5 A/cm^2 at low temperatures (4.2 K), but achieve only far lower current densities at the higher liquid-nitrogen temperatures (about 10^4 A/cm^2)(8).

Although superconductors do not exhibit losses in direct current (dc) operation, when alternating current (ac) travels through superconducting wire, ac losses invariably occur. Because the new superconductors require less refrigeration capacity, higher ac losses can be accepted without reducing a superconducting electrical machine's cost-effectiveness.

Brittleness and other poor mechanical properties of high-temperature superconductors also remain important obstacles to their use in practical devices. Although brittleness is less important in some applications, the conductors used in generators, motors, and electromagnets must be flexible enough to be bent into windings. This problem may be solved by the development of new techniques for fabricating thin fibers, or by the development of very thin superconductors (less than 25 micrometers) that may offer the flexibility necessary for most applications.

INCENTIVES FOR DEVELOPING HIGH-TEMPERATURE SUPERCONDUCTORS

Potential Energy Savings

In 1988, the electric energy sales in the United States totalled 2566×10^9 kWh. In many sectors of the economy, there exist significant opportunities to save energy through improved efficiency. Using high-temperature superconductors in motor applications is a leading example: Motors account for about two-thirds of the annual US electricity use, with motors sized 124 hp and above accounting for half that consumption (9) (see Figure 3).

The transportation sector accounts for about 27% of the total US energy consumption, and petroleum provides 97% of the energy used in this sector. Applying superconductivity to use electricity more efficiently and extensively in transportation—for example, with magnetic levitation train technology that provides sufficient speed (>500 km/hr) to compete with shorthaul aircraft—could help reduce overall oil consumption.

Superconductivity offers several advantages. Eliminating the resistive loads in the conductors of electrical equipment will have a significant impact

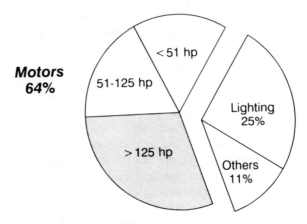

Figure 3   US electricity use—8.8 quads in 1988

on electric energy conservation. Superconductors also have advantages in the production of magnetic fields substantially greater than the normal 1–2 Tesla achievable with copper wire and iron structures. The ability to apply such fields would allow elimination of the iron core in devices such as motors, generators, transformers, and other magnet applications. This leads to reduced size and weight, in addition to increased efficiency due to the reduction of resistance and magnetic losses. If the operating magnetic field can be increased into the vicinity of 4 to 6 Tesla, then the volume of the electrical machine can be reduced by more than a factor of two. (Energy density is proportional to the square of the magnetic field strength.)

Reduced Cooling Requirements

One of the more important practical features of the new high-temperature superconductors is that the critical temperature of these materials is above the boiling point of liquid nitrogen. Many of the potential applications of superconductors that are impractical when cryogenic helium is required as a refrigerant may become feasible if liquid nitrogen is used. Both the availability and physical properties of nitrogen give it a significant advantage over helium.

To understand the importance of the higher critical temperature, it is first sufficient to recall that maximal achievable refrigeration efficiency for a heat engine is given by the Carnot efficiency (10). At cryogenic (liquid helium) temperatures the refrigeration efficiency goes to zero with T while the actual practical efficiency falls faster and is typically less than 20% of the Carnot value.

Operation of any superconducting machine at a higher temperature requires less refrigeration energy than operation at a lower temperature. The reduction in refrigeration energy is roughly proportional to the increase in temperature. For example, operation at 20 K requires five times less energy than at 4 K, and operation at 80 K would require nearly 100 times less energy. As a result, insulation can be less costly and more compact at the higher temperatures. Lower insulation requirements, along with less costly refrigeration systems, result in lower capital costs for higher operating temperatures. In general, the size at which a particular device becomes economic depends on how much these capital costs can be reduced.

In the sections that follow, the particular issues relevant to each of the major applications are discussed. The applications to electric systems are covered first, followed by sections describing six major classes of industrial and transportation applications. The interested reader is referred for more detail to several recent extensive reviews of the various power-system applications (11–13) and to the results of a major study by the US Department of Energy (DOE) and the Electric Power Research Institute (EPRI) of the end-use energy applications (14, 15), which forms the basis for a large fraction of this article.

POWER SYSTEM APPLICATIONS

The electric utility industry may eventually make widespread use of superconductors. This industry has researched superconducting transmission cables, generators, and energy storage for more than two decades, and the latest developments have renewed interest in superconductors. In looking toward power system applications of superconductivity, it is important to understand that the electrical efficiency of the modern power system is already quite high (>90%). Electrical losses that do occur have been carefully traded off against capital cost of the equipment. Rough estimates of the losses in a power system are:

Generator and transformer	<1%
Transmission	<2%
Distribution transformer	<1%
Distribution lines	<5%
Total	<8%

Once the electricity is delivered to a customer, the local internal wiring, typically at rather low voltages, can cause additional losses; however, overall electrical efficiency is clearly quite high.

Within a power station considerable electric energy is used, mainly in motors to drive the pumps and fans that blow air into the boiler, pump feedwater, cool water, and drive fuel handling and pollution control equipment. This internal use of electric energy depends on the nature of the power plant; it can approach 10% of the plant output. This usage is not included in the above numbers, but is properly incorporated in the power plant efficiency (heat rate).

Despite the high efficiency of the power system, the potential advantage from the use of superconductivity in the power system is substantial, because even a small percentage improvement is multiplied by large total energy costs. For example, a 1% improvement is worth more than $1 billion annually in the United States.

Transmission

Delivery of electric energy is accomplished through the transmission and distribution system. In current usage, *transmission* lines carry hundreds to thousands of megawatts at voltages typically above 100,000 volts. *Distribution* lines deliver the electricity from transmission system substations, where voltage is stepped down, to the end user. The basic relationship for electrical losses is i^2R while the power delivered is iV where i is the current, R is the resistance, and V is the voltage. If high power is to be transmitted great distances, then it is best to use high voltages so that current can be low. Normally, transmission is either overhead or underground. Typically, installation of underground transmission costs two to five times as much as overhead transmission. Additionally, as a result of specific details of the electrical properties of underground cable, overhead lines can normally carry more power over a particular right of way.

Superconducting transmission lines that can carry electricity large distances without resistive losses are one of the most promising potential applications for new superconductors. Transmission lines require relatively low current densities compared to other superconductor applications. Moreover, the flexibility requirement needed for wires wound into magnets can be relaxed because transmission cable can be installed underground in rigid or semi-rigid individual sections. However, only future research can determine if the enormous length and surface area of a superconducting transmission cable can be cooled and insulated economically. One approach is illustrated in Figure 4 (16).

Energy Storage

The concept of using an extremely large superconducting magnet to store electric energy for later use by an electric utility in meeting the demand for electricity during peak periods was developed in the early 1970s (17). Research on this concept continued at a modest level until it was recognized that such a magnet could also serve as a pulse power supply for "star wars" type

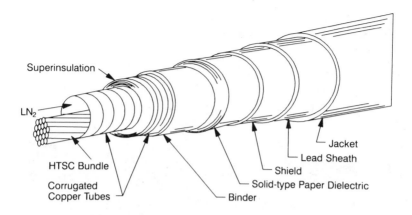

Figure 4   Schematic of liquid nitrogen–cooled transmission line

weapon systems—at which time funding increased dramatically. As a result of this expanded work, a 20-MWh superconducting magnetic energy storage system (SMES) may be constructed in the United States in the early 1990s using conventional NbTi superconducting wire.

In SMES, the energy is stored in the magnetic field of an extremely large solenoid in a subsurface structure, which allows a portion of the mechanical stress from the magnetic field to be taken up by the earth itself. The use of civil structures lowers the costs of the system. The SMES plant, shown in Figure 5, would be several hundred meters in diameter with 50,000–200,000 amperes of current running through a large solenoid (18).

Overall SMES efficiency from an ac input to ac output can be more than 90%, compared with pumped hydropower's 67–74%. Although low-temperature superconductors have been shown to be technically feasible for this very large-scale application, high-temperature superconductors will offer an advantage in small-scale systems where earthen reinforcement of the mechanical structure is not provided. Small-scale units storing perhaps 1–100 megawatt-seconds would serve not to meet utility energy demands but rather to ensure power quality or very short-duration peak supplies for industrial or transportation applications.

Generators

Electricity starts at a generator: two interacting electromagnets, one stationary and one rotating, that convert the mechanical energy of the generating station into electricity.

Generators with rotors wound from superconducting wire of adequate current-carrying capacity can produce magnetic fields much higher than those

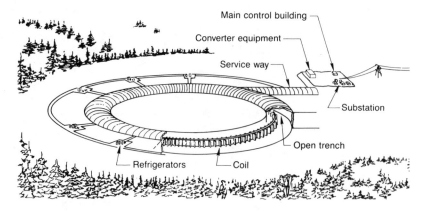

Figure 5 Superconducting magnetic energy storage (SMES) plant

produced by generators with copper windings. As shown in Figure 6, superconducting generators would thus be perhaps only half the size of ordinary generators and potentially less costly. This significant reduction in physical size is possible because of high current density and magnetic field strengths two to three times what is cost-effective in conventional nonsuperconducting designs based on copper and iron. Achievement of this size reduction offers the potential of making a generator from superconductors at a lower capital cost than that of a generator made from conventional materials.

Studies in the 1970s led to testing in the 1980s of prototype superconducting generators using NbTi superconductors (19–22). While the efforts in the United States have not led to full-scale application, designs were developed

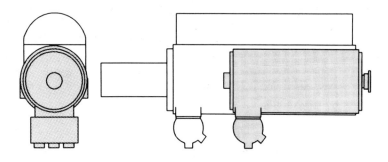

1200 MVA Electrical Rating
3600 RPM

Figure 6 Relative size of superconducting generator (shaded) vs conventional

that showed that application of the low-temperature superconductors is feasible and that significant operating and performance advantages exist for large generators (>300 MW).

MOTORS

None of the possible applications for superconductors would have more widespread impact on electricity use than electric motors, which account for about two-thirds of all electrical energy used in the United States (23). As with generators, superconducting windings could make it possible to increase significantly the magnetic field in the motor's air gap (and thus boost power density) without necessarily adding to the motor's weight and volume. Figure 7 illustrates how the reduction of losses from conventional windings could make superconducting motors extremely efficient.

The elimination of iron from the motor design not only eliminates iron losses, but also allows the use of substantially higher magnetic fields and a reduction in size and weight for a given performance rating. Also, the elimination of iron results in less variation of motor efficiency with load.

The potential size and weight reduction may be sufficient to allow development of a superconducting motor that is cheaper than a comparable conventional motor. Early projections of the advantage of small and compact generators and motors in ship propulsion systems, for example, may become reality once higher-temperature superconductors become available.

Earlier efforts indicate that there is no conceptual difficulty in applying

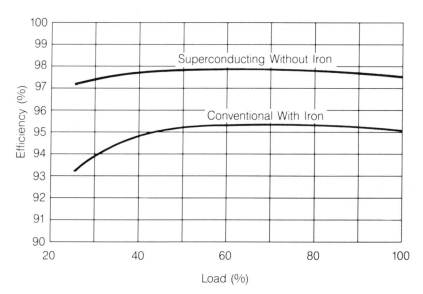

Figure 7 Load-efficiency curves for superconducting vs conventional motor

superconducting technology to any motors. The problems that limit the direct substitution of the older low-temperature superconductors for conventional conductors in motor applications— associated with ac losses, mechanical integrity of superconductor and cooling systems, the need for helium cooling and associated high cost, and electromechanical stability in the motor during startup and load variation—have been addressed in the development of prototype generators using low-temperature superconductors, and there is no reason to believe that they cannot also be resolved at higher temperatures.

ELECTROMAGNETIC PUMPING

Electromagnetic (EM) pumps operate by using the reaction between a magnetic field and a current passing through an electrically conducting fluid (24), as illustrated in Figure 8. EM pumps have been employed as an alternative to mechanical pumps. Because EM pumps do not require seals and have no moving parts in contact with the fluid, they are more reliable than mechanical pumps, especially when handling liquid metals. The pumps can also be oriented in any position.

Today's mechanical pumps generally have efficiencies of 80–85%, whereas EM pumps have efficiencies of 40–50%. The current use of EM pumps is limited to pumping electrically conducting fluids, such as aluminum, sodium, or lithium, which present special problems for conventional pumps. Many types of EM pumps have been designed and built, but none to date employs superconductors.

For pumping liquid metals, the use of superconducting windings would result in a reduction of pump size (and mass) and an increase in pumping efficiency compared to an EM pump with conventional windings. The pump efficiency can be improved from about 44% for the conventional EM pump to 68% for the EM pump with a superconducting winding. The efficiency increases with the magnetic field strength as indicated in Figure 9, taken from

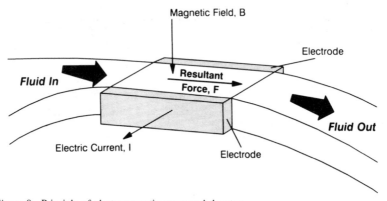

Figure 8   Principle of electromagnetic pump and thruster

a study (24a) of EM ship thrusters in which the EM pump substitutes for a ship propeller. Only the resistive losses were considered in this comparison and were set to zero for the superconducting winding. The analysis did not include the cryogenic refrigeration system for cooling of the rotor.

Superconducting EM pumps could find an economical application in transporting nonmetallic fluids that are difficult to handle because of their radioactivity, chemical reactivity, or corrosiveness and abrasiveness (e.g.

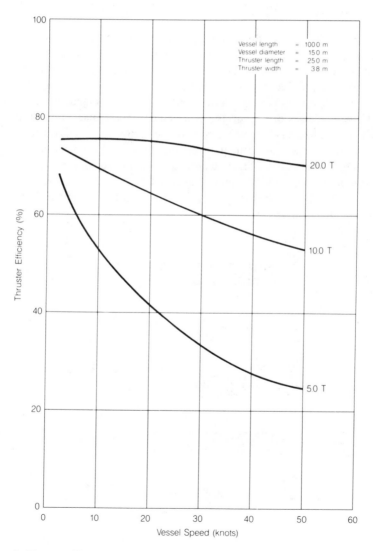

Figure 9 Thruster efficiency vs vessel speed for various field strengths

abrasive slurries). In these cases, the improved reliability of an EM pump could more than offset its low efficiency relative to a mechanical pump. However, if such a fluid has a relatively low electrical conductivity (less than that of sea water), the potential benefits gained by using superconducting EM pumps appear to be marginal. In closed systems it may be possible to improve performance by "doping" the primary fluid, but this process may not always be feasible.

MAGNETIC HEAT PUMPS AND REFRIGERATION

In the early part of this century, it was discovered that when certain paramagnetic or ferromagnetic materials were subjected to an intense magnetic field, alignment of their magnetic moments caused the materials to warm (25). Conversely, removal of the magnetic field caused cooling. This behavior is called the magnetocaloric effect (MCE) (see Figure 10) and forms the basis of a class of Carnot-like heat engines called magnetic heat pumps (MHPs). The MCE is highest for certain rare-earth compounds operated near their Curie temperature—the temperature at which a material loses its ferromagnetic properties. Material properties limit the achievable temperature changes in a single state to approximately 20 K, which requires roughly 10 Tesla (or 2 K/Tesla). Fields approaching 10 Tesla can be obtained by using superconducting magnets.

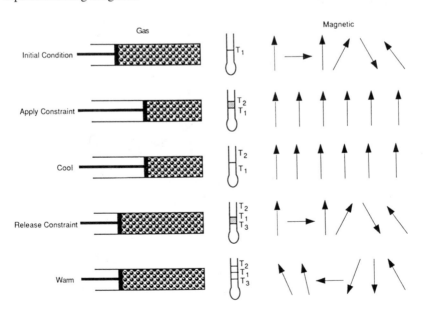

Figure 10   How the magnetocaloric effect relates to the steps of the Carnot Cycle

The magnetic heat pump is directly analogous to a conventional vapor-compression heat pump/refrigerator; the magnetic field corresponds to pressure and the magnetization corresponds to volume per unit mass. Manipulation of the magnetic field can be accomplished in one of three ways: 1. raising or lowering the magnetic field, 2. moving the magnetic material relative to the magnet, or 3. moving a shield between the magnetic material and the magnet.

Magnetic heat pumps have been used routinely in the laboratory during the past 50 years to provide small amounts of refrigeration at extremely low temperatures (under 1 K). Most MHP development efforts to date are for cryogenic refrigeration at temperatures below 20 K, with cooling powers of a few watts. For commercial interest, the MHP devices need to provide refrigeration and heat pumping around room temperature with capacities of tens to hundreds of kilowatts. A number of experimental devices have been built using gadolinium as a working material (293 K Curie temperature).

Economic comparisons were made between a conventional vapor-compression system and various MHPs to provide 50 kW of cooling at 0° F and 500 kW of heating at 240° F. The capital costs of the MHPs were estimated to be 2–3 times those of the conventional system for cooling and 3–4 times the conventional system's costs for heating. However, because of improved efficiency, the estimated operating costs (excluding cooling costs for the magnet) for the MHP are lower.

Gadolinium represents a significant cost item for the MHP, and it would be better to develop other materials. Designs and prototypes of MHPs suitable for a large market have only been conceptualized, and their feasibility remains to be determined.

MATERIALS PROCESSING

There are significant potential applications for superconductors in materials processing that utilize either modest magnetic fields of large volume or intense magnetic fields of small volume (26).

Magnetic Processing

Magnetic processing appears to be of potential value to a wide variety of material-driven industries: electronics, telecommunications, aerospace, automotive, etc. Magnetic fields can be used to manipulate the nature of growing crystals, the orientation of components such as crystals or fibers within a system, and the distribution of constituents within a system, such as dopants in semiconductors, conducting polymers, and glasses. This is best illustrated in the case of magnetic fields applied to the Czochralski process (27) for producing high-purity silicon shown in Figure 11. The application of the magnetic field reduces the thermal convection in the melt and results in a

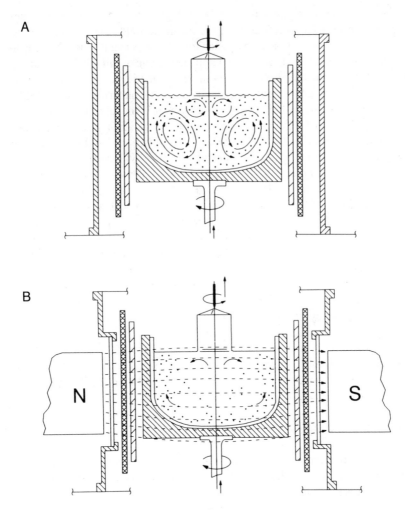

Figure 11   Magnetic processing of silicon without magnet (*A*) and with magnet (*B*)

higher-quality silicon. The generation of the large-volume magnetic fields required for material processing is only practical and economic with superconductivity. Large-volume conventional magnets can be built—but their energy consumption is very high.

Magnetic Separation

Magnetic separation is the use of magnetic forces to separate materials of various compositions on the basis of their magnetic properties (28). Magnetic separation is important in terms of economy, efficiency, and energy productivity for the processing of ferrous ores/metals and kaolin clay, and the recovery of magnetic materials from various types of solid waste and scrap. The use of superconductors can lead to substantial energy savings, which has led to the use of very large-scale superconducting magnetic operations in kaolin clay processing. Use in this application—which became commercial in 1986—is gradually expanding.

An important observation can be made that relates to the use of higher-temperature superconducting materials for magnetic separation. In the growing kaolin industry, liquid helium superconducting magnetic separators have been found to be reliable and economically superior to resistive electromagnetic separators on a substantial industrial scale. It follows that if higher-temperature superconducting separators (such as those that might be cooled with liquid nitrogen) can be made with comparable reliability and capital cost, then they would find a niche for substantial industrial application. By analogy, higher-temperature superconducting separators could displace resistive electromagnetic separators in other existing industrial applications and in new applications that are being developed.

Magnetic Fabrication

The availability of practical superconductors at liquid nitrogen temperatures can lead to discovery and development of new methods of fabrication involving very high magnetic fields. However, this area of high-field magnetic processing is largely unexplored (29). Important information concerning the technical barriers could be obtained by experiments using low-temperature superconductors with fields of 10–12 Tesla. This is pressure sufficient to deform a rod or tub of material of low-flow stress such as lead or tin. Experiments with these materials and with low-temperature superconducting solenoids (illustrated in Figure 12) are needed to clarify the pressure/

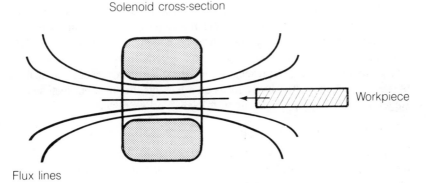

Solenoid cross-section

Workpiece

Flux lines

Figure 12 Solenoid magnet used to replace the die in wire drawing

deformation relationships as a function of solenoid inner diameter/workpiece diameter. They could also clarify the relationship between workpiece velocity and the rate of penetration of flux lines into the workpiece surface. Studies of the reactions of the solenoid to the magnetic pressure variations would provide information in two areas: the reaction of the solenoid body itself to the imposed forces, and the electrical reaction to the increased magnetic field as the workpiece penetrates the flux lines. Successful development of high-temperature superconducting high-field magnets will provide the motivation to explore new processes that are not currently practical.

TRANSPORTATION

Magnetic Levitation

Magnetic levitation (MAGLEV) vehicle technology is perhaps the most developed of all transportation applications of superconductors (30, 31). The Japanese repulsive-force-levitated vehicle, originally developed under the auspices of the Japanese National Railway, uses low-temperature supercon-ductors. High-temperature superconductors should be regarded as enhancing rather than enabling this technology. The use of superconductivity allows the MAGLEV vehicle to operate with a large space gap (10–15 cm) between the vehicle base and the guideway. With such a large gap, the vehicles can travel at speeds of 500 km/h or more without excessively tight tolerances on guideway alignment.

Use of high-temperature superconductors would reduce the cost and com-plexity of the cryogenic system considerably—as well as save operating energy and mass—and could significantly increase the reliability of on-board refrigeration systems. For a typical vehicle route, the energy saving is poten-tially on the order of 5–15%. However, the superconducting magnets represent only a fraction of the overall MAGLEV cost, and their use is predicated more on the costs of the right of way and the guideway. For the MAGLEV system to be economic, high traffic density is required. Recent investigations of MAGLEV (31) have identified it as suitable primarily for the mass transit niche served by short-distance commuter aircraft of 100–400 miles and linked to the existing air transport system. In the United States, this transportation segment represents about three quads of energy annually. Initial application will probably focus on a few intercity markets such as Los Angeles to Las Vegas, Washington to New York, etc.

Electromagnetic Ship Thruster

Electromagnetic thruster systems for ship propulsion have been investigated previously, and small models have been built to demonstrate the concept (32). In the EM thruster, an electrical current is passed directly through the sea water and interacts with an applied magnetic field to propel the vessel forward. The design of the EM thruster involves considerations similar to those associated with the construction of a dc conduction EM pump. (See Figures 8 and 9.)

POWER ELECTRONICS

Since the advent of modern power semiconductor devices, such as the transistors and thyristors introduced 30 years ago, power electronics have played an increasing role in the control of electric power devices. Two of the most common applications of power electronic devices are in motor drives and power supplies.

The major potential energy savings associated with superconducting power electronics devices will be in motor controllers. Any superconducting motor will require power electronic controllers. Motor controllers that use semiconductor devices consume about 2% of the energy used by the motor, whereas motor controllers using superconducting "transistors" would, in principle, be more efficient. However, the energy savings come largely from the better match between power delivered by the variable-speed motor and the system requirements for energy (33). (See Figure 13.)

SUMMARY

From this survey and assessment of applications, it is clear that significant opportunities for energy applications of superconductors exist today and that

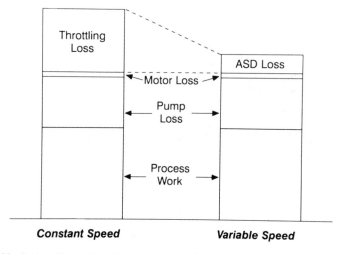

Figure 13   Savings from adjustable speed control of motors

the range of applications will greatly increase when new superconductors become practical.

Technology Readiness

Several energy applications are entering the marketplace today that use conventional low-temperature superconductors. In particular, low-temperature superconductors in magnetic separation, magnetic levitated trains, and magnetic processing of materials are establishing a record of success in both lowering energy costs and achieving technical results not readily accomplished without superconductivity.

Although further development of these technologies will occur even without the new superconductors, the new materials can certainly expand the range of economic application. Moreover, these applications represent a ready market for the new superconductors. Most of the long-term applications identified in Table 1 are limited by the critical current density values of today's superconductors. Performance targets (34) for these applications are indicated in Figure 14.

Economically Achievable Conservation Potential

Three of the superconductor technologies that hold the greatest potential for reducing energy waste are motors, magnetic heat pumps, and MAGLEV vehicles. Almost four quads per year are wasted in the United States because of the inefficiencies of the motors, heat pumps, and transport vehicles that could be displaced by superconductor technologies. Although superconducting technologies promise higher efficiencies, they will not be 100% efficient. If high-temperature superconductor technologies were to penetrate these three

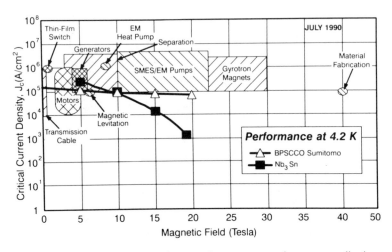

Figure 14   High-temperature superconductor performance targets for energy applications

Table 1 Technology readiness of superconductors

Near term (0-10 years)	Long term (10+ years)
Low-temperature	High-temperature
Medical diagnosis	Underground transmission lines
Magnetic separation	Motors
Magnetically levitated trains	Materials fabrication
Generators	Ship & railroad propulsion
Superconducting magnetic	Electromagnetic pumping
energy storage (SMES)	Magnetic heat pumps
	Supercomputers
High temperature	
Electronics and sensors	

applications completely, approximately two-thirds of this energy waste would be eliminated.

Assuming that these superconducting technologies can be commercially developed to meet the performance requirements of the applications, they would have a limited penetration in the market because of cost considerations. The primary energy savings that would likely result from commercialization of superconductor technologies cooled by liquid nitrogen in these three applications is about one-half quad or about 20% of what is technically feasible (35) (see Figure 15). A more optimistic view of the prospects of

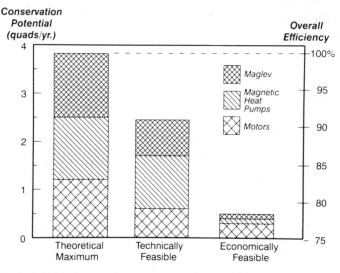

Figure 15 Technical and economic energy-conservation potential

efficiency improvements can, however, be constructed. Taking into account adoption of adjustable-speed drive technology for the superconducting motor applications (increasing the motor efficiency savings from perhaps 2–5% to as much as 25%) and allowing for further developments in MAGLEV (making it viable for replacing not only short-haul aircraft, but also intercity truck transport), perhaps the equivalent of a 25% savings could be achieved.

Although it is still too early to make reliable market forecasts, the successful development of the new higher-temperature superconductors is certain to affect electric energy use in many different ways. Direct substitution in current electrical apparatus will reduce energy use. New applications, typically utilizing magnetic fields, will add further value to the use of electric energy. The magnitude of the impact of this remarkable discovery will depend on our ability to translate this scientific breakthrough into engineering practicality.

ACKNOWLEDGMENTS

The author would like to acknowledge the large numbers of scientists and engineers who assisted in the DOE-EPRI project (36), which led to many of the results described in the article. More than 25 individuals at four national laboratories and several nongovernment laboratories and consultants performed the work. An outside advisory team provided constructive comments throughout the project. Thanks go to Drs. Brogan, Eberhardt, and Wolf of US DOE. Their support of major teams at Argonne National Laboratory, Idaho National Engineering Laboratory, and Lawrence Berkeley Laboratory made this project possible. Special thanks go to Dr. Steiner Dale of Oak Ridge National Laboratory, who was the overall project manager for DOE and EPRI. The interpretation of the results of this project and other references cited here are those of the author.

Literature Cited

1. Schurr, S. et al. 1991. Electricity in the American economy. *Contrib., Econ. Econ. Hist.* 11
2. Onnes, H. K. 1911. *Leiden Comm.* 120b, 122b, 124c
3. Bednorz, J. C., Muller, K. A. 1986. Possible high-T_C superconductivity in the Ba-La-Cu-O system. *Z. Phys. B* 64:189
4. Wu, M. K. et al. 1987. Superconductivity at 90 K in a new mixed phase Y-Ba-Cu-O compound system at ambient pressure. *Phys. Rev. Lett.* 58:908
5. Sheng, Z. Z., Hermann, A. M. 1988. Bulk superconductivity at 120 K in the Tl-Ca/Ba-Cu-O system. *Nature* 332: 128
6. Ono, A. et al. 1988. A 105 K superconducting phase in the Bi-Sr-Ca-Cu-O system. *Jpn. J. Appl. Phys.* 27:L1041.
7. Rose-Innes, A. C., Rhoderic, E. H. 1978. *Introduction to Superconductivity*, p. 204. Pergamon. 2nd ed.
8. Larbalestier, D. 1990. Critical currents pinned down. *Nature* 343:210–11
9. US Dept. Energy, Div. Buildings Community Systems. 1975. *Energy Efficiency and Electric Motors*. GPO, PCP/M50217–01. Reprinted in 1978
10. Refrigeration systems and applications. 1990. *ASHRAE Handbook*, Ch. 38
11. Sharma, D. K. et al. 1988. *Proc. EPRI Workshop on High-Temperature Super-*

conductivity. *EL/ER-5894P-SR*. Electr. Power Res. Inst. and Natl. Sci. Found.

12. Sharma, D. K. et al. 1989. *Proc: 1988 Conf. Electrical Applications of Superconductivity. EL-63325-D*. Electr. Power Res. Inst.

13. Assessing the impact of high-temperature superconductivity on electric power. 1988. *Proc. Experts' Meet*. Tokyo: Int. Energy Agency

14. Dale, S. J., Wolf, S. M., Schneider, T. R., eds. 1990. *Energy Applications of High-Temperature Superconductivity*, Vol. 1. US Dept. Energy and EPRI

15. Dale, S. J., Wolf, S. M., Schneider, T. R., eds. 1990. *Energy Applications of High-Temperature Superconductivity*, Vol. 2. US Dept. Energy and EPRI

16. Englehart, J. 1990. Private communication

17. Boom, R. W., Peterson, H. A. 1972. Superconductive energy storage for power systems. *IEEE Trans. Magn.*, pp. 701–3

18. Hassenzahl, W. V., Schainker, R. B., Peterson, T. M. 1990. Superconducting magnetic energy storage for utility applications. *CIGRE*, Section 1.1

19. Electr. Power Res. Inst. 1976. *Superconductors in Large Synchronous Machines TD-255*

20. Electr. Power Res. Inst. 1977. *Superconducting Generator Design EL-577*

21. Electr. Power Res. Inst. 1978. *Superconducting Generator Design ER-663*

22. Edmonds, J. S. 1979. Superconducting generator technology. *IEEE Trans. Magn.*, MAG.-15(1):373–79

23. US Dept. Energy. 1975. See Ref. 9

24. Dale, S. J., Wolf, S. M., Schneider, T. R., eds. 1990. See Ref. 15, Chap. 6

24a. Doragh R. A. 1973. Magnetohydrodynamic ship propulsion using superconducting magnets. *Trans. Soc. Nav. Architect. Mar. Eng.* 71:370

25. Dale, S. J., Wolf, S. M., Schneider, T. R., eds. 1990. See Ref. 15, Chap. 7

26. Weiss, P., Piccard, A. 1918. A new thermomagnetic phenomenon. *C R L'Acad. Sci.* 166:352–54

27. Ramachandran, P. A., Dudukovic, M. P. 1991. *Modeling of Czochralski Crystal Growth. RP8000–1*. Electr. Power Res. Inst.

28. Dale, S. J., Wolf, S. M., Schneider, T. R., eds. 1990. See Ref. 15, Chap. 10

29. Dale, S. J., Wolf, S. M., Schneider, T. R., eds. 1990. See Ref. 15, Chap. 9

30. Dale, S. J., Wolf, S. M., Schneider, T. R., eds. 1990. See Ref. 15, Chap. 8

31. Johnson, L. R., et al. 1989. *Maglev Vehicles and Superconductor Technology: Integration of High-Speed Ground Transportion into the Air Travel System. ANL/CNSV-67*. Argonne Natl. Lab., Cent. Transp. Res.

32. Dale, S. J., Wolf, S. M., Schneider, T. R., eds. 1990. See Ref. 15, Chap. 5

33. Poole, J. N. et al. 1990. *Commercial and Industrial Applications of Adjustable-speed Drives. CU-6883*. Electr. Power Res. Ins.

34. Dale, S. J., Wolf, S. M., Schneider, T. R., eds. 1990. See Ref. 14, page xx

35. Dale, S. J., Wolf, S. M., Schneider, T. R., eds. 1990. See Ref. 15, Chap. 11

36. Dale, S. J., Wolf, S. M., Schneider, T. R., eds. 1990. See Refs. 15 and 16

SUBJECT INDEX

557

CUMULATIVE INDEXES

CONTRIBUTING AUTHORS, VOLUMES 1–16

CHAPTER TITLES, VOLUMES 1–16

RISKS AND IMPACTS OF ENERGY PRODUCTION AND USE

ANNUAL REVIEWS INC.

A NONPROFIT SCIENTIFIC PUBLISHER

4139 El Camino Way
P.O. Box 10139
Palo Alto, CA 94303-0897 • USA

Annual Reviews Inc. publications may be ordered directly from our office; through booksellers and subscription agents, worldwide; and through participating professional societies. Prices subject to change without notice.   ARI Federal I.D. #94-1156476

- **Individuals:** Prepayment required on new accounts by check or money order (in U.S. dollars, check drawn on U.S. bank) or charge to credit card — American Express, VISA, MasterCard.
- **Institutional buyers:** Please include purchase order.
- **Students:** $10.00 discount from retail price, per volume. Prepayment required. Proof of student status must be provided (photocopy of student I.D. or signature of department secretary is acceptable). Students must send orders direct to Annual Reviews. Orders received through bookstores and institutions requesting student rates will be returned. You may order at the Student Rate for a maximum of 3 years.
- **Professional Society Members:** Members of professional societies that have a contractual arrangement with Annual Reviews may order books through their society at a reduced rate. Check with your society for information.
- **Toll Free Telephone orders:** Call 1-800-523-8635 (except from California) for orders paid by credit card or purchase order and customer service calls only. California customers and all other business calls use 415-493-4400 (not toll free). Hours: 8:00 AM to 4:00 PM, Monday-Friday, Pacific Time. **Written confirmation** is required on purchase orders from universities before shipment.
- **FAX: 415-855-9815   Telex: 910-290-0275**
- **We do not ship on approval.**

Regular orders: Please list below the volumes you wish to order by volume number.
Standing orders: New volume in the series will be sent to you automatically each year upon publication. Cancellation may be made at any time. Please indicate volume number to begin standing order.
Prepublication orders: Volumes not yet published will be shipped in month and year indicated.
California orders: Add applicable sales tax. **Canada:** Add GST tax.
Postage paid (4th class bookrate/surface mail) by **Annual Reviews Inc.** UPS domestic ground service available (except Alaska and Hawaii) at $2.00 extra per book. Airmail postage or UPS air service also available at prevailing costs. UPS must have street address. P.O. Box, APO or FPO not acceptable.

ANNUAL REVIEWS SERIES		Prices postpaid, per volume USA & Canada / elsewhere		Regular Order Please Send	Standing Order Begin With
		Until 12-31-90	After 1-1-91	Vol. Number:	Vol. Number:
Annual Review of ANTHROPOLOGY					
Vols. 1-16	(1972-1987)	$31.00/$35.00	$33.00/$38.00		
Vols. 17-18	(1988-1989)	$35.00/$39.00	$37.00/$42.00		
Vol. 19	(1990)	$39.00/$43.00	$41.00/$46.00		
Vol. 20	(avail. Oct. 1991)	$41.00/$46.00	$41.00/$46.00	Vol(s). _____	Vol. _____
Annual Review of ASTRONOMY AND ASTROPHYSICS					
Vols. 1, 5-14	(1963, 1967-1976)				
16-20	(1978-1982)	$31.00/$35.00	$33.00/$38.00		
Vols. 21-27	(1983-1989)	$47.00/$51.00	$49.00/$54.00		
Vol. 28	(1990)	$51.00/$55.00	$53.00/$58.00		
Vol. 29	(avail. Sept. 1991)	$53.00/$58.00	$53.00/$58.00	Vol(s). _____	Vol. _____
Annual Review of BIOCHEMISTRY					
Vols. 30-34, 36-56	(1961-1965, 1967-1987)	$33.00/$37.00	$35.00/$40.00		
Vols. 57-58	(1988-1989)	$35.00/$39.00	$37.00/$42.00		
Vol. 59	(1990)	$39.00/$44.00	$41.00/$47.00		
Vol. 60	(avail. July 1991)	$41.00/$47.00	$41.00/$47.00	Vol(s). _____	Vol. _____
Annual Review of BIOPHYSICS AND BIOPHYSICAL CHEMISTRY					
Vols. 1-11	(1972-1982)	$31.00/$35.00	$33.00/$38.00		
Vols. 12-18	(1983-1989)	$49.00/$53.00	$51.00/$56.00		
Vol. 19	(1990)	$53.00/$57.00	$55.00/$60.00		
Vol. 20	(avail. June 1991)	$55.00/$60.00	$55.00/$60.00	Vol(s). _____	Vol. _____

ANNUAL REVIEWS SERIES	Prices postpaid, per volume USA & Canada / elsewhere		Regular Order Please Send Vol. Number:	Standing Order Begin With Vol. Number:
	Until 12-31-90	After 1-1-91		

Annual Review of CELL BIOLOGY

Vols. 1-3	(1985-1987) .... $31.00/$35.00	$33.00/$38.00		
Vols. 4-5	(1988-1989) .... $35.00/$39.00	$37.00/$42.00		
Vol. 6	(1990) .... $39.00/$43.00	$41.00/$46.00		
Vol. 7	(avail. Nov. 1991) ...... $41.00/$46.00	$41.00/$46.00	Vol(s). _____	Vol. _____

Annual Review of COMPUTER SCIENCE

Vols. 1-2	(1986-1987) .... $39.00/$43.00	$41.00/$46.00		
Vols. 3-4	(1988, 1989-1990) ... $45.00/$49.00	$47.00/$52.00	Vol(s). _____	Vol. _____

Series suspended until further notice. SPECIAL OFFER: Volumes 1-4 are available at the special promotional price of $100.00 USA & Canada / $115.00 elsewhere, when all 4 volumes are purchased at one time. Orders at the special price must be prepaid.

Annual Review of EARTH AND PLANETARY SCIENCES

Vols. 1-10	(1973-1982) .... $31.00/$35.00	$33.00/$38.00		
Vols. 11-17	(1983-1989) .... $49.00/$53.00	$51.00/$56.00		
Vol. 18	(1990) .... $53.00/$57.00	$55.00/$60.00		
Vol. 19	(avail. May 1991) ...... $55.00/$60.00	$55.00/$60.00	Vol(s). _____	Vol. _____

Annual Review of ECOLOGY AND SYSTEMATICS

Vols. 2-18	(1971-1987) .... $31.00/$35.00	$33.00/$38.00		
Vols. 19-20	(1988-1989) .... $34.00/$38.00	$36.00/$41.00		
Vol. 21	(1990) .... $38.00/$42.00	$40.00/$45.00		
Vol. 22	(avail. Nov. 1991) ...... $40.00/$45.00	$40.00/$45.00	Vol(s). _____	Vol. _____

Annual Review of ENERGY

Vols. 1-7	(1976-1982) .... $31.00/$35.00	$33.00/$38.00		
Vols. 8-14	(1983-1989) .... $58.00/$62.00	$60.00/$65.00		
Vol. 15	(1990) .... $62.00/$66.00	$64.00/$69.00		
Vol. 16	(avail. Oct. 1991) ...... $64.00/$69.00	$64.00/$69.00	Vol(s). _____	Vol. _____

Annual Review of ENTOMOLOGY

Vols. 10-16, 18	(1965-1971, 1973)			
20-32	(1975-1987) .... $31.00/$35.00	$33.00/$38.00		
Vols. 33-34	(1988-1989) .... $34.00/$38.00	$36.00/$41.00		
Vol. 35	(1990) .... $38.00/$42.00	$40.00/$45.00		
Vol. 36	(avail. Jan. 1991) ...... $40.00/$45.00	$40.00/$45.00	Vol(s). _____	Vol. _____

Annual Review of FLUID MECHANICS

Vols. 2-4, 7	(1970-1972, 1975)			
9-19	(1977-1987) .... $32.00/$36.00	$34.00/$39.00		
Vols. 20-21	(1988-1989) .... $34.00/$38.00	$36.00/$41.00		
Vol. 22	(1990) .... $38.00/$42.00	$40.00/$45.00		
Vol. 23	(avail. Jan. 1991) ...... $40.00/$45.00	$40.00/$45.00	Vol(s). _____	Vol. _____

Annual Review of GENETICS

Vols. 1-21	(1967-1987) .... $31.00/$35.00	$33.00/$38.00		
Vols. 22-23	(1988-1989) .... $34.00/$38.00	$36.00/$41.00		
Vol. 24	(1990) .... $38.00/$42.00	$40.00/$45.00		
Vol. 25	(avail. Dec. 1991) ...... $40.00/$45.00	$40.00/$45.00	Vol(s). _____	Vol. _____

Annual Review of IMMUNOLOGY

Vols. 1-5	(1983-1987) .... $31.00/$35.00	$33.00/$38.00		
Vols. 6-7	(1988-1989) .... $34.00/$38.00	$36.00/$41.00		
Vol. 8	(1990) .... $38.00/$42.00	$40.00/$45.00		
Vol. 9	(avail. April 1991) ...... $41.00/$46.00	$41.00/$46.00	Vol(s). _____	Vol. _____

Annual Review of MATERIALS SCIENCE

Vols. 1, 3-12	(1971, 1973-1982) ...... $31.00/$35.00	$33.00/$38.00		
Vols. 13-19	(1983-1989) .... $66.00/$70.00	$68.00/$73.00		
Vol. 20	(1990) .... $70.00/$74.00	$72.00/$77.00		
Vol. 21	(avail. Aug. 1991) ...... $72.00/$77.00	$72.00/$77.00	Vol(s). _____	Vol. _____

<table>
<tr><td rowspan="3">**ANNUAL REVIEWS SERIES**</td><td colspan="2">Prices postpaid, per volume
USA & Canada / elsewhere</td><td>Regular Order
Please Send</td><td>Standing Order
Begin With</td></tr>
<tr><td>Until 12-31-90</td><td>After 1-1-91</td><td>Vol. Number:</td><td>Vol. Number:</td></tr>
</table>

Annual Review of **PSYCHOLOGY**

Vols. 4, 5, 8, 10, 13-24 (1953, 1954, 1957, 1959, 1962-1973)				
26-30, 32-38	(1975-1979, 1981-1987) . .	$31.00/$35.00	$33.00/$38.00	
Vols. 39-40	(1988-1989)	$34.00/$38.00	$36.00/$41.00	
Vol. 41	(1990)	$38.00/$42.00	$40.00/$45.00	
Vol. 42	(avail. Feb. 1991)	$40.00/$45.00	$40.00/$45.00	Vol(s). _____ Vol. _____

Annual Review of **PUBLIC HEALTH**

Vols. 1-8	(1980-1987)	$31.00/$35.00	$33.00/$38.00	
Vols. 9-10	(1988-1989)	$39.00/$43.00	$41.00/$46.00	
Vol. 11	(1990)	$43.00/$47.00	$45.00/$50.00	
Vol. 12	(avail. May 1991)	$45.00/$50.00	$45.00/$50.00	Vol(s). _____ Vol. _____

Annual Review of **SOCIOLOGY**

Vols. 1-13	(1975-1987)	$31.00/$35.00	$33.00/$38.00	
Vols. 14-15	(1988-1989)	$39.00/$43.00	$41.00/$46.00	
Vol. 16	(1990)	$43.00/$47.00	$45.00/$50.00	
Vol. 17	(avail. Aug. 1991)	$45.00/$50.00	$45.00/$50.00	Vol(s). _____ Vol. _____

Note: Volumes not listed are out of print.

SPECIAL PUBLICATIONS		Prices postpaid, per volume USA & Canada / elsewhere	Regular Order Please Send

The Excitement and Fascination of Science

Volume 1	(published 1965)	Clothbound	$25.00/$29.00	_____ Copy(ies).
Volume 2	(published 1978)	Softcover	$15.00/$19.00	_____ Copy(ies).
Volume 3	(published 1990)	Clothbound	$90.00/$95.00	_____ Copy(ies).

(Volume 3 is published in two parts with complete indexes for Volumes 1, 2, and both parts of Volume 3.
Sold in two-part set only.)

Intelligence and Affectivity:
Their Relationship During Child Development, by Jean Piaget
(published 1981)   Hardcover . . . . $8.00/$9.00 _____ Copy(ies).

TO: **ANNUAL REVIEWS INC.,** a nonprofit scientific publisher
4139 El Camino Way ● P.O. Box 10139
Palo Alto, CA 94303-0897 USA

Please enter my order for the publications checked above. **California orders, please add applicable sales tax.**
Prices subject to change without notice.

☐ Proof of student status enclosed.

Institutional purchase order No. _____

Amount of remittance enclosed $ _____

INDIVIDUALS: Prepayment required in U.S. funds or charge to
bank card below. Include card number, expiration date, and
signature.

Charge my account   ☐ VISA

☐ MasterCard   ☐ American Express

Acct. No. _____

Exp. Date _____ _____
                                    Signature

Name _____
          Please print

Address _____
          Please print

_____ Zip Code _____ Date _____

_____   Send free copy of current **Prospectus** ☐
   Area(s) of Interest   ARI Federal I.D. #94-1156476